Study Guide to Accompany Mulligan:

Introductory College Physics

SECOND EDITION

David A. Jerde

St. Cloud State University

McGraw-Hill Publishing Company

New York St. Louis San Francisco Auckland Bogotá Caracas Hamburg Lisbon
London Madrid Mexico Milan Montreal New Delhi Oklahoma City Paris
San Juan São Paulo Singapore Sydney Tokyo Toronto

Study Guide to Accompany Mulligan:
Introductory College Physics, second edition

1 2 3 4 5 6 7 8 9 0 MAL MAL 9 5 4 3 2 1 0

ISBN 0-07-044101-4

The editors were Anne C. Duffy and Denise T. Schanck;
the production supervisor was Richard A. Ausburn.
Malloy Lithographing, Inc. was printer and binder.

Cover photo: Fabry-Perot interferometer **fringes** from a low-voltage cadmium arc, produced by the PEPSIOS purely interferometric high-resolution scanning spectrometer in the physics department at the University of Wisconsin, Madison. (*Photograph by F. E. Barmore and F. L. Roesler; courtesy of Professor F. Roesler and Design, Inc.)*

TABLE OF CONTENTS

INTRODUCTION

In content and structure, this Study Guide is designed to accompany the Second Edition of INTRODUCTORY COLLEGE PHYSICS by Joseph Mulligan. The Guide continually attempts to anticipate commonly occurring points of confusion or misunderstanding and to clarify these both by pointing them out explicitly and by encountering them in the process of working out examples. In addition, the Guide emphasizes the primary importance of understanding the concepts which we call the principles of physics, and how these basic ideas, expressed in the precise language of mathematics, can be applied to analyze a great variety of specific problem situations. In its scope and methodology, this Guide is also meant to serve as a generic help to students in any algebra-based introductory physics class, regardless of the choice of primary text.

CONTENT OF THE GUIDE

Each chapter begins with a brief *summary*, including a *resume* of previous principles that will be involved and further developed within the chapter, in order that you clearly see the progressive synthesis of ideas from one chapter to the next. Then there is a listing of *basic data* and *physical constants* relevant to the chapter for easy reference. The main section of each chapter follows, in which each major *new concept or principle* introduced in the corresponding chapter in the text is restated. Specific *learning objectives* related to these concepts are listed and *problem examples* involving them are worked in detail. You will notice in the sections labelled *"Important Observations"* that emphasis is placed on getting a "feel" for the principles by pointing out assumptions underlying them, the measurement units involved, and some potential sources of confusion.

In the worked examples the solution is arrived at through dialogues between student and tutor. The questions asked in the dialogues are intended to be typical of those that should naturally occur to you in the process of pondering a given problem. Hopefully, in anticipating and actually verbalizing the critical questions which apply to the examples, this guide may quickly help you, when confronting a problem, to develop the insight and the ability to ask the right questions, make the right interpretations, and translate the questions and their answers into the language of mathematical statements. The rest of the problem then consists of straightforward mathematical manipulation.

You will notice in many cases that this process of solution involves mistakes which are easy to make, incorrect assumptions, or approaches that are more tedious than necessary. The focus is always in finding the right *questions* to ask before there is any hope of obtaining the right *answer*. In this way, the dialogue itself is intended to reflect the realities of the learning process, and becomes at least as important as the results themselves.

Since the text supplies an adequate number of first level worked examples whenever new concepts and principles are introduced, the examples in this guide are meant to be somewhat higher in difficulty, thereby avoiding duplication and providing the student insights into working the B and C level exercises in the text. In addition, there are a number of example problems which involve more than a single principle, corresponding to the level C or D exercises in the text. These are partially solved by giving hints in the form of critical questions, and answers to these are provided at the end of the guide.

Finally, a new feature in the second edition is a brief set of multiple choice questions which tests the student's ability to do the most straightforward calculations, involving each basic new concept in each chapter.

NOTE TO STUDENTS

Remember, all the mathematical statements in physics shouldn't be viewed as *formulas for* this or that, as a formula *for* energy, or a formula *for* heat transfer, etc. They represent *relationships* among various measurable physical properties, and as such can be manipulated precisely according to the rules of mathematics. In this way, the principles of physics reflect the logical behavior of the physical world. This is why your emphasis should always be on comprehending the basic concepts and their relationships.

Don't let the equations either scare or dominate you. Don't "throw" one equation after another at a problem, hoping that somehow one of them will find a solution. The facts, definitions, and theories or laws are "tools" which you have at your disposal with which to solve a wide variety of problems. To know which tools to use in a particular case, you have to first spend some time "diagnosing" the problem and finding out what quantities and relationships are involved. Once diagnosed, and the right "tools" selected, the "cure" is rather straightforward mathematics. A student who says "I understand all this stuff, but can't work problems" is just kidding himself or herself, assuming that he or she has the basic level of math skills. More often than not, it is precisely the lack of understanding of the basic ideas that is at fault.

More than anything else, be willing to make mistakes, and to learn from them. You'll notice that the students in the worked examples are always making some mistakes. Everytime you do, and then go on to correct them, you learn something. As with any other worthwhile skill, the command of problem solving in physics is achieved by practice, practice, and more practice. As in the first edition of this guide, I hope this edition will prove to be of some help, and I wish you the very best.

As a final note, I would like to thank Joe Mulligan and Anne Duffy for encouraging me to complete this new edition, and to express appreciation to David Damstra for overseeing the copy editing. My thanks also to Don Bruno for making professional illustrations from my rather primitive scrawls.

David A. Jerde
St. Cloud, Minnesota
November, 1990

CHAPTER 1. INTRODUCTION: IMPORTANT CONCEPTS AND TECHNIQUES.

BRIEF SUMMARY AND RESUME

This chapter is a brief introduction into what physics is, and how physicists proceed to gain experimental and theoretical knowledge of the physical world. Some discussion relevant to this is contained in the previous section of this guide. The problem of sources of error and precision in measurement and calculations is discussed. Physical *dimensions* are discussed, and the *International System* (SI) *of units* is introduced. Differences between *scalar* and *vector* quantities are defined, and examples involving the concepts of *distance* and *displacement* are given. Methods of *adding vectors*, *graphically* and *analytically*, are introduced.

NEW CONCEPTS AND PRINCIPLES

Units and Dimensions

Understanding this section should enable you to do the following:

1. *Identify the difference between dimensions and units.*
2. *Convert a quantity from one unit to another.*
3. *Find the dimensionality of terms in an equation.*

Dimension--A generic category of a physical property. These are categories which are fundamentally defined and not derived. In the study of motion only *three fundamental* physical dimensions are required. They are *mass* [M], *length* [L], and *time* [T]. When we get to the subject of electricity and magnetism, a *fourth* fundamental dimension will be added, that of *electrical current* [I].

Student: This seems quite different than referring to one-, two-, or three-dimensional spaces in geometry.

Tutor: It is a considerably more general use of the word. As we go along we shall derive many other measurable physical properties which are combinations of these fundamental dimensions.

Units--Specific scales of measurement in which quantities of a physical dimension may be expressed.

The units used in science today comprise the *International System*, or *SI*. In this system the *primary standards* are the *kilogram* of mass, the *second* of time, and the *meter* of length. The operational definitions of these quantities are given in Section 2.1 of your text. Notice that the precision with which we can measure length and time standards is nine and ten significant figures respectively, whereas measurement of mass is precise to eight figures at best.

Conversion Factors--Equivalencies between different units of the same physical property.

In order to change or convert from a quantity expressed in one system of units to another system, you must know the appropriate conversion factor.

Example 1-1. How many seconds are there in one year?

Student: Well, I know there are 60 seconds in a minute, 60 minutes in an hour and 24 hours in a day. But what should I use for the number of days in a year?

Tutor: Aha! That's an illuminating question! You can be very positive about the first three conversions because they are *exact* definitions. See Example 1-11 for reference. But we don't have the ability to make the earth go around the sun in an exact number of days, so the last conversion factor must be *measured* from nature, and the result will have a limited precision. Want to re-phrase your question?

S: How many significant figures should I use in converting days to years?

T: That's right to the point. The choice is entirely yours, and depends on how precisely you need to know the result. The exact conversion factors can't limit the precision of your calculation, only this measured one can. It's an experimental fact that, to eight significant figures, 1 year = 365.24219 days. If you choose a precision of three figures, your result will be:

$$1 \text{ yr} \times (365 \text{ d}/1 \text{ yr})\times(24 \text{ hr}/1 \text{ d})\times(60 \text{ min}/1 \text{ hr})\times(60 \text{ s}/1 \text{ min})$$

$$= 3.15 \times 10^7 \text{ s}.$$

If you choose 5 figures, you will get:

$$1 \text{ yr} \times (365.24 \text{ d}/1 \text{ yr})\times(24 \text{ hr}/1 \text{ d})\times(60 \text{ min}/1 \text{ hr})\times(60 \text{ s}/1 \text{ min})$$

$$= 3.1557 \times 10^7 \text{ s}.$$

Notice how the units are treated like algebraic quantities and can be cancelled. The only surviving unit is that of seconds. Each of the conversion factors in ratio form has the value unity, so the successive multiplications change the number and the units, but not the *basic quantity* of time.

Example 1-2. How many miles are there in one year?

S: That's crazy. There is no equivalence between miles and years, is there?

T: Absolutely correct. The example asks you to convert between two different dimensions, and there are no equivalencies between dimensions as there are between different units *of* a particular dimension.

Angle Measurement--The angle θ swept out by the radius R of a circle when the tip of the radius has moved an arc length S is: $\theta = S/R$, or $S = R\theta$. This defines the measurement of angular quantities in *radians*.

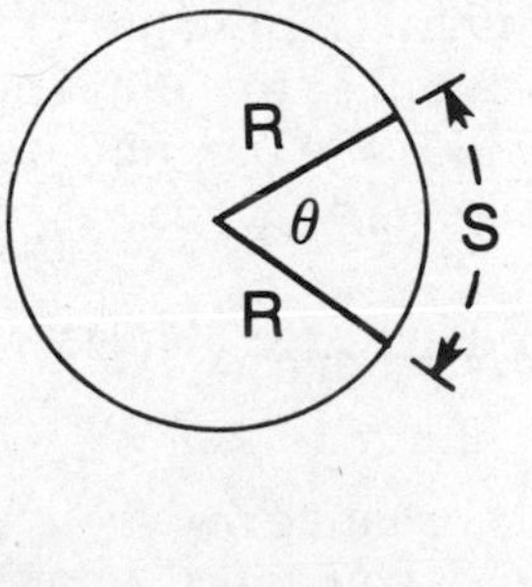

Important Observations

1. The conversion between radians and degrees is 1 radian = $180/\pi$ degrees = 57.3^o.

2. The radian is a "natural" way to define angles, since it does not involve an *arbitrary* division of a circle as the degree does. Since it involves the ratio of lengths, the radian is *dimensionless*. However, whenever an angle enters a calculation *directly* as an *algebraic factor*, it *must* be expressed in radians. No other angle measure has this fundamental meaning.

Derived Units--Algebraic combinations of units which result from calculations with quantities involving fundamental dimensions.

Example 1-3. The following table gives a few examples of derived physical properties, along with their SI units and derived dimensions.

Physical Property	Derived Units (SI)	Derived Dimension
Speed (distance/time)	meters/sec (m/s)	[L]/[T]
Acceleration (velocity change/time)	m/s/s, or m/s^2	$[L]/[T]^2$
Force (mass x acceleration)	kg m/s^2 or newton	$[M][L]/[T]^2$
Work (force x distance)	kg m^2/s^2 or newton-meter or joule	$[M][L]^2/[T]^2$

The last two are examples in which the derived unit is given a specific SI name. Notice the way that any derived unit can be reduced to a combination of the three fundamental dimensions.

Dimensional Analysis--The process of examination of the fundamental dimensionality of each term in an equation.

Since we can only add or subtract terms which represent like things, dimensional analysis can be used to check the correctness of the basic structure of equations and formulas, and to determine the dimensions that constants of proportionality must have in order to be consistent with other terms in an equation.

Example 1-4. In the following let s = distance, v = speed, t = time, a = acceleration, and F = force. Using the table in Example 1-3, check each for dimensional consistency.

(a) $$s = vt + \tfrac{1}{2}at^2$$

Check: $[L] = [L]/[T] \times [T] + [L]/[T]^2 \times [T]^2$.

All the terms have the dimension [L], so the equation is structurally correct. This does not prove that it is specifically correct with regard to numerical coefficients.

(b) $$Ft + mv = m(as)^{\frac{1}{2}}.$$

Check: $[M][L]/[T]^2 \times [T] + [M] \times [L]/[T]$
$= [M]\{[L]/[T]^2 \times [L]\}^{1/2}.$

All terms have the dimension $[M][L]/[T]$.

(c) $$v = 2as.$$

Check: $[L]/[T] = [L]/[T]^2 \times [L]$

This is not consistent. Furthermore, the analysis shows us what is necessary in order to make the equation even *possibly* correct. The dimensions of the right side are the square of the dimensions of the left side. Thus dimensional consistency would require $v^2 = 2as$.

Example 1-5. What dimensions must the proportionality constant G have in the following expression?

$$F = Gm^2/d^2$$

Dimensional consistency requires

$$[M][L]/[T]^2 = [G] \times [M]^2/[L]^2, \text{ or } [G] = [L]^2/[M][T]^2$$

Accuracy of Physical Data

Understanding this section should enable you to do the following:

1. *Identify the number of significant figures in quantities expressed in decimal or powers of ten notation.*
2. *Calculate the appropriate number of significant digits in the sum or product of quantities having various precision.*
3. *Identify the difference between accuracy and precision in physical measurement.*

Significant Figures--This is the number of digits in a measured or calculated quantity that are definitely known.

Example 1-6. 3.649 cm. has 4 significant figures.

Example 1-7. 137.249 sec. has 6 significant figures.

Example 1-8. 20.030 pounds has 5 significant figures.

Student: Why does the last zero in Example 1-8 count as significant?

Tutor: Its presence there signifies that we *know* this last digit is zero, and that the number is *not* 20.029 or 20.031. If the fifth digit was not known, the number should have been written 20.03.

Example 1-9. 0.00167 grams has 3 significant digits.

S: Why don't the first zeros count as significant?

T: All zeros before a non-zero digit merely locate the decimal place. They have nothing to do with the precision with which the number is known.

S: What determines the number of significant figures in a measurement?T: There are two types of measurement error which limit this. *Systematic errors* are due to errors in the measuring instrument or in the design or execution of the experiment. *Statistical errors* refer to the spread of individually measured values around the average value of a group of measurements. Measurements with small systematic errors are said to be *accurate*, while those with small statistical errors are *precise*.

S: What limits precision in measurements?

T: Any physical instrument has an inherent limit on the fineness or sensitivity of its measuring mechanism. Even if you could eliminate systematic error, this inherent limit would ultimately determine the possible precision of the measurement and how many significant figures the measured quantity could have.

Powers-of-Ten Notation--Also known as scientific notation, this is the useful practice of expressing all numbers as a number between 1 and 10, containing the correct number of significant figures, times some power of ten. A main reason for doing this is that in physics we encounter a tremendous range in the size of numbers, and using straight decimal system notation becomes very cumbersome and confusing. A second convenience of this notation is that it makes it easier to keep track of the significant figures in any number, however large or small.

Example 1-10. 437,189 miles becomes 4.37189×10^5 miles. Notice that both notations show 6 significant figures.

Example 1-11. 0.000000927 grams becomes 9.27×10^{-7} grams. It is clear that there are only three significant figures. See Example 1-9 for reference.

For this section and the rest of this chapter, if you are a bit rusty about the manipulation of exponents and powers of ten, you should consult the review in Appendix III of the text.

Calculation with Significant Figures

It should be quite obvious that the mere manipulation of measured data in calculations cannot increase the precision associated with an experiment. Yet very often students will try to claim all eight or ten figures in their calculator readout as being significant, even though the input data had only two or three significant figures! It is important therefore that you keep in mind the following simple rules for determining the number of significant figures you can claim as a result of your calculations.

Addition and Subtraction. The sum or difference of numbers cannot have more *precision* than the least precise number entering the calculation.

Multiplication and Division. The product or quotient of a series of factors can have no *greater number of significant figures* than does the factor containing the fewest significant figures.

Example 1-12. What is the sum of 39.37 inches, 36 inches, and 1.2×10^2 inches?

S: Which number is the least precise?

T: 1.2×10^2 inches. It has the same number of significant figures as does 36 inches, but *less precision*. 36 inches is known to the nearest inch, whereas 1.2×10^2 inches is known only to the nearest 10 inches.

S: We should get all the numbers in powers-of-ten notation.

T: Yes, and furthermore we should round them all off to the same precision. Thus

$$39.37 \text{ becomes } 0.4 \times 10^2$$

and

$$36 \text{ becomes } 0.4 \times 10^2.$$

The sum then becomes

$$1.2 \times 10^2 \text{ in.} + 0.4 \times 10^2 \text{ in.} + 0.4 \times 10^2 \text{ in.} = 2.0 \times 10^2 \text{ in.}$$

S: Is the zero after the decimal point significant?

T: Yes. Can you tell why?

S: Haven't I thrown away information that I really know?

T: You know an *individual* measurement to four significant figures and a precision to the nearest .01 inch. However, when you add this to a much less precise number, the sum has the uncertainty of the latter. The precision of 39.37 in. is lost except for the leading digit (in this case) when it enters the sum.

Example 1-13. Calculate

$$\frac{3.987 \times 4.3}{12.143}$$

S: The calculator shows 1.4118504. The rule says I must round this off to the same number of figures as contained in the factor 4.3. This would make the answer 1.4. It seems a shame to throw all that other information away. Do I really have to?

T: The precision of the calculator is an illusion. It does not represent valid information, so you are not throwing away anything. To show this, let's calculate the largest and smallest value this calculation could produce, based on the largest and smallest values of the individual factors. Notice:

3.987 can be anything between 3.9865 and 3.9875,
4.3 can have any value between 4.25 and 4.35, and
12.143 must be between 12.1425 and 12.1435.

Then the *largest* possible result of the calculation would be

$$\frac{3.9875 \times 4.35}{12.1425} = 1.4285052 = 1.43$$

The *smallest* possible result would be

$$\frac{3.9854 \times 4.25}{12.1435} = 1.3952011 = 1.40$$

You can see that these results do *not* agree to *three* significant figures. Only the first two digits from the calculation are *certain*. The result is therefore <u>1.4</u>, as we obtained by using our rule.

<u>Vectors and Scalars</u>

Understanding this section sbould enable you to do the following:

1. *Identify scalar and vector quantities.*
2. *Identify the difference between distance and displacement.*

<u>Scalar</u>--A property which is completely specified by its quantity or size alone.

<u>Vector</u>--A property whose complete specification needs both a quantity and a direction.

The size or quantity of either a scalar or a vector is referred to as its *magnitude*. To distinguish between scalars and vectors symbolically, either a small arrow is drawn over the symbol, or as is the case throughout this guide, the vector symbol is printed in bold face type. The symbol without the arrow or in standard type represents the vector's magnitude. To illustrate these concepts we will discuss the difference between *distance* and *displacement* and between *speed* and *velocity*.

<u>Distance</u>--The scalar length between two points measured along a given path.

<u>Example 1-14</u>. In the figure below, the lengths of the three paths between A and B are obviously different. Thus the distance between A and R depends on which path you choose.

S: Isn't path #3 the *actual* distance?

T: It is the *straight-line*, or *minimum* distance between A and B, but as the definition shows, the distance between two points is not specified until you describe a specific path.

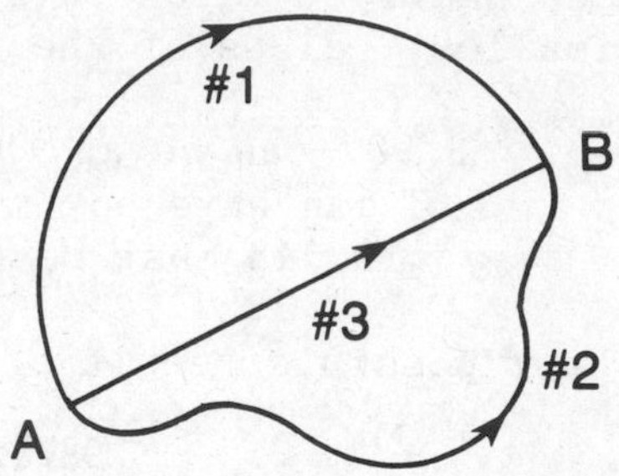

Displacement--The displacement from A to B is the straight-line distance between points A and B, directed from A to B.

This is a *vector* quantity, which is represented by the arrow **AB** in the figure in Example 1-14. Notice that the displacement from B to A, **BA**, would be just the opposite, or *negative*, of **AB**. Thus

$$\mathbf{AB} = -\mathbf{BA}.$$

S: So the displacement depends only on the two end points?

T: That's right. It doesn't depend on how you get there, as distance does.

S: Since in one dimension only straight paths are allowed, aren't distance and displacement the same in those cases?

T: In one dimension, distance is the magnitude of the displacement. In two or three dimensions, the *straight-line* distance is the magnitude of displacement.

Vector Addition: Graphical Method

Understanding this section should enable you to do the following:

1. *Find the sum of a number of vectors graphically.*
2. *Find the sum and difference of two vectors by the parallelogram method.*

Any vector can be represented on paper by an arrow whose direction shows the vector's direction, and whose length represents the vector's magnitude according to a suitable choice of scale.

Example 1-15. Displacements:

a. A displacement **AB** of 10 meters to the right.

Scale: 1 inch = 5 meters.

b. A displacement AB of 30 miles southwest.

Scale: 1 inch = 20 miles.

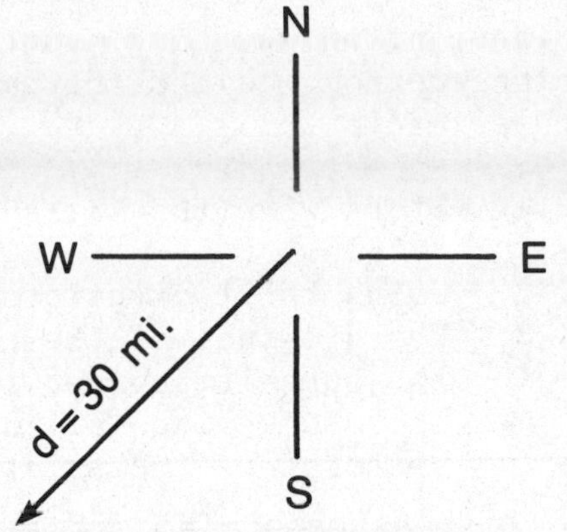

Two or more vectors expressed in the same physical units can be added graphically by successively joining the vectors head to tail *in any order*. The arrow drawn from the starting tail to the final head then represents the sum of the vectors. Its length, referred to the same scale used for the individual vectors, gives the magnitude of this sum, and its direction is the direction of the sum.

Example 1-16. Suppose you have four displacements as follows: $\mathbf{D}_1$ = 2 meters right, $\mathbf{D}_2$ = 3 m down, $\mathbf{D}_3$ = 1 m 45° above the x axis, and $\mathbf{D}_4$ = 4 m left. Find the vector sum: $\mathbf{D} = \mathbf{D}_1 + \mathbf{D}_2 + \mathbf{D}_3 + \mathbf{D}_4$.

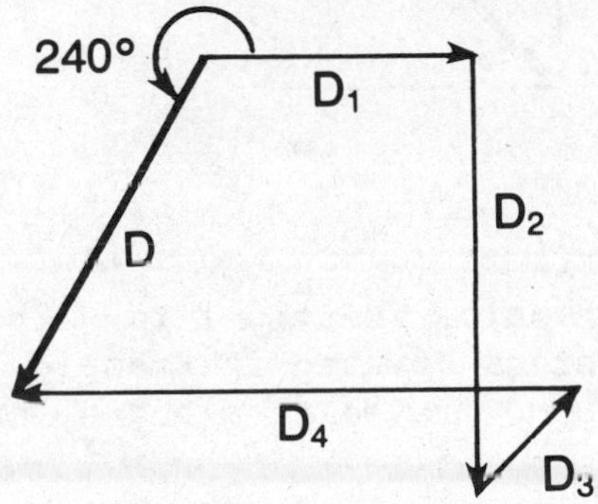

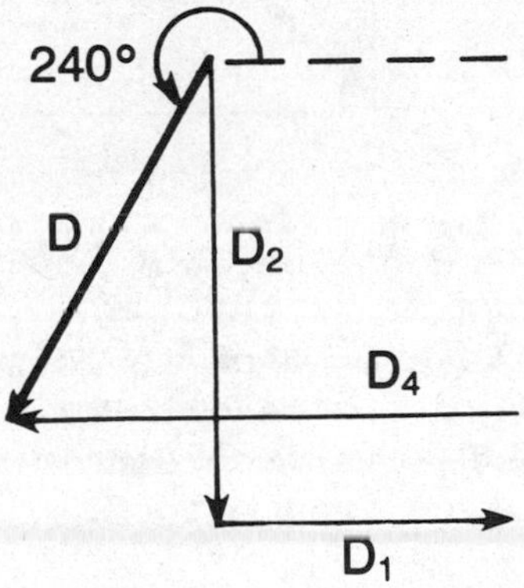

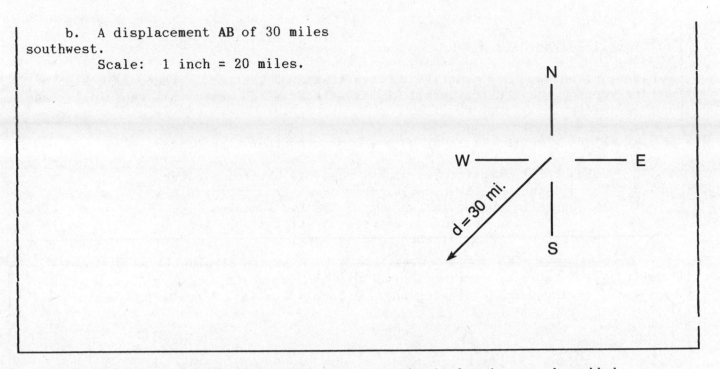

S: Which vector shall I start with?

T: You may take them in *any* order. Vector addition is associative and commutative. Also, any choice of scale will do. Let's choose 1 inch = 2 meters for convenience.

S: Moving vectors around on the page is O.K.?

T: As long as you keep the *length* and *direction* the same, you have not changed the vector, so you can do this in calculating the sum of vectors.

In both cases, D is 1 5/8 inches long, representing D = 3.25 m , at an angle of 240° measured from the +x axis.

Parallelogram Method.

Very often you have just two vectors to add or subtract. There is a simple graphical method, called the parallelogram method, which makes this especially easy. The rules are:

1. Start with the tails of both vectors together at a common origin.
2. Using the two vectors as sides, complete a parallelogram.
3. The diagonal of the parallelogram drawn from the origin to the opposite corner is the *sum* of the two vectors.
4. The other diagonal, drawn between the tips of the two vectors, is the *difference* between them.

Example 1-17. Take two displacements, $\mathbf{D}_1$ = 5 miles north and $\mathbf{D}_2$ = 10 miles northeast. Using a scale drawing where 1 inch = 5 miles, find $\mathbf{D}_1 + \mathbf{D}_2$ and $\mathbf{D}_2 - \mathbf{D}_1$.

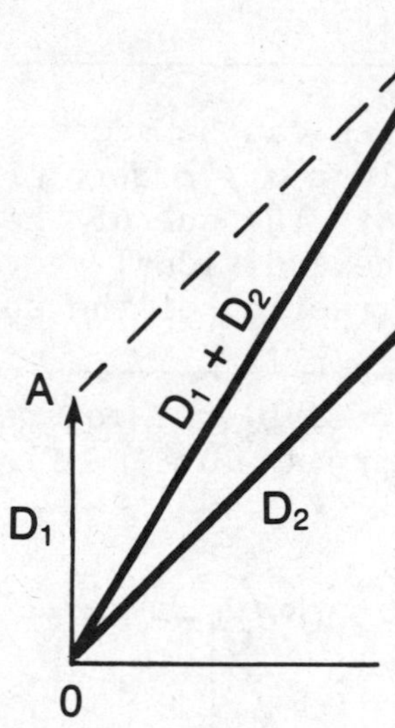

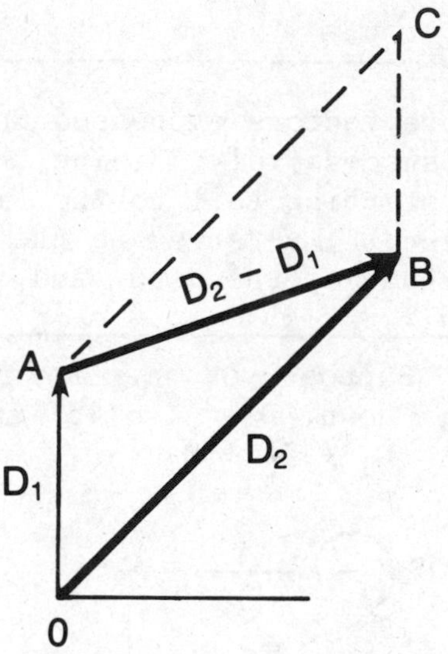

S: I don't see why the diagonal is $\mathbf{D}_1 + \mathbf{D}_2$.

T: Move $\mathbf{D}_2$ over to coincide with side AC, which it is identical to. Then the vectors are head to tail, and you have the situation covered in Example 1-16. Or move $\mathbf{D}_1$ over to coincide with side BC. Once again, traveling along $\mathbf{D}_2$ and then $\mathbf{D}_1$ gets you from O to C.

S: I'm even more confused about the second diagram. I don't recognize $-\mathbf{D}_2$ anywhere.

T: Think of what $-\mathbf{D}_1$ means. From Chapter two we remember that the negative of a vector is just the same magnitude but *in the opposite direction*. In the above diagram, the vector **AB** is seen to equal the sum of vectors **AO** and **OB** by our graphical method. But **AO** is just $-\mathbf{D}_1$ and OB is the same as $\mathbf{D}_2$, so $\mathbf{AB} = \mathbf{D}_2 + (-\mathbf{D}_1) = \mathbf{D}_2 - \mathbf{D}_1$.

S: How do you remember which direction to draw the diagonal? It could represent either $\mathbf{D}_2 - \mathbf{D}_1$ or $\mathbf{D}_1 - \mathbf{D}_2$.

T: Good point. The rule is: the vector the diagonal *leaves* is the one you're subtracting from the other, to which the diagonal *goes*. Now review the previous rules of the parallelogram method.

Vector Addition: Rectangular Components Method

Any vector can be resolved into components along the x and y (and z) axes of a chosen rectangular coordinate system. Likewise, a vector's magnitude and direction can be composed from its rectangular components. From the following figure, we have the basic equations for resolving or composing:

Resolution: $V_x = V \cos\theta$, $V_y = V \sin\theta$

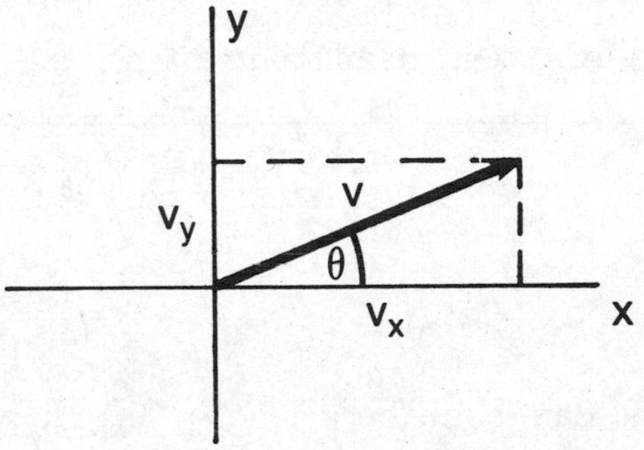

Composition:

$$\mathbf{V} = (V_x^2 + V_y^2)^{\frac{1}{2}}$$

and $\theta = \tan^{-1}(V_y/V_x)$.

When a number of vectors are to be added, they can *each* be resolved into their x and y components. By adding all the x components together and all the y components together, which is a one dimensional *algebraic* process, you get the x and y components of the vector sum, or *resultant*.

Example 1-18. Find the resultant, **V**, of the three vectors shown in the following figure, where $\mathbf{V_1}$ = 1 inch, $\mathbf{V_2}$ = 1 inch, and $\mathbf{V_3}$ = 2 inches.

S: I have to watch out for the signs of the components, right?

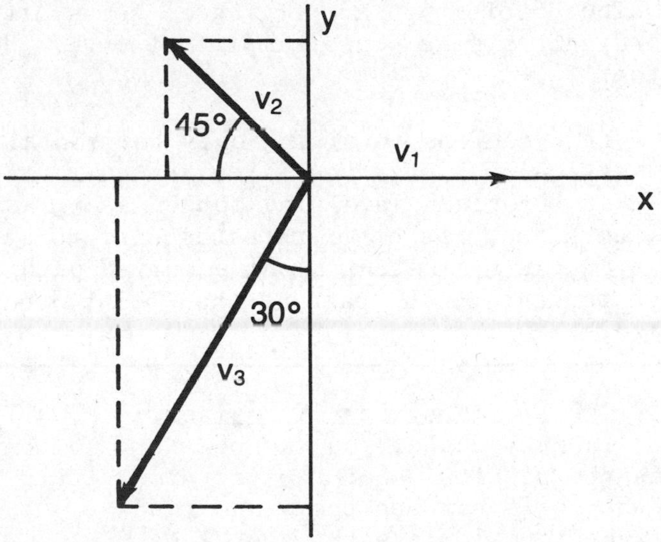

T: Yes. Either you do it by inspection of the diagram and using your judgment, or you may measure all angles relative to the +x axis, in which case the sign of components in all quadrants will be given correctly by the sign of the sine and cosine functions. Resolving these vectors gives $V_{1x} = 1$, $V_{1y} = 0$, and

$$V_{2x} = V_2 \cos 135^\circ = (1)(-.707) = -.707$$

$$V_{2y} = V_2 \sin 135^\circ = (1)(.707) = +.707$$

$$V_{3x} = V_3 \cos 240^\circ = (2)(-.500) = -1.00$$

$$V_{3y} = V_3 \sin 240^\circ = (2)(-.866) = -1.73$$

These are indicated by dashed lines on the figure.

S: How do we get **V** from this?

T: The x component of **V** is the algebraic sum of the individual x components, and similarly for the y component of **V** in terms of the individual y components. Thus to three figures,

$$V_x = V_{1x} + V_{2x} + V_{3x} = 1.00 - .707 - 1.00 = -.707 \text{ inches}$$

$$V_y = V_{1y} + V_{2y} + V_{3y} = 0 + .707 - 1.73 = -1.02 \text{ inches}$$

S: And the next step is to compose **V** from V_x and V_y?

T: Right. The magnitude of V is

$$V = \sqrt{V_x^2 + V_y^2} = \sqrt{(-.707)^2 + (-1.02)^2} = 1.25 \text{ inches}$$

The angle is

$$\theta = \arctan(V_y/V_x) = \arctan(-1.02/-.707) = 55.4^\circ.$$

S: How can that be right? This would put **V** in the 1st quadrant.

T: This is a good example of not taking the mathematical result blindly. Looking at the signs of V_x and V_y will give you a feel for interpreting the result. The only quadrant that has both x and y components negative is the third, so it should be obvious that the angle you calculated is 55.4 below the x axis in the third quadrant, or θ = 235.4 measured from the +x axis. It would be a good idea for you to check this result by applying the graphical method to the same vectors.

S: Is there some way to know how to make the best choice of a coordinate system?

T: The problem you are trying to solve often gives you a clue as to which coordinate system would simplify things. Here are some cases which occur quite often:

1. Problems involving only horizontal and vertical forces or only horizontal or vertical motion. Use a horizontal x axis and a vertical y axis.
2. Problems involving motion along an inclined plane. Use one axis parallel to the plane, the other perpendicular to it.
3. At an instant along a curved path of motion, one axis is taken along the line tangent to the path at that point, and the other perpendicular to the path there.

Law of Sines, Law of Cosines.

In many cases, you may be given the magnitudes of two vectors and the angle between them, and you may want to calculate the sum $V_1 + V_2$ directly without the inconvenience of resolving or graphing the vectors. You can do this by means of the Law of Sines and the Law of Cosines. If you have forgotten these, you might want to refer to an introductory book on trigonometry for review.

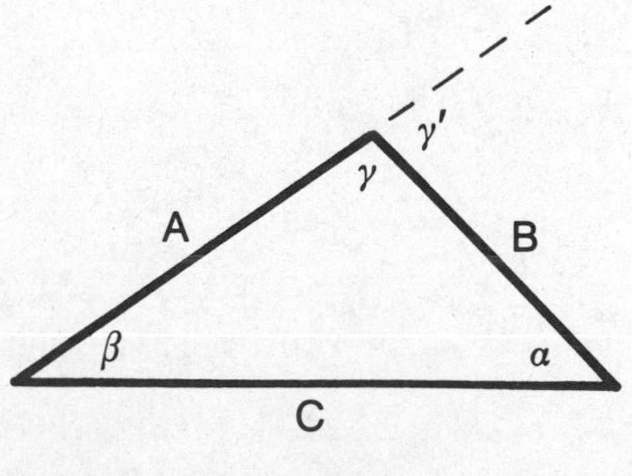

A general triangle is shown in the figure above, with sides A, B, and C. The angles opposite these sides are labelled α, β, and γ, respectively.

Law of Sines:

$$\frac{\sin\alpha}{A} = \frac{\sin\beta}{B} = \frac{\sin\gamma}{C}$$

Law of Cosines:

$$C^2 = A^2 + B^2 - 2AB\cos\gamma$$

<u>Example 1-19</u>. Suppose you have two forces, $\mathbf{F}_1$ = 10 N and $\mathbf{F}_2$ = 20 N, and there is an angle of 80° between them. What is their sum?

<u>S</u>: Some kind of drawing would help.

<u>T</u>: Yes, it can just be a labelled sketch as below. Notice I've drawn it so that we have the $\mathbf{F} = \mathbf{F}_1 + \mathbf{F}_2$ triangle.
The angle of 80° between $\mathbf{F}_1$ and $\mathbf{F}_2$ is *exterior* to the triangle. The *interior* angle is therefore 100°.

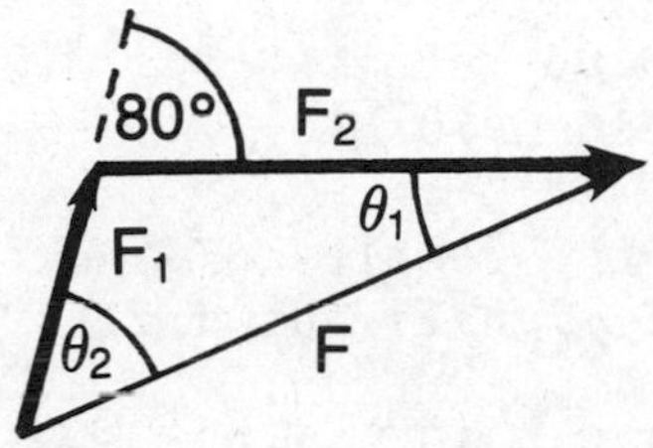

<u>S</u>: We know sides F_1 and F_2 and the one angle. What will give us the length of side F?

<u>T</u>: At this point, any of the Law of Sines equations would contain two unknowns. But the Law of Cosines contains only the unknown F:

$$F^2 = F_1^{\,2} + F_2^{\,2} - 2F_1F_2 \cos 100^\circ$$

$$= (10\text{ N})^2 + (20\text{ N})^2 - 2(10\text{ N})(20\text{ N}) \cos 100^\circ$$

Thus

$$F^2 = 569\text{ N}^2,\text{ and } F = 23.9\text{ N}$$

<u>S</u>: Which of the angles do we want to know?

<u>T</u>: Either one will do. The other is immediately then known from the relation $\theta_1 + \theta_2 + 100^\circ = 180^\circ$. The Law of Sines gives

$$(\sin\theta_1)/F_1 = (\sin 100^\circ)/F .$$

Then

$$\sin\theta_1 = (F_1/F)\sin 100^\circ = (10\text{ N})(.985)/(23.9\text{ N}) = 0.412$$

$$\theta_1 = \arcsin .412 = 24.3^\circ,\text{ and then } \theta_2 = +55.7^\circ .$$

TEST ON FUNDAMENTALS

The following questions are meant to test your understanding of the basic fundamentals contained in this and each succeeding chapter.

1. A car traveling 60 miles per hour is going how fast in SI units?

   (a.) 96 km/hr (b.) 9.6×10^3 m/hr (c.) 1.6 km/s (d.) 26.7 m/s (e.) .0267 km/s

2. What physical quantity is represented by 1 Joule/Newton?

   (a.) 1 meter (b.) 1 m/s (c.) 1 km (d.) $(\text{meters})^{-1}$ (e.) there is no other name than J/N for this combination

3. How many significant digits are there in the following:

   (a.) 40.06 inches (b.) 0.031 seconds (c.) 3×10^2 kg (d.) 0.16×10^{-3}

4. What are the results of the following calculations to the correct number of significant digits?

   (a.) $(4.8 \times 10^6 \text{ m})/(3.25 \text{ s})$ = ?
   (b.) $(3.75 \text{ m})(14.59 \text{ kg})/(3.01 \times 10^3 \text{ s})$ = ?
   (c.) 13.67 ft + 0.771 ft - 12.2 ft = ?
   (d.) 3.83 ft + 14.65 inches = ?

5. What is the vector sum of $\mathbf{V}_1$ = 2.00 meters North and $\mathbf{V}_2$ = 4.70 meters West?

   (a.) 6.7 m, 23.1° N of W (b.) 26.1 m, 23.1° N of W, (c.) 5.11 m, 23.1° N of W (d.) 26.1 m, 23.1° W of N (e.) 5.11 m

6. Using the two vectors in the above question, calculate the vector difference $\mathbf{V}_2 - \mathbf{V}_1$.

   (a.) 5.11 m, 23.1° S of W (b.) 2.7 m, 23.1° N of W (c.) 26.1 m, 23.1° S of W (d.) 5.11 m, 23.1° N of E (e.) 26.1 m, 23.1° N of E

7. What are the x and y components of the vector $\mathbf{V}$ = 40.0 meters, directed 25° above the -x axis?

   (a.) V_x = 16.9 m ; V_y = 36.3 m (b.) V_x = -36.3 m ; V_y = 16.9 m (c.) V_x = 36.3 m ; V_y = 16.9 m (d.) V_x = 16.9 m ; V_y = -36.3 m (e.) V_x = -16.9 m ; V_y = 36.3 m

8. A triangle has sides A and B of 1.87 m and 2.74 m, respectively. The angles opposite these sides are $\alpha = 33^\circ$ and $\beta = 72^\circ$. Find the length of the third side, C.

   (a.) 11.0 m (b.) 3.32 m (c.) 3.69 m (d.) 13.7 m (e.) 2.89 m

GENERAL PROBLEMS

Problem 1-1. How many significant figures are there in 2100 °C?

Q: How can you tell if the last zeros are *known*, or whether they are just locating the decimal point?

A: This is a tricky situation, and illustrates the advantage of powers-of-ten notation. If only two figures were known, it would be written 2.1×10^3 °C. If all four were certain the number would be 2.100×10^3 °C. An alternative if four figures were known, would be to show the decimal point explicitly, as in 2100 °C. Still another possibility would be to show the range of experimental error, as 2100 ± 50 °C (2 significant figures) or 2100 ± 0.5 °C (4 significant figures). As it is stated in the example, the precision of the number is ambiguous.

Problem 1-2. Find the number of significant figures in 10^3 meters.

Q: What does 10^3 mean? What would it be in powers-of-ten notation?

A: Another tricky one. 10^3 meters *implies* 1×10^3 meters. If any more precision was claimed, the number should have been written 1.0×10^3 or 1.00×10^3, etc. As it is stated, the number of significant figures should be interpreted as *one*. This and the previous example demonstrate the care you should use in expressing the precision of numbers clearly.

Problem 1-3. How many significant figures are there in the number: 15 apples?

Q: This is different from measuring, isn't it? This is counting!

A: Very true! And counting, or enumerating, is basically an *exact* process. Don't confuse it with estimating--we're talking about actual counting, regardless of difficulty. Such numbers should be regarded as having an *unlimited* number of significant figures, in the sense that they cannot be the limiting factor in a calculation which also involves *measured* numbers. The same also holds for purely *irrational* numbers such as π, which can be calculated to any desired precision.

Problem 1-4. Calculate

$$\frac{4.5\times(3.34\times10^{3})}{2.465\times10^{-1}}-9.834\times10^{3}$$

Q: This involves both rules. In what order do I apply them?

A: In the same order as you do the calculation. The term on the left must contain only two significant figures after calculation. Thus it must be

$$\frac{4.5\times(3.34\times10^{3})}{2.465\times10^{-1}}=6.0974\times10^{4}=6.1\times10^{4}$$

Next you must compare this precision with that of the number to be subtracted from it. 6.1×10^4 is *less* precise than 9.834×10^3. If you refer back to Example 1-7, you will see that we have to round off as follows:

$$9.834 \times 10^3 = 0.9834 \times 10^4 = 1.0 \times 10^4$$

to the same precision as the number 6.1×10^4. The final answer is thus

$$6.1 \times 10^4 - 1.0 \times 10^4 = \underline{5.1 \times 10^4}$$

Problem 1-5. City B is located 100 miles due northeast of city A, but the road connecting them is 135 miles long.

(a) If a car takes three hours to make the trip from A to B, what is the average speed of the car? What is the average velocity?

(b) Suppose the car makes a round trip in 5 hours. What are the car's average speed and average velocity for the round trip?

Q1: Going back to the definitions, average speed = distance/time and average velocity = displacement/time. What are the values of distance and displacement for the trip?

Q2: When you make a round trip, what is your displacement? How much distance has been traveled?

Problem 1-6. Initially your car is going east at 100 km/hr (about 62 mi/hr). Two seconds later, you are going southeast at 80 km/hr. Use the Law of Cosines and the Law of Sines to find the size and direction of your average acceleration vector during this time.

Q1: The Law of Cosines and the Law of Sines are used to add two vectors together when the angle between them is known. But $\langle a \rangle$ involves the *difference* in the velocity vectors. How can the laws apply here?

Q2: What do the Law of Cosines and the Law of Sines look like when applied to this case?

Q3: Once I've found Δv, then $\langle a \rangle = \Delta v/\Delta t$. Do I have a units problem here?

ANSWERS TO QUESTIONS

FUNDAMENTAL QUESTIONS:

1. (d.)
2. (a.)
3. (a.) 4 (b.) 2 (c.) 1 (d.) 2
4. (a.) 1.5×10^6 m/s (b.) 1.82×10^{-2} kg-m/s (c.) 2.2 ft (d.) 5.05 ft
5. (c.)
6. (a.)
7. (b.)
8. (e.)

GENERAL PROBLEMS:

Problem 1-5. A1: Distance = 135 miles; Displacement = 100 miles northeast

A2: Round trip displacement = 0; Distance = 270 miles.

Problem 1-6. A1: To get v, just think of adding $-v_1$ to v_2, as in the diagram below.

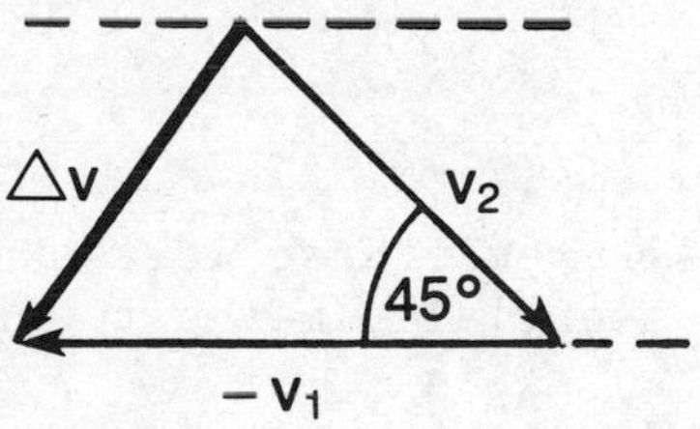

A2: $(\Delta v)^2 = v_1^2 + v_2^2 - 2v_1v_2 \cos 45^\circ$

$\sin \theta/100 = \sin 45^\circ/v$

A3: Yes. To get to SI units, you must convert km to meters and hours to seconds.

CHAPTER 2. RECTILINEAR MOTION

BRIEF SUMMARY AND RESUME

Here the description of motion in one dimension is introduced, and the concepts of *speed*, *velocity*, and *acceleration* are operationally defined. Equations describing one dimensional motion with constant acceleration are developed and applied. The specific example of *gravitational free fall* near the earth's surface is explored. Motion on an inclined plane is studied as an example of resolution of vectors in a convenient choice of reference frame.

Material from the previous chapter which is used in this chapter includes:

1. Calculating with operationally defined physical quantities.
2. Addition and subtraction of vectors.

NEW DATA AND PHYSICAL CONSTANTS

Free Fall Acceleration. Acceleration of freely falling objects near sea level due to earth's gravity has the symbol g and a value of

$$g = 9.80\ \mathrm{m/s^2}.$$

Air resistance is assumed to be negligible. This is a vector pointing "down," which essentially means toward the center of the earth.

NEW CONCEPTS AND PRINCIPLES

Speed, Velocity, and Acceleration.

Understanding this section should enable you to do the following:
1. *Identify the difference between speed and velocity.*
2. *Calculate the average velocity for a given time interval from position versus time data in one dimension.*
3. *Calculate the average acceleration of an object from the value of its velocity at various times.*

Speed--The *rate* of covering distance, i.e., the distance covered in a period of time divided by that time period. This is a *scalar*, with symbol v.

Velocity--A *vector* quantity in which the *direction* of motion and speed are both specified. The speed is thus the magnitude of the velocity. The symbol for velocity is **v**.

S: What if an object's speed or velocity *changes* during the time period being considered?

T: Then of course we would be measuring the *average* speed or velocity in that period. If we take a short time interval, short enough so that no change takes place, then we can claim to have measured the *instantaneous* speed or velocity within that brief interval. Let's be more specific and look at an example. First, a couple of definitions:

Average Velocity--The *displacement* an object undergoes during a time period, divided by that time period. We will denote averages by brackets, so this would be $\langle \mathbf{v} \rangle$.

The direction of the average velocity is the same as the direction of the displacement vector from which it was calculated.

Instantaneous Velocity--The infinitesimal displacement an object undergoes in an infinitesimal time period, divided by that period. This is the object's velocity at the middle of that small time interval, symbolized by **v**.

From a practical standpoint, notice that we are here calculating an *average* over a very *small* interval of time. If that interval is small enough so that **v** does not change during it, then this average value will represent the actual velocity at the instants of time in that interval.

Example 2-1. The following table shows a list of position coordinates at various times for an object moving along the x axis. What is the average velocity during the time period t = 1 sec. to t = 5 sec.?

t (sec)	0	1	2	3	4	5	6	7	8
x (cm)	2	6	7	12	20	22	23	20	15

S: We'll need the displacement during this period and divide by the 4 second time interval.

T: The displacement is the difference in the positions, 22 cm - 6 cm = 16 cm, in the positive x direction. So

$$\langle \mathbf{v} \rangle_{1-5} = 16 \text{ cm}/4 \text{ sec} = \underline{4 \text{ cm/sec, toward } +x}.$$

Now let's find the instantaneous velocity at t = 3 seconds.

S: We'll need information on the positions close to that instant, won't we?

T: Yes. The following table shows additional data. Notice that the measurements necessarily become more precise in position as well as in time.

t (sec)	2.90	2.95	3.00	3.05	3.10
x (cm)	11.3	11.8	12.0	12.7	13.5

t (sec)	2.980	2.990	3.000	3.010	3.020
x (cm)	11.83	11.89	12.02	12.08	12.21

Between t = 2.95 sec. and 3.05 sec. we have a displacement of 12.7 cm - 11.8 cm = 0.9 cm. Thus

$$\langle \mathbf{v} \rangle_{2.95-3.05} = 0.9 \text{ cm}/0.10 \text{ sec} = \underline{\underline{9 \text{ cm/sec}}}.$$

Between t = 2.99 sec. and t = 3.01 sec. we have a displacement of 12.08 - 11.89 = 0.19 cm. Thus

$$\langle \mathbf{v} \rangle_{2.990-3.010} = 0.19 \text{ cm}/0.020 \text{ cm} = \underline{\underline{9.5 \text{ cm/sec}}}.$$

You should verify that the above results display the correct number of significant digits.

S: But we still haven't calculated the instantaneous value of **v**, have we?

T: No, and we never will, if you expect an exact answer. Remember, these are measurements. We have sampled a time interval as small as .02 seconds, centered on t = 3 sec., and derived ⟨v⟩ to two significant figures. If our data had sampled more precisely, we could pursue the calculation to greater precision.

Acceleration--The rate of change of the *velocity* of an object, i.e., the change of velocity during a time interval divided by that time interval.

Strictly speaking, this is the *average* acceleration during the time interval. As was the case with velocity, you have to let the time interval shrink to an infinitesimal amount to define the *instantaneous* acceleration. If the acceleration is *constant*, as is often the case, the average and instantaneous values are the same. Mathematically,

$$\langle \mathbf{a} \rangle = \Delta \mathbf{v}/\Delta t = (\mathbf{v}_2 - \mathbf{v}_1)/(t_2 - t_1)$$

Important Observations

1. "$\Delta \mathbf{v}$" reads "change in **v**", or $\mathbf{v}_2 - \mathbf{v}_1$. Similarly, "$\Delta t$" means "change in t", or the time interval $t_2 - t_1$. The earlier value is always subtracted from the later value.
2. *Instantaneous* values of velocity are required. $\mathbf{v}_2$ occurs at t_2: $\mathbf{v}_1$ at t_1.
3. Acceleration is a *vector*. Its direction is not $\mathbf{v}_2$ or $\mathbf{v}_1$ but is the direction of the vector $(\mathbf{v}_2 - \mathbf{v}_1)$.
4. Keep in mind that a vector such as **v** can change either in magnitude or direction. In one dimension, change of direction can only mean the reversal of the sign of **v**, but in Chapter 4 we'll see that *turning* an object, even at constant speed, is an acceleration according to the above definition.
5. The SI units of acceleration are m/s/s, or m/s^2. This is verbalized either as "meters per second per second" or more commonly "meters per second squared." Notice again how the units have been treated as algebraic symbols.

Example 2-2. Suppose we have $\mathbf{v}_1$ = +2 m/s at t_1 = 2 s and $\mathbf{v}_2$ = +10 m/s at t_2 = 4 s. Find ⟨**a**⟩.

T: From the definition we have

$$\langle \mathbf{a} \rangle = \frac{+10\ \mathrm{m/s} - (+2\ \mathrm{m/s})}{4\ \mathrm{s} - 2\ \mathrm{s}} = \frac{8\ \mathrm{m/s}}{2\ \mathrm{s}} = +4\ \mathrm{m/s^2}$$

S: In what direction is **a**?

T: The + sign of **a** shows it to be in the same direction as $\mathbf{v}_1$ and $\mathbf{v}_2$, representing an increase of speed in the original direction.

Example 2-3. Often the initial time is taken to be t = 0, and the corresponding velocity $\mathbf{v}_1$ is referred to as the initial velocity, with the symbol $\mathbf{v}_0$. Then any later time is simply expressed as t and the corresponding velocity as $\mathbf{v}$. Then we have $\langle \mathbf{a} \rangle = (\mathbf{v} - \mathbf{v}_0)/t$.

Suppose $\mathbf{v}_0$ = -2 m/s and $\mathbf{v}$ = +10 m/s when t = 2 s. What is $\langle \mathbf{a} \rangle$?

$$\langle \mathbf{a} \rangle = \frac{+10 \text{ m/s} - (-2 \text{ m/s})}{2 \text{ s}} = (12 \text{ m/s})/2 \text{ s} = +6 \text{ m/s}^2$$

S: This is a greater acceleration than the previous example, yet the numbers seem the same.

T: Notice the importance of the signs. Here there is a change in the direction of motion, from a negative $\mathbf{v}_0$ to a positive $\mathbf{v}$. This is a bigger change in velocity than before, when the initial velocity was already positive. If you're careful with the algebra, the result will be consistent with your choice of signs.

Example 2-4. Suppose $\mathbf{v}_0$ = -10 m/s, $\mathbf{v}$ = -2 m/s, and t = 2 s. Find $\langle \mathbf{a} \rangle$.

$$\langle \mathbf{a} \rangle = \frac{-2 \text{ m/s} - (-10 \text{ m/s})}{2 \text{ s}} = (+8 \text{ m/s})/2 \text{ s} = +4 \text{ m/s}^2$$

S: How can $\langle \mathbf{a} \rangle$ have the opposite sign of both velocities? What does it mean?

T: Think of what is happening physically here. The object starts with large negative velocity (towards the -x direction) and later has a smaller negative velocity. This means a slowing of the object's speed, represented by an acceleration vector opposite in direction to the velocities. This is often referred to as a deceleration. Again, using signs carefully in the definition yields the correct interpretation of direction in the result.

One Dimensional Motion With Constant Acceleration

Understanding this section should enable you to do the following:

1. *Interpret word problems involving one dimensional motion with constant acceleration.*
2. *Use the equations which describe such motion, both singly in the case of one unknown and simultaneously in cases of more than one unknown.*
3. *Calculate times of fall, distances, and speeds for the case of objects freely falling under constant gravitational acceleration.*

By restricting ourselves to one dimension, i.e. , a straight line, calculation with vectors is considerably simplified. In Chapter 4 we shall take up the more difficult situation of adding vectors in two and three dimensions.

Since on any straight line there are only two possible directions (up or down, front or back, right or left, etc.), and these are *opposite* to each other, *we can completely specify the direction of a vector in one dimension by either assigning it a positive or a negative sign.* Once that choice of sign is made, then any vectors in the same direction will have the same sign, and opposite vectors will have opposite signs. Vector addition is then the same as the *algebraic* addition of positive and negative numbers.

S: How do I know which direction should be considered to be + and which to be -?

T: That's a completely arbitrary choice for you to make. There is no right or wrong choice, but often you can sense that one would be more convenient than the other. Once made, however, you must retain that choice consistently throughout a problem. This means *all* vectors in the problem, i.e. velocities, accelerations, forces, etc., must share the same choice of sign.

The following equations fully describe one dimensional motion *when the acceleration is constant*. Here $\mathbf{s}$ stands for position, such as the x or y coordinate. $\mathbf{s}_0$, the initial coordinate, is very often taken to be zero for convenience. The other symbols have been introduced and used before. The vector signs remind you that you must be aware of the need to use signs consistently to represent direction.

a. $\langle \mathbf{v} \rangle = (\mathbf{v}_0 + \mathbf{v})/2$. (Mathematical definition of average)

b. $\mathbf{s} = \mathbf{s}_0 + \langle \mathbf{v} \rangle t$. (Operational definition of $\langle \mathbf{v} \rangle$).

c. $\mathbf{v} = \mathbf{v}_0 + \mathbf{a}t$. (Operational definition of $\langle \mathbf{a} \rangle$).

d. $\mathbf{s} = \mathbf{s}_0 + \mathbf{v}_0 + \frac{1}{2}\mathbf{a}t^2$. (Mathematical derivation using a., b., and c. above)

e. $v^2 = v_0^2 + 2a(s - s_0)$. (Mathematical derivation using a., b., and c. above)

S: This looks pretty confusing. How do I know which equation to use in a problem?

T: You must identify by symbol what information is given in the problem and what is asked for. Choose the equation or equations which involve the quantities you have identified in this way.

S: I always have trouble changing a word problem into equation form.

T: That translation problem is really the crucial step to problem solving. Read the words and phrases carefully. The following translation guide should help:

"When?" --------------------------- "What value of t?"
"Where?" -------------------------- "What value of $\mathbf{s}$?"
"Starts from rest" ----------------- "$\mathbf{v}_0 = 0$"
"How fast?" -------------------------"What value of $\mathbf{v}$?"
"How long does it take?" ---------- "What is t?"
"How far?" ------------------------- "What is $\mathbf{s} - \mathbf{s}_0$?"
"Comes to rest" ------------------- "$\mathbf{v}$ is 0"

Example 2-5. If an object starts from rest and undergoes a constant acceleration of 8 m/s^2, how fast is it moving when it has gone 36 meters?

S: The example tells us that v_0 = 0 and a = 8 m/s. It wants to know when $s - s_0$ = 36 meters, right? It doesn't make any mention of time. So what equation do I pick?

T: The one equation that contains **v**, **s**, and **a**, but not t, is equation (e.). Using it gives

$$v^2 = 0 + 2(8\ m/s^2)(36\ m) = 576\ m^2/s^2.$$

Thus v = 24 m/s.

Notice that if the problem had given **v** but left out one of the other quantities as unknown, the equation could just as well have been used. This points up the fact emphasized in the introduction--that this equation is *not* just a *formula for* **v**, but a *relationship* between **v**, **a**, and **s**.

Example 2-6. If a car traveling 20 m/s comes to rest 10 seconds after the brakes are applied, how far does it travel in coming to a stop?

S: v_0 and t are given, and $s - s_0$ is asked for. There is no equation which connects these. Is there enough information given?

T: You are also given that v = 0 at t = 10 s. You're right that no *single* equation is sufficient to give the answer. Equations c., d., and e. all involve **a**, which is another unknown. But remember, it's alright to have two unknowns as long as you can find two equations involving them. For instance, equation c. will give us the value of **a**, and then d. can give us $s - s_0$. Watch:

$$v = v + at$$

So

$$a = (v - v_0)/t = (0 - (20\ m/s))/10\ s = -2\ m/s^2.$$

Then

$$s - s_0 = v_0 t + \tfrac{1}{2}at^2 = (20\ m/s)(10\ s) + \tfrac{1}{2}(-2\ m/s^2)(10\ s)^2$$

$$= 200\ m - 100\ m = 100\ m.$$

S: Why couldn't I use equations a. and b.?

T: You can! It represents an alternate solution, without having to find **a**.

$$\langle v \rangle = (v_0 + v)/2 = (20\ m/s + 0)/2 = 10\ m/s.$$

Then

$$s - s_0 = \langle v \rangle t = (10\ m/s)(10 s) = 100\ meters.$$

Example 2-7. A police car is sitting a quarter of a mile off a main highway. He receives a report of a car, A, speeding along the highway at 70 miles per hour. The police car can accelerate at 28 $feet/s^2$. How near to the intersection can car A get and still have the police car head it off with a minute to spare? See the figure below for reference. Assume the velocities and acceleration are known to three significant figures.

S: What kind of motion does car A have? What equations describe it?

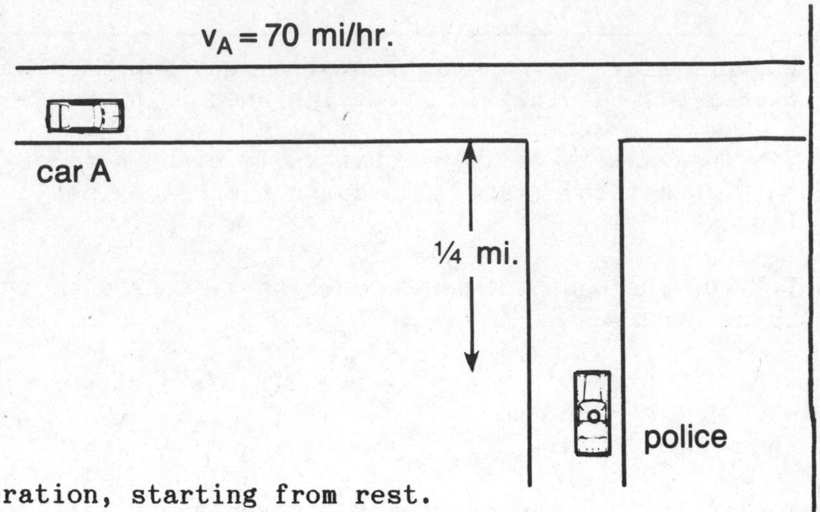

T: Since it moves with constant velocity, it has a = 0. Then

$$v = v_0 = 70.0 \text{ mi/hr, and}$$

$$s - s_0 = v_0 t = (70.0 \text{ mi/hr})t$$

S: What equations describe the police car's motion once it starts?

T: It undergoes constant acceleration, starting from rest.

$$v = 0 + at = (28.0 \text{ ft/s })t, \text{ and}$$

$$s - s_0 = 0 + \tfrac{1}{2}at^2 = \tfrac{1}{2}(28.0 \text{ ft/s}^2)t^2.$$

S: How long does it take the police to get to the intersection?

T: This means finding out what value of t corresponds to $s - s_0$ = 1/4 mile = 1320 feet. From above,

$$1320 \text{ ft} = \tfrac{1}{2}(28.0 \text{ ft/s}^2)t^2, \text{ or}$$

$$t^2 = (2 \times 1320 \text{ ft})/28.0 \text{ ft/s}^2 = 94.3 \text{ s}^2.$$

Thus $\underline{t = 9.71 \text{ s.}}$

S: How far does car A travel in this time?

T: Using the equations which apply to car A, we have

$$\begin{aligned} s - s_0 &= (70.0 \text{ mi/hr})t \\ &= (70.0 \text{ mi/hr})(9.71 \text{ s}) \times (1 \text{ hr/ } 3600 \text{ sec}) \\ &= 0.189 \text{ mi.} \times (5280 \text{ ft/1 mi}) = \underline{997 \text{ feet}}. \end{aligned}$$

S: So if the police started when car A was 997 feet from the intersection, they would arrive at the same time. How much further back would car A be if the police started one minute earlier?

T: This means, "how far does car A travel in one minute"?

$$\begin{aligned} s - s_0 &= 70.0 \text{ mi/hr} \times 1 \text{ min} \times (1 \text{ hr/60 min}) \\ &= 1.17 \text{ mi.} = \underline{6180 \text{ ft.}} \end{aligned}$$

The last digit is not significant. Do you agree?

S: How far from the intersection is car A when the police started?

T: 6180 feet + 997 feet. However, remember the rule of addition as it applies to significant figures. To get both numbers to the same precision, we must round off 997 feet to 1000 feet, the first two zeros being significant. The answer, to three significant figures, is 6180 + 1000 = <u>7180 ft</u>.

<u>Graphical Interpretation</u>. The equations of motion can be interpreted by displaying the relations between s and t, and v and t graphically:

<u>Zero Acceleration</u>: s increases *linearly* with t, while v is a constant.

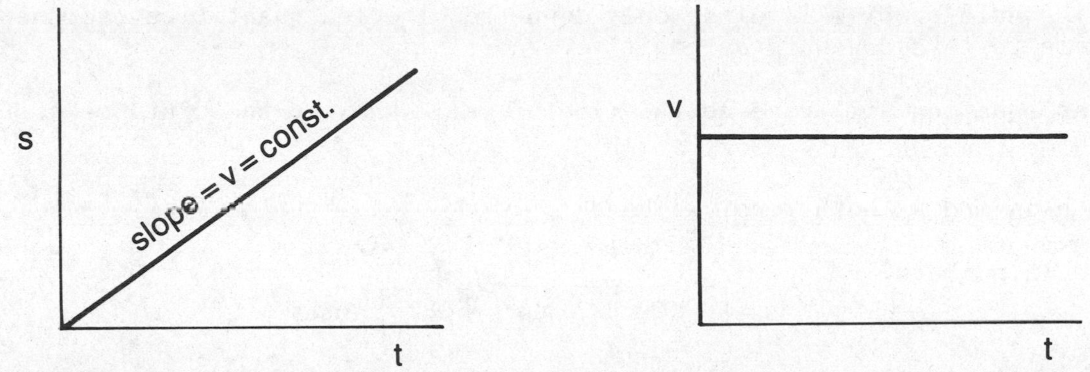

<u>Constant Acceleration</u>: s increases *quadratically* with t (in a parabolic curve). v increases linearly with t with a slope equal to the acceleration a.

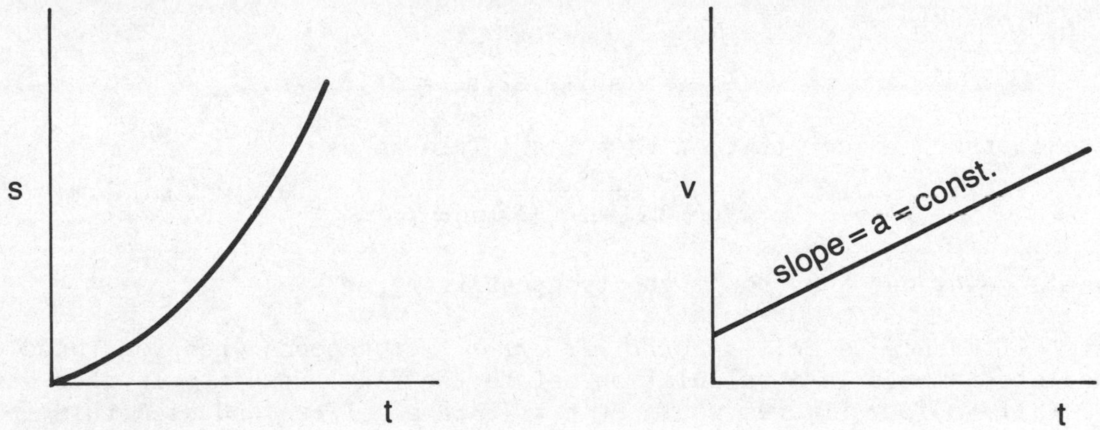

<u>Gravitational Free Fall</u>.

Understanding this section should enable you to do the following:

1. *Apply the principles of one dimensional motion under constant acceleration to problems involving objects falling under the influence of earth's gravity near the suface of the earth.*

In many cases involving dense objects, it is observed that their motion while falling near Earth's surface can be described by a constant downward acceleration of 9.80 m/s^2. This is given the symbol g and is referred to as the *gravitational free fall acceleration*. The equations of motion of the preceding section apply here by putting g in for a and carefully choosing signs in each problem according to convenience. s and s_0 will of course represent the vertical position coordinates of the falling object.

<u>Example 2-8.</u> You are standing on an observation deck 300 meters above the city street and drop a rock from rest. How long does it take to fall to the street, and with what speed does it hit?

S: What shall I take for the origin of coordinates, the street or where I'm standing?

T: Either choice will do, but if you choose the origin to be where you're standing, you can put $s_0 = 0$, which is a convenience.

S: What gives me a clue as to choice of positive direction?

T: This whole problem involves only downward directed quantities, so the best choice is to take downward as positive.

S: What equation will give me the time of fall, knowing the distance of fall is $s = 300$ m?

T: With s_0 and v_0 both zero, we have

$$s = 0 + \tfrac{1}{2}gt^2, \quad \text{or } t = (2s/g)^{\frac{1}{2}}$$

Putting in numbers,

$$t = \sqrt{2(300\text{ m})/9.80\text{ m/s}^2} = \sqrt{61.2\text{ s}^2} = 7.82\text{ s}$$

S: The next question is to find out speed at this time. I'll use $v = 0 + gt$.

T: O.K. This gives

$$v = (9.80\text{ m/s}^2)(7.82\text{ s}) = 76.6\text{ m/s}.$$

You have another choice, that of $v^2 = 2gs$. This gives

$$v = \sqrt{2(9.80\text{ m/s}^2)(300\text{ m})} = 76.7\text{ m/s}$$

S: What's going on? Why don't the two results agree?

T: This difference is called *round off error*. It happens when you round off intermediate numbers in a calculation before the final result. In this case, the time of fall to *four* figures would be $t = 7.825$ s. If we had used this in $v = gt$, our answer would have been 76.7 m/s to three figures, in agreement with the second method. This would have been the correct way to treat t when it was an intermediate number in the calculation before the final result.

Example 2-9. Vary the above problem by giving the rock an initial upward velocity of 10 m/s. Answer the same questions.

S: Now I have to be more careful with signs, don't I?

T: Yes, because you have both upward and downward motion in the problem. It's still best to choose $s_0 = 0$ as before. Except in the case where *everything* is downward, the usual choice for signs is upward positive, although the opposite choice would work also.

S: How has the distance vs. time equation changed?

T: Now we *can't* set $v_0 = 0$, so the equation looks like

$$s = (+10\text{ m/s})t + \tfrac{1}{2}(-9.80\text{ m/s}^2)t^2.$$

Notice carefully the + sign for the upward v_0 and the - sign for the downward g. The question is now to find t when s = -300 m. Again, this - sign is necessary since the street where the rock hits is 300 meters *below* the origin. Thus we solve for t from

$$-300 \text{ m} = (10 \text{ m/s})t - 4.90t^2.$$

S: I hate quadratic equations.

T: They are no more *difficult* than linear equations, just a bit more *tedious*. First get it into the standard form, $At^2 + Bt + C = 0$. Omitting the units temporarily, we have

$$-4.90t^2 + 10t + 300 = 0.$$

So A = -4.90, B = 10, and C = 300.

The form of the quadratic solution is

$$t = \frac{-B \pm \sqrt{B^2 - 4AC}}{-2A} = \frac{-10 \pm \sqrt{100 - 4(-4.90)(300)}}{2(-4.90)}$$

which gives t = +8.91 s. or -6.87 s.

S: Which is the right answer? What does the negative time mean?

T: Obviously, since the problem started at t = 0, only the positive time, t = 8.91 s, can be physically meaningful. Let's find the final velocity before we comment on the negative time solution. We have

$$v = v_0 + gt = +10 \text{ m/s} + (-9.80 \text{ m/s}^2)(8.91 \text{ s}) = -768 \text{ m/s}.$$

Notice the consistency of signs, representing a downward final velocity. As to the interpretation of the negative time solution, if a person standing at street level had thrown a rock upward with +768 m/s initial velocity 6.87 seconds before you had thrown yours from above (6.87 seconds *before* your t = 0), it would have reached your height at the instant of your throw, moving upward at 10 m/s, and would duplicate the motion of your rock from then on until they hit the street simultaneously with a velocity of -768 m/s. The total time of flight of this second rock would have been 6.87 + 8.91 sec = 15.78 seconds.

Example 2-10. In the previous example, find the maximum height of the rock.

S: What physical condition does "maximum height" imply?

T: It is the position, s, at the instant the velocity changes direction, or when v = 0. We call this a *turning point* of the motion. The equation relating s directly to v is

$$v^2 = v_0^2 + 2gs,$$

or

$$0 = (10 \text{ m/s})^2 + 2(-9.8 \text{ m/s}^2)s.$$

Thus s = 0.510 m at the top of the flight. The plus sign correctly shows this is above the point of throw.

Gravitational Acceleration on an Inclined Plane.

Inclined planes are flat surfaces inclined at an angle of θ relative to the horizontal, as shown in this figure.

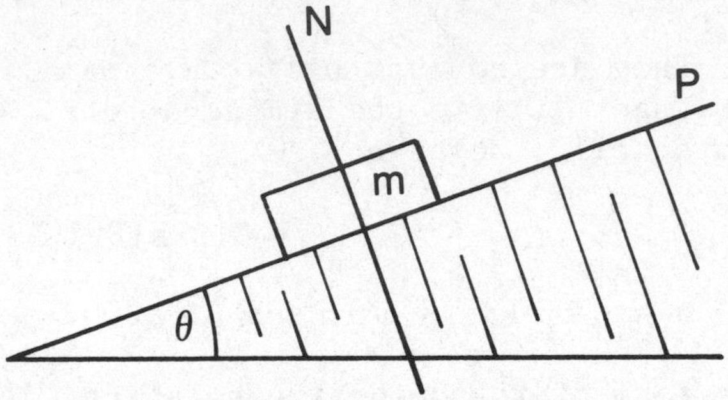

Objects on such planes are constrained to move along the plane. The most convenient choice of a coordinate system in which to describe motion in this situation is to take one axis (P) parallel to the plane's surface and the other axis (N) perpendicular, or *normal*, to the plane. All vectors in the problem are then *resolved* into components along these axes. For example, **the gravitational acceleration g** is directed *vertically downward*. Its components along and perpendicular to the plane are g sin θ and g cos θ, respectively. The rigidity of the plane exerts enough force to prohibit an object from moving perpendicular to the plane, so the only motion of an object on the plane will be an acceleration *g sin θ* downward along the plane (assuming no friction).

Example 2-11. An object is released from rest 2 meters from the end of a plane inclined 40 from the horizontal. There is a negligible amount of friction between the object and the plane. How long does it take for the object to reach the bottom of the plane?

S: The discussion in the text says that the component of acceleration down the surface of the plane is g sin θ. I still don't see this.

T: Let's look at the diagram showing the optimal choice of coordinate system. Notice that the line representing g cos θ is perpendicular to the x-axis and the line marked g is perpendicular to the horizontal. The angle between the x-axis and horizontal is θ.

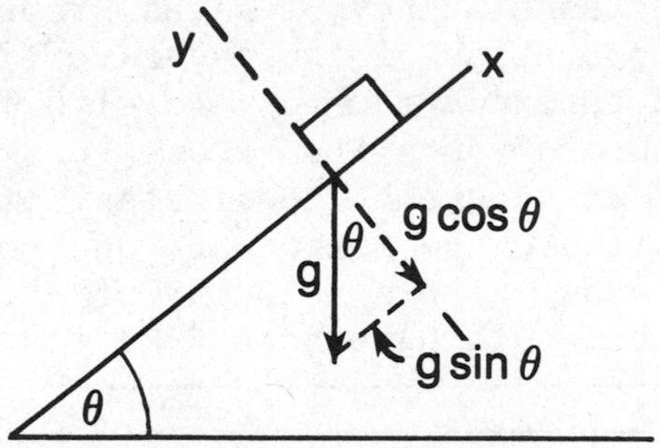

S: I seem to remember something from geometry saying that if two lines are perpendicular to two other lines which enclose an angle θ, the angle between the first pair of lines is also θ.

T: That's precisely the point. Then notice that g cos θ is perpendicular to the plane and that the component g sin θ is parallel to it. The rigidity of the plane allows motion only along the plane, so the only component of g that describes the allowed motion is g sin θ.

S: The one-dimensional equations of motion then apply to the x-axis then?

T: Exactly. Since the acceleration along x, g sin θ, is constant, you can apply them in the same way.

S: So we have a situation where $v_o = 0$, and we want to find the time t for the object to cover a given distance. The angle in this problem is θ = 40°, so g sin 40° = (9.8 m/s^2)(.643) = 6.30 m/s^2. This is the value of a to put into the equation $x = \frac{1}{2}at^2$, right?

T: That's right, where we want the value of t when x = 2 meters, measured along the x-axis:

$$2 \text{ m} = \tfrac{1}{2}(6.30 \text{ m/s}^2)t^2,$$

or

$$t^2 = 2(2 \text{ m})/(6.30 \text{ m/s}^2); \qquad \underline{t = 0.797 \text{ s}}$$

to 3 significant digits.

QUESTIONS ON FUNDAMENTALS

1. A person rides 12 miles north in 30 minutes and then 3 miles west in 15 minutes. What is her average *speed*?

(a.) 18 mi/hr (b.) 20 mi/hr (c.) 0.33 mi/hr (d.) 12.4 mi/hr

2. In question 1, what is the magnitude of her *displacement*?

(a.) 9 mi (b.) 15 mi (c.) 19.3 mi (d.) 12.4 mi (e.) 11.3 mi

3. In question 1, what is the magnitude of her *average velocity* for the whole trip?

(a.) 16.5 mi/hr (b.) 20 mi/hr (c.) 18 mi/hr (d.) 12.3 mi/hr
(e.) 15 mi/hr

4. In question 3, what is the *direction* of her velocity?

(a.) 75.5° W of N (b.) 14° W of N (c.) 14° N of W
(d.) 75.5° N of W

5. If an object moves with constant acceleration of 5 m/s^2, starting from rest, what will its average velocity be during the first 10 seconds?

(a.) 25 m/s (b.) 0.5 m/s (c.) 50 m/s (d.) 250 m/s

5a. A particle is dropped from rest. After 3 seconds, how fast is it going?

(a.) 44.1 m/s (b.) 9.8 m/s (c.) 14.7 m/s (d.) 5.4 m/s
(e.) 29.4 m/s

5b. How far has the above object fallen after 3 seconds?

(a.) 44.1 m (b.) 176 m (c.) 29.4 m (d.) 14.7 m (e.) 58.8 m

6. If a ball is thrown with a speed of 15 m/s at an angle of 50° above the horizontal, how long before it hits the same level from which it was thrown?

(a.) 1.2 s (b.) 3.1 s (c.) 1.97 s (d.) 2.4 s (e.) 1.55 s

GENERAL PROBLEMS

Problem 2-1. A lead weight is dropped from rest into a lake from a platform 10 meters above the water. Once it hits the water, it sinks with a constant velocity equal to 1/10 the velocity it had acquired up to impact with the water. It reaches the bottom of the lake 5 seconds after impact with the water. How deep is the lake?

Q1: How fast is the weight going just before it hits the water? By that time it has fallen 10 meters. How fast does it sink?
Q2: How far does the weight travel in 5 seconds at its sinking velocity?

Problem 2-2. You are on the observation deck of Example 2-14, and drop a rock from rest. A friend stands on the street directly below and throws a rock upward at the same instant, with an initial speed of 50 m/s. Assuming they are moving head on toward each other, where will they collide? When? Will your friend's ball be rising or descending when they collide?

Q1: I have to set up equations of motion for both objects. They should be in the same coordinate system. If the deck is at $s = 0$, what is the initial position s_0 for the rock thrown from the street? What are the initial velocities for both objects?

Q2: What are the specific equations for s_1, v_1, and for s_2 and v_2? Are all units and signs consistent?

Q3: What is the mathematical condition for the collision? Can I find from this at what time the collision occurs?

Q4: What are the positions s_1 and s_2 at this time? Does that make sense?

Q5: How long does the rock thrown upward rise? Is this time longer or shorter than the time at collision?

Problem 2-3. Suppose a sports car has the ability to speed up with an acceleration $a_+ = 24\ \mathrm{ft/s^2}$, and to brake with a maximum deceleration of $a_- = -32\ \mathrm{ft/s^2}$. Starting from rest, what minimum time would it take this car to reach a stop sign 1/4 mile away by accelerating as hard as possible, followed by a period of maximum deceleration to a stop at the sign?

Q1: There will be a period of max acceleration followed by a period of max deceleration. I'll call these t_1 and t_2. The distances covered during these times will be s_1 and s_2. What does the problem say about s_1 and s_2?

Q2: By the time t_1 has occurred, the car will be at s_1 with some speed v_1. What are the specific equations connecting these?

Q3: For the deceleration period, what is the initial speed? I know the velocity must go to zero at the end of this time, t_2. What are the specific equations for this period?

Q4: Do I have a_- as a negative sign? Are all other signs and units consistent?

Q5: Just what are my unknowns? How many are there? Do I have enough equations?

Q6: The rest is mathematics. How do I solve simultaneous equations?

ANSWERS TO QUESTIONS

QUESTIONS ON FUNDAMENTALS:

1. (b.)
2. (d.)
3. (a.)
4. (d.)

5a. (e.)
5b. (a.)
6. (d.)

GENERAL PROBLEMS:

Problem 2-1.

A1: $v = 2as = 14$ m/s just before impact.
It sinks at $.1v = 1.4$ m/s.

A2: $s = 1.4$ m/s x 5 s = 7 meters.

Problem 2-2.

A1: $s_0 = -300$ m for the rock thrown upward.
$(v_1)_0 = 0$; $(v_2)_0 = +50$ m/s.

A2: $s_1 = -4.9t^2$; $v_1 = -9.8t$
$s_2 = -300 + 50t - 4.9t^2$; $v_2 = +50 - 9.8t$.
All numbers are in the SI units.

A3: The collision occurs where $s_1 = s_2$. Yes, $t = 6$ s.

A4: $s_1 = s_2 = -176.4$ m. Yes. Both equations should give the same position at this time.

A5: 5.1 s. Collision time is later than this.

Problem 2-3.

A1: $s_1 + s_2 = 1/4$ mile = 1320 ft.

A2: $s_1 = 12t_1^2$; $v_1 = 24t_1$

A3: Initial speed for deceleration is v_1. $s_2 = v_1t_2 - 16t_2^2$;
$v_2 = 0 = v_1 - 32t_2$.

A4: All numbers are expressed in feet and seconds.

A5: There are 5 unknowns, s_1, v_1, s_2, t_1, and t_2. Counting the equation from Q1, there are 5 equations.

A6: Carefully.

CHAPTER 3. FORCES AND NEWTON'S LAWS OF MOTION

BRIEF SUMMARY AND RESUME

Continuing under the restriction to one dimension, the concept of *force* is introduced and its vector nature is discussed. The idea of *inertia* is developed in Newton's *First Law of Motion*, and action-reaction pairs of forces are discussed in the *Third Law of Motion*. The introduction of Newton's *Second Law of Motion* connects the acceleration of objects to the forces acting on them, and provides the basis of what is called *dynamics*, or the study of the physical *cause* of various types of motion. *Mass* is operationally defined by the Second Law. *Friction* and other forces resistive to motion are examined, and with them the concepts of *normal force* and *coefficients of friction* are introduced.

Material from previous chapters which is used directly in this chapter includes the following:

1. Equations of motion when acceleration is constant.

   $v = v_0 + at$; $v^2 = v_0^2 + 2a(s - s_0)$; $s - s_0 = v_0t + \frac{1}{2}at^2$

2. Free fall acceleration at Earth's surface.

   $g = 9.80$ m/s

3. Methods of addition of vectors.

NEW DATA AND PHYSICAL CONSTANTS

1. <u>SI Unit of Force</u>

   1 newton (N) = 1 kg-m/s^2 = .225 pounds

NEW CONCEPTS AND PRINCIPLES

<u>Forces</u>

Understanding this section should enable you to do the following:

1. *Identify various types of forces commonly encountered.*
2. *Calculate the resultant of a number of individual forces acting along one dimension.*
3. *State the operational definition of mass.*
4. *For each force acting on an object, identify the reaction force.*
5. *Knowing two of the following: mass, net force, and acceleration, calculate the third.*

Force--A physical interaction which if applied alone to an object, will produce an acceleration of that object.

We constantly encounter a variety of forces in our daily lives. Table 3-1 is a brief list of a few commonly experienced types of forces, all of which will be explained in detail throughout the text.

Table 3-1. Some Common Types of Forces

Type	Examples
Forces of *compression*, involving rigid objects.	Forces that support and bear weight. Exerted by floors, beams, shelves, etc.
	Forces occurring during collisions between solid objects.
	Forces that occur between solid objects which are pushed together.
	Forces arising from fluid pressure.
Forces of *tension*	Pulling forces exerted through strings, ropes, cables, or other objects attached to a given object. The opposite of compression.
Forces of *friction* or *viscosity*	Forces which resist the slipping or sliding motion between solid surfaces in contact.
	Forces which resist the movement of an object through gases or fluids.
Fundamental forces which act between physically separated objects	Gravity acting between any material objects.
	Electrical force between objects possessing electrical charge.
	Magnetic forces between electrical currents.

Forces are *vectors*, since the direction of an object's acceleration will depend on the direction,in which the force is applied. The symbol used generically to denote force is **F**. The SI unit of force is the *newton*, symbolized by N, and defined as follows:

Newton--The SI unit of force. An amount of force which will give an acceleration of 1 m/s^2 when acting alone on a 1 kg mass. The dimensionality of force is given in the section involving the Second Law of Motion.

When a number of separate forces act simultaneously on an object, their *vector sum* is called the *net force* or *resultant force*. If all the various forces were replaced by a *single* force equal in magnitude and direction to this net force, it would produce the *net* effect of the combination.

In one dimension the problem of finding the vector sum of forces again reduces to designating signs to represent the two possible directions, so that forces in the same direction add and those in opposite directions subtract.

Example 3-1. Suppose a force $\mathbf{F}_1$ = 10 N is acting upward on an object and forces $\mathbf{F}_2$ = 3 N, $\mathbf{F}_3$ = 2 N, and $\mathbf{F}_4$ = 7 N are acting downward. What is the net force on the object?

S: We have to make a choice of signs first. How about upward + and downward - ?

T: That's the usual choice. Then we have

$$\mathbf{F}_{net} = \mathbf{F}_1 + \mathbf{F}_2 + \mathbf{F}_3 + \mathbf{F}_4 = 10\text{ N} + (-3\text{ N}) + (-2\text{ N}) + (-7\text{ N}) = -2\text{ N}.$$

This tells us the net force is 2 N downward, and the object would react as if this *single* force was acting on it.

Newton's First Law of Motion. An object at rest will remain at rest and an object in motion will continue to move in a straight line at constant speed unless some net external force acts to change this motion.

Important Observations

1. This property of an object tending to maintain its state of motion is referred to as the object's *inertia*. The more inertia, the stronger this tendency. The *quantitative* measure of inertia is the object's *mass*.
2. Changing the state of motion requires a net *external* force. An object cannot produce changes in its speed or direction by *internal* forces.

Newton's Second Law of Motion. When a net force $\mathbf{F}_{net}$ acts on an object having a mass m, the object undergoes an acceleration **a** in the direction of the net force, given by

$$\mathbf{a} = \mathbf{F}_{net}/m$$

Corollary to the Second Law. For a given force acting on various objects, the mass of an object is inversely proportional to its acceleration.

$$m_1/m_2 = a_2/a_1$$

Important Observations

1. The additional fundamental dimension of *mass* is introduced here. The *standard* SI mass of 1 kilogram (kg) is defined in Section 2.1 of your text.
2. You should appreciate the impressive *generality* of the 2nd Law. *Any* kind of forces or combination of forces are included. Whatever their source, you have only to find the vector sum of all forces on an object. This is the net force which determines the acceleration of the object.
3. The *dimensionality* of force can be derived from the 2nd Law in the form F = ma. Thus dimensions of force are $[M]x[L]/[T]^2$. In SI units, 1 kg m/s^2 is called 1 *newton* (N).
4. The 2nd Law is a *vector* equation. Since in a rectangular coordinate system the x, y, and z axes are mutually perpendicular, it is possible to write a one dimensional version of this equation along each axis separately. Thus

$$a_x = \frac{(F_{net})_x}{m}; \quad a_y = \frac{(F_{net})_y}{m}; \quad a_z = \frac{(F_{net})_z}{m}$$

5. At present we are still considering motion in only one dimension, for example along the x axis. This means that, even though some forces may be acting perpendicular to that axis, the *net* perpendicular force to the direction of motion must be *zero*. Otherwise, an unbalanced perpendicular force would produce a change in the *direction* of motion, which would then no longer be in one dimension.
6. The motion of objects with constant acceleration which we described in the last chapter now can be explained physically. They are situations where the net force on the object remains constant.

Newton's Third Law of Motion. Whenever an object A exerts a force on object B, B exerts an equal force on A in the opposite direction.

Important Observations

1. This law deals with interactions between *pairs* of objects, stating that in *any* kind of interaction, when one object exerts a force, which we'll call an *action*, on another object, this second object exerts an *equal* and *opposite* force of *reaction* on the first object.
2. These two forces are *not* applied to the same object. Each is applied to *one* of the interacting objects.
3. The *effects* of these equal forces may be very unequal, depending on the masses of the interacting objects.

Example 3-2. In a collision between a large truck and a subcompact car, which vehicle is hit harder?

S: What does "hit harder" mean?

T: Good point. It's an example of the lack of precision in much of our common language. We should probably translate this as "hit with greater force." Then the Third Law tells us that the car hits the truck with just as much force as the truck hits the car, but in the opposite direction.

S: That doesn't seem to be the case when you observe the result!

T: Don't confuse the *result* with the *force* that acted. The result depends on how much mass is reacting to the force. Obviously the smaller mass--the car--will be affected more than the heavy truck by the same amount of force.

Newton's Third Law is very important. In the exercises which follow, you should try to identify the pairs of interaction forces involved. Table 3-2 contains a list of common examples which should be of help. The choice of which force is action and which is reaction is entirely arbitrary.

Table 3-2. Examples of Action-Reaction Pairs

Action	Reaction	Comments
Your weight pressing down on a chair	The rigid chair pushes *up* on you, supporting your weight	If the seat of the chair rips or breaks, down you go!
The *rearward* frictional force of a car's tires *on the road* when the car accelerates	The *forward* frictional force of the road *on the tires* (and hence on the car) causing the car to accelerate	If there is glare ice (i.e., no friction), the wheels spin, but no acceleration of the car results
The *forward* force of the car seat *on you* causing you to accelerate with the car	The *backward* force you exert *on the car seat* causing you to "sink" into it	If it is a reclining seat that is not locked into place, you'll end up in a horizontal position as the car accelerates
The force of a bat *on a ball* causes the ball to fly over the fence for a home run	The ball hits the bat *backward* with an equal force	Sometimes this force is large enough to break the bat
When you throw a heavy anchor horizontally over the stern of a boat, you exert a *rearward* force *on the anchor*	The anchor exerts a *forward* force *on you* (and hence on the boat), causing you and the boat to lunge forward	This is the principle by which jet engines and rockets operate. They are called "reaction engines"

Force Vectors

Understanding this section should enable you to do the following:

1. *Identify the forces acting on an object and draw its free body diagram.*

In most problems, a number of forces from different causes are acting simultaneously on a given object. What happens to the object depends on the resultant, or net, force according to Newton's 2nd Law. It is of basic importance, then, to be able to identify the existence of all forces acting on an object by inspecting which influences such as gravity, friction, reaction forces, etc. may apply to a given problem situation. Often you can identify the *direction* of the forces even though their magnitude is not immediately known.

Free Body Diagram--A diagram which abstracts an object from all extraneous factors in its physical environment, identifying only the *forces* acting on it from all influences in its surroundings.

In order to illustrate, we will draw free body diagrams for (a.) a falling ball, (b.) a picture hanging on the wall, and (c.) a box on a table with someone exerting a pull on it.

(a.)

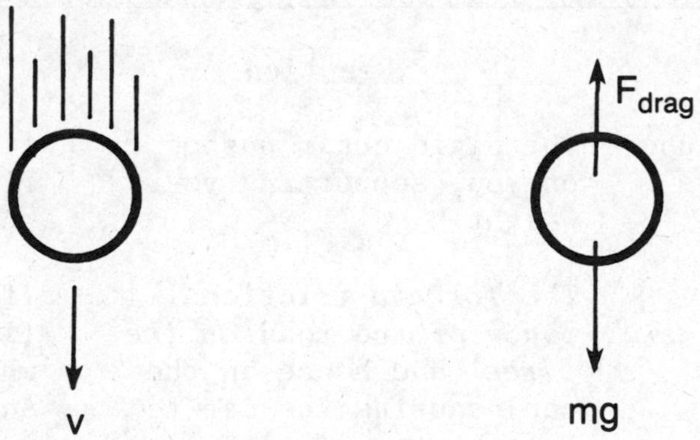

(b.)

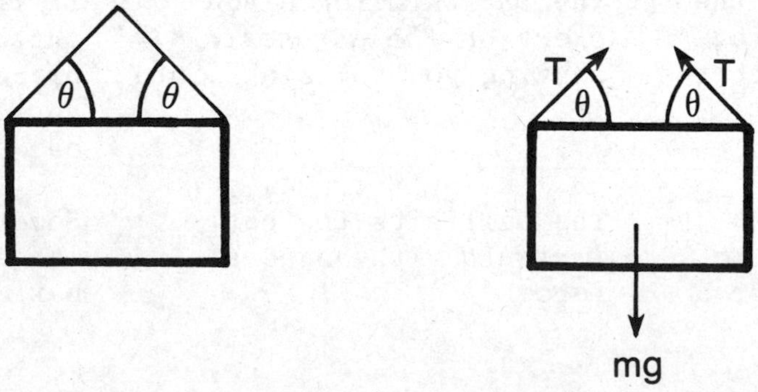

Here T represents the force of tension in the strings.

(c.)

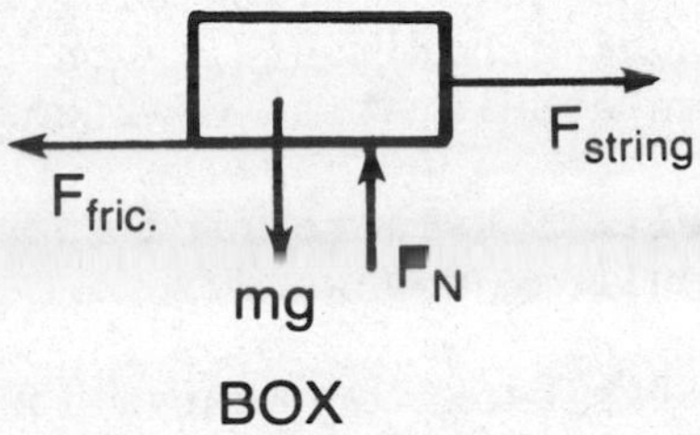

Below are the free body diagrams for the boy and the table. Notice how these diagrams reveal the sometimes hidden reaction forces.

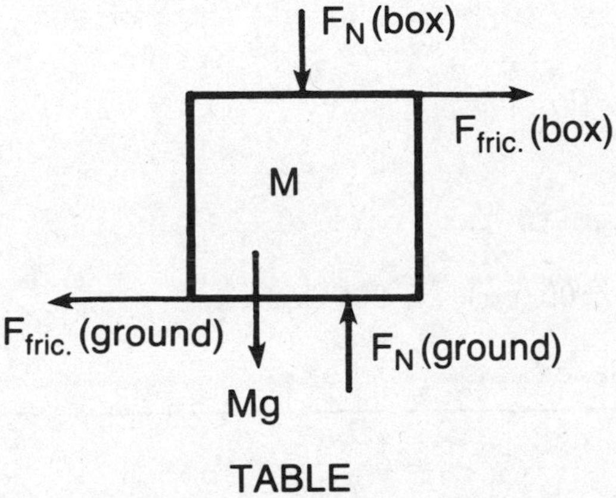

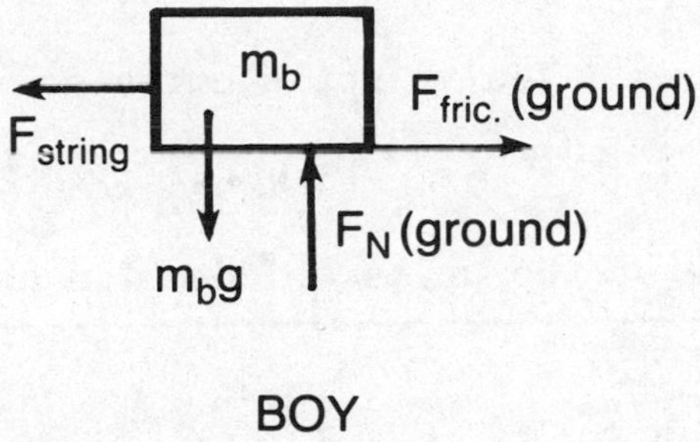

These diagrams are meant merely to show the existence of each force and its direction. Nothing is meant to be implied about their magnitudes in the diagrams. Remember, when you want to know how a given object is reacting, only those forces *acting on that object* are of interest. In this example, try to categorize each force into one of the following:

1. Forces of compression, or forces that push objects together.
2. Forces of tension, or pulling forces.
3. Gravitational force.
4. Resistive forces.

<u>Example 3-3</u>. A 30 kg object is being pulled upward by a crane with a force of 300 N. The earth's gravity is pulling down on the object with a force of 294 N. What is the acceleration of the object?

<u>S</u>: There are two forces acting. What is the *net* force?

<u>T</u>: The net force is 300 N - 294 N = 6 N *upward*. Then the acceleration is

$$a = (6\ \text{N})/(30\ \text{kg}) = 0.2\ (\text{kg-m/s})/\text{kg} = 0.2\ \text{m/s}^2$$

Notice how the derived unit (N) can be written in fundamental form for cancellation purposes.

Example 3-4. I am pushing a box along a horizontal table with a force of 15 N while you pull horizontally in the same direction with a force of 10 N. Friction is dragging horizontally on the box with a force of 9 N. We observe the box to have an acceleration of 2 cm/s^2. What is the box's mass?

S: Well, I have to find F_{net} again. But what do I do about the *weight* of the box, pulling downward?

T: Refer back to observation #5 above. The table is *constraining* the box to move horizontally. The contact force of support provided by the rigidity of the table is upwards and *must* balance the box's weight, so there is a net force only in the horizontal direction. We have

$$F_{net} = 15\text{ N} + 10\text{ N} + (-9\text{ N}) = 16\text{ N},$$

in the direction we are exerting our forces.

Then $m = (16\text{ N})/(2\text{ cm/s}^2)$.

S: This gives me m = 8 kg, right?

T: No way! You've got a units conversion problem to take care of.

$$m = (16\text{ kg-m/s}^2)/\{(2\text{ cm/s}^2)/(1\text{ m}/100\text{ cm})\} = 800\text{ kg}$$

Be aware of the units "hidden" in derived quantities.

Friction

Understanding this section should enable you to do the following:

1. *Identify the direction of the frictional forces in a given situation.*
2. *Calculate the relationship between normal force and frictional forces for various surfaces.*
3. *State the condition for a falling object to attain terminal velocity.*

Frictional forces are those between two surfaces in contact which resist the relative motion of the surfaces parallel to themselves. There are two categories of friction:

Static friction--A force which resists an attempt to *start* the surfaces sliding.

Sliding friction--A force resisting an existing relative motion between the surfaces in contact.

Experimentally, the magnitude of these forces is found to depend on the nature of the surfaces, such as composition and degree of roughness, and on the force perpendicular to a surface. A perpendicular to a surface is called the *normal* to the surface, so the perpendicular force of compression between the surfaces is called the normal force, F_N.

The composition and condition of the surfaces is summarized by experimentally measured *coefficients of friction*, μ_k and μ_s, which are tabulated in Table 3.1 of your text.

In mathematical form, the forces of friction are:

$$F_{\text{static friction}} \le \mu_s F_N$$

$$F_{\text{sliding friction}} = \mu_k F_N.$$

Important Observations

1. The *direction* of friction forces is always *parallel* to the surfaces, and opposed to the sliding motion or the tendency to slide.
2. The *amount* of friction force is proportional to the compressional force *perpendicular* to the surfaces.
3. The *maximum* static friction force is $\mu_s F_N$. The static force can take on *any* value up to this maximum in order to oppose an attempt to make the surfaces slip.
4. μ_s is usually larger than μ_k. They are dimensionless numbers.
5. $F_{\text{sliding friction}}$ does *not* depend on the *speed* of sliding.

Example 3-5. How much horizontal force must you exert on a 2 kg block of oak held against a vertical oak wall in order that the block doesn't fall? Assume the wood grain in the block and wall are parallel. The figure below shows the situation.

S: How do I translate "doesn't fall?"

T: The weight of the block will cause it to fall unless there is an equal amount of upward force. This weight is w = mg = (2 kg)x(9.80 m/s^2) = 19.6 N.

S: I know that, but how can I create an *upward* force by pushing *horizontally*?

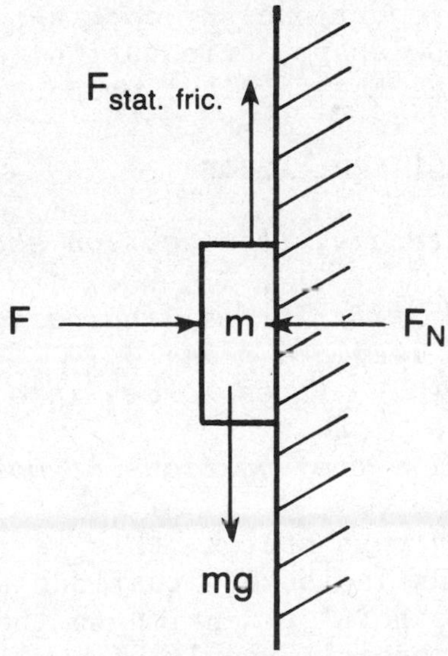

T: You are creating a normal force between the block and the wall. That creates a static frictional force *parallel* to the wall, opposing any attempt to move vertically. Since gravity attempts to pull the block down, the frictional force will be *upward*. From Table 3.1 in your text, μ_s = 0.62 for this case, and thus

$$F_{\text{static fric.}} \le \mu_s F_N = 0.62\ F_N.$$

You need $F_{\text{static fric.}}$ = 19.6 N for the block to stay. This means

$$19.6\ \text{N} \le 0.62\ F_N, \qquad \text{or} \quad F_N \ge \underline{31.6\ \text{N}}.$$

By the way, the above figure shows the forces acting *on the block*. Can you identify the forces acting on the wall? They are reactions to two of the forces shown. On what do the reactions to the other two forces act?

Air Resistance and Terminal Velocity. As an object moves through the air or a fluid, the molecules exert a force resistive to the motion, called the *force* of *air resistance* or *viscosity*. For falling objects this produces an upward force, opposing the force of gravity. Experimentally it is found that, for low speeds, this force is proportional to the object's speed. This can be written as

$$F_{vis} = -kv \ .$$

Important Observations

1. The minus sign indicates the force is *opposite* in direction to the velocity, i.e., a *resistive* force.

2. A graph of F_{vis} vs. v shows that the magnitude of the force increases linearly with v. On the same graph, F_g is shown as a constant force, independent of v. When the two forces are equal, $F_{net} = 0$, the object no longer has an acceleration, and falls thereafter with a constant speed. This is called the *terminal speed*, v_{term}, and is given by

$$v_{term} = mg/k$$

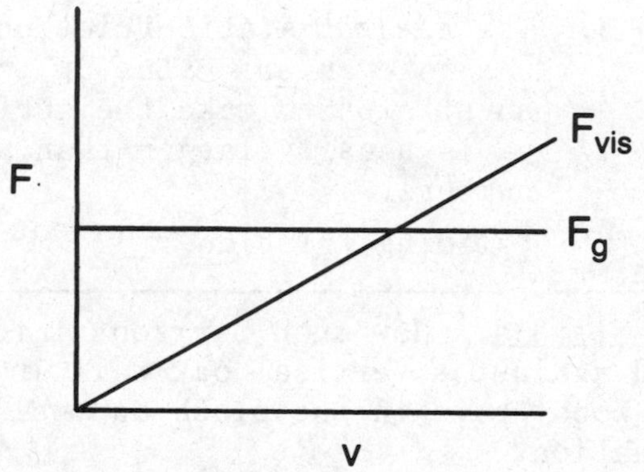

3. k is an experimentally determined proportionality constant, and depends on the shape, orientation, etc. of the object and the physical properties of the air or fluid.

Motion on Inclined Planes

Understanding this section should enable you to do the following:

1. *Identify forces, including friction, acting on an object on an inclined plane.*
2. *Resolve these forces into components parallel and normal to the plane.*
3. *Solve the equations of motion for the object along the plane.*

As your text states, this is really one-dimensional motion again, except that not all forces in these situations are purely parallel or normal to the motion, so we must pay careful attention to the choice of coordinate systems and resolution of vectors. A general example of inclined plane is shown in Chapter 2, with its surface making an angle θ relative to horizontal.

<u>No Friction</u>. The free body diagram of mass m on the plane with no friction or applied forces other than gravity is shown in the figure to the right.

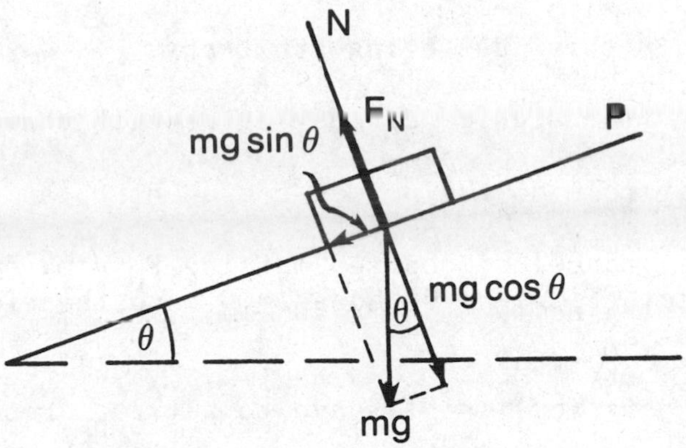

<u>Important Observations</u>

1. The convenient coordinate system is shown, with the P-axis parallel to the plane and the N-axis *normal* to the plane.
2. All motion must be *along the plane*, which means **v** and **a** must lie *entirely* along the P-axis.
3. This means that all force components along the N-axis must cancel, and any net force which exists must be entirely along the P-axis.

<u>Example 3-6</u>. Let's add various applied forces to the above figure. These may be tensions in attached strings, direct pushes, or whatever. In each case, find an expression for the acceleration in terms of mg, F_{app}, and θ.

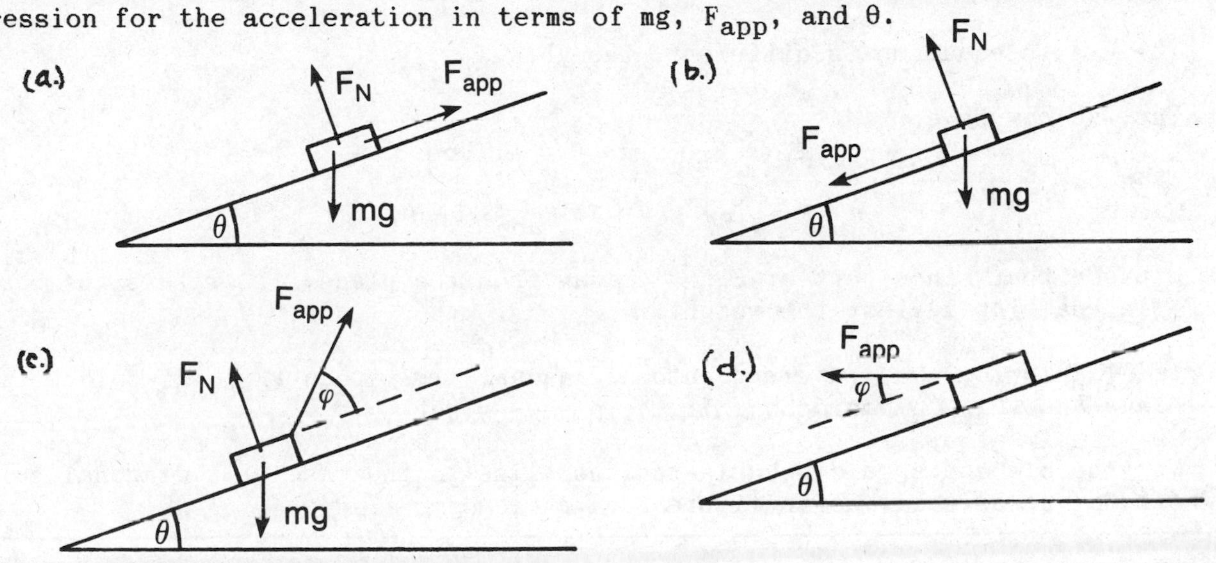

<u>S</u>: In (a.) and (b.) F_{app} is purely along the P-axis, so it will directly add or subtract from the P component of weight, mg sin θ, right? The net force will be
F_{app} - mg sin θ for (a.) and
F_{app} + mg sin θ for (b.).

<u>T</u>: *Partly* right. There's no question of the direction of motion in (b.), down the plane. But be careful in (a.). You can't assume that F_{app} is necessarily greater than mg sin θ. So Newton's 2nd Law would give

$$F_{app} - mg \sin\theta = ma_p, \quad \text{if } F_{app} > mg \sin\theta$$

or

$$mg \sin\theta - F_{app} = ma_p, \quad \text{if } F_{app} < mg \sin\theta$$

<u>S</u>: Isn't this just being arbitrary?

<u>T</u>: No. The sign of a_p must agree with the sign of the net force.

<u>S</u>: What role do the normal components play?

<u>T</u>: They cancel of course, meaning $F_N = mg \cos\theta$ in both (a.) and (b.). This says that $(F_{net})_N = 0$, so $a_N = 0$, and therefore $a_{total} = a_p$.

<u>S</u>: In part (c.), we'll have to get F_{app} resolved into our coordinate system.

<u>T</u>: Right. $(F_{app})_p = F_{app} \cos\theta$, directed up the plane

and $(F_{app})_N = F_{app} \sin\theta$, directed away from the plane.

<u>S</u>: Except for cos θ, the equations along the P-axis are the same as before.

<u>T</u>: That's right. Depending on whether the motion is up or down the plane, we have

$$F_{app} \cos\theta - mg \sin\theta = ma_p$$

or

$$mg \sin\theta - F_{app} \sin\theta = ma_p$$

<u>S</u>: The N-axis equation is different, though.

<u>T</u>: Yes, we now have

$$F_N + F_{app} \sin\theta = mg \cos\theta,$$

or

$$F_N = mg \cos\theta - F_{app} \sin\theta .$$

You can see that since part of F_{app} is away from the plane, it *reduces* the burden of F_N in reacting against the weight.

<u>S</u>: If F_{app} in (c.) was *reversed* into a diagonal *push*, then $(F_{app})_p$ would be *down* the plane and $(F_{app})_N$ would be *into* the plane, right?

<u>T</u>: Very good! You should be able to generalize to the case of a diagonal pull down the plane as well, as in figure (d.).

<u>Example 3-7</u>. Suppose we take a specific case, where θ = 25°, m = 3 kg, and a force is applied as in part (c.) of the previous example with θ = 40 and F_{app} = 20 N. If the block starts from rest 10 meters from the end of the ramp, how long does it take to reach the end?

<u>S</u>: It sounds like a problem using the one dimensional equations of motion. This *is* one dimensional motion, but the equations require a to be constant. Is it?

<u>T</u>: Let's look. From before, noting by inspection that in this case $mg \sin\theta > F_{app} \cos\theta$, we can write

$$(F_{app} \cos\theta)/m - g \sin\theta = a_p.$$

Everything determining a_p is constant, so the equations of motion we've used previously *do* apply. Putting in the numbers, we get

$$a_p = (20\ N)(\cos 40^\circ)/(3\ kg) - (9.80\ m/s^2)\sin 25^\circ = 0.97\ m/s^2 .$$

S: Now we can let $s - s_0 = 10$ meters along the plane and find t.

T: And also set $v_0 = 0$. So with $s - s_0 = \frac{1}{2}(a_p)t^2$,

$$10\ m = \tfrac{1}{2}(0.97\ m/s^2)(t^2), \quad \text{giving } t^2 = 20.6\ s^2$$

or

$$t = 4.54 \text{ seconds} .$$

S: We calculated (F_{app}) without friction before, and it was *up* the plane,

$$(F_{net})_p = F_{app} \cos \theta - mg \sin \theta$$

$$= (20\ N)\cos 40^\circ - (3\ kg)(9.8\ m/s^2)\sin 25$$

$$= 2.9\ N, \text{ up the plane.}$$

T: So the *tendency* is to slide *up* the plane. Can static friction hold the mass?

S: You mean can $F_{stat.frict.}$ *down* the plane be as great as the force we just calculated *up* the plane?

T: Exactly. Look at the free body diagram:

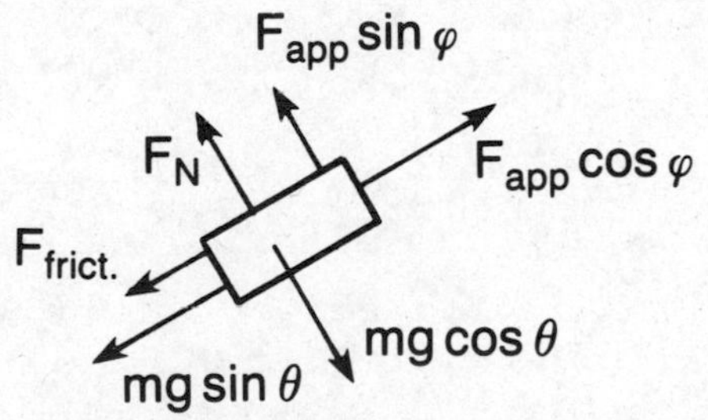

S: We have to know F_N in order to calculate maximum static friction.

T: Good: Balancing forces in the N-direction gives us

$$F_N + F_{app} \sin \theta = mg \cos \theta,$$

or

$$F_N = mg \cos \theta - F \sin \theta$$

$$F_N = (3\ kg)(9.8\ m/s^2) \cos 40^\circ - (20\ N) \sin 40^\circ = 13.7\ N.$$

S: This is a lot bigger than 2.9 N.

T: But it is not the force of friction. $F_{stat.\ frict.} \leq \mu s\ F_N$. In this case, $F_{stat.\ frict.} \leq 0.20(13.7\ N) = 2.7\ N$.

S: So friction *can't* make the block stick. The net force will be 2.9 - 2.7 N = 0.2 N up the plane.

T: *Watch out*! Many people have fallen into that trap! We had to find out *whether* the block would move, so we calculated the max static friction. *When we find out it does move, the resistive force is not static friction, but sliding friction, with a different coefficient.* Thus

$$F_{\text{sliding fric.}} = \mu_k F_N = 0.10(13.7\ \text{N}) = 1.4\ \text{N},$$

so

$$(F_{net})_p = 2.9 - 1.4\ \text{N} = 1.5\ \text{N}\ , \textit{ up the ramp}$$

including sliding friction. Then

$$a_p = (F_{net})_p/M = (1.5\ \text{N})/(3\ \text{kg}) = 0.5\ \text{m/s}^2.$$

QUESTIONS ON FUNDAMENTALS

1. If a 2 kg object is observed to be accelerating at $a = 4\ m/s^2$, the net force on the object must be

(a.) 2 N (b.) 0.5 N (c.) 8 N (d.) 6 N

2. If one person pulls on a 10 kg mass with a force of 5 N and another person pulls in the opposite direction with a force of 2 N, what is the acceleration of the mass?

(a.) 30 m/s^2 (b.) 0.7 m/s^2 (c.) 1.4 m/s^2 (d.) 0.3 m/s^2
(e.) zero

3. A force of 3 N acts on an object along the +x axis and a second force of 3 N acts along the -y axis. If the object moves at constant velocity, there must be an additional force F of

(a.) 4.24 N, 45° below the +x axis
(b.) 6 N, 45° above the -x axis
(c.) 18 N, 45° above the -x axis
(d.) 4.24 N, 45° above the -x axis
(e.) 6 N, 45° below the +x axis

4. If a force of 15 N acts on a 6 kg object in a direction 30^c above the +x axis, what is the x-component of the object's acceleration?

(a.) 2.17 m/s^2 (b.) 2.5 m/s^2 (c.) 5 m/s^2 (d.) 9.8 m/s^2
(e.) 0.4 m/s^2

5. If the coefficient of sliding friction between a 3 kg mass and the floor is 0.4, what is the force of sliding friction the floor exerts on the mass?

(a.) 1.2 kg (b.) 0.75 N (c.) 11.8 N (d.) 1.2 N (e.) 38.4 N

6. If a force of 12 N acts alone on a 2 kg mass that is initially at rest, how far will the mass have traveled in 5 seconds?

(a.) 150 m (b.) 300 m (c.) 120 m (d.) 30 m (e.) 75 m

7. How fast will the mass in #6 be going at t = 5 seconds?

(a.) 75 m/s (b.) 120 m/s (c.) 300 m/s (d.) 30 m/s (e.) 15 m/s

8. If you have to exert 50 N horizontally to move a 5 kg box at constant speed against friction, what horizontal force would you have to exert in order to give the box an acceleration of 2 m/s^2?

(a.) 100 N (b.) 60 N (c.) 52 N (d.) 125 N (e.) 75 N

GENERAL PROBLEMS

Problem 3-1. Two masses are connected to each other by a string which passes over a massless, frictionless pulley as shown in the figure. When M_1 is on the ground, M_2 is held 10 meters above. $M_1 = 5$ kg and $M_2 = 10$ kg. They are let go from rest.

(a) Where and when will they pass each other?

(b) If you make $M_2 = 15$ kg instead, how will these answers change?

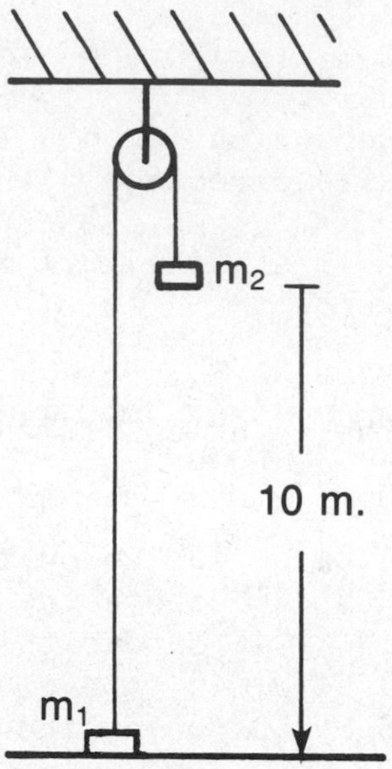

Q1: What are the free body diagrams for each mass?

Q2: How does Newton's 2nd Law apply to each diagram? I can see they will have the same a, but I have to find out what it is.

Q3: What are the equations of motion for each mass? Both have $v_0 = 0$, but what is s_0 in each case?

Q4: What is the mathematical condition for "passing each other"? Is there a direct clue for where this must happen?

Problem 3-2. In the following figure, assume $M_2 > M_1$, and that there is no friction either in the pulleys or on the surface bearing M_1. The pulleys are also massless. Find the algebraic expression for the acceleration of M_2 in terms of M_1, M_2, and g. If $M_1 = 2$ kg and $M_2 = 5$ kg, what is the numerical result?

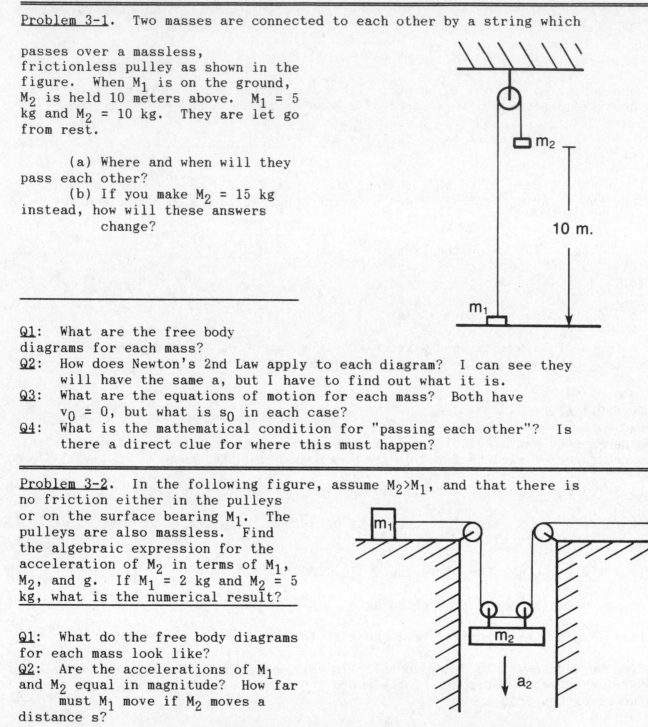

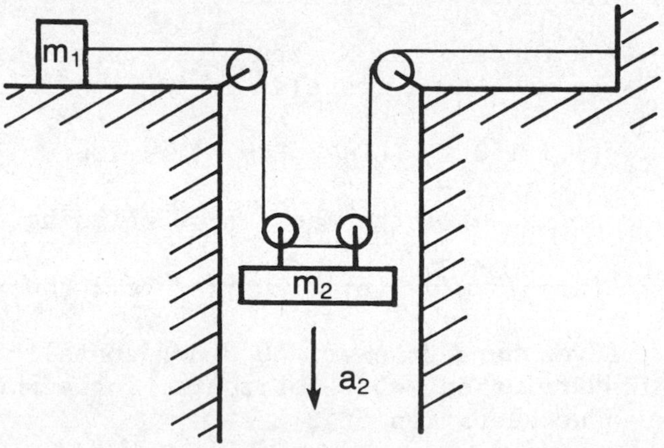

Q1: What do the free body diagrams for each mass look like?

Q2: Are the accelerations of M_1 and M_2 equal in magnitude? How far must M_1 move if M_2 moves a distance s?

Q3: What are the 2nd Law equations for each mass?

Q4: Are these equations sufficient? What are the unknowns?

Problem 3-3. The planets Earth and Mars have radii of 4000 miles and 2100 miles respectively. The mass of Mars is 1/10 that of Earth. On Earth, the surface acceleration of gravity is g = 9.80 m/s^2. What is the free fall acceleration at the surface of Mars?

Q1: What determines Earth's surface gravity?
Q2: Can I generalize Q1 to any given planet?
Q3: Can I compare Mars to Earth by ratios of their physical properties?

Problem 3-4. Two blocks are in contact on a horizontal frictionless table, as shown in the figure. M_1 = 1 kg and M_2 = 2 kg. The blocks are accelerating with a = 3 m/s^2 in response to a horizontal force F applied to M_1.

(a) How big is F?

(b) What is the force of contact between the blocks?

(c) Suppose now there is friction between the table and the blocks, and μ_k = 0.4. The blocks again are made to have a = 3 m/s^2. Answer (a) and (b) again for this case.

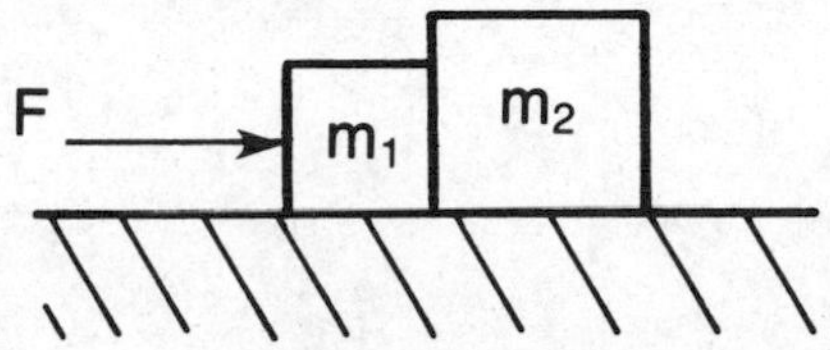

Q1: Both masses have the same a. Can't I treat the two as one 3 kg mass in order to find F?
Q2: I don't see any contact forces. What do the free body diagrams show?
Q3: What is the net force on each block?
Q4: In part (c), what is the *total* friction force opposing F?
Q5: From the free body diagrams, what is the net force on each object when friction is present?

Problem 3-5. Take the two masses in the previous problem and stack the 1 kg mass on top of the 2 kg mass. Suppose the coefficient of static friction between their surfaces is 0.5. M_2 is on a frictionless table. A force F is applied horizontally to M_2.

(a) How big can F be in order that M_1 does not slip?
(b) If F is applied to M_1 instead, how big can it be before slipping occurs?

Q1: What are the free body diagrams for the two masses?
Q2: What is the algebraic expression for the net force on M_1?
Q3: What determines the condition for M_1 to slip on M_2?
Q4: What is the expression for the net force on M_2?
Q5: How big is the normal force at the surface between the masses?
Q6: How big can the force of friction be?

Repeat the above questions for part (b).

ANSWERS TO QUESTIONS

FUNDAMENTALS:

1. (c.)
2. (d.)
3. (d.)
4. (a.)
5. (c.)
6. (e.)
7. (d.)
8. (b.)

GENERAL PROBLEMS:

Problem 3-1.

A1:

A2: $M_2g - T = M_2a$; $T - M_1g = M_1a$
A3: Taking the floor as $s = 0$, $s_1 = 1/2at^2$; $s_2 = 10 - 1/2at^2$; $v_1 = at$ and $v_2 = -at$.
A4: "Passing each other" means the instant when $s_1 = s_2$. Yes, the length of the string is fixed, so whatever distance M_2 comes down, M_1 must rise.

Problem 3-2.
A1:

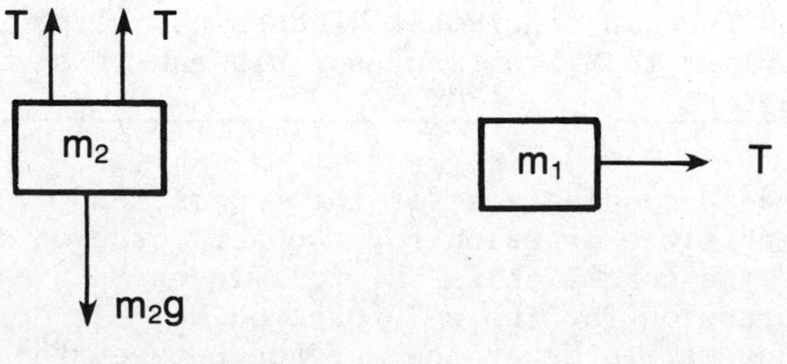

A2: No. If M_2 moves a distance s, M_1 must move 2s.
A3: $M_2g - 2T = M_2a_2$; $T = M_1a_1$
A4: No. The unknowms are T, a_1, and a_2. You must find a relationship between a_1 and a_2 from the hint in A2 above.

Problem 3-3.

A1: From the Universal Law of Gravitation, we saw that $g = G(M_E)/r_E^2$.

A2: Certainly. The law claims to be *universal*, and our planet doesn't have "special" gravitational properties. On planet "P," $g_P = GM_P/r_P^2$.

A3: Sure. Write down g for each of them, and form their ratio.

Problem 3-4.

A1: Yes.

A2: We'll label the compression force between their surfaces F_N. This is not the same as the normal force between them and the floor, of course.

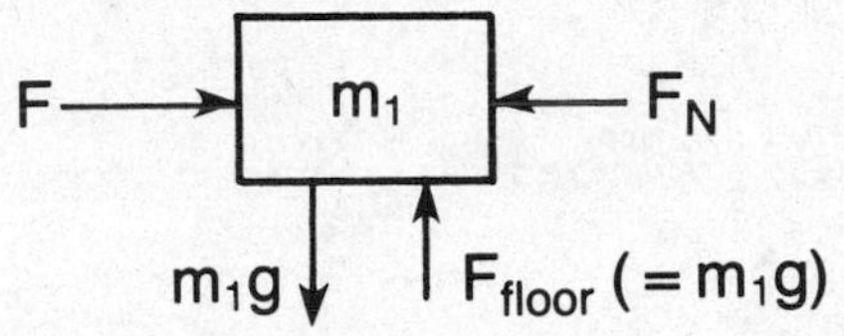

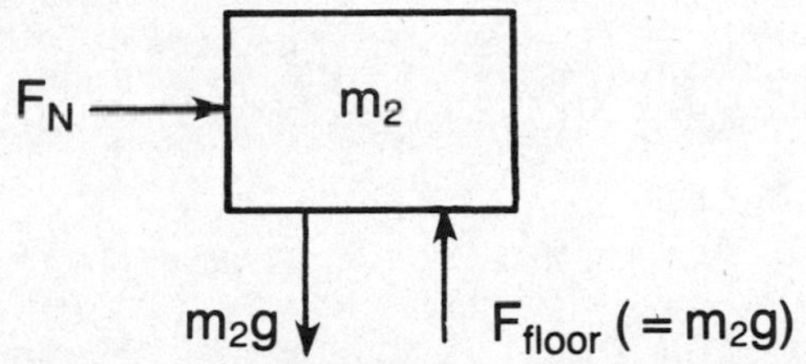

A3: For M_1, $F - F_N$; for M_2, F_N.

A4: $\mu_k(M_1 + M_2)g$.

A5: For M_1, $F - \mu_k M_1 g - F_N$; for M_2, $F_N - \mu_k M_2 g$.

Problem 3-5.

A1:

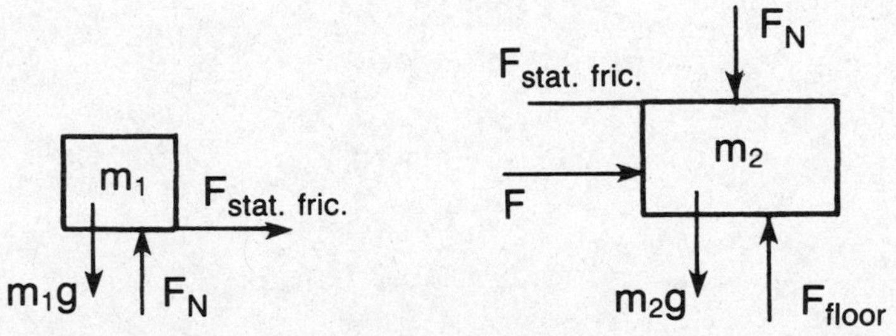

A2: $F_{static\ friction} = F_{net}$, to the right.

A3: That both masses accelerate equally, $a_1 = a_2$. In the case of M_1, friction is the only source of a_1.

A4: $F_{net} = F - F_{static\ friction}$.

A5: $F_N = m_1 g$.

A6: $F_{static\ friction}$ can at most be 0.5 F_N, or .5 $m_1 g$.

Part (b)

A1:

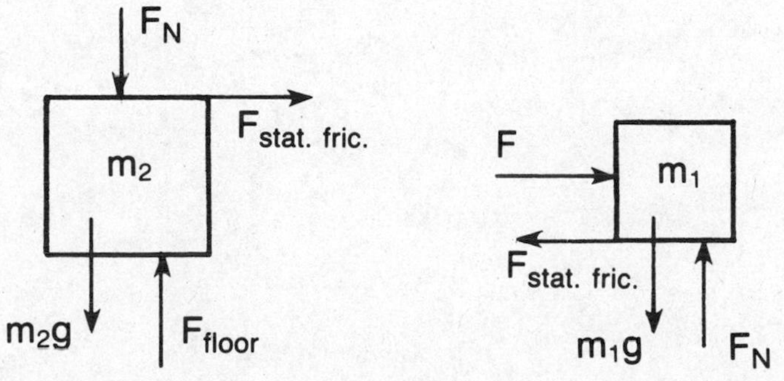

A2: $F - F_{stat.\ fric.} = F_{net}$, to the right.
A3: That both masses accelerate equally. This time friction is the only source of the acceleration of M_2.
A4: $F_{net} = F_{stat.\ fric.}$
A5: $F_N = M_1 g$.
A6: $0.5\ M_1 g$.

CHAPTER 4. MOTION IN A PLANE

SUMMARY AND RESUME

The principles developed in preceding chapters are applied more generally to *two dimensional* motion. Newton's *Universal Law of Gravitation* is given as an example of a fundamental force. The relation between mass and *weight* is discussed. Acceleration related to *turning* or *change of direction* is introduced. Motion on *inclined planes* and *projectiles* is used to exemplify two dimensional motion. The *kinematics* or description of circular motion is developed. Newton's Laws of Motion are applied to *circular* motion, and the ideas of centripetal and centrifugal force are discussed. *Kepler's Laws* describing *planetary motion* are given and are interpreted by Newton's Laws. *Earth satellites* and *apparent weight*, including *weightlessness*, are discussed.

Material from previous chapters which is used in this chapter includes:

1. Kinematic equations of motion with constant acceleration.
2. Newton's Laws of Motion.

NEW CONCEPTS AND PRINCIPLES

1. Universal Gravitational Constant

$$G = 6.67 \times 10^{-11}\ m^3/kg\text{-}s$$

Newton's Law of Gravitation

Understanding this section should enable you to do the following:

1. *Calculate the gravitational force between two given masses separated by a given distance.*
2. *Calculate the difference between the weight and mass of an object.*
3. *Explain the free fall acceleration of an object in terms of the Universal Law of Gravitation.*

Law of Universal Gravitation. Every mass attracts every other mass by a force which is directly proportional to the product of the two masses and inversely proportional to the square of the distance between them. The mathematical form of this law is

$$F_g = G(m_1 m_2)/r^2 .$$

Important Observations

1. The above is the expression for the magnitude of the gravitational force between the two masses m_1 and m_2. The distance between the masses is r. G is the *constant of proportionality* between the units of kg and meters on the right side and the units of newtons on the left. It is called the *Universal Gravitational Constant*, and has a measured value of $6.67 \times 10^{-11}\ m^3/kg\ s^2$. You should convince yourself by dimensional analysis that these are the correct units for G.
2. $\mathbf{F_g}$ is an *attractive* force, directed along the line joining m_1 and m_2. This means the force acting on m_1 is directed toward

m_2, and the same amount of force (notice F_g remains the same if you interchange the subscripts on the masses) acting on m_2 is directed toward m_1. This is another example of Newton's 3rd Law.

3. Strictly speaking, the correct separation distance is obvious only for *point* masses, i.e., masses without any extension in space. An important special case, however, is that, for masses possessing spherical symmetry, the law applies if you interpret r as the distance between their centers.

Example 4-1. How much gravitational attraction would a couple of 220 pound football linemen exert on each other at a distance of 2 meters? On earth an object that weighs 220 pounds has a mass of 100 kg.

S: Since they are not point masses nor spheres, how can we do this? What does the 2 meter distance refer to?

T: Very often in physics a meaningful approximation can be produced by greatly idealizing a complicated situation. In this case, when we are given a rather undefined distance of 2 meters separation, we'll assume it is the effective distance between their centers. Thus we are making a kind of "spherical lineman" approximation. It's useful!

$$F_g = 6.67 \times 10^{-11}\ m^3/(kg\ s^2) \times (100\ kg)(100\ kg)/(2\ m)^2$$

$$= 1.7 \times 10^{-7}\ kg\ m/s^2 = 1.7 \times 10^{-7}\ N$$

Notice how *weak* the gravitational force is. Since 1 N = 0.225 pounds, this amounts to only 4×10^{-8} pounds, or about 40 *billionths* of a pound!

Example 4-2. Earth's mass is about 6.0×10^{24} kg. That of the moon is .0123 as much, while that of Jupiter is 317 times as much. When Jupiter is closest to Earth, it is about 6.3×10^8 kilometers (km) away, while the moon maintains a distance of about 3.8×10^5 km from Earth. How does Jupiter's *maximum* pull on Earth compare with the moon's?

S: I need to know the masses, and then plug them and the distances into the formula for F_g for each case, right?

T: That's a direct way to do it, and is also rather lengthy. When you are asked to do a *comparison*, i.e., a *ratio*, you often find simplification by looking at the ratio in *algebraic* form, before doing any numerical work. Notice that

$$F_g\ (\text{from Jupiter}) = G\ (M_J M_E)/(r_{J-E})^2$$

and

$$F_g\ (\text{from moon}) = G\ (M_m M_E)/(r_{m-E})^2 \quad .$$

The symbols ought to be self-explanatory. If we now take the *ratio* of these expressions, G and M_E will cancel, and we have

$$\frac{F_g(\text{Jupiter})}{F_g(\text{moon})} = \left(\frac{M_J}{M_m}\right) \times \left(\frac{r_{m-E}}{r_{J-E}}\right)^2 = (317/.0123) \times \left(\frac{3.8\text{x}10^5}{6.3\text{x}10^8}\right)^2$$

Thus the ratio = 9.4×10^{-3}, or = 1×10^{-2} to one significant figure.

Thus Jupiter's maximum pull (at closest distance) on earth is about .01 or 1% as much as the moon's. This illustrates the technique of reducing a problem in comparisons to the process of calculating simple ratios. Also notice that in ratios, we didn't even have to worry about converting everything into SI units. The km cancelled, and we put masses in terms of "earth masses" instead of kg.

Weight--The weight of an object is the gravitational force acting on the object.

Important Observations

1. The weight of an object varies according to the object's gravitational environment. Weight on Mars or the moon would be less than on earth because they have weaker gravity. In space, far away from any stars or planets, weight would become negligibly small. Mass, on the other hand, remains the same, measuring the inertia of the object, regardless of where it is.
2. In some dynamic situations encountered later, such as earth satellites and elevators, we shall refer to "apparent weight," where weight reduction and weightlessness can occur because of acceleration even when F_g is not diminished.
3. The dimensions of mass and weight are different. Weight has units of *force*, measured in newtons or pounds, whereas mass is a fundamental dimension, measured in kg.

Example 4-3. What is the weight of a 1 kg mass at the earth's surface?

S: To use the law of gravity, I have to figure out the appropriate distance r.

T: Right. If we assume the earth is a uniform sphere, we can use the distance from the object at the surface to the earth's center, which is just the mean radius of the earth, r_E, as shown on this figure. r_E has a value of 6.4×10^6 m. Then

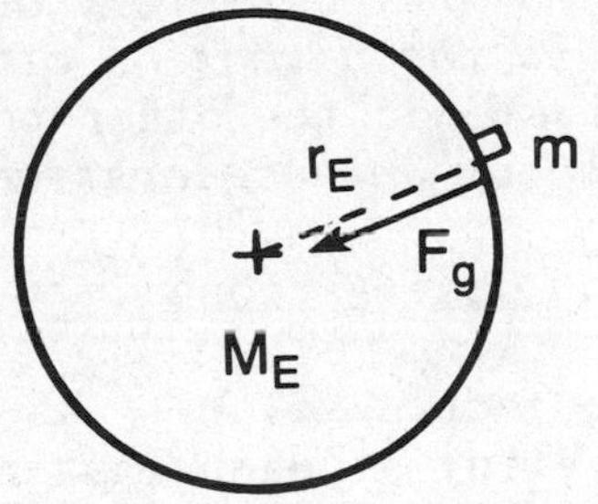

$$w = F_g = G\frac{mM_E}{r_E^2} = 6.67\text{x}10^{-11}\frac{m^3}{(kg/s^2)} \text{x} \frac{(1kg)(6\text{x}10^{24}kg)}{(6.4\text{x}10^6 m)^2}$$

Thus w = 9.8 N .

In British units, this is 9.8 N/(4.45 N/lb) = 2.2 pounds.

S: If we let this object free fall, this force should give it an acceleration of g, shouldn't it? How can objects with different weight all have the same free fall acceleration?

T: The key lies in the fact that the gravitational mass in the expression for F_g is the same as the inertial mass in the 2nd Law, so in free fall, the mass of the object cancels out:

$$G(mM_E)/(r_E)^2 = ma = mg$$

Notice that this shows an easy way to convert between mass and weight *at Earth's surface*, namely <u>w = mg</u> .

<u>S</u>: This is true even if the mass is not falling with acceleration g?

<u>T</u>: Sure. Whether it falls or not depends on whether other forces support the object against gravity. The same downward gravitational force is present regardless.

<u>Velocity and Acceleration in a Plane</u>

Understanding this section should enable you to do the following:

1. *Calculate the parallel and normal acceleration conponents of an object whose velocity vector changes in magnitude and direction.*

To describe motion in a plane (two dimensions), we must use the vector definitions of velocity and acceleration to represent changes of *direction* as well as changes of speed. An important point to remember is that the instantaneous velocity of an object moving along a curved path is always in a direction *parallel* or *tangent* to the curve at each instantaneous position, as shown in the figure below. This is quite intuitive, and just says that your direction of motion is always along the path you're taking.

Recall that in defining acceleration in Chapter 2, we observed that if **a** is going to represent the rate of change of the velocity vector, it would have to describe changes of *direction* as well as changes of speed. In one dimension, *changes of speed are completely described by an acceleration either parallel or anti-parallel to the direction of motion.* This remains true for two dimensions as well, but we must add the other possibility: that *changes of direction of motion are represented by an acceleration vector perpendicular to that direction.*

<u>Important Observations</u>

1. An object's velocity is always parallel to the path it is traveling at each instantaneous position.
2. If the object's speed is changing, the rate of change of speed is represented by an acceleration component a_T tangent to the path.
3. If the object is turning, the rate of change of direction is represented by an acceleration component a_N normal to the path.

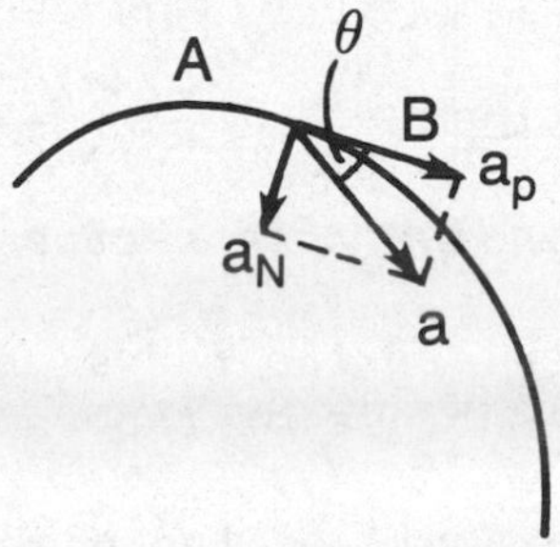

The following figure shows the situation with both acceleration components. You should imagine points A and B to be very close to each other, since we are considering a to be the *instantaneous* acceleration. It ought to be obvious that the *total* acceleration is

$$a = (a_N{}^2 + a_T{}^2)^{1/2} ,$$

and the direction of **a** relative to the instantaneous direction of motion is given by

$$\theta = \tan^{-1}(a_N/a_T)$$

Example 4-4. Suppose that in 0.5 sec, the velocity of an object changes direction by 15° clockwise and its speed goes from 10 m/s to 15 m/s. Find ⟨**a**⟩ and its components relative to $\mathbf{v_1}$ graphically. Check your results using trigonometric methods.

S: The graphical approach sounds like it should involve the parallelogram method, with Δ**v** being drawn *from* $\mathbf{v_1}$ to $\mathbf{v_2}$.

T: Exactly. The sketch shows this, on a scale of 1 inch = 10 m/s. Measuring the length of **v** gives 0.6 inches, or 6 m/s, and the angle relative to $\mathbf{v_1}$ is about 42.5°, measured by a protractor.

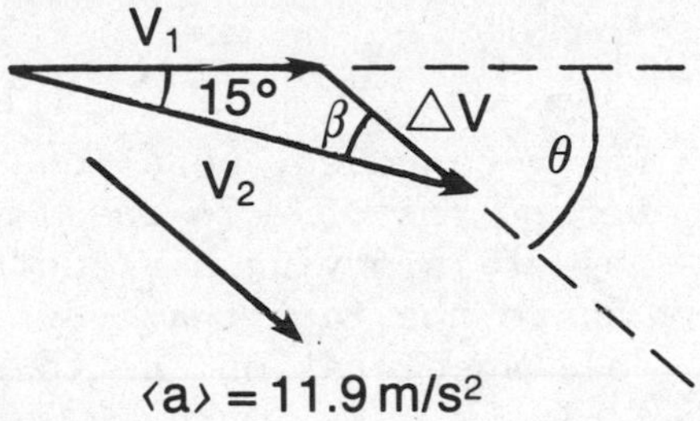

S: Now let's do it analytically with trigonometry.

T: Wait a minute! We've found Δ**v**, but that's *not* the same as ⟨**a**⟩.

$$\langle a\rangle = \Delta v/\Delta t = (6 \text{ m/s})/0.5 \text{ s} = 12 \text{ m/s}^2 ,$$

in the same direction as Δ**v**. Also, the components are

$$\langle a_T\rangle = \langle a\rangle\cos 42.5^\circ = 8.85 \text{ m/s}^2$$

and

$$\langle a_N\rangle = \langle a\rangle\sin 42.5^\circ = 8.11 \text{ m/s}^2$$

S: O.K. *Now* let's use the Laws of Sines and Cosines.

T: From the diagram, we can write

$$(\Delta v)^2 = v_1{}^2 + v_2{}^2 - 2v_1v_2 \cos 15^\circ = 35.2 \text{ m}^2/\text{s}^2,$$

with Δv = 5.93 m/s.

Thus $\langle a\rangle = (5.93 \text{ m/s})/0.5 \text{ s} = 11.9 \text{ m/s}^2$.

Also, $\Delta v/\sin 15^\circ = v_1/\sin \beta$,

which gives $\beta = 25.9^\circ$ and $\theta = \beta + 15^\circ = 40.9^\circ$.

These are reasonably close to the graphical results.

Projectile Motion

Understanding this section should enable you to do the following:

1. *Calculate the range and time of flight of a projectile near the surface of the earth for various launch conditions.*

A good example of two dimensional motion in a curved path is projectile motion near the surface of the earth. We may assume g is constant throughout the path of the projectile, which is called its *trajectory*. We also make the simplifying assumption of *neglecting air resistance*. This is really unjustified, except that unless we do assume this, the problem becomes unduly complicated. Our coordinate system will have the x-axis horizontal and the y-axis vertical, as before. Under these assumptions, we have

$$a_x = 0 \text{ and } a_y = -g.$$

Important Observations

1. The x-component equations of motion will be those of unaccelerated motion, i.e.,

$$V_x = (V_0)_x = \text{constant, and } x = x_0 + (V_0)_x t.$$

2. The y-component equations of motion *will be the same as before for objects falling vertically*. The fact that the object is moving horizontally while it is falling doesn't change the fact that vertically it is still undergoing free fall acceleration due to gravity. Thus

$$V_y = (V_0)_y - gt; \text{ and } y = y_0 + (V_0)_y t - \tfrac{1}{2}gt^2.$$

3. The equations of motion along the x- and y-axes are *uncoupled*, i.e., they are *independent* of each other. No y -dependent variables appear in the x equations and vice versa.

There are relatively few separate categories of projectile problems. They can be summarized by the following:

I. Horizontal launch from height h above the impact point.

Special Conditions: At launch the initial velocity is V_0. Its components are: $(V_0)_x = V_0$; $(V_0)_y = 0$. The launch point is at the coordinates $x_0 = 0$; $y_0 = h$. The impact point is at the coordinates $x = R$ (the *range*); $y = 0$.

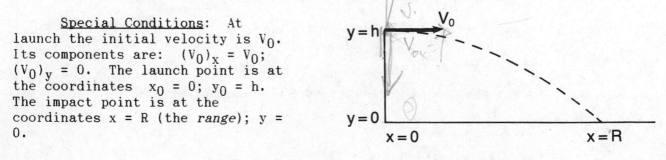

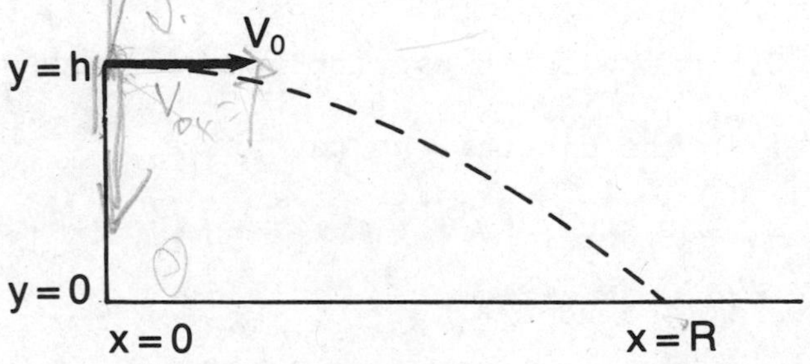

II. Launch at angle θ_0 above horizontal.

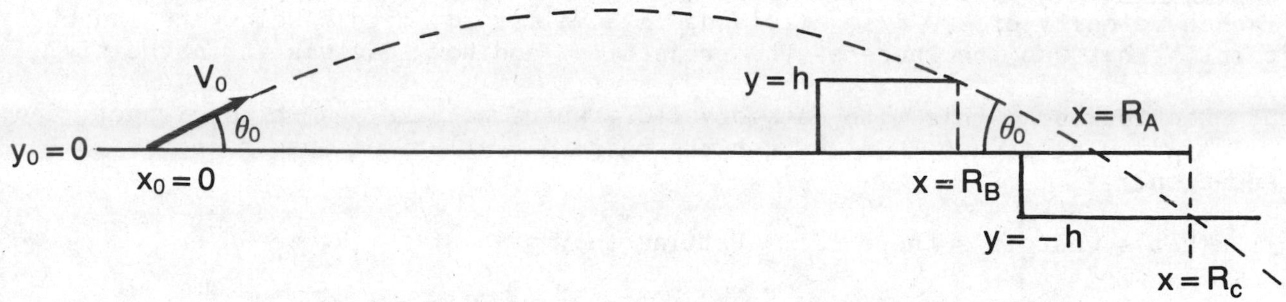

Special Conditions: At launch the components of initial velocity V_0 are $(V_0)_x = V_0 \cos \theta_0$; $(V_0)_y = V_0 \sin \theta_0$. The launch point is at $x_0 = 0$; $y_0 = 0$. The projectile reaches its maximum height when $V_y = 0$.

Case A. Impact at same level as launch.

Special Conditions: At impact $y = 0$; $x = R$ and $V_y = -(V_0)_y$.

Case B. Impact at height h *above* launch point.

Special Conditions: At impact $y = +h$; $x = R$.

Case C. Impact at height h *below* launch point.

Special Conditions: At impact $y = -h$; $x = R$.

III. Launch at angle θ_0 below horizontal with impact at height h below launch point.

Special Conditions: At launch the velocity has components $(V_0)_x = V_0 \cos \theta_0$; $(V_0)_y = -V_0 \sin \theta_0$. The coordinates at launch are $x_0 = 0$; $y_0 = 0$. At impact they are $x = R$; $y = -h$.

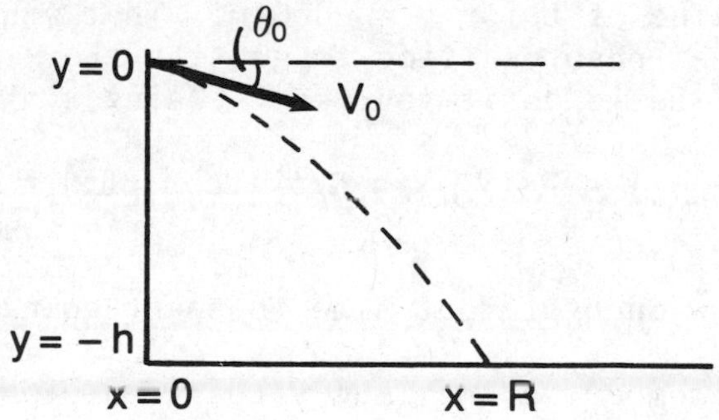

Trajectory Equation. By eliminating t from the x and y equations of motion, we get the equation of the projectile's *trajectory*, or path:

$$y = y_0 + (\tan\theta_o)x - \frac{g}{2(V_o)_x^2}x^2$$

You can perhaps recognize this as describing a parabolic curve in the x-y plane. Notice that $\tan \theta_0 = (V_0)_y/(V_0)_x$. If θ_0 is below the horizontal, this has to be negative.

Time of Flight--This is obviously the time between launch and impact.

Range--This is the value of x at impact, i.e., the horizontal distance traveled between launch and impact.

Example 4-5. Suppose you fire a projectile 350 above the horizontal with a launch velocity of 200 m/s. It lands in a valley 300 meters below the launch point. What was the range of the projectile, and how long was it in the air?

S: Looking back, this fits category IIC., where we take y = -h for impact, and $x_0 = y_0 = 0$ at launch. Can we get the range directly from the trajectory equation?

T: That's what it's there for. Putting in y = -h at x = R, we have

$$-h = 0 + (\tan 35^\circ)R - (9.8\ m/s^2)R^2/2(20\ m/s \cos 35^\circ)^2$$

With h = 300 m, this can be put into standard quadratic form:

$$(1.83 \times 10^{-4})R^2 - 0.700R - 300 = 0$$

The solutions are: $R = +4.22 \times 10^3$ m, or $R = -3.91 \times 10^2$ m.

S: Again we have a positive and a negative answer. Only the positive one makes sense here.

T: That's right. Negative values of x refer to positions *behind* the launch point. If you traced the trajectory *backward* from launch down to the -300 meter level, you would find yourself at x = -391 meters. That's the meaning of the "phantom" result. If you inspect the physical situation you can always interpret the "real" solution.

S: All right, what about the time of flight? Now that we know R, we could find the t corresponding to R.

T: That would do it. Or, you could get it from the y vs. t equation, by finding the value of t for y = -300 m. That would be a quadratic again, and you would get the "phantom" time to go with the x = -391 m point, as well as the correct time. Let's do it your way. Using $x = (V_0)_x t$, the time of flight is given by

$$t = R/(V_o)x = R/(V_1 \cos 35^\circ) = (4.22 \times 10^3\ m)/(200\ m/s) \cos 35^\circ$$
$$= \underline{25.8\ seconds}$$

S: How much of that time is spent going upwards?

T: Remember, at maximum y the trajectory is *horizontal*, so $V_y = 0$ at that instant. So the time to top-of-flight comes from setting $V_y = 0$ in the V_y vs. t equation:

$$0 = V_0 \sin \theta_0 - gt$$

Thus

$$t = (200\ m/s) \sin 35^\circ/(9.8\ m/s^2) = \underline{11.7}\ s \text{ to top-of flight.}$$

Circular Motion

Understanding this section should enable you to do the following:

1. *Calculate the relation between angular velocities and accelerations and their linear counterparts a distance R from the axis of rotation.*

2. *Calculate the centripetal acceleration of an object moving in a circle at constant speed, and the centripetal force on the object.*
3 . *Convert between radians, degrees, and revolutions of rotational motion.*

When an object moves along a circular path, its position can be located by specifying the *angle* which a line connecting it with the circle's center makes relative to some reference line. Likewise, the object's *angular velocity* is related to how fast this angular position is changing, and the *angular acceleration* describes how fast the angular velocity is changing. This angular description of circular motion is entirely analogous to our previous description of translational motion, with angular variables substituting for their translational counterparts.

Angular Velocity--The angular displacement $\Delta\theta$ occurring in a time period divided by that period. Symbolized by (omega).

$$\omega = \Delta\theta/\Delta t, \quad \text{or} \quad \Delta\theta = \omega\Delta t$$

Angular Acceleration--The time rate of change of angular velocity. Symbolized by α (alpha).

$$\alpha = \Delta\omega/\Delta t, \quad \text{or} \quad \omega = \alpha\,\Delta t$$

Important Observations.

1. The units of angular velocity are radians/second. Dimensionally, this is just $(s)^{-1}$ but it is often useful to use the radian label to remind us of the angle measure being used.
2. Similarly, the units of angular acceleration are rads/s^2.
3. Only if Δt is very small does $\Delta\theta/\Delta t$ approach the *instantaneous* angular velocity or $\Delta\omega/\Delta t$ approach instantaneous angular acceleration. Otherwise we have the *average* values, as with the translational definitions.
4. The connection between the linear velocity along the circle's rim and the angular velocity about the center is $V = R\omega$.
5. The component of acceleration parallel to the circular path, a_T, is related to the angular acceleration about the center by $a_T = R\alpha$.
6. One assumption in all the above is that the plane in which the circular motion occurs is *fixed in its orientation*. In this way the definitions of θ, ω, and α involve only their *magnitudes*.

Frequency of Circular Motion--The number of complete *revolutions*, or *cycles*, of motion per second. Symbolized by f.

Period of Circular Motion--The time it takes to make one complete revolution, or cycle, of motion. Symbolized by T.

Important Observations.

1. f and T are related by $f = 1/T$.
2. The units of f are revolutions/second, or just $(s)^{-1}$ dimensionally. You see why it is important to remember whether you're expressing quantities in radians or cycles,

numbers. Counting revs per second or cycles per second is common emough to warrant a special name, called the *hertz* (Hz). Thus 1 Hz = 1 rev/sec or 1 cycle/sec.

3. Since there are 2π radians per revolution, the obvious connection between frequency and angular velocity is $\omega = 2\pi f$

Example 4-6. If a car has 15.0 inch radius tires and is going along a road at 55.0 miles/hour,

(a.) What is the linear speed of a point on the rim of the tire relative to the axle?
(b.) What is the angular velocity of a point on the rim?
(c.) What is the frequency of rotation?

S: It seems that in order not to slip on the road, the linear speed of points on the rim relative to the axle would have to be the same as the speed of the road relative to the axle. You can see that by looking at the point momentarily in contact with the road.

T: Exactly right. And the axle is a fixed point of the car, whose speed relative to the road is 55 mi/hr.

S: For the angular velocity, the radius of the wheel connects v with ω.

T: You mean

$$V = R\omega \text{ , or } \omega = V/R = (55.0 \text{ mi/hr})/(15.0 \text{ in})(1 \text{ mi}/66360 \text{ in}),$$

giving

$$\omega = 2.32 \times 10^5 \text{ rads/hr} = 64.5 \text{ rads/sec.}$$

S: For part (c.), we need $f = \omega/2\pi$, in Hz. What are Hz again?

T: Hz refers to revolutions/s, so,

$$f = (64.5 \text{ rads/s})/2 \text{ rads/rev} = 10.3 \text{ revs/s, or Hz.}$$

By the way, this is 10.3 x 60s/min = 616 revs/min, called r.p.m.

Example 4-7. A certain flywheel is set into rotation by a motor. Starting from rest, it takes 20 seconds to reach 3000 r.p.m.

(a.) What is the average angular acceleration of the flywheel?
(b.) What is the linear acceleration of a point 50 cm. from the axis?
(c.) Assuming α to be constant, how many revolutions has the wheel turned just in getting up to speed?

S: For part (a.), I need to find the angular velocity which corresponds to 3000 rpm.

T: The conversion is (rads/s) = $(2\pi/60)$ (revs/min). Thus

$$= (2\pi/60)(3000) = 100\pi \text{ rad/s.}$$

Then from $\omega = \omega_0 + \alpha t$, so

$$\alpha = (100\pi \text{ rad/s})/20\text{s} = 5\pi \text{ rad/s}^2.$$

S: We'd better convert these to decimal form.

T: Why? There are reasons why you shouldn't be too quick to replace an irrational number by its decimal approximation:

1. You introduce round-off error prematurely.
2. In a later part of a series of calculations, π may cancel *exactly*.

S: Anyway, for (b.) we should use $a_T = R\alpha$. What units do we get?

T: $$a_T = R\alpha = (.50 \text{ m})(5\pi \text{ rad/s}^2) = 2.5\pi \text{ m/s}^2.$$

Remember, radians don't have a dimension, but nevertheless, the equations connecting linear and angular quantities *demand* that the latter be in radian measure.

S: We haven't mentioned the connection between revolutions and angular acceleration yet, have we?

T: Not directly, but because of the analogies mentioned earlier, and from the definitions of θ, ω, and α, it should be quite obvious that there are *angular equations of motion*,

$$\theta - \theta_0 = \omega_0 t + 1/2\alpha t^2, \quad \text{and} \quad \omega - \omega_0 = \alpha t.$$

These require that α is constant. In our case, we get

$$\theta - \theta_0 = 0 + 1/2(5\pi \text{ rad/s}^2)(20 \text{ s})^2 = 1000\pi \text{ radians}$$
$$= 1000\pi/2\pi = 500 \text{ revolutions}.$$

Centripetal Acceleration--We have observed that changing the *direction* of the velocity vector is related to the acceleration *normal* or perpendicular to the velocity. In Section 4.6 of your text, the mathematical expression for this acceleration component, called *centripetal* ("center-seeking") acceleration is derived. The result is that the motion of an object traveling in a circle of radius R with linear speed V can be described by an acceleration *directed toward the center* of the circle, having magnitude $a_C = v^2/R = R\omega^2$. This is called *centripetal acceleration*, which represents the rate of turning of the velocity vector. It is an example of a *normal* acceleration, but for purely circular motion we adopt the symbol a_C to replace a_N.

Centripetal Force--From Newton's 2nd Law, a net force is required to cause any acceleration. For an object to follow a circular path, there must be a net component of force *normal* to the circle, such that

$$(F_{net})_C = ma_C = mv^2/R = mR\omega^2.$$

This component of force, directed *inward toward the center*, is called the centripetal force.

Important Observations

1. Notice that the units of v^2/R are m/s^2, as they must be to represent an acceleration.
2. If there is no change of *speed*, then a_C and $(F_{net})_C$ are the *total* acceleration and net force vectors.

Table 4.1 Some Common Examples of Centripetal Force

Situation	Centripetal Force
Car driving along a level circular road.	Friction between tires and road, not allowing sideways slipping.
Car driving along a banked circular road.	Horizontal components of both friction and the normal force of the road on the car.
Ball on a string whirled in a horizontal circle.	Tension in the string attached to the ball.
Ball on a string whirled in a vertical circle.	Tension in the string (varies with position) plus the component of weight normal to the circle (also varies).
Moon or satellite in orbit around Earth.	The force Earth exerts on the moon or satellite according to Newton's Law of Gravity.
A charged particle moving in the presence of a magnetic field.	The magnetic force on a moving charge is always normal to its velocity, causing it to move in a circular path.

Example 4-8. A girl is whirling a 500 gram ball in a horizontal circle at the end of a 1.5 meter "massless" string. The breaking strength of the string is 50 pounds.

(a.) Is this physically possible, strictly speaking?
(b.) Assuming it is, find the maximum speed with which she can whirl the ball before the string breaks.

S: What do you mean, "is it possible"? That if she has the strength she can?

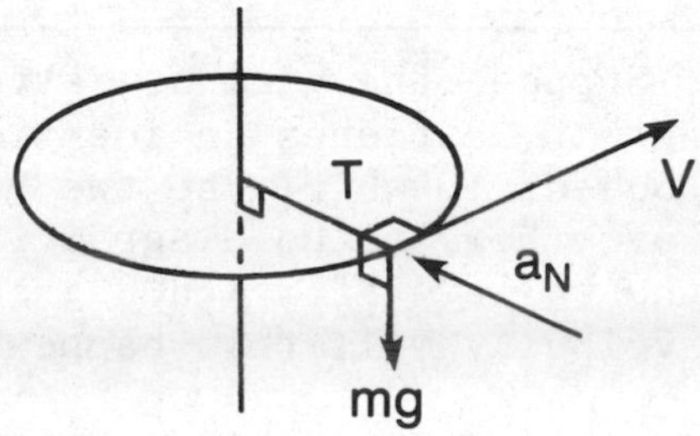

T: Look carefully at the following free body diagram. a_N is horizontal, pointing toward the center of the circular motion. Do you see anything wrong?

S: Yeah. There's no force component balancing the weight mg vertically, yet the acceleration is completely horizontal. The tension must have *some* upward component to cancel mg.

T: You're getting good! So no matter how hard she tries, the string must slant downward somewhat. Let's neglect this for now, and just treat the horizontal approximation. Tension is the *only* horizontal force, so

$$T = ma_C = mv^2/R, \quad \text{giving} \quad v^2 = TR/m.$$

Maximum v happens when T is maximum, which is 50 pounds, or 50 x 4.45 = 222 N. Then

$$v_{max}^2 = (222\ N)(1.5\ m) = 666\ m^2/s^2, \text{ giving } v_{max} = 25.8\ m/s.$$

S: What about the real situation? Can we find out the angle the string must make with the horizontal?

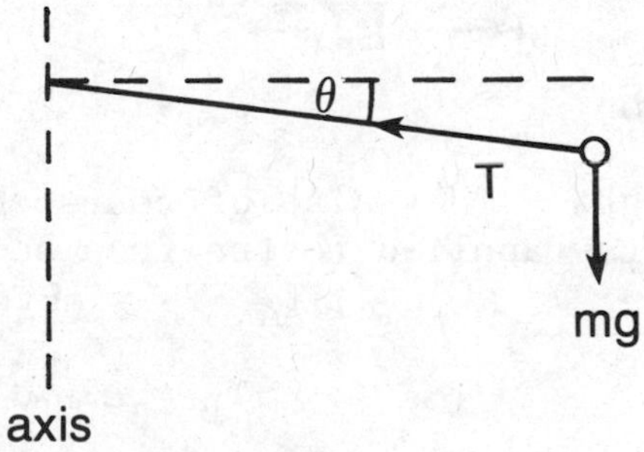

T: Quite easily. A vertical cross-section at an instant of time would look like this sketch.

S: So $T \sin \theta = mg$.

T: Right, and the net force toward the center of the circle is $T \cos \theta$. Thus

$$T \cos 0 = ma_C = mv^2/R.$$

At maximum tension,

$$\sin \theta = mg/T_{max} = (.5\ kg)(9.8\ m/s^2)/222\ N = 0.022$$

So

$$\theta = \sin^{-1}(.022) = 1.26^\circ, \text{ almost horizontal.}$$

S: But now *less* of T_{max} can provide the centripetal force, it seems, since part of the tension supports the weight.

T: Which means she can't swing the ball quite as fast as we thought.

$$v_{max}^2 = T(\cos 1.26^\circ)R/m = 665.8\ m^2/s^2, \text{ or}$$

$$v_{max} = 25.8\ m/s \text{ again, to three figures.}$$

<u>Example 4-9.</u> Suppose the girl now attaches two balls of mass m together with a string of length L, attaches an identical string to the other side of one of the balls, and proceeds to whirl the two balls together in a horizontal circle as shown from above them in the diagram. Which string will break first as she accelerates the balls? If m = 500 grams, L = .5 meters, and T = 50 pounds, at what angular velocity will this happen?

S: With just one string, you'd expect the tension to be the same throughout.
T: Not so. These strings are attached to different objects, are separate, and are exerting different forces. Look at the free body diagrams.

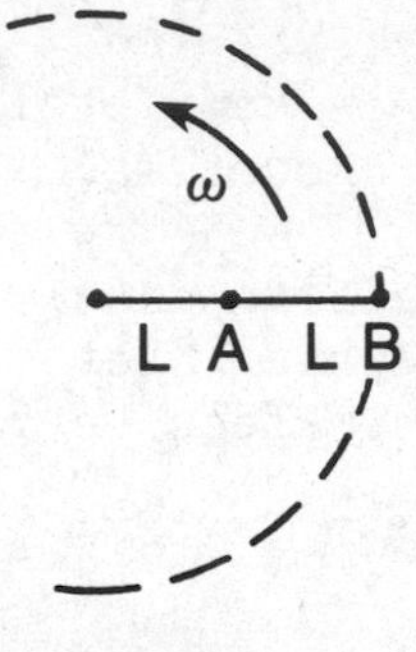

S: I also see that the accelerations are different because, even though both masses have the same , they have different radii of

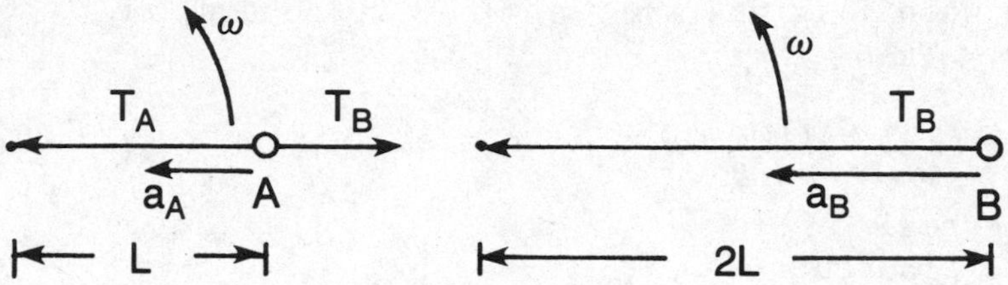

circular motion.

T: You're right. They also of course have different linear velocities. Newton's 2nd Law applied to the free body diagrams gives us

$$T_A - T_B = mR_A\omega^2 = mL\omega^2, \text{ and}$$

$$T_B = mR_B\omega^2 = m(2L)\omega^2.$$

S: So T_A is considerably larger than T_B?

T: Yes. T_A has to balance T_B *and in addition* provide the centripetal acceleration for mass A.

$$T_A = mL\omega^2 + T_B = mL\omega^2 + m(2L)\omega^2 = 3mL\omega^2 = (3/2)T_B.$$

S: So string A will break first, when $3mL\omega^2 = T_{max}$.

T: Right. This means

$$\omega_{max}^2 = T_{max}/3mL = 222 \text{ N}/(3)(.5 \text{ kg})(.5 \text{ m}) = 296 \text{ rad}^2/\text{s}^2.$$

$$\text{Then } \omega_{max} = 17.2 \text{ rad/s.}$$

<u>Planetary Motion</u>

Understanding this section should enable you to do the following:

1. *Apply the principles of circular motion and the Law of Gravitation to problems involving planetary motion.*

2. *Explain apparent weight and weightlessness in terms of the acceleration of an object in the presence of a gravitational force.*

The motion of planets around the sun and satellites around the earth are explainable by the principles just developed. In these cases the force providing the centripetal acceleration of orbiting bodies is *gravity*. We will treat one of the masses in the gravitational interaction as remaining stationary because of its much greater mass than that of the orbiting object. We will also consider only *circular* orbits, although in actuality the most general orbital paths are *ellipses*.

Conditions for Circular Planetary Orbits.

Free body diagram:

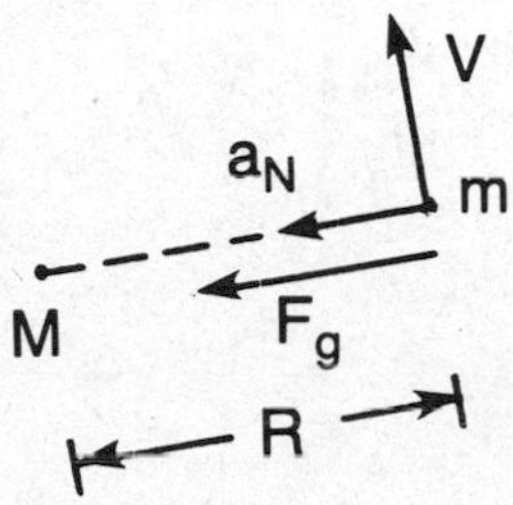

Actual Orbit:

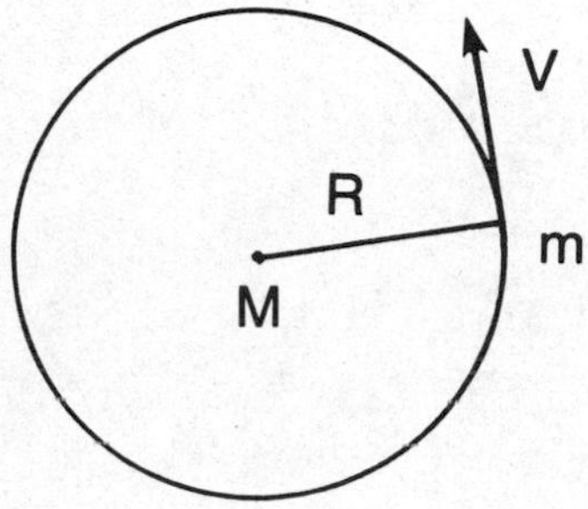

Force equation (Newton's 2nd Law)

$$F_g = ma_C, \quad \text{which becomes} \quad GmM/R^2 = mv^2/R$$

Orbital velocity. From the force equation we get the velocity required for orbit,

$$v = (GM/R)^{\frac{1}{2}} .$$

Orbital period. The orbiting body travels one *circumference*, or $2\pi R$, in one orbital *period*, T. Thus

$$T = 2\pi R/v = (4\pi^2 R^3/GM)^{\frac{1}{2}}$$

Important Observations

1. We are considering a mass m in circular orbit around a much greater and stationary mass M.
2. Notice that the conditions for orbital velocity and hence period are *independent* of m. *All* masses share the same orbital conditions. For example, a satellite, an astronaut, a toothbrush, and a book can all be in the same orbit around the earth.
3. The orbital speed is *constant* as long as there is no force *parallel* to the orbital path. This illustrates the *inertia* of the body, where the speed was caused by some previous force which originally accelerated the object into the orbit. All gravity does is *turn* the object.

Example 4-10. The planet Mercury is about .39 times as far from the sun as Earth is. Assuming a circular orbit, about how long would Mercury's "year" be?

S: Don't we have to have more information, such as the masses of Mercury and the sun?

T: Remember, the requirements for a gravitational orbit do not depend on the orbiting mass. If you wanted to use the basic expression for T you're right in thinking you need the sun's mass. We assume we know Earth's year.

S: It sounds like we should form a ratio between Earth and Mercury.

T: That's the easiest way. We'll let M_s be the mass of the sun, and R_{E-s} and R_{M-s} be the distance between the sun and Earth and Mercury. Then

$$T_M = \sqrt{\frac{4\pi^2 R^3_{m-s}}{GM_s}} \text{ and } T_E = 365 \text{ days} = \sqrt{\frac{4\pi^2 R^3_{E-s}}{GM_s}}$$

So

$$\frac{T_M}{T_E} = \sqrt{\frac{R^3_{M-s}}{R^3_{E-s}}}$$

This is just Kepler's 3rd Law of Planetary Motion referred to in your text, which was developed by Kepler in 1624 A.D. Thus

$$T_M/T_E = (.39)^{3/2} = .243,$$

So Mercury's year is .243(365 days) = 89 earth days.

Example 4-11. The average distance from Earth to the sun is about 93 million miles (1.5×10^{11} m). Taking the earth's year to be 365 days (3.15×10^7 s), find the mass of the sun and the orbital velocity of the earth around the sun.

S: The mass of the sun is involved in the orbital period equation. We know R and T and everything else is just constants.

T: Good. With a little manipulation,

$$T = (4\pi^2 R^3/GM)^{1/2} \quad \text{becomes} \quad M = 4\pi^2 R^3/GT^2$$

S: We really have to watch our units now. Is everything in SI?

T: That's the only system we've expressed G in. So we plug in the numbers:

$$M = \frac{4\pi^2(1.5 \times 10^{11} \text{ m})^3}{(6.67 \times 10^{-11} \text{ m}^3/\text{kg s}^2)(3.15 \times 10^7 \text{ s})^2} = 2.0 \times 10^{30} \text{ kg}$$

S: It's neat the way all the extraneous units cancelled. The orbital velocity is easy:

$$v = 2\pi R/T = 2\pi(1.5 \times 10^{11} \text{ m})/(3.15 \times 10^7 \text{s}) = 3 \times 10^4 \text{ m/s, or } 30 \text{ km/s}$$

T: Just for the sake of reference, this is about 67,500 mi/hr!

<u>Apparent Weight and Weightlessness</u>. Often you hear of an astronaut being in a "weightless" condition while in orbit. One of the inadequately known effects of a prolonged stay in orbit is the effect of such a condition on the muscles of the body. Yet the astronauts, with the exception of the Apollo missions to the moon, have orbited close enough to the earth that the force of gravity on them was only slightly less than if they were on the ground. In Chapter 3 we indicated that changes in "apparent" weight could be due to certain dynamic situations.

<u>Example 4-12</u>. What is the value of the force of gravity on an astronaut in orbit 500 miles above the earth's surface, compared to mg on the surface?

<u>S</u>: We could go back and calculate F_g from the Law of Gravity, but it seems we can get the answer easier by using ratios again.

<u>T</u>: Yes, We have that

$$F_g(\text{surface}) = GmM_E/R_E{}^2, \text{ where } R_E = 6.4 \times 10^6 \text{ m.}$$

At an altitude of 500 miles (800 km),

$$F_g(500 \text{ mi. up}) = GmM_E/R^2 ,$$

where R would be 7.2×10^6 m. So the ratio of weights would be

$$w(500 \text{ mi. up})/w(\text{surface}) = (6.4 \times 10^6)^2/(7.2 \times 10^6)^2 = 0.79$$

The force of gravity on the astronaut would thus be 79% of the surface value.

<u>Apparent Weight</u>--The force which, in any given situation, opposes the force of gravity on an object.

<u>Important Observations</u>

1. When you stand on a scale, the normal force exerted by the scale in supporting you against gravity is the reading on the scale. If you are not accelerating in the vertical direction, this force equals F_g, your "real" weight.
2. If you *do* have an acceleration *in the direction of gravity*, then this opposing force must be *less* than F_g by an amount equal to ma:

$$F_g - F_{opposing} = ma.$$

Thus

$$F_{opposing} = \text{apparent weight} = F_g - ma < \text{weight}$$

3. If you have an acceleration *opposite* to the direction of gravity, the opposing force must be *greater* than F_g :

$$F_{opposing} - F_g = ma.$$

So in this case,

$$\text{apparent weight} = F_{opposing} = F_g + ma > \text{weight}$$

4. In the case of *vertical* motion, this acceleration means your *speed* is changing when you experience an apparent weight

different than F_g. For *circular* motion, a change of apparent weight is associated with your change of *direction*.

5. If in either case there is *no* force opposing gravity, so that your acceleration is F_g/m, you are *apparently weightless*. This means either that you are in *free fall* vertically or that you have sufficient tangential velocity to be in an *orbital* condition.

Example 4-13. During a rocket lift-off, suppose the ship reaches a maximum upward acceleration of 40 m/s^2. During this period, what would the apparent weight of a 180 pound astronaut be? Assume this acceleration occurred within the first 20 miles of flight.

S: What does that last statement tell me?

T: That the ship is still close enough to Earth's surface that the force of gravity on the astronaut is essentially its surface value of 180 pounds.

S: What force is accelerating the astronaut?

T: He and the ship are experiencing a 40 m/s^2 upward acceleration. The ship is deriving this from the thrust produced by the rocket engines. I imagine the astronaut is in some seat whose rigidity produces sufficient force to give him the 40 m/s^2 acceleration along with the ship.

S: This force is opposite to gravity. Is this his apparent weight?

T: That's right. If the astronaut was sitting on some sort of scale, it would read his "weight," equal to this force. Look at the free body diagram, and remember that 40 m/s^2 amounts to 4.1g.

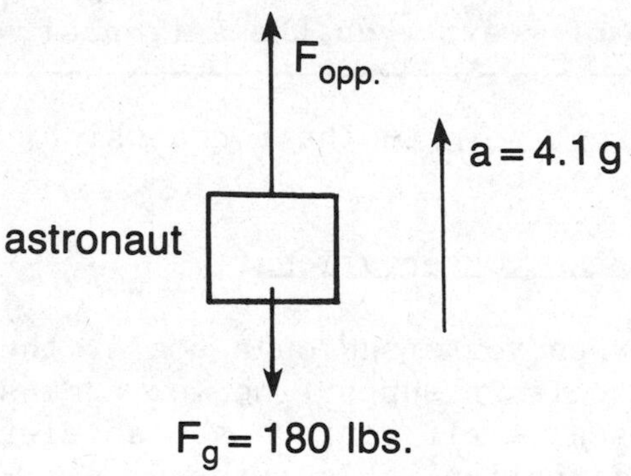

S: Then $F_{opposing}$ would be 4.1(mg), and the astronaut would have an apparent weight of (4.1)(180) = 738 pounds!

T: That's *not* right. It's even worse for him. Be more careful in using the 2nd Law. We have $F_{opposing} - mg = ma = m(4.1g)$, so

$$\text{Apparent weight} = (4.1)mg + mg = (5.1)mg = 918 \text{ pounds!}$$

Let me ask a slightly different question: *what if this acceleration occurred when the ship was 500 miles above the earth*?

S: We previously found that F_g at that height is only .79 of surface value, so the force creating apparent weight can be less and still provide 40 m/s^2 acceleration.

T: Precisely. We would have

$$\text{apparent weight} - .79(180 \text{ pds.}) = m(4.1g),$$

or

$$\text{app. weight} = 739 + 142 = 880 \text{ pounds .}$$

QUESTIONS ON FUNDAMENTALS

1A. An object is shot from a gun with an initial velocity of 200 m/s at an angle of 40° above the horizontal. What maximum height does the object achieve?

(a.) 656 m (b.) 2040 m (c.) 204 m (d.) 843 m (e.) 1200 m

1B. If the object in 1A lands at the same level, how far away from the gun is its impact?

(a.) 8.16 km (b.) 4.02 km (c.) 2.01 km (d.) 13.6 km
(e.) 980 meters

2. How much gravitational force is exerted by a 1000 kg mass on another 1000 kg mass a distance of 2 meters away?

(a.) 1000 kg (b.) 980 N (c.) 102 N (d.) 4900 N (e.) 9800 N

3. How much does a 1000 kg object weigh at the surface of the earth?

(a.) 1000 kg (b.) 980 N (c.) 102 N (d.) 4900 N (e.) 9800 N

4A. If the object in #3 were taken to a distance of 5 earth radii from the center of the earth, how would its weight compare with that in #3?

(a.) 5 times more (b.) 5 times less (c.) 25 times more
(d.) 25 times less (e.) weight would be zero

4B. Referring to the object in #4A, what would its mass be at that height?

(a.) 1000 kg (b.) 200 kg (c.) 40 kg (d.) 5000 kg (e.) zero

5. If an object is spinning at 360 revs/min, what is its angular velocity, in rad/s?

(a.) 12π rad/s (b.) $3/\pi$ rad/s (c.) 6 rad/s (d.) $180/\pi$ rad/s
(e.) 360 rad/s

6. If a ball is at the end of a 2 meter string and is being whirled horizontally at a rate of 2 revs/s, what is its centripetal acceleration?

(a.) $32\pi^2$ m/s^2 (b.) 8π m/s^2 (c.) $8\pi^2$ m/s^2 (d.) 4π rad/s^2
(e.) zero

7. If the ball in #6 has a mass of 1.5 kg, how much tension would there be in the string?

(a.) 686 N (b.) 546 N (c.) 140 N (d.) 826 N (e.) zero

8. If you normally weigh 686 N, what would you appear to weigh in an elevator which was giving you an upward acceleration of 2 m/s^2?

(a.) 686 N (b.) 546 N (c.) 140 N (d.) 826 N (e.) zero

GENERAL PROBLEMS

Problem 4-1. A jet fighter is flying with a velocity V_j = 600 mi/hr, climbing at the rate of 9000 ft/mi. The moment it is directly over an anti-aircraft gun it is at an altitude of 1000 ft. If the gun is fired at this instant, at what angle (relative to ground) should it be aimed in order to hit the aircraft, and at what height does impact occur? Assume the muzzle velocity of the gun's projectile is V_P = 2000 ft/sec. (You can do this in British units, with g = 32.2 ft/s^2.)

Q1: What is necessary for impact to occur?
Q2: What is a good choice of coordinate system?
Q3: What are the equations of motion of the aircraft?
Q4: What are the aircraft's velocity components?
Q5: What are the projectile's equations of motion?
Q6: How do you find the aiming angle?
Q7: What happens when I equate the equations for y?

Problem 4-2. An 800 kg sports car with a 160 pound driver hits a dip in the road whose radius of curvature is 60 meters. This causes the car's springs to "bottom out," or compress completely. It normally takes a load of 5000 N in addition to the car's weight to fully compress the springs when standing still. How fast was the car going when it hit the dip? What was the apparent weight of the driver at that instant?

Q1: What category of motion is this?
Q2: At the bottom of the dip, what is the free body diagram for the car alone?
Q3: What is Newton's 2nd Law for this situation?
Q4: What is the condition that makes the springs compress fully?
Q5: What is the free body diagram and the 2nd Law for the driver?

Problem 4-3. An astronaut is flying 1 mile above the surface of the moon at orbital speed. She wants to jettison a container to hit a target area on the surface. The jettison tube is directed horizontally to the rear of the craft, and can give the container an initial speed of 1000 m/s relative to the spacecraft. How long will it take the container to impact the surface? How far before the target horizontally must she jettison it? With what *velocity* will the container impact? Use the following data: M_{moon} = 7.38 x 10^{22} kg; R_{moon} = 1.74 x 10^6 meters; g_{moon} = (1/6)g_{earth}. For the falling projectile, you can treat the moon as being "flat" over the range of its trajectory.

Q1: What are the crucial changes here from our previous projectile problems?
Q2: What is the expression which gives the orbital speed?
Q3: Relative to the moon's surface, what is the initial velocity of the container?
Q4: What are the equations of motion of the falling container?
Q5: What is the time of impact?

Problem 4-4. Consider a car going around a banked circular curve with friction between the road and the tires. Find an expression for the maximum speed that car can have without slipping. Take the angle of bank to be θ, the mass of the car m, and the radius of the curve R. Also, the coefficient of static friction is μ_s.

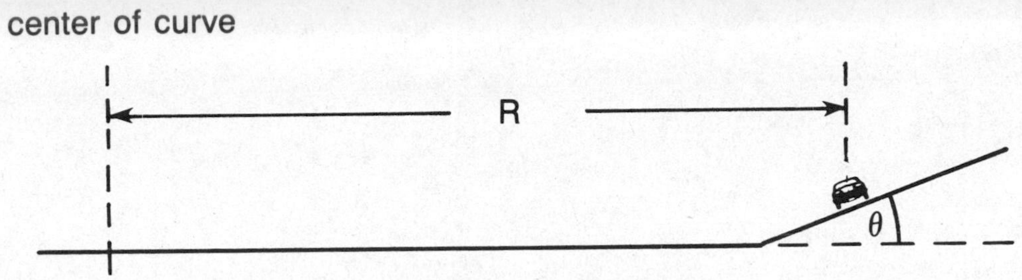

Q1: What is the free body diagram of forces on the car?
Q2: What are the best coordinate axes to choose?
Q3: What is the 2nd Law equation normal to the surface?
Q4: What is the 2nd Law equation parallel to the surface?
Q5: What is the condition for the *maximum* possible speed?
Q6: What is the first step in combining these equations?

Problem 4-5. A space station is constructed in the form of a wheel with two corridors as showm in the diagram below. One corridor is 50 meters from the hub, the other one 100 meters from the hub. The station rotates around the hub in order to produce "artificial gravity."

(a.) What angular speed would produce normal weight for a person in the outer corridor?

(b.) At that rate of rotation, what "gravity" would be produced at the inner corridor?

(c.) What linear speed would a person have at the outer corridor?

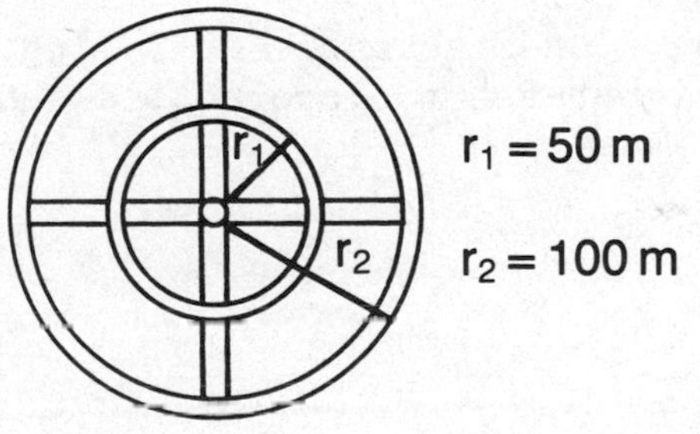

Q1: For a person in the outer corridor, what would a free body diagram look like, and what would the 2nd Law state?
Q2: What would the condition for normal weight require?
Q3: How does the condition on the inner corridor compare with the outer one?
Q4: What relates the linear speed at the outer corridor to the angular speed?
Q5: In the presence of this "artificial gravity," would you stand with your head or your feet toward the hub?

ANSWERS TO QUESTIONS

FUNDAMENTALS:

1A. (d.)
1B. (b.)
2. (c.)
3. (e.)
4A. (c.)
4B. (a.)
5. (a.)
6. (a.)
7. (c.)
8. (d.)

GENERAL PROBLEMS:

Problem 4-1.

A1: The x and y coordinates of the aircraft and the projectile must be equal at the same time.

A2: Take the origin at the position of the gun. The aircraft then has $x_0 = 0$, $y_0 = +1000$ ft. Obviously $t = 0$ when the gun fires.

A3: $y = 1000 \text{ ft} + (V_a)_y t$; $x = (V_a)_x t$.

A4: The aircraft has a velocity of 600 mi/hr (880 ft/s), whose upward component is $(V_a)_y = 9000$ ft/min (150 ft/s). Thus the horizontal component is $(V_a)_x = [880^2 - 150^2]^{\frac{1}{2}} = 867$ ft/s.

A5: $y = 2000(\sin\theta)t - 1/2(32.2)t^2$; $x = 2000(\cos\theta)t$.

A6: Set the x coordinates equal. Since they both have $a_x = 0$, these coordinates must be equal at all times. You'll notice that t cancels out.

A7: Having found the angle of aiming, this gives you the time of impact between the projectile and the aircraft.

Problem 4-2.

A1: Circular motion in a vertical plane, at least for an instant.

A2:

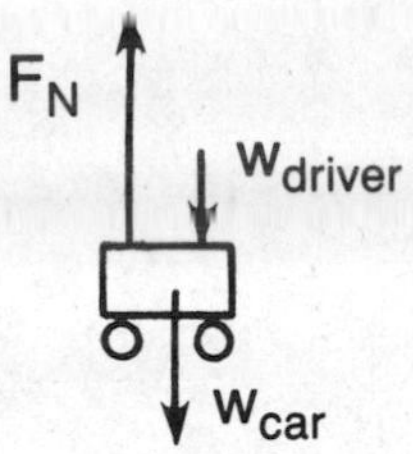

A3: $F_N - w_{car} - w_{driver} = (M_{car})V^2/R$

A4: That $F_N = 5000\ N + w_{car} = 1.28 \times 10^4\ N$

A5: $(F_N)_{driv} - w_{driv} = m_{driv}\ V^2/R$

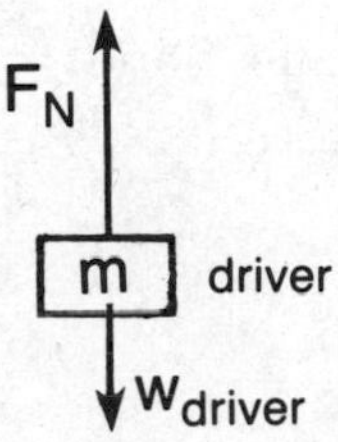

Problem 4-3.

A1: The force of lunar gravity dictates a different orbital speed, and produces a different vertical acceleration on falling objects, g_{moon}.

A2: $F_g = GmM_{moon}/R^2_{moon} = mV^2/R_{moon}$ for circular orbit close to the moon's surface.

A3: $V_0 = V_{orbit} - 1000$ m/s, in the horizontal forward direction.

A4: $y = 1600$ meters $- (1/2)g_{moon}t^2$; $V_y = -g_{moon}t$
$x = V_0t$; $V_x = V_0$.

A5: When $y = 0$.

Problem 4-4.

A1:

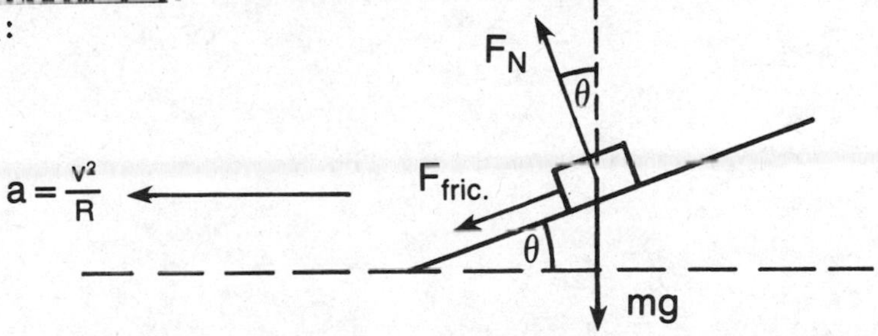

A2: There are two choices:

1. Axes parallel and normal to the bank. You'll have to find the components of a and mg in that system.
2. Axes vertical and horizontal. You'll have to find the components of F_{fric} and F_N in that one.

A3: $F_N - mg \cos\theta = ma_N = mV^2 \sin\theta/R$.

A4: $F_{fric} + mg \sin\theta = ma_P = mV^2 \cos\theta/R$.

A5: That F_{fric} be maximum, i.e., $F_{fric} = \mu_s F_N$.
A6: To substitute $\mu_s F_N$ for F_{fric} and eliminate F_N from the two component equations. The mass will cancel and give an expression for V_{max} in terms of μ_s, θ, g, and R.

Problem 4-5.
A1: $F_N = mR\ \ ^2$

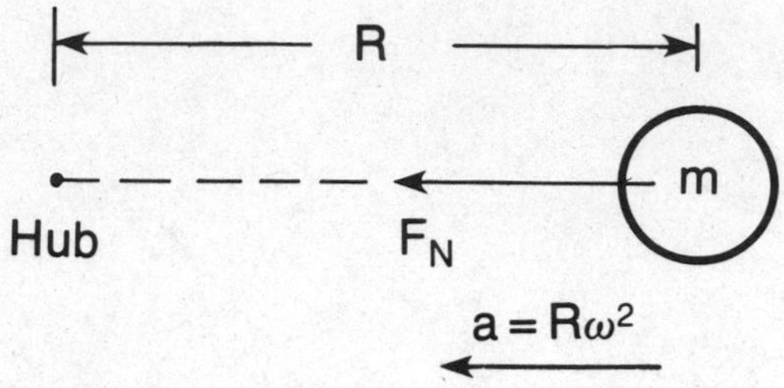

A2: That $F_N = mg$, where $g = 9.8\ m/s^2$.
A3: That radius is half as much, and the angular speed is the same. A4: $v = R_2\omega$, with $R_2 = 100$ meters.
A5: Head toward the hub.

CHAPTER 5. EQUILIBRIUM

BRIEF SUMMARY AND RESUME

The central subject of this chapter is that of *balanced forces*, responsible for the physical condition called *equilibrium*. *Tension* and *compression* forces are discussed. The conditions under which an object is in *translational equilibrium* is derived. The idea of *torques* is introduced, and the condition for the *rotational equilibrium* of an object is derived. Finally, the *center of gravity* of an object is defined.

Material from previous chapters which is used in this chapter includes:

1. The concept of force vectors.
2. Methods of addition of vectors.
3. Newton's three laws of motion.

NEW DATA AND PHYSICAL CONSTANTS

1. The SI unit of torque: the newton-meter, N-m.

NEW CONCEPTS AND PRINCIPLES

Translational Equilibrium

Understanding this section should enable you to do the following:

1. *Calculate the force conditions on an object in translational equilibrium.*

If the net force on an object at rest is zero, then the object will remain at rest, according to Newton's 1st Law. If the object is in motion, it will change neither its speed or direction of motion. In either case, the object is said to be in *translational equilibrium*. Since very often there are a number of forces acting on an object, these forces must exactly cancel in order to have this condition. If we resolve all the forces into their x and y (and z) components, we can write this mathematically, called the *First Condition of Equilibrium*:

$$(F_{net})_x = \sum_{i-1}^{N} (F_i)_x = 0 \; ; \quad (F_{net})_y = \sum_{i=1}^{N} (F_i)_y = 0$$

These are two scalar equations.

Example 5-1. Refer to the free body example (c) on page 39 of this guide. If the mass of the box is 5 kg and the box and the table are both steel, how hard can the boy pull on the string before the box begins to slide?

S: The boy can pull on the box with a force as large as the maximum force of static friction, but I don't know how much force that is.

T: So let's find out. Here the magnitude of a horizontal force is determined by the amount of a vertical force, so we expect there to be a *linkage* or *coupling* between the equations in the x and y directions. As long as it doesn't slide, the box is in equilibrium, so the 1st Condition equations give us x components: $F_{string} - F_{friction} = 0$

y components: $F_N - mg = 0$

S: There are *three* forces we don't know: $F_{friction}$, F_N and F_{string}. We need another equation.

T: Good! The missing equation is how $F_{friction}$ is determined by the compression between surfaces, F_N. That is

$$F_{static\ friction} \leq \mu_s F_N.$$

From Table 3-1 of your text we find that $\mu s = 0.78$ for steel on steel. Then

$$F_N = mg = (5kg)(9.8\ m/s^2) = 49\ N$$

So

$$F_{friction} \leq (9.78)(49\ N) = 38.2\ N$$

and finally

$$F_{string} = F_{friction} \leq 38.2\ N.$$

Example 5-2. A helium filled balloon weighing 4000 N experiences a lifting or buoyant force of 5000 N. As shown in the figure, two tethering ropes of negligible weight are attached to opposite sides. Both ropes make an angle of 75° with the ground. How much tension must each rope exert to keep the balloon stationary?

S: I need to draw a free body diagram, but I don't know whether the tensions in the ropes are equal or not.

T: From the diagram, you can see that the ropes are in a symmetric situation, so you should expect them to be doing the same thing. But we don't *have* to assume this at the start. Lets call them T_1 and T_2 and have the following free body diagram:

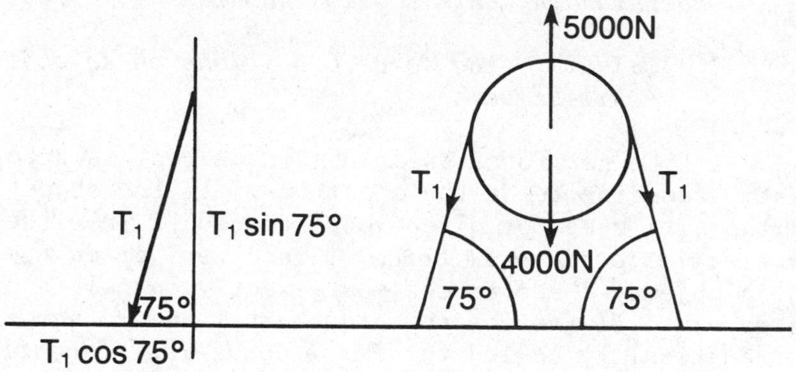

S: There are two unknown tensions, so we need two equations?

T: Yes. One of them is that the horizontal components cancel.

S: You mean the horizontal components of T_1 and T_2.

T: Right. $T_2 \cos 75^o$ is to the right and $T_1 \cos 75^o$ must balance it to the left. That shows that $\underline{T_1 = T_2}$, as we suspected. Now let's just call the individual tensions T.

S: The second equation is that the vertical forces add up to zero.

T: Mathematically that means

$$5000\ N - 4000\ N - 2T \sin 75^o = 0$$

Now the physics is over. It's straight algebra from here on.

$$2T \sin 75^{\circ} = 1000 \text{ N}$$

Or

$$T = 500 \text{ N}/\sin 75^{\circ} = 518 \text{ N}$$

Rotational Equilibrium

Understanding this section should enable you to do the following:

1. *Calculate the torques produced by forces acting at given locations on an object.*
2. *Calculate the torque conditions required for an object to be in rotational equilibrium.*
3. *Calculate the location of the center of mass of a distribution of point masses.*
4. *Include the force of gravity on an object in calculating its rotational equilibrium.*

Forces That Produce Rotation: Torques. For translation equilibrium, we didn't need to pay attention to *where* the forces were attached to the object. We needed only their direction, or *line of action*, in addition to their magnitude. However, if the object is free to rotate about some axis, a force whose line of action does not pass through that axis will tend to rotate the object.

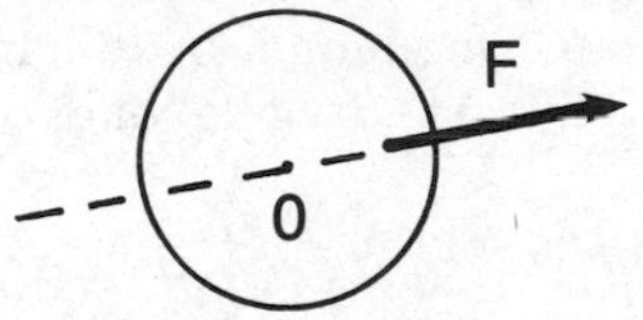

no tendency to rotate

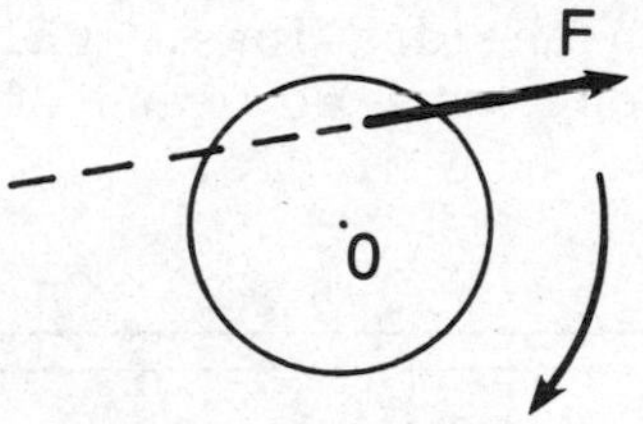

tendency to rotate clockwise

The effectiveness of a force in producing a rotation depends on how large the force is and how far away its line of action is from the axis of rotation.

Torque--The product of the magnitude of a force and its lever arm, l.

Lever Arm--The perpendicular distance from the line of action of a force to an axis of rotation. This is also called the *moment* of the force.

The symbol for torque is τ (tau). The mathematical form of the definition of torque is

$$\tau = Fl$$

Important Observations

1. There are *two* directions that rotation about an axis can have: clockwise or counterclockwise. To distinguish between these opposites, we shall assign a + sign to torques producing counterclockwise rotations and a - sign to those producing clockwise rotations.

2. The SI units of torque are the newton-meter (N-m).
3. Remember that l is always perpendicular to the line of action of F.

<u>Couple</u>--The special situation where two equal but oppositely directed forces are applied equidistant from and on opposite sides of an axis.

<u>Important Observations</u>

1. Since $F_{net} = 0$ for a couple, they cannot produce a translational acceleration of an object.
2. The total torque produced by a couple about the axis is F times L, where L is the perpendicular distance between the lines of action of the two forces.

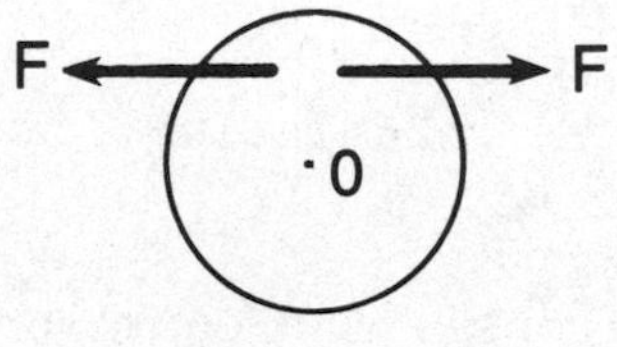

<u>not</u> a couple
($\tau = 0$)

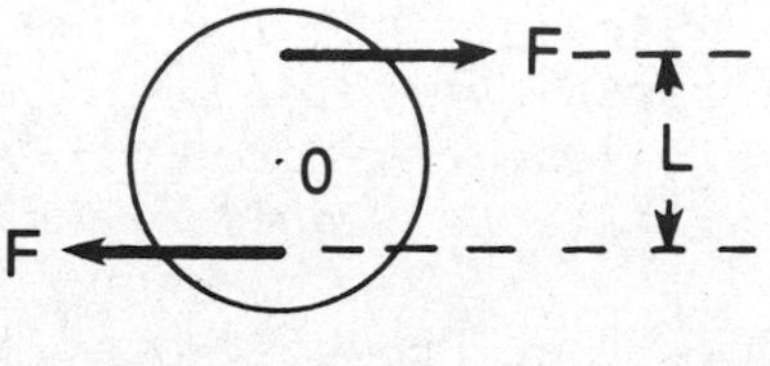

a couple,
($\tau = -F \times L$)

<u>Example 5-3</u>. In the following, forces are applied to a meter stick at various points as shown in the drawings. Calculate the torque produced by each force and the net torque about axes at 0 and 0' for each case. 0 is at the end of the stick, and 0' is at the middle.

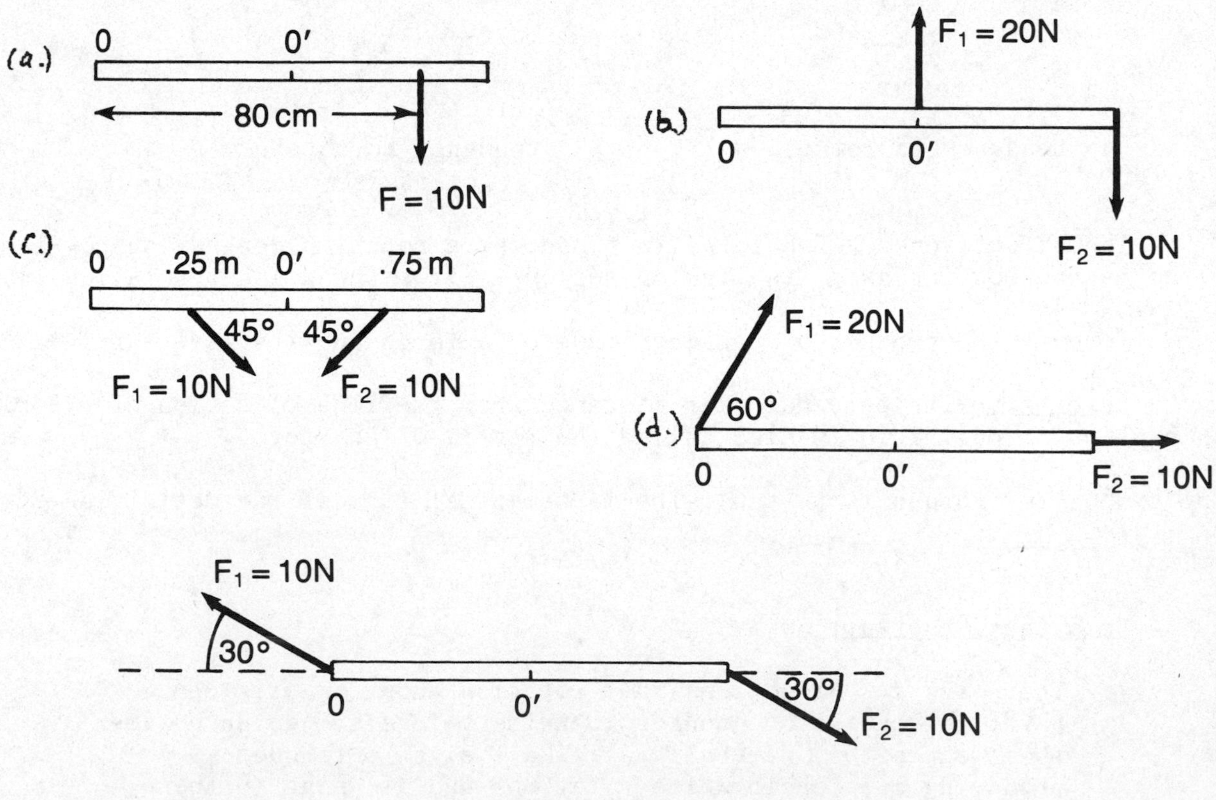

S: Part (a.) is easy. l = .80 m from O and .30 m from O', so

$$\tau = (10\ N)(.8m) = 8\ N\text{-}m \text{ about O, and}$$

$$\tau = (10\ N)(.3m) = 3\ N\text{-}m \text{ about O'.}$$

T: Right, except you forgot the *sign*. F would cause *clockwise* rotation about both O and O', so both your quantities should be *negative*.

S: In (b.) the torques *oppose* each other around O. But F_1 doesn't produce a torque at all about O', I don't think.

T: You're right, since F_1 passes right through O', so l = 0 for that one. Otherwise, we have

$$\tau_{net} = +(20\ N)(.5\ m) - (10\ N)(1\ m) = 0 \text{ about O,}$$

and

$$\tau_{net} = 0 - (10\ N)(.5\ m) = -5\ N\text{-}m \text{ about O'.}$$

See how changing the axis can change τ_{net}?

S: In (c.) the lever arms aren't so obvious. But I can see by *symmetry* that τ_{net} = 0 about O'.

T: Good! Symmetry inspection is *very* helpful in simplifying physics problems. As you saw, the forces were equal, angles were equal, points of application were equidistant from O', and the forces tended to rotate the stick oppositely about O'. As for the lever arms about O, look at the diagram as follows:

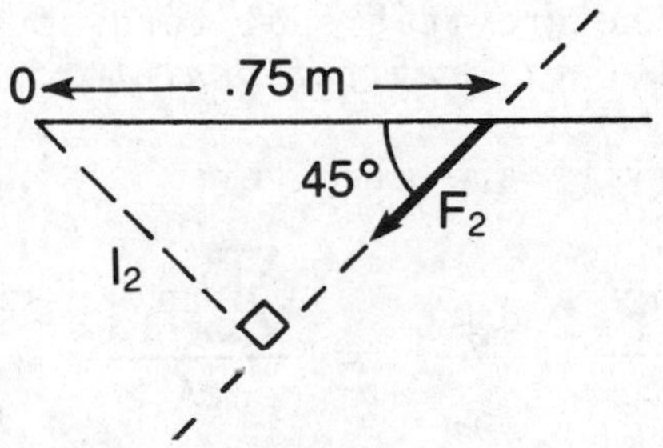

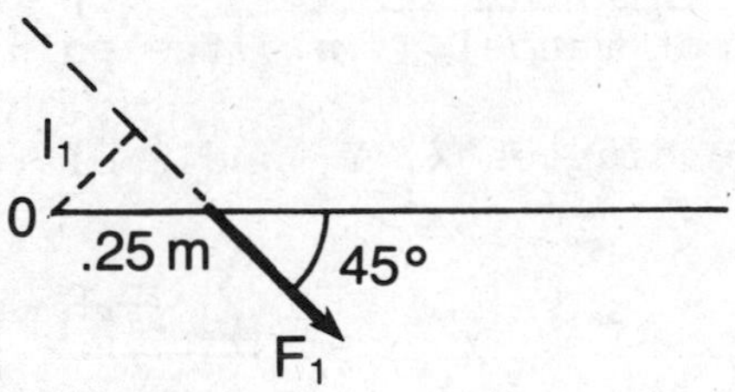

It should be obvious that

$$\tau_1 = -\ (10\ N)(.25 \sin 45^o\ m),$$

and

$$\tau_2 = +\ (10\ N)(.75 \sin 45^o\ m).$$

So about O,

$$\tau_{net} = +5.30 - 1.77\ N\text{-}m = +3.53\ N\text{-}m\ .$$

S: In (d.) it looks like both lines of action pass through O, so

$$\tau_1 = \tau_2 = \tau_{net} = 0 \text{ about O.}$$

T: Very good. Furthermore, F_2 produces no torque about O' either. So the only torque is the one produced by F_1 about O', with l_1 = (.50 m) sin 60° = 0.25 m. So

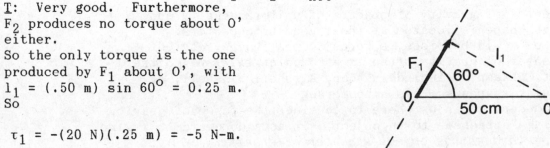

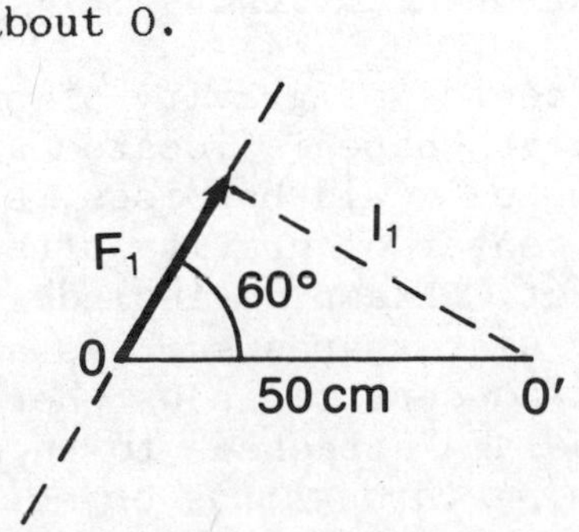

$$\tau_1 = -(20\ N)(.25\ m) = -5\ N\text{-}m.$$

S: In (e.) we again have a symmetry situation about O', but different than part (c.). F_1 and F_2 are opposite--they form a *couple* about O'.

T: They sure do. You can see that this is the only one of the cases which is in *translational equilibrium*. Can you show that the distance L between the lines of action of F_1 and F_2 is .50 m? Then, about O',

$$\tau_{net} = F \times L = -(10\ N)(.5\ m) = -5\ N\text{-}m.$$

This exactly the same as τ_{net} about O, by the way. Verify this. Can you *generalize* this result?

The Second Condition for Equilibrium--When the algebraic sum of clockwise and counterclockwise torques on an object (i.e., the *net* torque) about any arbitrary axis is zero, the object is in *rotational equilibrium*.

Important Observations

1. A key point is that *any* axis can be chosen for the application of the 2nd Condition, because if the object is in rotational equilibrium, it cannot have unbalanced torques about any stationary axis. The chosen axis can even be *external* to the object, as well as one passing through it.
2. The above observation means you should, wherever possible, choose an axis through which some of the forces pass, in order to simplify the torque calculations.

Center of Gravity--The point about which the gravitational torques acting on an object cancel, i.e., the point about which the *net gravitational torque is zero.*

The coordinates X, Y, and Z of the center of gravity are given by:

$$X_{cg} = \frac{\sum m_i x_i}{M} \; ; \; Y_{cg} = \frac{\sum m_i y_i}{M} \; ; \; Z_{cg} = \frac{\sum m_i z_i}{M}$$

as explained in your text. M is the total mass of the object. Without the methods of calculus, these calculations can only be done for situations consisting of a finite number of individual masses, m_i.

Strictly speaking, these equations define the object's *center of mass.* As long as the gravitational force on the object is *constant* and *uniform* across its volume, this is the same as the center of gravity.

Important Observations

1. The center of gravity of objects of uniform density and regular shape is located at their geometric centers. Examples would be discs, spheres, cubes, hoops, etc.
2. The center of gravity often is at a point *external* to the object. Examples include a hoop, a spherical shell, or irregular shapes such as a boomerang.
3. This concept tells us where to consider the force of gravity as being "attached" to an object. It acts downward *on a line passing through the object's center of gravity.*

Example 5-4. A uniform rod 1 meter long having a mass of 500 grams is suspended by two vertical strings. One is attached at one end and the other is attached 80 cm away, as showm in the figure below. If the rod is in equilibrium, find the tensions in the strings.

S: Only vertical forces are acting, so the tensions up must equal the weight down. Thus

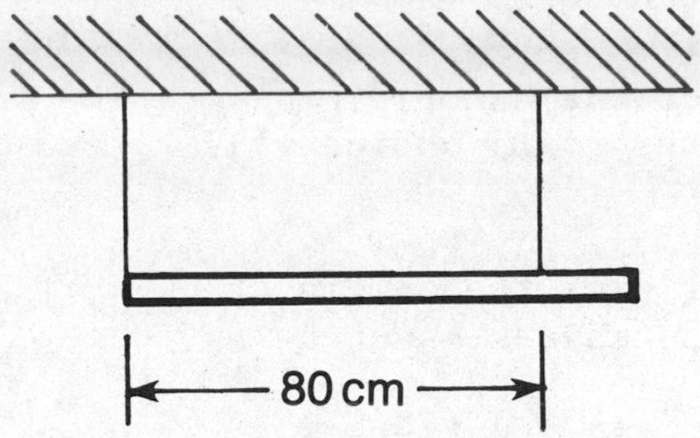

$$T_1 + T_2 = mg = (.5 \text{ kg})(9.80 \text{ m/s}^2) = 4.9 \text{ N}$$

I need another equation.

T: You've applied the lst Condition. Now apply the 2nd Condition.

S: I have to make two choices. What axis to use for the calculation of torques, and where to put the weight vector.

T: You have no choice in where w goes. The rod is uniform, so its center of gravity is at the 50 cm mark. Look at the free body diagram. Let's choose two points, the end and the midpoint, and apply the 2nd Condition in each case. They should give the same result, as should any other point you might choose.

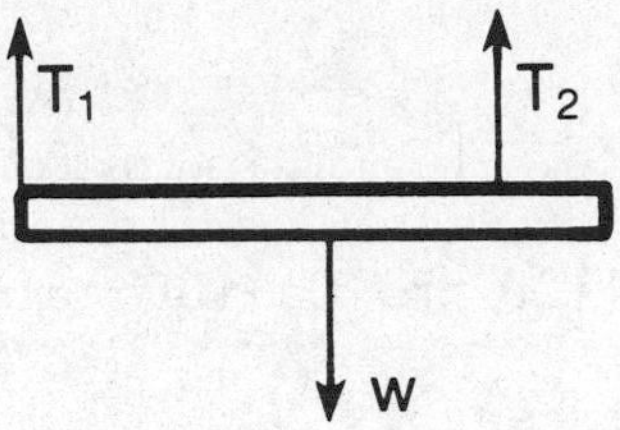

About the end $\tau_1 = 0$, so

$$T_2(.8 \text{ m}) -(4.9 \text{ N})(.5 \text{ m}) = 0$$

This gives $T_2 = 3.06$ N, and so from the lst Condition $T_1 = 1.84$ N

Calculating torques about midpoint: $\tau_w = 0$, and

$$-T_1(.50 \text{ m}) + T_2(.30 \text{ m}) = 0$$

This gives $T_2 = 1.67\ T_1$. Then

$$T_1 + 1.67\ T_1 = 4.9 \text{ N},$$

giving $T_1 = 1.84$ N and $T_2 = 3.06$ N , the same results as before.

<u>Example 5-5</u>. You have three masses arranged in an equilateral triangle as shown. Where is the center of mass of this arrangement?

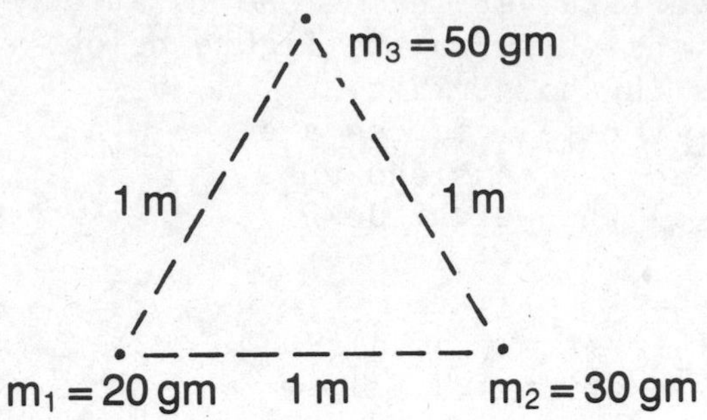

<u>S</u>: I see the formulas, but I'm not clear on what the x_i and y_i mean here.

<u>T</u>: *You* choose an origin of coordinates. *Any* choice will do.

<u>S</u>: Can I pick the origin at one of the masses, such as m_1?

<u>T</u>: Sure. As a matter of fact, that simplifies things, making $x_1 = 0$ and $y_1 = 0$.

<u>S</u>: But then m_1 seems to disappear from the calculations.

<u>T</u>: It makes itself felt in $M = m_1 + m_2 + m_3 = 100$ grams. The rest of the calculation is then

$$X_{cg} = (m_1x_1 + m_2x_2 + m_3x_3)/M$$

$$= \{(20 \text{ gm})(0)+(30 \text{ gm})(1 \text{ m})+(50 \text{ gm})(.5 \text{ m})\}/100 \text{ gm}$$

Thus $X_{cg} = 0.550$ meters .

Similarly,

$$Y_{cg} = (m_1y_1 + m_2y_2 + m_3y_3)/M$$

$$= \{(20 \text{ gm})(0)+(30 \text{ gm})(0)+(50 \text{ gm})(.707 \text{ m})\}/100 \text{ gm},$$

giving $Y_{cg} = 0.354$ meters .

QUESTIONS ON FUNDAMENTALS

1. A frame is hanging from two identical cords attached at its two top corners. If the frame weighs 10 N and the cords each make a 60° with the vertical, what is the tension in each cord?

   (a.) 5.8 N (b.) 20 N (c.) 10 N (d.) 5 N (e.) 11.0 N

2. A cord is wound around an axle whose radius is 6 cm. If you pull on the cord with a force of 7 N, how much torque do you exert about the axis?

   (a.) 0.42 N-m (b.) 7/6 N-cm (c.) 252 N-cm (d.) 4.3 N-m

3A. A uniform 500 gram meter stick is resting on two knife edge supports. One (A) is at the 10 cm mark, and the other (B) is at the 65 cm mark. Which is exerting more force on the stick?

   (a.) A (b.) B (c.) both are exerting the same

3B. Referring to #3A, how much more force is one exerting than the other?

   (a.) 6.5 times as much (b.) both the same (c.) 2.67 x as much
   (d.) 3.5 x as much

4. Where is the center of gravity of the following collection of objects: 3 kg at x = -2 cm; 1 kg at x = +2 cm; and 2 kg at x = +5 cm.

   (a.) x = +3 (b.) x = +6 (c.) x = +2 (d.) x = 0 (e.) x = +1

GENERAL PROBLEMS

Problem 5-1. Imagine that the masses in Example 5-5 are somehow fixed in position by thin rods of negligible mass. If you suspend this rigid arrangement by a string attached to m_3, at what angle relative to vertical will the side connecting m_1 to m_3 hang when in equilibrium? If the three masses were all equal, what would this angle be?

Q1: What property determines how it will hang?
Q2: What does this look like in a diagram?
Q3: The angle looks difficult to calculate. How best to determine it?

Problem 5-2. The crane shown in the figure consists of a uniform 2 meter long beam allowed to pivot around a fixed end A. A cable is attached to the other end B and runs to a winch at C, 2m directly above A. Suppose the crane is in equilibrium when it is bearing a 100 kg mass hanging from B, and the angle between cable and beam is 50°. The beam itself has a mass of 20 kg. Find the force components that pivot A must exert on the beam in this situation, and find the total force of compression on the pivot.

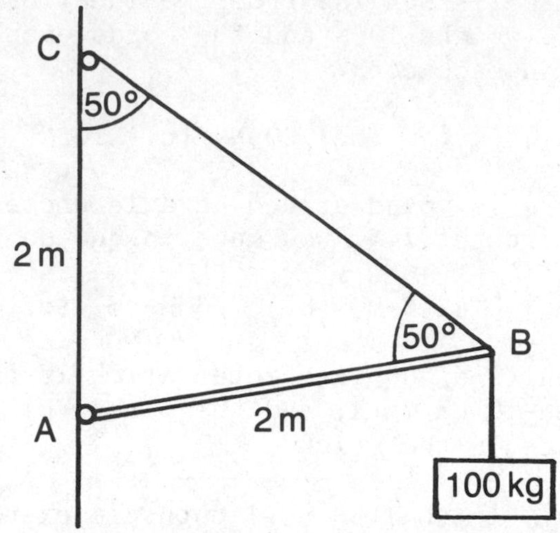

Q1: Let's see. The beam is in equilibrium, so we'll draw a free body diagram for it. How many forces are acting on it and where?

Q2: How many unknowns do I have in the diagram?

Q3: What equations does the 1st Condition of Equilibrium yield?

Q4: I'm never sure about being right in choosing the point about which to calculate torques.

Q5: Then what does the 2nd Condition for Equilibrium give?

Q6: What does "total force of compression" mean?

Problem 5-3. Your car has a wheelbase of 10 ft. When the rear wheels are on a scale, the scale reads 1800 pounds, and when the front wheels are on the scale, it reads 220 pounds. How far in front of the rear axle is the car's center of gravity located?

Q1: What do these scale readings mean?

Q2: What is the total weight of the car?

Q3: What does the free body diagram for the car in equilibrium look like? Have I included the weight of the car?

Q4: What information do the conditions for equilibrium give me?

Q5: Where does the length of the car come in?

Problem 5-4. Consider a square block made of uniform material, which is 1 meter on a side, shown in the diagram. The block has a mass of 1 kg. A hole of radius 0.2 meters, centered in the middle of the upper right quadrant of the block, is then cut out of the block. Where is the center of mass of the block after it has the hole cut in it?

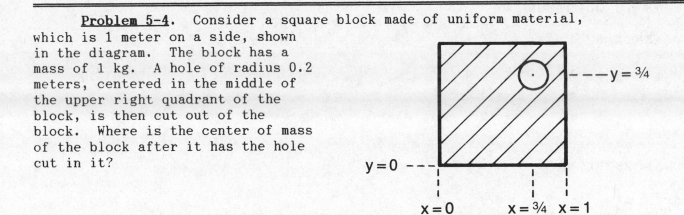

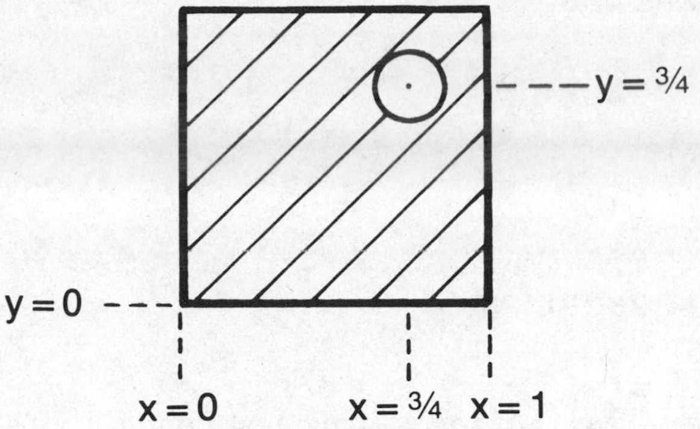

Q1: The block with the hole doesn't have any simple symmetry left. I can't calculate it from the basic definition unless I have an arrangement of point masses or other symmetric masses whose centers of mass I know. If the square was whole, its center of mass would be at its geometric center.

Q2: A circular block would have its center of mass at *its* center, but here a circular mass was *removed*, not added to the square.

Q3: Removing the circular block from the square is just the *opposite* of adding a similar block to the whole square. What does this mean?

Q4: How can I find the mass of the material removed from the square?

ANSWERS TO QUESTIONS

FUNDAMENTALS

1. (c.)
2. (a.)
3A. (b.)
3B. (c.)
4. (e.)

GENERAL PROBLEMS:

Problem 5-1.

A1: That gravitational torques cancel. This means the vertical line through the point of suspension (m_3) must also pass through the center of gravity. The location of that point was found in Example 5-5.

A2:

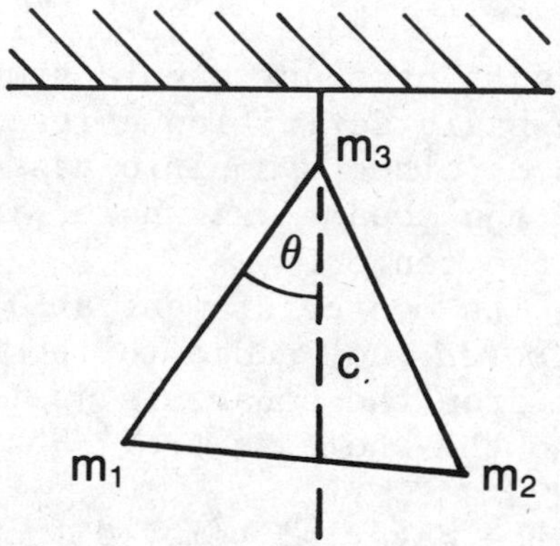

A3: You can solve for it using geometry and trigonometry, but it is somewhat involved. If you make a careful scale drawing, you can easily measure the angle with a protractor.

Problem 5-2.

A1: There are 5 forces: T, F_x, F_y, (100 kg) x g, and the weight of the beam, w.

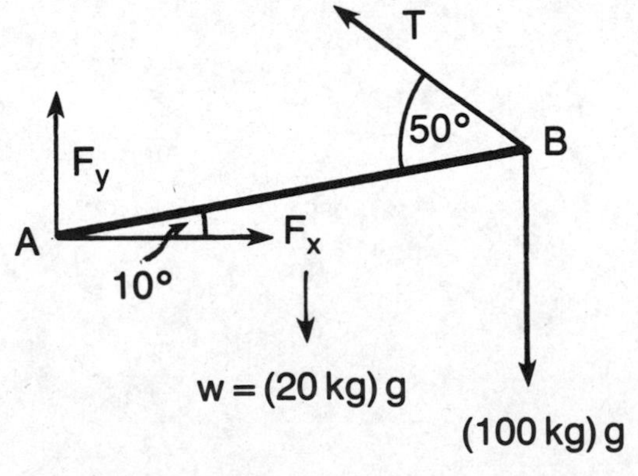

A2: Three.

A3: $T_x = F_x$; $T_y + F_y = (120\text{ kg})g$. The diagram gives you the angles for calculating T_x and T_y in terms of T.

A4: Any point will do. But both components of F pass through A, so this would be the most convenient point.

A5: $T(2\text{ m})\sin 50 - (20\text{ kg})g(1\text{ m})\cos 10 - (100\text{ kg})g(2\text{ m})\cos 10^{\circ} = 0$. This gives T directly. Going back to A3 then gives F_x and F_y.

A6: The vector composed from F_x and F_y.

Problem 5-3.
A1: They measure the normal force between the car and ground at the rear and front wheels.
A2: 4000 pounds.
A3:

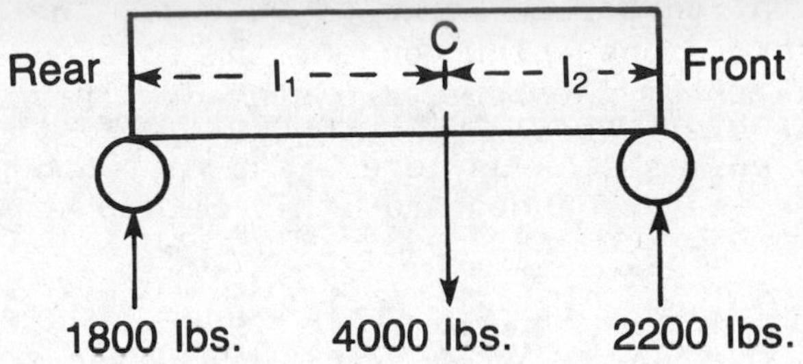

A4: τ_{net} about the center of mass is zero.
A5: The sum of the two lever arms l_1 and l_2 is 10 ft.

Problem 5-4.
A1: That's a good observation, even though it may seem irrelevant at the moment. If you had a circular block of the material like the one that was cut out, where would its center of mass be?
A2: If you had the *whole* square again, and *added* the circular block on the top of the square at the same spot as described before for the hole, the coordinates of the center of mass for this combination would be

$$X_{cm} = (M_{square}X_{square} + M_{circle}X_{circle})/(M_{square} + M_{circle})$$

and

$$Y_{cm} = (M_{square}Y_{square} + M_{circle}Y_{circle})/(M_{square} + M_{circle})$$

Here (X_{square}, Y_{square}) would be the coordinates of the square's center, and (X_{circle}, Y_{circle}) are those of the circle's center.
A3: The *opposite* of *adding* is *subtracting*, so we can use the A2 equations with $-M_{circle}$ instead of $+M_{circle}$, in both numerator and denominator.
A4: 1 sq. meter of the block has a mass of 1 kg. The 1 kg/m^2 can be called the mass density per area of the material. The area of the circle is πr^2 where r = 0.2 m. Thus $M_{circle} = \pi r^2 \times 1$ kg/m^2.

CHAPTER 6. WORK AND ENERGY

BRIEF SUMMARY AND RESUME

Previous material on force and acceleration is developed into the concepts of *work* and *energy*. The energy of motion, *kinetic energy*, and energy of position, *potential energy*, are introduced as forms of energy. *Thermal energy* is also discussed briefly. Expressions which connect the work done by forces with changes of energy, called *work-energy theorems*, are stated. This leads to the fundamental principle of *conservation of energy*. The concepts of *mechanical advantage* and efficiency of *simple machines* such as levers and compound pulleys are developed. The definition of *power* is introduced and its relation to force and speed is examined.

Material from previous chapters relevant to this chapter includes:

1. Force and Newton's 2nd Law of Motion.
2. Displacement.
3. Gravitational Force.

NEW DATA AND PHYSICAL CONSTANTS

1. The SI unit of work and energy.

   1 N-m = 1 joule (J)

2. The elastic force constant of a spring.

   $k = F/x$, with units of N/m.

3. The SI unit of power.

   1 watt (W) = 1 joule/sec = 1 N-m/s.

NEW CONCEPTS AND PRINCIPLES

Work

Understanding this section should enable you to do the following:

1. Given a constant force on an object and the object's displacement, calculate the work done by the force.

In previous chapters we have seen how the basic concept of force is related to the description of the motion of objects. This was a *vector* description, and we had to be very careful in giving attention to the choice of coordinate system and the calculation of vector components. We now turn to a *scalar* description of what forces do, by defining *work* and *energy*. This gives us an advantage, because scalar mathematics is easier than working with vectors.

Work--The magnitude of a force component *in the direction* of motion of the object on which the force acts, times the *magnitude* of the object's *displacement*.

If θ is the angle between **F** and the displacement **s**, the mathematical form of this is

$$W = F\,s\cos\theta$$

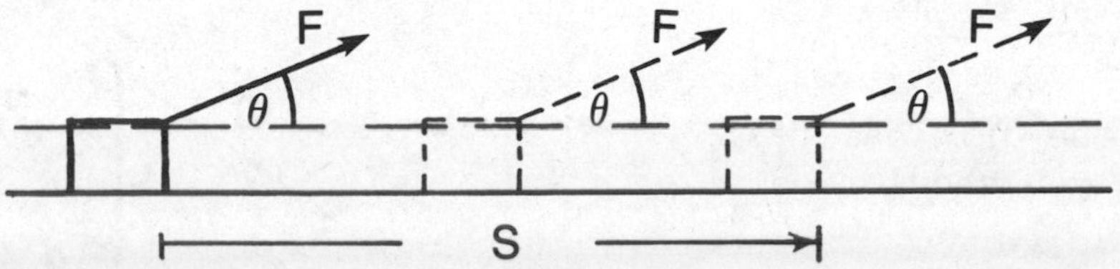

Important Observations

1. Since $\cos\theta$ is dimensionless, the derived SI unit of work is the N-m. This is given the special name of *joule*:

   1 joule (J) = 1 N-m.

   Although this looks the same as the unit of *torque*, remember in that case F is *perpendicular* to the lever arm. Work and torque obviously represent *totally* different physical situations.
2. Since only the *magnitudes* of F and s are involved, *work is a scalar* quantity.
3. Work can be *positive* or *negative*, depending on the sign of $\cos\theta$. In the case of negative work, $\theta > 90$. Examples of this would be a moving object to which a retarding force is applied, or sliding friction, which is always opposite to the direction of motion.
4. If many forces are acting on an object, you can use the definition to calculate the work done by each force individually. The *net work* done on the object will be the algebraic sum of these individual amounts. This will give you the same result as you would get by calculating the work done by the *net force*.
5. The work done by a force can be *zero* in either of *two* ways:
   (a.) No displacement occurs, i.e., the force is applied to a stationary object.
   (b.) $\cos\theta = 0$, which means that $\theta = 90^\circ$. This tells us that *normal forces do no work*.

This latter point can be summarized by saying that *turning* an object requires no work; changing its speed does.

Example 6-1. A 7.00 kg box slides down a 35.0° ramp from the second story of a building to the street 6.00 meters below (vertical height). How much work did the force of gravity do?

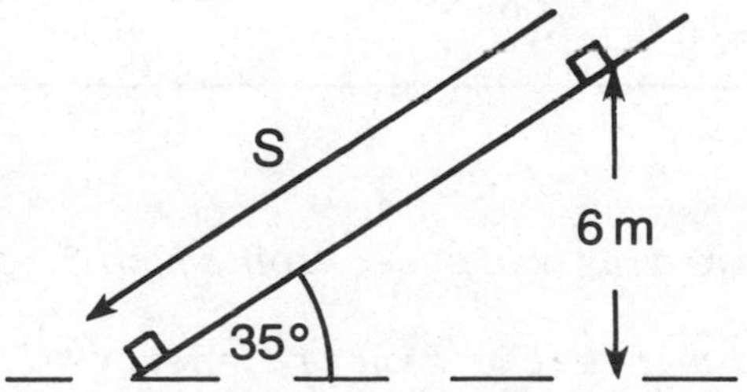

S: We need F_g, s, and $\cos\theta$. $F_g = mg = 68.6$ N. We have to identify θ. Is it 35°?

T: No. You'd better sketch F_g and s on a free body diagram: You can see that θ is complementary to 35°, i.e., $\theta = 55^\circ$, and so $\cos 55^\circ = .574$.

S: Finding s is a little confusing.

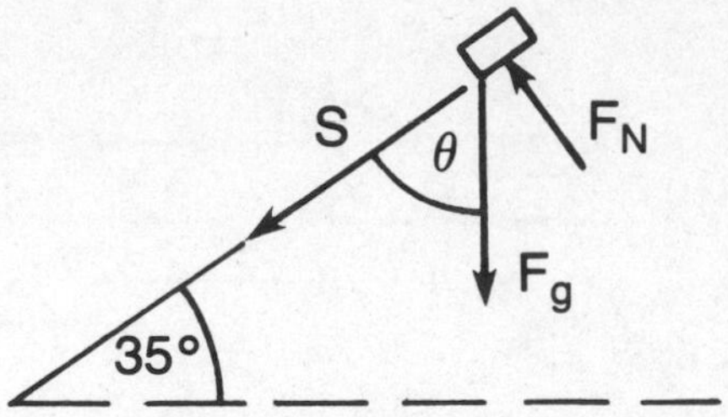

T: Go back to the original real diagram. There you should see that

$$(6\text{m})/s = \sin 35^\circ \text{ , so } s = (6\text{m})/\sin 35^\circ = 10.5 \text{ meters.}$$

Now we have the required numbers:

$$W = F\, s \cos 55^\circ = (68.6 \text{ N})(10.5 \text{ m}) \cos 55^\circ = 413 \text{ N-m} = 413 \text{ J.}$$

S: There was no mention of friction. Wouldn't that make a difference?

T: Not in the calculation of the work done by F alone. If friction *was* present, the result we got wouldn't then be the *total* or *net* work done on the object. Want to try that, with $\mu_k = 0.20$?

S: I'm sorry I asked. But we've already done all those inclined plane situations. With no other forces, $F_N = mg \cos 35^\circ$.

T: Good. You can see why in physics it's important to keep relating to previous material. We now have

$$F_{frict.} = \mu_k F_N = (.20)(68.6 \text{ N}) \cos 35^\circ = 11.2 \text{ N.}$$

S: This time s is *anti-parallel* to the force, so

$$W_{frict.} = -F_{frict.}\, s.$$

T: Right. Friction is always 180° from the displacement. So

$$W_{frict.} = F_{frict.}\, s \cos 180^\circ = (11.2 \text{ N})(10.5 \text{ m})(-1) = -118 \text{ J.}$$

The net work is then

$$W_{total} = W_g + W_{frict.} = 305 \text{ J} - 118 \text{ J} = 187 \text{ J.}$$

S: And the third force, F_N, did no work?

T: By definition, F_N is perpendicular to s, and $\cos 90^\circ = 0$. Thus $W(F_N) = 0$ *always*.

Energy

Understanding this section should enable you to do the following:

1. *Calculate the kinetic energy of an object.*
2. *Calculate the gravitational potential energy of an object near the Earth's surface.*
3. *Calculate the elastic potential energy of a stretched or compressed spring.*

As we shall see, what we define as energy has many forms, all convertible one into another. It is perhaps the single most *unifying* concept in physics, since various forms of energy occur in all the topics which comprise physics.

Energy--The ability to perform work.

This is an astonishingly brief and general definition, and probably leaves you wondering "what does *that* mean?". Have patience. The best way to find out is to define some specific examples of energy and see after that that they do agree with this broad definition. We could turn this statement around and say that work is the process of transferring energy.

It should be clear from everyday experience that a moving object can exert a force and do work on another object that it hits. In keeping with the above definition, we define an energy of motion, or kinetic energy, of a moving object as follows:

Translational Kinetic Energy (KE)--The energy associated with the *motion* of an object, equal to one-half the product of its mass times the square of its speed. In mathematical form,

$$KE = \tfrac{1}{2}mv^2.$$

Important Observations

1. The SI units of KE are $(kg)(m/s)^2$ = N-m = joules, the same as the unit of work.
2. KE is a *scalar* quantity, and is *always positive*, since neither m nor v^2 can ever be negative.

Example 6-2. What is the kinetic energy of a 40 grain .22 caliber rifle bullet as it leaves the muzzle at 1400 ft/s? (A grain is a measure of weight, with 7000 grains = 1 pound).

S: We need to calculate $\tfrac{1}{2}mv^2$. You sure have crazy units here.

T: But *realistic* ones. They are precisely how the performance of firearms are described. First we'd better convert everything into SI.

S: First we'll convert grains to newtons:

$$40 \text{ grains} \times (1 \text{ pound}/7000 \text{ grains}) \times (4.45 \text{ N}/1 \text{ pound}) = .0254 \text{ N}$$

Now ft/s into m/s:

$$1400 \text{ ft/s} \times (.305 \text{ m}/1 \text{ ft}) = 427 \text{ m/s}.$$

So then KE = $\tfrac{1}{2}(.0254 \text{ N})(427 \text{ m/s})^2$?

T: You've just made a very common mistake, using *weight* instead of *mass*.
S: Tricky, tricky. The bullet's *mass* is

$$m = w/g = (.0254 \text{ N})/(9.8 \text{ m/s}^2) = 2.59 \times 10\text{-}3 \text{ kg}.$$

T: *Now* you can plug numbers in:

$$KE = \tfrac{1}{2}(2.59 \times 10^{-3} \text{ kg})(427 \text{ m/s})^2 = 2.36 \times 10^2 \text{ J}.$$

It is clear that as a falling object loses height it gains speed and hence kinetic energy. This kinetic energy can subsequently do work. So just by virtue of occupying a higher vertical position, an object has greater *potential* for doing work than in a lower position. Hence we define a *gravitational potential energy* as follows:

Gravitational Potential Energy--The energy associated with the vertical position of an object experiencing a gravitational force. In mathematical form, this is

$$(PE)_g = mgh$$

Important Observations

1. The units are again joules.
2. This definition is appropriate only *close to Earth*, where F_g can be written as mg over the range of heights being considered in a given problem.
3. h is the *vertical* position measured from some *reference* level, which can be arbitrarily chosen for convenience in each given situation.
4. If an object is *below* the reference level, h and hence its potential energy are *negative*. If *above* the reference, they are *positive*.
5. When an object moves downward, it loses PE, i.e., $\Delta PE < 0$. When it moves upward, $\Delta PE > 0$.

A second example of energy of position is related to *elastic springs*. An object on the end of a stretched or compressed spring potentially has more energy than when the spring is at its equilibrium position.

Elastic Spring--A spring whose *restoring force* when stretched or compressed an amount x from its equilibrium length is linearly proportional to x and directed back toward equilibrium. In mathematical terms,

$$F_{elast.} = -kx \ .$$

Important Observations

1. x is the amount of stretch (positive x) or compression (negative x) from the spring's equilibrium length. Notice that in either case the negative sign in the force expression means F is directed toward the equilibrium position.
2. k is called the *force constant* (sometimes *spring constant*), and as can be seen from the equation, has units of N/m.
3. Not all "stretchy" materials, such as rubber bands, have this specific elastic property.
4. As you stretch an elastic spring, you must apply a force at least equal to its restoring force. This increases with stretched (or compressed) position, and is an example of a force that is *not constant* over a displacement.

Elastic Potential Energy--The energy stored in a spring due to an increase or decrease in its length relative to its unstretched or equilibrium length. The mathematical form is derived in your text, and is

$$(PE)_{elast} = \tfrac{1}{2}kx^2.$$

Important Observations

1. Both k and x^2 are always positive, so this energy is *positive* for both stretching and compression.
2. Notice that once again the SI units for this form of energy are $(N/m)x(m)^2 = J$.

Example 6-3. It takes 30 N to compress a certain spring by 10 cm. If you placed the spring in a vertical position, how much mass would cause the spring to compress .50 meters? How much energy would then be stored in the spring?

S: For F = 30 N, the compression is x = .10m. The force constant k is found by

$$k = F/s = 30 \text{ N}/.10 \text{ m} = 300 \text{ N/m}.$$

T: That's the most important quantity here. We can answer the second question immediately. For a compression of .50 m,

$$(PE)_{elast} = 1/2kx^2 = 1/2(300 \text{ N/m})(.50 \text{ m})^2 = 37.5 \text{ J}.$$

S: Where does the mass enter in?

T: Only that here F_g is the force which is doing the compressing. mg will compress the spring until its restoring force upward equals mg. For x = .50 m, the required mass is found from

$$kx = mg, \quad \text{or } m = kx/g = (300 \text{ N/m})(.50\text{m})/(9.80 \text{ m/s}^2) = 15.3 \text{ kg}.$$

Example 6-4. A 110 pound girl is standing on the top of a cliff 200 meters above the sea.

(a.) How much gravitational potential energy does she have relative to sea level?
(b.) How much PE_{grav} does she have relative to a helicopter hovering 500 meters above the sea?
(c.) If the girl climbs halfway down the cliff, what is the *change* in her potential energy in both cases above?
(d.) How fast would the girl have to be moving to have an amount of kinetic energy equal to the potential energy she had in part (a.)?

S: "Relative to sea level" must mean that we measure h from there, so h = +200 m. I think we also need to know the girl's *mass*.

T: You always have to have mass to calculate KE, but you'll note that PE_{grav} involves (mg)h, so we don't need to know mass as long as we're given the *weight* directly. In this case mg = 110 pounds = 490 N. Then

$$PE_{grav} = (490 \text{ N})(200 \text{ m}) = 9.80 \times 10^4 \text{ J}$$

S: In (b.), is the heliocopter *directly* above her?

T: It makes no difference in calculating PE_{grav}. h indicates relative vertical position *regardless of horizontal displacement*. Even if the helicopter was 2 miles out to sea, the girl is still 300 meters *below* it, so h = -300 m.

S: So then PE_{grav} = (490 N)(-300 m) = -1.47×10^5 J. How can she have two values of PE? What is her *true* potential energy?

T: Since choice of reference level is completely arbitrary, there is no *absolute* value of PE here. Both answers are "true," as long as you specify the reference level. Physically, only *changes* in potential energy are meaningful. Part (c.) will show this.

S: O.K., for sea level as a reference, the change in PE will be

$$\Delta PE = PE_{final} - PE_{initial} = mg(100\ m) - mg(200\ m)$$
$$= -4.90 \times 10^4\ J.$$

T: Correct. And relative to the helicopter, her final position will be -400 m. So

$$\Delta PE = mg(-400\ m) - mg(-300\ m) = -4.90 \times 10^4\ J.$$

S: Yeah, I can see now that the choice of reference didn't make any difference in PE, so I can always make the choice that's the most convenient. Going on to part (d.), we should set $\frac{1}{2}mv^2 = 9.80 \times 10^4$ J. *Now* I'll need the girl's *mass* in order to get v.

T: This is a point that bears repeating: PE_{grav} involves weight, whereas KE involve mass. The girl's mass is

$$m = w/g = (490\ N)/(9.80\ m/s^2) = 50\ kg.$$

S: Then

$$v = [(2)(9.80 \times 10^4\ J)/(50\ kg)]^{1/2} = 62.6\ m/s$$

T: Right. Did you check the cancellation of units hidden in joules? By the way, let's calculate, using previous methods, what speed a free falling object would acquire if it fell off the cliff 200 meters to the sea.

S: You mean using $v^2 = v_0^2 + 2gs$? With $v_0 = 0$, this gives us

$$v = [(2)(9.8\ m/s^2)(200\ m)]^{1/2} = 62.6\ m/s\ .$$

That's the same!

T: This is an example of *conversion* of potential into kinetic energy. As the object falls, its *loss* of PE is equalled by the *gain* in KE. At the bottom, PE has gone to *zero*, and the KE just before impact is equal to the original PE.

Work-Energy Theorems and the Conservation of Energy

Understanding this section should enable you to do the following:

1. *Calculate changes in the speed of an object by using the work-energy theorem for net force.*
2. *Define mechanical energy and conservative force.*
3. *Calculate the relation between speed, position, and dissipated mechanical energy using the extended work-energy theorem.*

As your text shows, the 2nd Law of Motion can be reinterpreted in terms of work-energy language. The result is what we call a *work-energy* theorem.

Work-Energy Theorem for Net Force--The work done by the *net force* acting on an object is equal to the change in the *kinetic* energy of the object.

Important Observations

1. Notice that this statement applies only to the *net* force, not to individual forces which may be acting together on an object.
2. If the work is *positive*, there is a *gain* in KE as the speed increases. If the net work is *negative*, KE *decreases*, and the object moves slower.

Example 6-5. A 5 kg box is being dragged by a force of 40 N applied 30^0 above horizontal. A 20 N frictional force is present. After the box has been dragged 15 meters,

(a.) How much work has the *net* force done?
(b.) If the box started from rest, that is its KE at the 15 meter mark?
(c.) If the applied force is released at the 15 meter mark, how far will the box travel before it loses all its KE, and stops?

S: The net force is horizontal, and in the direction of the displacement

$$F_{net} = (40\text{ N})\cos 30^0 - 20\text{ N} = 14.6\text{ N},$$

so

$$W_{net} = (14.6\text{ N})(15\text{ m}) = 220\text{ J}.$$

This is equal to the change in KE, right?

T: Yes, and $KE = \frac{1}{2}mv^2 - \frac{1}{2}mv_0{}^2$. We have $v_0 = 0$, so

$$W_{net} = \tfrac{1}{2}mv^2 = 220\text{ J}.$$

This means the box's speed is $v = [(2)(220\text{ J})/(5\text{ kg})]^{1/2} = 9.4\text{ m/s}$.

S: Part (c.) sounds familiar from our equations of motion in previous chapters.

T: It's the same type of problem, but now we want to get used to using the methods of work and energy. At the 15 meter mark the box has 220 J of kinetic energy. Friction is the only force doing work thereafter, and it does *negative* work. When it has done 220 J of negative work, the KE will have become zero.

S: So we want to set $W_{frict.} = -(F_{frict})s = -220$ J?

T: Exactly, because KE = -220 J by the time the box has stopped. This gives

$$s = 220 \text{ J}/20 \text{ N} = 11 \text{ meters.}$$

A more general work-energy theorem can be stated with the help of the following definitions.

Mechanical Energy (ME): The sum of kinetic plus potential energies an object has. For the energy forms considered so far,

$$ME = KE + PE_{grav.} + PE_{elast}.$$

Of course one or more of these energies may be zero in a given situation, and they will be changing from instant to instant in a dynamic situation.

Conservative Forces--Forces which do work which leaves the mechanical energy of a system unchanged, or *conserved*. Work done by these forces can be represented by potential energies. Examples are gravity, elastic restoring forces, and electrostatic forces between charges, which we will encounter later.

Dissipative Forces--Forces which do work which converts mechanical energy into *thermal* energy, or *heat*, symbolized by Q. $W_{diss.} = -Q$. These forces cannot be represented by a potential energy. Examples include friction and viscosity or air drag.

The Extended Work-Energy Theorem--The work done on a system by an outside agent, W_{ext}, is equal to the change in the mechanical energy of the system plus the amount of heat energy produced by dissipative forces in the system. In mathematical form, this is

$$W_{ext} = \Delta KE + \Delta PE + Q.$$

Important Observations

1. Since the work done by dissipative forces is always negative, and we have written $W_{diss} = -Q$, the symbol Q itself represents a positive quantity in the equation.
2. If W_{ext} is *negative*, we usually say that work is being done *on the external agent*.
3. Perhaps you are beginning to better understand our definition of energy. If an object or a system has some potential or kinetic energy, it can expend part or all of it to *do work* on its surroundings (negative W_{ext}). If it has no KE or PE, it does not have the ability to do such work.
4. The potential energy terms represent work done by all the conservative forces present. Thus W_{ext} does *not* include such work.

Special Case: Conservation of Mechanical Energy.

In many cases of interest, we can neglect friction or other heat producing forces (Q = 0), and the system has only conservative forces acting on it ($W_{ext} = 0$). Then the work-energy theorem says

$$\Delta KE + \Delta PE = \Delta ME = 0 .$$

This is a statement that, under the above conditions, *mechanical energy* of the system *remains constant*, i.e., is *conserved*. This can also be written as

$$\Delta KE = - \Delta PE, \quad \text{or} \quad (KE + PE)_{initial} = (KE + PE)_{final}$$

We encountered this in Example 6-4.

Example 6-6. A 10 kg object is dropped from a height of 3 meters onto a vertical spring whose force constant, k, is 300 N/m. How far does the spring compress before the object comes to a momentary stop?

S: We're assuming $Q = 0$ and $W_{ext} = 0$, so we can find the increase of kinetic energy as it falls, then convert this KE into elastic potential energy of the spring as it compresses.

T: That's a good description of what happens. But our solution can even be simpler. You can write down (KE + PE) at *any* two instants you choose and equate the results, *regardless of what happened in between.* What are the two instants that are crucial in the problem?

S: I would say the instant the object was dropped and either the instant it hits the spring or the instant it comes to rest. I think it is the latter, but don't we need to know the speed or energy with which it hits the spring?

T: The main point here is that the mechanical energy the system started with is going to remain with it. So even though the original gravitational energy is converted to KE, all of it will reappear as potential energy again when the spring momentarily stops the object. Both original and final instants have KE = 0.

S: I see. You're saying that

$$(PE)_{initial} = mgh = (PE)_{final} = \tfrac{1}{2}kx^2_{max} ,$$

so that we can get x_{max} from

$$x_{max} = (2mgh/k)^{1/2} \ ?$$

T: Once again, you must pay more attention to detail in translating the basic ideas into mathematics. What are the references for each potential energy?

S: The reference for gravity is the position of the end of the spring when uncompressed. The same for elastic PE. Why?

T: I think you should re-evaluate your final potential energies.

S: Uh-oh. I forgot that as the spring compresses, the object goes *below* its gravitaional reference, and has negative PE_{grav}.

T: Exactly! It's not hard, but you have to be alert! So

$$(PE)_{final} = \tfrac{1}{2}kx^2_{max} - mgx_{max}$$

Then the conservation theorem allows us to write

$$mgh = \tfrac{1}{2}kx^2_{max} - mgx_{max}.$$

In standard quadratic form,

$$\tfrac{1}{2}kx^2_{max} - mgx_{max} - mgh = 0.$$

Putting in numbers,

$$150x^2_{max} - 98x_{max} - 294 = 0.$$

Then

$$x_{max} = \frac{98 \pm \sqrt{(98)^2 - 4(150)(-294)}}{300} = +1.76 \text{ m } (or\ -1.11 \text{ m})$$

When we wrote $-mgx_{max}$ as the final PE_{grav}, we chose to interpret x_{max} as a *positive* quantity, hence our choice of the positive root of the equation.

Example 6-7. Starting from rest, a couple of boys exert 400 N of force pushing a 50 kg sled 30 meters on level ice toward a downhill slope. They let go just as the sled plunges over the bank. On the way downhill the sled encounters some gravel on the ice. It reaches the bottom 50 meters below the top (vertical distance), with a speed of 30 m/s. How much energy was dissipated by the friction with the gravel? (Neglect friction on the level ice at the top.)

S: We don't know the angle of the slope or the coefficient of friction. So we can't figure out frictional force or displacement. How can we find work without these?

T: The *definition* for work is not the *only* way to calculate work. The work energy theorem keeps track of energy transactions, and one of them is Q, the work done by dissipative forces.

S: There still seems to be a rather bewildering array of material. The problem is in knowing where to start. I can recognize some things, like W_{ext} = (400 N)(30 m) = 12,000 J is the work done at the top by the boys, and mgh is the PE lost down the hill, with
h = 50 meters. But how do I put the pieces together?

T: You've made a good start. The *crucial* step in applying the work-energy theorem is again to *choose the two instants*, initial and final, that span the situation. Then write down each term in the equation as it applies to those times. In our case, initial time was when the boys *started* pushing, and the final time is when the sled reached the bottom of the slope. During that time span, W_{ext} = 12,000 J as you said, and friction did an unknown amount of work, -Q.

S: And $\Delta PE = -mgh = -(50 \text{ kg})(9.8 \text{ m/s}^2)(50 \text{ m}) = -24{,}500$ J. They started at rest, so $KE_{initial} = 0$. $KE_{final} = 1/2(50 \text{ kg})(30 \text{ m/s})^2 = 22{,}500$ J, and Δ KE is just = KE_{final}.

T: Good! The theorem now looks like this:

$$12{,}000 \text{ J} = 22{,}500 \text{ J} - 24{,}500 \text{ J} + Q$$

Solving for Q: $Q = 14{,}000$ J .

If you turn the equation around a bit, it looks like this:

$$(KE)_{final} = W_{ext} + mgh - Q$$

In words, the KE that survives at the bottom is the difference between the energy supplied by W_{ext} and gravity and that taken away as heat by friction, Q.

Simple Machines

Understanding this section should enable you to do the following:

1 . Calculate the ideal and actual mechanical advantages and efficiency of simple machines.

Simple Machine--A simple device which provides a useful output of work when provided an input of work by multiplying the force applied to the machine. Examples are levers, pulley systems, screws, inclined planes, and the wheel and axle.

Important Observations

1. In each simple machine, the output force that the machine can exert is some multiple of the applied input force.
2. The *work* output can be *no more* than the work input done by the applied force, since *energy* cannot be created or multiplied.
3. In real machines, because of friction, the work output is actually *less* than the work input. If the output force F_{out} moves through a distance l, the input force f_{in} must move through a greater distance L, and we have that

$$f_{in}L \geq F_{out}l$$

Ideal Mechanical Advantage (IMA)--In a simple machine, the ratio of the distance the input force moves to the distance the output force moves,

$$IMA = L/l$$

Actual Mechanical Advantage (AMA)--In a simple machine, the ratio of output force to input force.

$$AMA = F_{out}/f_{in}$$

Efficiency--The ratio of output work to input work in a simple machine.

$$\text{Efficiency} = F_{out}l/f_{in}L = AMA/IMA$$

By multiplying by 100, the efficiency is expressed as a percentage.

Important Observations

1. Because of friction or other dissipative forces in real machines, some of the input work, $f_{in}L$, goes into heat energy rather than output work. In other words,

$$f_{in}L = F_{out}l + Q.$$

Thus

$$AMA < IMA.$$

2. The efficiency of an ideal frictionless machine would be 1.0 (or 100%). A real machine, with AMA < IMA, has an efficiency *less than* 1.0.

Example 6-8. A winch (wheel and axle) is one type of simple machine, shown in the diagram. If a bucket of water weighs 40 pounds, how much force would you have to apply at the handle to lift the water, assuming an ideal machine? If the efficiency is actually 80%, how much force would you have to apply?

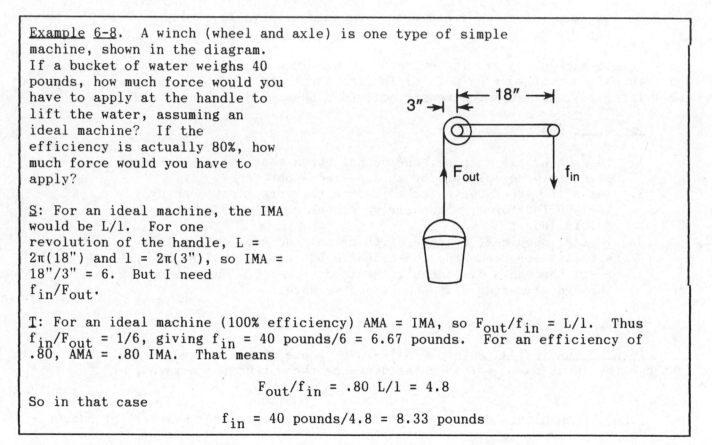

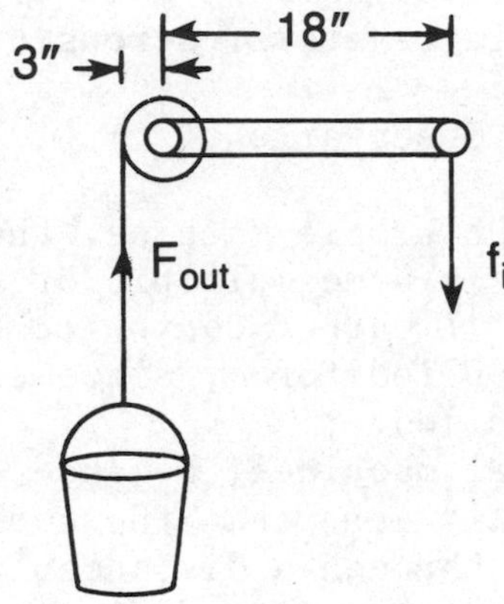

S: For an ideal machine, the IMA would be L/l. For one revolution of the handle, L = 2π(18") and l = 2π(3"), so IMA = 18"/3" = 6. But I need f_{in}/F_{out}.

T: For an ideal machine (100% efficiency) AMA = IMA, so $F_{out}/f_{in} = L/l$. Thus $f_{in}/F_{out} = 1/6$, giving f_{in} = 40 pounds/6 = 6.67 pounds. For an efficiency of .80, AMA = .80 IMA. That means

$$F_{out}/f_{in} = .80\ L/l = 4.8$$

So in that case

$$f_{in} = 40 \text{ pounds}/4.8 = 8.33 \text{ pounds}$$

Example 6-9. A pulley system is used in a garage to enable a mechanic to lift the engine out of a car. The IMA of the machine is 40. The mechanic needs to hoist a 400 pound engine. He finds he has to exert a 15 pound force to move the engine. If he does 2670 J of work, how high has he raised the engine? What is the efficiency of the machine?

S: The IMA gives us L/l = 40, but we don't know either L or l.

T: But the *forces* are given, which gives us the AMA:

$$AMA = F_{out}/f_{in} = 400 \text{ pounds}/15 \text{ pounds} = 26.7$$

S: The efficiency of the machine is defined as W_{out}/W_{in}, but we don't know W_{out}.

T: Remember that you don't always have to go to the definition directly to calculate some quantity. The efficiency is also AMA/IMA = 26.7/40 = 0.67. So

$$W_{out} = (.67)W_{in} = .67(2670 \text{ J}) = 1790 \text{ J}$$

S: Then

$$1790 \text{ J} = F_{out}\, l = (400 \text{ pounds})\times(4.45 \text{ N}/1 \text{ pound})\times(l)$$

So $$l = 1790 \text{ J}/1780 \text{ N} = 1.01 \text{ meters} .$$

T: Right. And to do this, the mechanic had to pull a lot of rope through his hands:

$$L = \text{IMA} \times l = 40 \times 1.01 \text{ m} = 40.2 \text{ meters!}$$

Power

Understanding this section should enable you to do the following:

1. *Calculate the power represented by work done by a force in a given time.*
2. *Calculate the power delivered by a force acting on an object with given average speed.*
3. *Calculate the energy consumed by a process operating at a certain power level for a given time.*

In common usage the word power is very often confused with energy. As with every other term in science however, it has a very specific meaning.

Power--The rate at which work is done or energy transformed. Mathematically,

$$\langle P \rangle = W/t$$

An equivalent formulation is derived in your text as

$$\langle P \rangle = F\langle v \rangle ,$$

where F is a force (or component of force) in the direction of the average velocity.

Important Observations

1. Power is a *scalar*, with SI units of joules/second. This is given the specific name of *watt*.

$$1 \text{ watt (W)} = 1 \text{ J/s}.$$

2. A corollary of this definition is *Work = Power x time*. If a machine operates at constant power output or input, the work it does or the energy it consumes is the product of its power times the time it operates. Electrical appliances are examples of this, and a common unit of electrical energy is the *kilowatt-hour*, or kWh.
3. A common unit of power still used in the U.S.A. is the *horsepower* (h.p.). The SI equivalence is

$$1 \text{ h.p.} = 745 \text{ W}.$$

4. The second expression for P indicates that a constant force develops more power when acting on a rapidly moving object than on a slow object.

Example 6-10. What power does the force of gravity produce on a 10 kg object falling from rest through a distance of 10 m? Would it be different if the object already had an initial downward velocity?

S: Which expression for power shall I use?

T: Both.

S: O.K. I'll start with $P = W/t$. W is just mgh here, but I'll have to find the time of fall.

T: Yes, given by $s = \frac{1}{2}gt^2$. Thus $t = [2(10 \text{ m})/(9.8 \text{ m/s}^2)]^{1/2} = 1.43 \text{ s}$.

S: Then

$$P = mgh/t = (10 \text{ kg})(9.8 \text{ m/s}^2)(10 \text{ m})/(1.43 \text{ s}) = 685 \text{ W}.$$

T: This is about $685/745 = 0.92$ h.p. Now try the alternate definition.

S: I need to find $\langle v \rangle$.

T: Since acceleration is constant, we can use

$$\langle v \rangle = \tfrac{1}{2}(v_0 + v_{final}) = \tfrac{1}{2}v_{final}, \quad \text{since } v_0 = 0,$$

and

$$v_{final} = (2gs)^{1/2} = 14 \text{ m/s}.$$

S: Then

$$P = F\langle v \rangle = mg\langle v \rangle = (98 \text{ N})(7 \text{ m/s}) = 686 \text{ W}.$$

T: For the last question, although W remains the same, namely mgh, the *time* it would take the object to fall the 10 meters would be less if it had an initial velocity, so P would be greater. Another view would be that $\langle v \rangle$ would be greater, so $P = mg\langle v \rangle$ would be larger.

QUESTIONS ON FUNDAMENTALS

1. The length of an incline is 18 m and its height is 6 m. How much work must you do to pull a 4 kg mass all the way up the incline at constant speed in the absence of friction?

(a.) 39.2 J (b.) 706 J (c.) 24 J (d.) 235 J (e.) 72 J

2. How much gravitational PE does a 10 kg mass have relative to the ground when it is at a height of 1500 m?

(a.) 147,000 J (b.) 15,000 J (c.) 1530 J (d.) 480,000 J

3. If the object in #2 is let fall from rest and strikes the ground with a speed of 100 m/s, how much work has been done by air resistance?

(a.) $+430 \times 10^3$ J (b.) -97×10^3 J (c.) -47×10^3 J
(d.) -1030 J (e.) +5000 J

4. In catching a 500 gram object, you exert a constant force of 20 N over a 40 cm distance, bringing the object to a stop. How much work did you do on the object?

(a.) +0.8 J (b.) -7.84 J (c.) -0.8 J (d.) -20 J (e.) +1960 J

5. In stretching a spring by 30 cm, you do 9 J of work. What is the force constant of the spring?

(a.) 200 N/m (b.) 30 N/m (c.) 15 N/m (d.) 2000 N/m
(e.) 300 N/m

6. If a 3 kg object is shot upward with a KE of 3750 J, how high will it rise?

(a.) 1250 m (b.) 254 m (c.) 625 m (d.) 375 m (e.) 127 m

7. In using a manual winch to raise a 250 kg engine out of a car to a height of 2 m, you exert a 245 N force as you pull the rope. What is the actual mechanical advantage of this winch?

(a.) 1.02 (b.) 5 (c.) 20.4 (d.) 10
(e.) can't tell from the above

8. In question #6, if you have to pull 25 m of rope through your hands to lift the engine 2 m, what is the efficiency of the winch?

(a.) 67% (b.) 100% (c.) 80% (d.) 8% (e.) 50%

GENERAL PROBLEMS

Problem 6-1. Take another look at Example 6-6. At what position does the object experience the greatest acceleration? Where does it experience zero acceleration?

Q1: What is the physical condition determining maximum acceleration?
Q2: What is the net force before hitting the spring?
Q3: What is the net force after hitting the spring?
Q4: What is the condition for zero acceleration?

Problem 6-2. Find the power available at the bottom of a waterfall 160 feet high when the flow rate is 4000 ft^3/min. Use the fact that water has a density of 1.00 x 103 kg/m^3.

Q1: What does the power depend on?
Q2: What work can be done by a mass M of water falling through a distance h?
Q3: How much mass is in 1 ft^3?
Q4: How much potential energy is lost by 4000 ft^3 of water falling 160 feet?

Problem 6-3. Suppose a 1000 kg car is going 50 km/hr on a level circular road whose radius of curvature is 500 meters. The driver hits the accelerator and maintains an acceleration of 3 m/s^2 parallel to the road for the next 200 meters. Find an expression for the net force acting on the car during this period, and calculate the change in KE and the final speed of the car.

Q1: What is the general expression for F_{net}.
Q2: Is this constant during the first 200 meters?
Q3: How can I use the work-energy theorem when F_{net} is not constant? I can't calculate the work.

Problem 6-4. A 50 kg skydiver falls from rest from a height of 2000 meters above the ground. By the time she has fallen the first 1000 meters, she has reached her terminal speed of 240 km/hr. How much work has been done by the friction of air resistance during the first 1000 meters of fall? How much during the last 1000 meters?

Q1: The work done by friction is $\langle F_{fric} \rangle$ x 1000 meters. Can we find $\langle F_{fric} \rangle$ during the first 1000 meters?
Q2: What does the work-energy theorem state?
Q3: Is there any W_{ext}?
Q4: In the first 1000 meters, what are ΔPE and ΔKE?

Problem 6-5. The jackscrew shown in the picture is used to lift a load of15,000 N including the weight of the moving screw mechanism. This requires the application of a 150 N force at the end of the lever handle 1.2 meters from the axis. The pitch of the screw, which is the vertical separation of the screw threads, is 1.5 cm. What are the ideal and actual mechanical advantages of this machine? In raising this load a vertical distance of 30 cm, how much heat energy was produced?

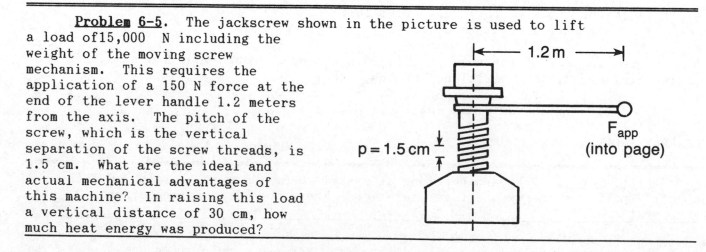

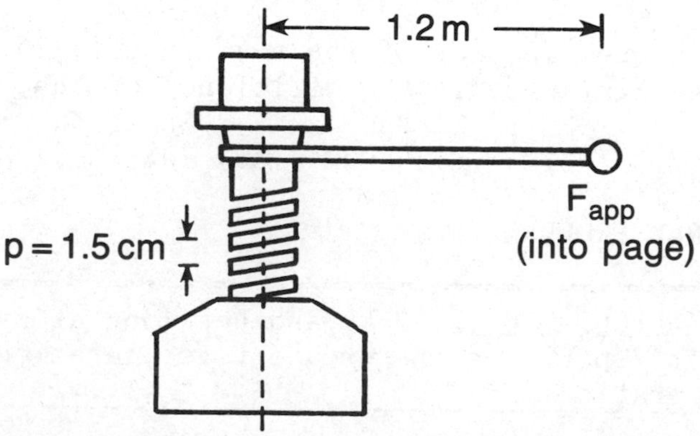

Q1: How far does F_{app} have to move to raise the load by 1.5 cm?
Q2: How does the work-energy theorem apply here?
Q3: Is there another approach to finding Q?

ANSWERS TO QUESTIONS

FUNDAMENTALS:

1. (d.)
2. (a.)
3. (b.)
4. (c.)
5. (a.)
6. (e.)
7. (d.)
8. (c.)

GENERAL PROBLEMS:

Problem 6-1.
A1: That F_{net} is a maximum.
A2: $F_{net} = -mg$.
A3: $F_{net} = kx = mg$
A4: That $F_{net} = 0$.

Problem 6-2.
A1: Power = work/time.
A2: The work available is ΔPE, or $Mg(\Delta h)$.

A3: 1.00×10^3 kg/m^3 x $(1/3.28)^3$m^3/ft^3 = 28.3 kg.
A4: 5.41×10^7 J.

Problem 6-3.
A1: $F_{net} = ma$, and a has two components, $a_p = F_p/m$ and $a_N = F_N/m = v^2/F$.
A2: *No.* V is increasing during this time, hence a_N and F_N are also.
A3: The component F_p *is* constant, and the changing component, F_N, does no work. So $W_{tot} = F_p s$.

Problem 6-4.
A1: Not directly. Use a different approach to find the work.
A2: $W_{ext} = \Delta KE + \Delta PE + Q$, where Q = work done by air resistance.
A3: No. The only force other than friction is that of gravity, which is accounted for in PE.
A4: $\Delta PE = -4.90 \times 10^5$ J; $\Delta KE = +1.12 \times 10^5$ J.

Problem 6-5
A1: One revolution of the screw takes it through a distance equal to the circumference of a circle of radius 1.2 m.
A2: $W_{ext} = \Delta PE = Q$. ($\Delta KE = 0$).
$W_{ext} = F_{app} s$, where $s = 2\pi R$ for one revolution, and
PE = 15,000 N x 1.5 cm for one revolution.
A3: Yes. Efficiency = AMA/IMA = W_{done}/W_{app},
and $W_{done} = W_{app} - Q$.

CHAPTER 7. LINEAR MOMENTUM AND COLLISIONS

BRIEF SUMMARY AND RESUME

The *linear momentum* of a particle is defined and Newton's 2nd Law of Motion is re-examined in momentum terminology. The *impulse* created by a force is defined. The fundamental *Principle of Conservation of Linear Momentum* is introduced and applied in problems involving *collisions* between objects. *Elastic and inelastic* collisions are defined. The conservation of linear momentum is applied to problems involving *recoil* and the basic physics of *rocket propulsion* is discussed.

Material from previous chapters relevant to this chapter includes:

1. Newton's 2nd and 3rd Laws of Motion.
2. The center of mass of a system.
3. Kinetic and gravitational potential energy.

NEW DATA AND PHYSICAL CONSTANTS

1. The SI unit of linear momentum is the kg-m/s. No specific name is used for this.

NEW CONCEPTS AND PRINCIPLES

Linear Momentum

Understanding this section should enable you to do the following:

1. *Calculate the momentum of individual objects and the total momentum of a system.*
2. *Calculate the velocity of the center of mass of a system from its total momentum.*
3. *State the momemtum form of Newton's 2nd Law.*

To Newton, the "quantity of motion" was one of the most important physical properties of a moving object. This term, which he used in first formulating his 2nd law of motion, refers to what we now call linear momentum.

Linear Momentum--A *vector* quantity equal to the product of the *mass* of a particle and its *velocity*, symbolized by **p**. For an extended mass or a system of masses, the total linear momentum is the vector sum of the individual momenta or equivalently, the product of the *total mass* times the *velocity of the center of mass* of the system. Mathematically,

$$\mathbf{p} = m\mathbf{v}, \quad \text{or} \quad \mathbf{P}_{tot} = M_{tot}\mathbf{V}_{cm}.$$

Important Observations

1. The SI units are derived to be kg-m/s, and are *not* given any other specific name.
2. Momentum is a *vector* quantity.

Momentum Statement of Newton's 2nd Law--The time rate of change of the linear momentum of a system is equal to the net external force applied to the system. In mathematical terms,

$$\mathbf{F}_{net} = \Delta\mathbf{p}/\Delta t = m(\Delta\mathbf{v}/\Delta t) + (\Delta m/\Delta t)\mathbf{v}.$$

Important Observations

1. This is a more general statement of the 2nd Law than we used before. It extends to systems where the mass may change. For *constant mass* (like a single object), m/ t = 0, so then F_{net} = ma as before.

Example 7-1. A 10,000 pound truck is traveling 30 mi/hr east and a 4000 pound car is traveling 60 mi/hr north. What momentum does each have, and what is their total momentum as a system? Use SI units.

S: First, the conversion of units gives

$$10{,}000 \text{ pounds} = 44{,}500 \text{ N}, \quad \text{and } 4000 \text{ pounds} = 17{,}800 \text{ N},$$

also,

$$30 \text{ mi/hr} = 13.3 \text{ m/s}, \quad \text{and } 60 \text{ mi/hr} = 26.7 \text{ m/s}.$$

Then the truck's momentum is (44,500 N)(13.3 m/s) = 5.92×10^5 kg m/s and the car's is (17,800 N)(26.7 m/s) = 4.75×10^5 kg m/s.

T: That's kind of sloppy! You used *weight* instead of *mass*, and you said nothing about *direction*. Remember, we're back to vectors again. If you divide by g to get masses, you'll have the magnitudes right.

S: Sorry. The truck's *mass* is (44,500 N)/(9.8 m/s^2) = 4540 kg, and the car's is (17,800 N)/(9.8 m/s^2) = 1820 kg. Then

$$P_{truck} = (4540 \text{ kg})(13.3 \text{ m/s}) = 6.04 \times 10^4 \text{ kg m/s } \textit{East},$$

and

$$P_{car} = (1820 \text{ kg})(26.7 \text{ m/s}) = 4.86 \times 10^4 \text{ kg m/s } \textit{North} .$$

T: That's right. To get P_{tot} you have to add these *vectorially*. Since the two momenta are at right angles to each other, that's pretty easy.

$$P_{tot} = [P^2_{truck} + P^2_{car}]^{\frac{1}{2}} = 7.75 \times 10^4 \text{ kg m/s}.$$

and

$$\theta = \tan^{-1} (P_{car}/P_{truck}) = \tan^{-1} .805 = 3.38^\circ \text{ N of E}.$$

S: What is the importance of the total momentum?

T: It remains *constant* for a given system, unless there are *external* forces acting on the system, even if the *individual* momenta change from *any* sort of *internal* reactions. Also, you can find the velocity of the center of mass of the system from

$$V_{cm} = P_{tot}/M_{tot} .$$

For this case,

$$V_{cm} = (7.75 \times 10^4 \text{ kg m/s})/(6360 \text{ kg}) = 12.2 \text{ m/s}$$

in the direction of P_{tot}.

Impulse.

Understanding this section should enable you to do the following:

1. *State the definition of the impulse delivered by a force and its relation to the momentum of a system on which the force acts.*

Another familiar word commonly used without any particular precision is *impulse.* We say that we do things on impulse, or that we are impulsive buyers, etc. The scientific definition is:

Impulse--The product of the *average net force* acting on an object of a system and the *time* t during which it acts. Mathematically,

$$\text{Impulse} = \langle F_{net} \rangle \Delta t.$$

Important Observation

1. Impulse is a *vector* in the same direction as the average net force, but with units of N-s. These are the same units as momentum, since 1 N-s = 1 kg m/s.

Impulse-Momentum Theorem--Combining the definition of impulse with Newton's 2nd Law, we find that the impulse delivered to an object or a system is equal to *the change of linear momentum* of that object or system. Mathematically,

$$\text{Impulse} = \langle F_{net} \rangle \Delta t = \Delta \mathbf{P}.$$

Important Observation

1. The direction of *change* of momentum, $P_2 - P_1$, will be in the direction of the impulse.

Example 7-2. A 2.5 kg lead mass is dropped from rest into a lake from a platform 10 meters above the water. In 0.15 sec. after hitting the water, the mass reaches a sinking speed 1/10 of its speed just before impact. What average force did the water exert on the mass during this period?

S: I have to find the speed at impact. Then I'll be able to get the change in momentum, knowing the speed is cut to 1/10 that value by the impact.

T: Good. You've done a lot of free fall problems, and probably remember that

$$v_1 = (2gs)^{\frac{1}{2}} = 14 \text{ m/s}$$

just before impact. Thus $p_1 = mv_1 = (2.5 \text{ kg})(14 \text{ m/s}) = 35$ kg m/s.

S: And just .15 sec. later it has 1/10 of this, or $v_2 = 1.4$ m/s, making $p_2 = 3.5$ kg m/s. Do I just subtract these to get p?

T: You subtract the *vectors.* If they are in the same direction, as they are here, that means you subtract the magnitudes. If they were in opposite directions, the opposite signs would lead you to *add* the magnitudes. Just remember to always take extra care when handling vectors.

S: So $\Delta \mathbf{p} = \mathbf{p}_2 - \mathbf{p}_1$ = 3.5 kg m/s - 35 kg m/s = -31.5 kg m/s.

The negative must mean that $\Delta \mathbf{p}$ is opposite to $\mathbf{p}_1$ and $\mathbf{p}_2$, which we wrote as positive, right?

T: Yes. The rest is easy. This momentum change is caused by the impulse of the force during the impact.

$$\langle \mathbf{F}_{net} \rangle \Delta t = \Delta \mathbf{p}, \quad \text{with } \Delta t = .15 \text{ sec.}$$

So

$$\langle \mathbf{F}_{net} \rangle = (-31.5 \text{ kg m/s})/(.15 \text{ s}) = -210 \text{ N}$$

Conservation of Linear Momentum

Understanding this section should enable you to do the following:

1. *Calculate the relation between momentum changes of various parts of isolated systems.*

A direct corollary to the impulse-momentum theorem is the following:

Principle of Conservation of Linear Momentum--If no *net external force* acts on a system, the *total linear momentum* of the system will remain constant.

Important Observations

1. The total momentum of the system can either be written as the sum of the individual momenta,

$$\mathbf{P}_{tot} = m_1\mathbf{v}_1 + m_2\mathbf{v}_2 + \ldots + m_n\mathbf{v}_n ,$$

or as the momentum of the center of mass,

$$\mathbf{P}_{tot} = M_{tot}\mathbf{V}_{cm}.$$

2. Individual momenta of *parts* of the system may change, but always in such a way that the momentum changes cancel out.
3. The key to using this principle lies in choosing the system so that forces in the problem are all *internal* to the system.

Example 7-3. A 4 kg block of plastic is sitting on a horizontal frictionless surface. A 10 gram bullet with horizontal velocity of 800 m/s hits the block and penetrates all the way through, coming out the other side with horizontal velocity $\mathbf{v'}_b$. The block is seen to have acquired a speed of 40 cm/sec. What is the bullet's speed, $\mathbf{v'}_b$?

S: What is the system we should consider? The only forces are between the bullet and the block.

T: Then if you choose the bullet and the block as your system, there are no external forces. Find the total momentum of the system.

S: At first the block isn't moving, so only the bullet has momentum.

$$P_{bull} = m_b v_b = (.01\ kg)(800\ m/s) = 8\ kg\ m/s.$$

This must be the initial P_{tot}.

T: And since no external force acts?

S: This amount of momentum must still be present in the system. So what?

T: So write down an algebraic expression for the system momentum after impact and equate it to the number you just calculated.

S: Okay.

$$P'_{tot} = m_b v'_b + M_{block} v'_{block} = P_{tot} = 8\ kg\ m/s.$$

I see. v'_b is the only unknown here, so

$$v'_b = [8\ kg\ m/s - (4\ kg)(.40\ m/s)]/.01\ kg = 640\ m/s$$

But couldn't we have gotten the same from the conservation of energy?

T: Think of what is necessary to claim that mechanical energy is conserved. No energy can be transformed into heat. Intuitively, do you think that is true for a bullet to pass through a block? We can check by calculating the KE of the system before and after the impact.

$$(KE) = \tfrac{1}{2} m v_b^2 = \tfrac{1}{2}(.01\ kg)(800\ m/s)^2 = 3200\ J.$$

$$(KE)' = \tfrac{1}{2} m_b v'^2_b + \tfrac{1}{2} M_{block} V^2_{block} =$$

$$= \tfrac{1}{2}(.01\ kg)(640\ m/s)^2 + \tfrac{1}{2}(4\ kg)(.40\ m/s)^2 = 2048\ J + 0.32\ J.$$

Thus $\Delta KE = -1050\ J$!

S: How can the block with so much mass have so little energy? What happened?

T: It shows how strongly KE depends on *speed*, since it goes as the *square* of v. The frictional force between bullet and block has created about 1050 J of heat energy.

Example 7-4. A 50 kg girl and a 90 kg boy are in a boat at rest whose mass is 300 kg. The boy dives off the stern horizontally with a speed of 2 m/s and the girl dives horizontally off the left side at a right angle to the stern. Her speed is 2.5 m/s. Neglecting any viscosity between the boat and the water, what happens to the boat?

S: I think from the last statement that you can treat the boy, girl, and boat as a system with no net external force acting on it.

T: Precisely, which means the total momentum is conserved *no matter what the parts of the system do individually*. What is the total momentum initially?

S: Initially, everything's at rest, so we have $\mathbf{P}_{tot} = 0$. It must *remain* zero when they dive.

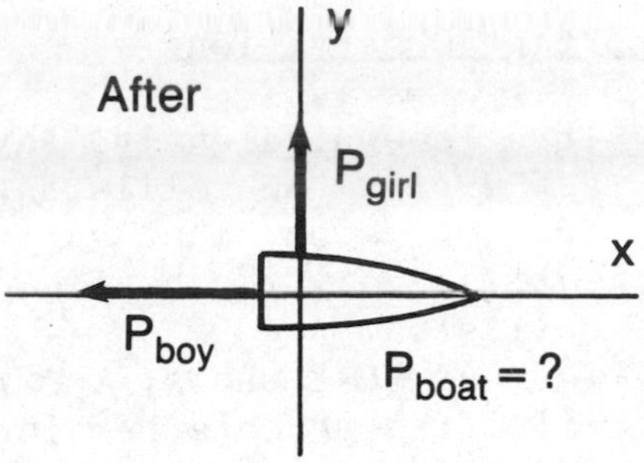

T: Let's look at the situation in a diagram seen from above:

S: That makes it look simple. The boat must have an an x component of momentum that is opposite and equal to the boy's momentum, and a y component that balances the girl's momentum. So

$$(P_{boat})_x = -P_{boy} = -(90\ \text{kg})(-2\ \text{m/s}) = +180\ \text{kg m/s},$$

and

$$(P_{boat})_y = -P_{girl} = -(50\ \text{kg})(2.5\ \text{m/s}) = -125\ \text{kg m/s}.$$

Should I get the total momentum vector of the boat first and its speed, or get the velocity components right away?

T: Either way will do. I'm glad to see you were careful with signs and your choice of coordinate system. It's easy to see that

$$P_{boat} = [(P_{boat})^2_y + (P_{boat})^2_y]^{1/2} = 219\ \text{kg m/s.} = M_{boat}V_{boat}$$

So

$$V_{boat} = (219\ \text{kg m/s})/(300\ \text{kg}) = 0.73\ \text{m/s}.$$

S: I still have to find the angle of the momentum vector. By looking at the components, I see that with P_x positive and P_y negative, P_{boat} is going to be in the 4th quadrant.

T: Right. We have

$$\theta = \arctan\ (P_y/P_x) =$$
$$= (-125\ \text{kg m/s})/(180\ \text{kg m/s}) = \arctan\ (-.694) = -34.80^o\ ,$$

which is an angle of 34.8^o *below* the +x axis.

Collisions in One Dimension

Understanding this section should enable you to do the following:

1. Calculate the final velocities of objects undergoing head-on collisions of various types.

The previous example showed one kind of collision between two objects. In *all* collisions, the forces involved are *internal* to the system, so the conservation of linear momentum *always* applies. We have also seen that conservation of mechanical energy *can't* always be assumed.

Elastic Collision--A collision in which total mechanical energy *is* conserved, in addition to linear momentum.

Important Observations

1. This is equivalent to saying that only conservative forces act during the collision. The material, when deformed during the collision, must spring back to its *original shape elastically.* No heat can be produced in an elastic collision.
2. As your text shows, the *relative velocity* between the colliding objects remains unchanged except for *reversal of direction* during an elastic collision:

$$v_2 - v_1 = -(v'_2 - v'_1)$$

3. Special Case: When the target object m_2 is at rest ($v_2 = 0$) and the incident particle m_1 has impact velocity v_1, the velocities *after* impact are given by

$$v'_1 = (m_1-m_2)v_1/(m_1+m_2) = v'_2 = (2m_1)v_1/(m_1+m_2).$$

Furthermore, if the masses are equal, this gives

$$v'_1 = 0, \quad \text{and } v'_2 = v_1.$$

The objects have just changed roles.

Example 7-5. A ball (A) of mass 1 kg moving westward at 8 m/s overtakes and collides with a ball (B) of mass 3 kg moving westward at 4 m/s. The collision is *elastic*. What are the velocities v'_A and v'_B afterward?

S: Momentum must be conserved, and we can calculate how much there is initially. Also, we can find the initial KE, which we know is conserved. This gives us two equations for two unknowns.

T: Very good. The energy equation, involving the squares of the speeds, can give some trouble if you're not careful. We don't always have to start from scratch every time, either. The *general* solution to this type of collision was derived in your text, and gave the result mentioned in observation (2.) above: that the relative velocity stays the same except for reversal of direction.

S: I'm not clear on what that means precisely.

T: The relative velocity before collision can be written

$$V_{rel} = V_B - V_A .$$

For elastic collisions, $V'_{rel} = -V_{rel}$, or

$$V'_B - V'_A = V_A - V_B$$

S: So that plus the momentum conservation equation will be enough to solve the problem.

T: Yes. Starting from the relative velocity equation is like not re-inventing the wheel all the time, but you should remember where that equation comes from.

S: Then

$$V'_B - V'_A = -(V_B - V_A) = 8 \text{ m/s} - 4 \text{ m/s} = 4 \text{ m/s}$$

and

$$P_{tot} = m_A V_A + m_B V_B = (1 \text{ kg})(8 \text{ m/s}) + (3 \text{ kg})(4 \text{ m/s}) = 20 \text{ kg m/s}$$

$$= m_A V'_A + m_B V'_B.$$

How do I know whether V'_A and V'_B are in the same direction or not?

T: You *don't* at this stage. You know B is going to move in the original direction, so you can put V'_B positive. For V'_A you have to guess, but *guess consistently* in both equations. The worst that can happen then is that you'll get a negative number for V'_A if you guessed its direction wrong.

S: So I might as well treat it as positive for now. Then the equations are

$$V'_B - V'_A = 4 \text{ m/s}, \quad \text{and} \quad m_A V'_A + m_B V'_B = 20 \text{ kg m/s}.$$

T: Two simultaneous equations again. Using substitution,

$$V'_B = +2 \text{ m/s} + V'_A,$$

so

$$(1 \text{ kg})V'_A + (3 \text{ kg})(4 \text{ m/s} + V'_A) = 20 \text{ kg m/s}$$

Collecting terms involving V'_A, we get

$$V'_A = +2 \text{ m/s}, \quad \text{and } V'_B = +6 \text{ m/s} .$$

Your guess was right!

Inelastic Collisions--Collisions in which some mechanical energy is converted into heat.

Completely Inelastic Collisions--Collisions in which the colliding objects stick together after collision.

Important Observations

1. Because momentum implies kinetic energy, and total momentum is conserved in *all* collisions, the amount of mechanical energy that can be converted into heat is limited by the requirement that enough kinetic energy must survive after collision to account for the system's linear momentum. *Only if the initial system momentum is zero can all* the mechanical energy be converted into heat.
2. The completely inelastic collision represents the *maximum* conversion of mechanical energy into heat.
3. In all inelastic collisions, the relative velocity after collision is *less* than before collision. For *completely* inelastic collisions, the relative velocity after collision is *zero*.

Example 7-6. A man on ice skates is gliding at 2 m/s toward a group of friends at the edge of the rink. His mass is 60 kg. Ten friends each throw a 200 gram snowball at him with a horizontal velocity of 15 m/s. He somehow manages to catch all of them.

(a.) What is his speed after this barrage?

(b.) If this group can collectively throw ten such snowballs at him every 3 seconds, how long would it take to stop the skater? Again assume he catches them all.

S: The collisions are going to be completely inelastic. I can't quite see what the best choice of system is here.

T: I'd say the skater plus 10 snowballs in flight.

S: Then using the direction of the skater as positive, the system momentum would be

$$P_{tot} = P_{skater} - 10P_{ball} = (60\text{ kg})(2\text{ m/s}) - 10(.2\text{ kg})(15\text{ m/s})$$
$$= 90\text{ kg m/s.}$$

After hitting, they're all moving together.

T: Right. So the expression for the momentum after is

$$P'_{tot} = (M_{skater} + 10M_{ball})V' = P_{tot} = 90\text{ kg m/s.}$$

Thus

$$V' = (90\text{ kg m/s})/62\text{ kg} = 1.45\text{ m/s}$$

Would the skater be moving faster if he had dropped the balls instead of catching them?

S: Let's see. The only thing that would change is that the balls would not be moving faster, with

$$V' = (90\text{ kg m/s})/60\text{ kg} = 1.50\text{ m/s}$$

For part (b.), we want to find how many snowballs it would take to stop the skater, that is, to make $V' = 0$.

T: Can you develop a *general* equation for V after n snowball hits?

S: General derivations often puzzle me. I never seem to see which way to go, or what to look for.

T: Use the same ideas, but in general terms instead of treating a specific numerical case. For instance, go back to the beginning and answer the question for *any* number of snowballs, n.

S: Well, if you say n balls instead of 10, we would have

$$P_{tot} = P_{skater} - nP_{ball}.$$

Then what?

T: Good so far. After n balls have hit, write down P'_{tot}.

S: It's the same amount, by the conservation principle.

T: Sure, but it's *composed* differently.

S: Oh, you mean $P'_{tot} = M_{tot}V'$. And I see that

$M_{tot} = M_{skater} + nM_{ball}$. So then

$$V' = (P_{skater} - nP_{ball})/(M_{skater} + nM_{ball})$$

T: There you have it! Now the advantage is that we can examine it in a general way. What does it take to make $V' = 0$?

S: It takes $P_{skater} = nP_{ball}$, or

$$n = P_{skater}/P_{ball} = (120 \text{ kg m/s})/(3 \text{ kg m/s}) = 40 \text{ balls}$$

So at 10 balls per 3 seconds, it would take 40/3 = 13.3 seconds to stop the skater.

T: Right. Look at one more aspect of the general equation. Does n depend on whether he catches or drop the balls?

S: When you asked that before, V *did*, because the mass of the balls adds to his when he catches then. So I'd say *yes*.

T: You *guessed* and guessed *wrong*. The general expression for *V'* is different for the two cases:

$$\text{(catching)} \quad V' = (P_{skater} - nP_{ball})/(M_{skater} + nM_{ball})$$

$$\text{(dropping)} \quad V' = (P_{skater} - nP_{ball})/(M_{skater})$$

But *the condition to make $V' = 0$ is the same*, i.e., $P_{skater} = nP_{ball}$. Thus n is the same in both cases. Once you develop the general algebraic equation it tells you about many different situations.

Example 7-7. A 4 kg mass (m_1) is moving to the right with $v_1 = +16$ m/s, and is hit by a 1 kg mass (m_2) moving left with $v_2 = -10$ m/s. The relative velocities of the objects after collision is 0.4 of that before collision.

(a.) What is the velocity of m_2 relative to m_1 before and after collision?
(b.) What is the velocity of m_1 after collision?
(c.) What happens to the kinetic energy as a result of the collision?

S: The relative velocity of m_2 and m_1 is $V_{rel} = v_2 - v_1$.

$$v_2 - v_1 = -10\text{m/s} - (16 \text{ m/s}) = -26 \text{ m/s}.$$

T: Very good! A lot of people get signs mixed up in a problem like this.

S: After collision is $V'_{rel} = v'_2 - v'_1$, or is it $v'_1 - v'_2$?

T: Right the first time,

$$V'_{rel} = v'_2 - v'_1 = -(.4)V_{rel} = -(.4)(-26 \text{ m/s}) = +10.4 \text{ m/s}.$$

This clearly shows the *reversal* of direction of V_{rel}.

S: Next, how do I know what signs to give v'_1 and v'_2?

T: You really don't know, but again a wrong choice will just give you a negative sign in the solution, so you'll always know in the end whether you were right or not. I'll demonstrate in a moment, but let's write the momentum equation first.

S: Initially,

$$P_{tot} = (4 \text{ kg})(16 \text{ m/s}) + (1 \text{ kg})(-10 \text{ m/g}) = +54 \text{ kg m/s}$$

T: And after collision, it's

$$P'_{tot} = m_1 v'_1 + m_2 v'_2 = +54 \text{ kg m/s}.$$

Now let's make two choices of signs and see how each works out:

$v'_1 = +$ and $v'_2 = +$	$v'_1 = +$ and $v'_2 = -$
$v'_2 - v'_1 = 10.4$ becomes $v'_2 - v'_1 = 10.4$ $m_1 v'_1 + m_2 v'_2 = 54$ becomes $4v'_1 + 1v'_2 = 54$ Solving gives $v_1' = +8.7$ m/s and $v'_2 = +19.1$ m/s. These are consistent with the designated signs, so our choice was correct.	$v'_2 - v'_1 = 10.4$ becomes $-v'_2 - v'_1 = 10.4$ $m_1 v'_1 + m_2 v'_2 = 54$ becomes $4v'_1 - 1v'_2 = 54$ Solving gives $v'_1 = +8.7$ m/s, showing this sign was correct, and $v'_2 = -19.1$ m/s. This is the right number, but our sign choice was wrong.

S: The original KE was

$$(KE)_i = \tfrac{1}{2}m_1 v_1^2 + \tfrac{1}{2}m_2 v_2^2 = \tfrac{1}{2}(4 \text{ kg})(16 \text{ m/s})^2 + \tfrac{1}{2}(1 \text{ kg})(-10\text{m/s})^2$$

$$= 562 \text{ J}.$$

T: The final KE is

$$(KE)_f = \tfrac{1}{2}m_1(v_1^1)^2 + \tfrac{1}{2} M_2(v_2^1)^2$$

$$= \tfrac{1}{2}(4 \text{ kg})(8.7 \text{ m/s})^2 + \tfrac{1}{2}(1 \text{ kg})(19.1 \text{ m/s})^2 = 334 \text{ J}.$$

Thus 562 J - 334 J = 228 J of kinetic energy has been converted to heat during the collision.

Collisions in Two Dimensions

Understanding this section should enable you to do the following:

1. *State what information must be known in order to solve collision problems in two dimensions.*
2. *Apply the conservation of momentum to collision problems in two dimensions.*

If collisions do not occur head-on, or if the collisions result in fragmenting one or both particles, then momentum conservation has to be applied in two or three dimensions. In brief, conservation of momentum requires

$$(P'_{tot})_x = (P_{tot})_x; \quad (P'_{tot})_y = (P_{tot})_y$$

for two dimensions.

Special Case: Target particle at rest ($V_2 = 0$) . Applying the above eqations to the case shown in the diagram, we have:

$$\text{x-components: } m_1 v_1 = m_1 v_1' \cos \alpha + m_2 v_2' \cos \beta$$

$$\text{y-components: } 0 = m_1 v_1' \sin \alpha - m_2 v_2' \sin \beta.$$

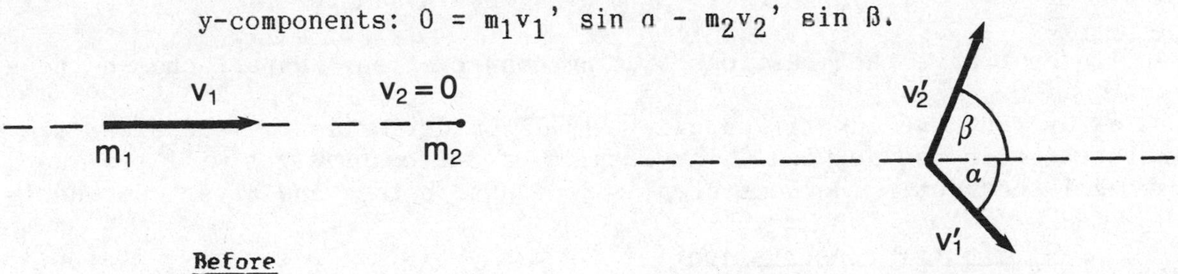

Before

After

Important Observations

1. These are *two* equations containing *seven* quantities (m_1, m_2, v_1, v'_1, v'_2, α and β). To have a *solvable* problem, either *5* of these quantities have to be known, or *4* have to be known and an additional statement about the mechanical energy before and after has to be given.

Example 7-8. Suppose you are told that a 2 kg mass (m_1) and a 750 gram mass (m_2) are moving along paths that intersect at an angle of 40°. m_1 is moving along the +x-axis, and m_2 is moving down and to the left. Suppose $v_1 = 5$ m/s and $v_2 = 12$ m/s just before they collide. After collision, each moves off at a right angle to its former path.

(a.) Find v'_1 and v'_2.
(b.) Find out whether or not mechanical energy has been conserved.

S: We can find the two angles from a diagram, so the two momentum component equations ought to be sufficient for getting v'_1 and v'_2.

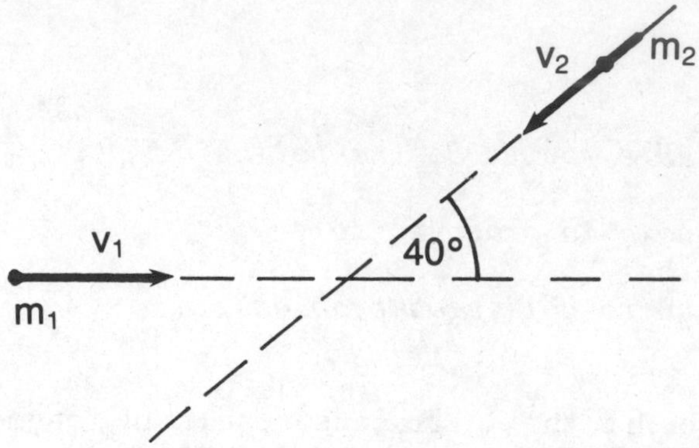

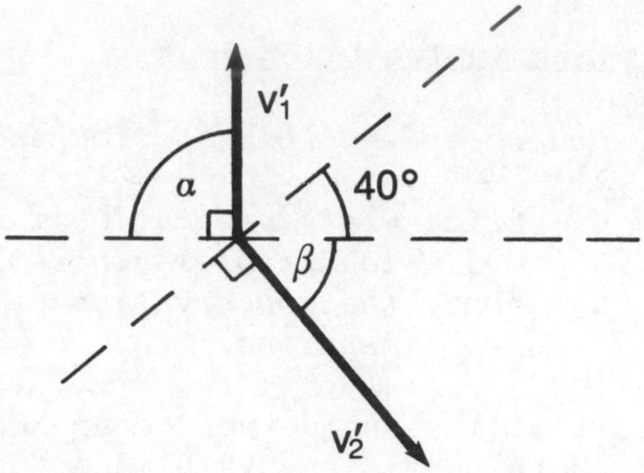

T: That's right. From your diagram, $\alpha = 90^o$ and $\beta = 90^o - 40^o = 50^o$.

S: Then *before* collision,

$$\text{x-components:} \quad (P_1)_x = m_1v_1 = +10 \text{ kg m/s}$$
$$(P_2)_x = -m_2v_2 \cos 40^o = -6.9 \text{ kg m/s}$$

$$\text{y-components:} \quad (P_1)_y = 0; \; (P_2)_y = -m_2v_2 \sin 40^o = -5.76 \text{ kg m/s}$$

Afterward, $(P_1)_x' = 0; \; (P_2)_x' = m_2v_2' \cos 50^o$
$(P_1)_y' = m_1v_1'; \; (P_2)_y' = -m_2v'_2 \sin 50^o$

I still don't have the equations stating conservation. Have I chosen the signs right?

T: So far you have identified everything from definition. Your signs are consistent with the drawing. The *results of the calculation* will tell us if the drawing is correct. The next step is to equate before and after components:

<u>x-component conservation</u>: $10 \text{ kg m/s} - 6.9 \text{ kg m/s} = m_2v_2'\cos 50^o$

<u>y-component conservation</u>: $0 - 5.76 \text{ kg m/s} = m_1v_1' - m_2v_2'\sin 50^o$

S: I notice the 1st equation has only one unknown, so

$$v_2' = (3.1 \text{ kg m/s})/(.75 \text{ kg})\cos 50^o = +6.4 \text{ m/s}.$$

Then from the other equation,

$$v_1' = \{(-5.76 \text{ kg m/s}) + (.75 \text{ kg})(6.4 \text{ m/s})\sin 50^o\}/(2 \text{ kg})$$

$$= -1.04 \text{ m/s}.$$

I thought we chose v_1' to be *positive*!

T: We chose the *upward y-direction* to be positive, consistent with the diagram. As I mentioned earlier, our calculations will show us whether this is correct or not. As it turns out, the diagram *should* have been drawn like this, with m_1 moving *downward* at an angle of 90° to its original direction. Now how do we determine if KE has been conserved?

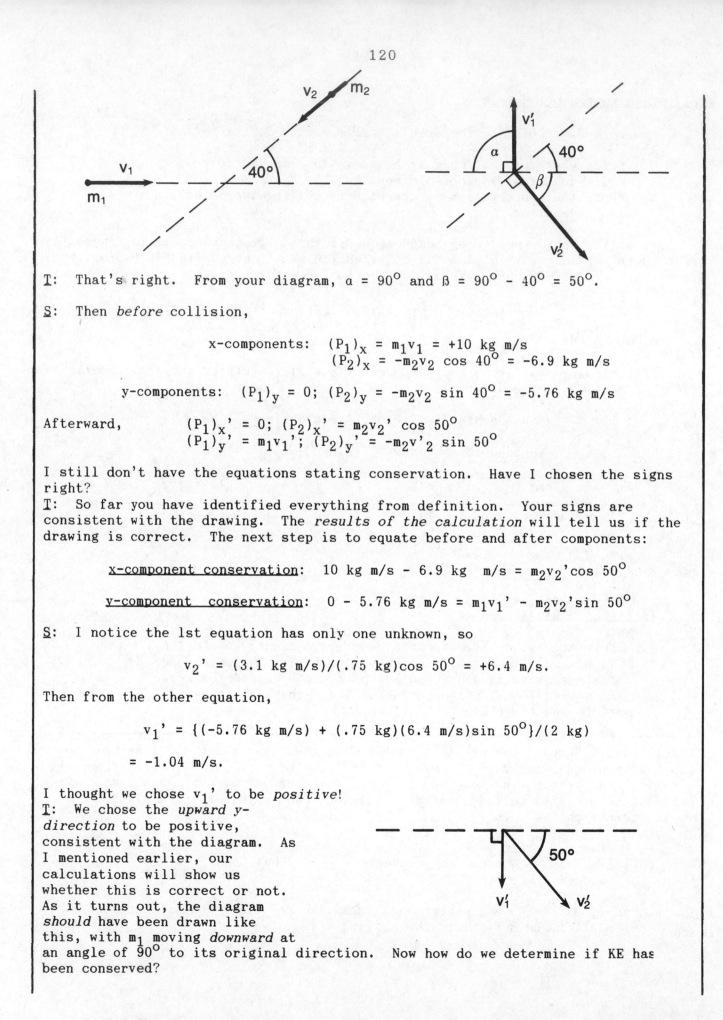

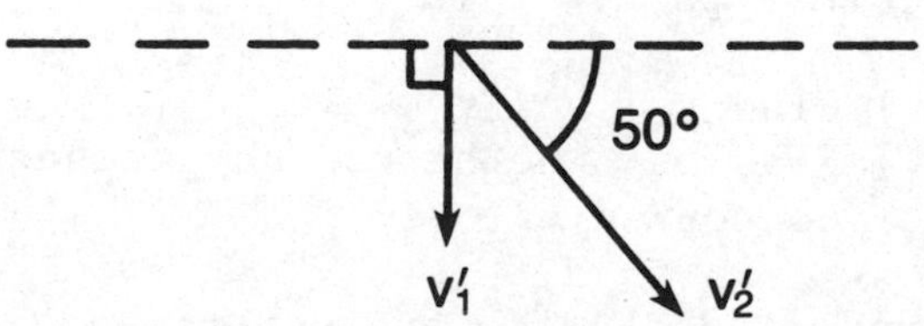

S: We have to calculate KE before and after, and compare the values.

$$KE_{tot} = \tfrac{1}{2}m_1v_1^2 + \tfrac{1}{2}m_2v_2^2 \; ; \quad KE'_{tot} = \tfrac{1}{2}m_1v'_1{}^2 + \tfrac{1}{2}m_2v'_2{}^2$$

Q: When we put in the numbers we get

$$KE_{tot} = 25 \text{ J} + 54 \text{ J} = 79 \text{ J} \text{ , and}$$

$$KE'_{tot} = 41 \text{ J} + 0.4 \text{ J} = 41.4 \text{ J}$$

T: You see that we have an example where momentum was conserved, but *mechanical energy was not.* During the collision, 37.6 J of kinetic energy was converted into heat.

Newton's 3rd Law and Momentum Conservation

Understanding this section should enable you to do the following.

1. *Apply the conservation of momentum to recoil problems.*
2. *Calculate a rocket's thrust, and determine the rocket's acceleration.*

Given the 2nd Law in momentum form, the principle of conservation of linear momentum is precisely a corollary of the 3rd Law, which says that forces only occur in interaction pairs. When you define a system so that all forces are internal, that means that both action and reaction forces are contained within. As seen from outside, *all* internal forces must be cancelled by their reaction counterparts. A system *cannot* exert a net force *on itself from within.* So unless some *external* force acts, the total momentum of the system must remain constant.

Recoil and Rocket Propulsion--When a system originally at rest fragments into a small mass m and a relatively larger mass M, conservation of linear momentum demands that

$$MV = -mv \; .$$

The momentum and velocity acquired by the larger mass is referred to as *recoil.* Examples include firearms, cannon, and disintegrating nuclei. The equations describing recoil are very simple:

Recoil speed: $V = (m/M)v$ (V in opposite direction from v)

Recoil KE: $(KE)_M = (m/M)(KE)_m$

Example 7-9. Refer to Example 6-2. Calculate the recoil speed and kinetic energy that a 3 kg rifle would have immediately after firing the .22 cal. bullet.

S: We have from before that $m = 2.59 \times 10^{-3}$ kg. Then

$$V_{recoil} = (2.59 \times 10^{-3} \text{ kg})(1400 \text{ f/s})/(3 \text{ kg}) = 1.2 \text{ ft/s}$$

and

$$KE_{recoil} = \{(2.59 \times 10^{-3} \text{ kg})/(3 \text{ kg})\} \times 2.36 \times 10^2 \text{ J} = .204 \text{ J} \; .$$

T: Very easy. And revealing, too! You have to absorb only .2 J of recoil energy, but the bullet hits with 236 J despite its tiny mass.

Rockets--Recoil can be extended to explain the operation of *rockets* as well. Return to the skater in Example 7-6 and ask yourself what would happen to him if he started with an armload of snowballs and threw them toward the side of the rink. That is, what if the physical process had been exactly reversed? If you feel he would have increased his speed in the direction opposite to his throws each time he released a snowball, you're quite right. This amounts to an *acceleration*, and the force causing it is the continuing *recoil* from throwing the snowballs. This is precisely the principle of rocket propulsion, except that rockets involve a more continuous *flow* of mass, ejected at a high *exhaust velocity*.

Rocket Thrust--The *thrust force* produced by a rocket is the product of the *rate of mass consumed* in the energy process and ejected as exhaust times the *exhaust velocity*, $\mathbf{V}_{GR}$. Mathematiclly

$$\mathbf{F}_{thrust} = -(\Delta m/\Delta t)\mathbf{V}_{GR}$$

Important Observations

1. The negative sign in the vector equation above indicates that the thrust is *opposite* in direction to the exhaust velocity.
2. A rocket consists of *payload* and other *hardware mass* plus *fuel mass*. As the engine burns, the fuel mass decreases at a constant rate and fuel is used up and ejected. For this and other reasons discussed in Section 7.7 of your text, the *acceleration* of a rocket is *not* constant, even though the thrust is.

Example 7-10. The earliest true rocket used as a long range weapon in war was the German V-2 toward the close of World War II. Its engine burned approximately 600 kg of fuel each second, and had an exhaust velocit of 6000 ft/sec. If the rocket had a total mass of 9×10^4 kg at launch,

(a.) What initial acceleration would it have at launch?
(b.) If after 2 minutes the rocket was at an altitude of 67 miles and still climbing vertically, what acceleration would it have?

S: We know its initial mass M_o, and hence its initial weight, $M_o g$. We have to find the thrust developed by the engine.

T: And that is given by

$$F_{thrust} = (\Delta m/\Delta t)V_{GR} = (600 \text{ kg/s})(6000 \text{ ft/s})(1 \text{ m}/3.28 \text{ ft})$$

$$= 1.10 \times 10^6 \text{ N}$$

S: Then the initial acceleration is given by the 2nd Law:

$$a_o = (F_{thrust} - M_o g)/M_o$$

$$= \{1.10 \times 10^6 \text{ N} - (9 \times 10^4 \text{ kg})(9.8 \text{ m/s}^2)\}/(9 \times 10^4 \text{ kg}) = 2.40 \text{ m/s}^2$$

T: This is approximately (1/4)g. How about part (b.)?

S: Well, the rocket has lost mass, so weight and inertia will *both* be less.

T: Good! Glad to see you distinguished between the two. Anything else?

S: Does thrust stay the same?

T: As long as the engine is burning, $\Delta m/\Delta t$ and V_{GR} remain constant. How about the effect of the altitude?

S: Air drag will be less, but we've neglected that, anyway.

T: Yes, and that really is not negligible at lower altitudes. But gravity is somewhat weaker. How much?

S: Gravitational force at 67 miles? We did something like that in Example 4-5.

$$g(67\ \text{mi})/g(\text{surface}) = (6.4 \times 10^6\ \text{m})^2/(6.51 \times 10^6 \text{m})^2 = .966$$

I see! So the weight of the rocket would decrease *both* because of less mass *and* weaker F_g.

T: Exactly. Of course the mass loss is going to be the *dominant* effect, but I wanted you to see the smaller effect of a 3.3% reduction in F_g. Now put it all together.

S: O.K. At 67 miles altitude,

$$a = F_{thrust} - mg(\text{at } 67\ \text{mi.})\ ,$$

where $m = M_o - (\Delta m/\Delta t)t$. Thus at the 67 mile altitude the mass has diminished to

$$m = 9 \times 10^4\ \text{kg} - (600\ \text{kg/s})(120\ \text{s}) = 1.8 \times 10^4\ \text{kg}$$

So

$$a = \{1.1 \times 10^6\ \text{N} - (1.8 \times 10^4\ \text{kg})(.966)(9.8\ \text{m/s}^2)\}/(1.8 \times 10^4\ \text{kg})$$

$$= 51.6\ \text{m/s}^2!$$

Wow!

T: Wow, indeed! This is 5.3 "g's," comparable to accelerations experienced for short times by Apollo astronauts.

QUESTIONS ON FUNDAMENTALS

1. A 4 kg object is moving to the right with a velocity of 12 m/s, and a 5 kg object is moving to the left with a velocity of 10 m/s. What is the total momentum of this system?

(a.) +98 kg-m/s (b.) +2 kg-m/s (c.) -2 kg-m/s (d.) zero
(e.) -98 kg-m/s

2. An impulse of 10 N-s is given to a 2 kg object initially at rest. What speed does the object acquire?

(a.) 20 m/s (b.) 5 m/s (c.) 8 m/s (d.) 0.5 m/s (e.) 0.2 m/s

3. A 500 gm mass is moving with a velocity of 5 m/s in a direction 30° above the +x-axis. After a force acts on the mass for 0.1 sec, it moves along the +y-axis with a speed of 2.5 m/s. What force must have acted?

(a.) 12.5 N, left (b.) 25 N, left (c.) 12.5 N, up
(d.) 12.5 N, 30° below the -x axis (e.) 25 N, 30° below the -x axis

4. A 500 gm rock moving to the right at 3 m/s strikes a stationary 2 kg mass, sticking to it. What speed does the combined mass have after collision?

(a.) zero (b.) 0.75 m/s (c.) 1.2 m/s (d.) 1.5 m/s
(e.) 0.6 m/s

5. What is the relative velocity between a 1 kg mass moving up at 30 cm/s and a 4 kg mass moving down at 50 cm/s?

(a.) 80 cm/s (b.) 20 cm/s (c.) 20 cm/s, down
(d.) 170 cm/s, down (e.) 80 cm/s, down

6. A 4 kg gun shoots a 20 gm bullet at a muzzle velocity of 1000 m/s. What is the recoil speed of the gun if no other forces act?

(a.) .05 m/s (b.) 4 m/s (c.) 12.5 m/s (d.) 5 m/s
(e.) 50 m/s

7. If the gun in #6 is an automatic which shoots 6 rounds per second, what is the average recoil force exerted on the gun?

(a.) 240 N (b.) 600 N (c.) 400 N (d.) 120 N (e.) 80 N

GENERAL PROBLEMS

Problem 7-1. A 50 gram racquetball is hit normal to the front wall of the court, hitting with a speed of 50 m/s at a spot .5 meters above the floor. The ball rebounds with an initially horizontal velocity and lands 18 meters from the front wall. What is the speed with which the ball rebounds off the wall? If the duration of impact was .01 seconds, what average force did the wall exert on the ball?

Q1: What horizontal rebound velocity would give the ball a range of 18 meters?
Q2: What was the momentum change of the ball during impact?

Problem 7-2. Go back to Example 2-15. Suppose when the two rocks collide, they stick together. How long after the collision will they hit the street? Take the masses of the rocks to be 100 grams each.

Q1: We need the velocities before collision.
Q2: What is the combined velocity just after impact? Momentum conservation should tell.
Q3: With this as an initial downward velocity, how far do the rocks have to fall?
Q4: What is the equation giving the time to impact?

Problem 7-3. Suppose a mass m_2 is stationary, and is struck elastically by mass m_1 whose speed is v_1. Take m_1 as some general multiple of m_2, i.e., $m_1 = km_2$, where k is any number or fraction. Find the fraction of the original KE of the system which is transferred from m_1 to m_2 by the collision. This fraction will be in terms of k. Find the value of this fraction when k = 100, 10, 1, 0.1, and 0.01. Can you generalize these results?

Q1: What are the equations for the velocities after collision, especially the one involving v_2'?
Q2: When we put in $m_1 = km_2$, what do we get?
Q3: Precisely what is the question of energy transfer, stated mathematically?

ANSWERS TO QUESTIONS

FUNDAMENTALS:

1. (c.)
2. (b.)
3. (a.)
4. (e.)
5. (c.)
6. (d.)
7. (d.)

GENERAL PROBLEMS:

Problem 7-1.

A1: The initial height was .5 meters, and from Chapter 4 the range equation for this is $R = (2y_0/g)v_0^2$, giving $v_0 = 18.8$ m/s.

A2: Taking the direction of the rebound to be positive,

$$(\Delta p)_{ball} = (.05 \text{ kg})(18.8 \text{ m/s}) - (.05 \text{ kg})(-50 \text{ m/s}) = +3.44 \text{ N-s}$$

Problem 7-2.

A1: From Problem 2-3, the collision time was 6 seconds after release of the rocks. At that time,

$$v_1 = -gt = -58.8 \text{ m/s (downwards)}, \quad \text{and}$$
$$v_2 = 50 - gt = 8.8 \text{ m/s (downwards)}$$

A2: This is given by $P'_{tot} = P_{tot}$.

$$P_{tot} = (.1 \text{ kg})(-58.8 \text{ m/s}) + (.1 \text{ kg})(-8.8 \text{ m/s}) = -6.76 \text{ N-s}$$

Thus

$$P'_{tot} = M_{tot}V' = -6/76 \text{ N-s}$$

A3: From Problem 2-3, the impact took place at s = -176 m, and street level is at s = -300 m.

A4: $s = s_0 + v_0t - \frac{1}{2}gt^2$, or $-300 = -176 - 33.8t - 4.9t^2$

Problem 7-3.

A1: The previously derived result for this special kind of collision gave

$$v_2' = (2m_1)v_1/(m_1 + m_2); \quad v_1' = (m_1 - m_2)v_1/(m_1 + m_2)$$

A2:

$$v_2' = (2k)v_1/(k + 1); \quad v_1' = (k - 1)v_1/(k + 1)$$

A3: What is the ratio of the KE of mass m_2 after collision to the original KE? (i.e., what is:)

$$KE_2'/KE_1, \quad \text{or} \quad m_2(v_2')^2/m_1(v_1)^2$$

CHAPTER 8. ROTATIONAL MOTION

BRIEF SUMMARY AND RESUME

Most of this chapter reveals the analogies between the descriptions and laws of translational motion and those of *rotational motion*. *Rotational equations of motion* with constant angular acceleration are developed and applied. The *work* done by *torques* is derived. The concept of *moment of inertia* of a rigid body is introduced, and rotational kinetic energy is defined. Conservation of mechanical energy is extended to include rotational energy. Moments of inertia of various solids are discussed and the *parallel axis theorem* is introduced. Newton's 2nd Law of Motion connecting *torques* with *angular acceleration* is examined and combined with the effect of translational forces to treat problems involving simultaneous rotation and translation. Finally, the *impulse-momentum theorem* is applied to rotation, and the *Principle of Conservation of Angular Momentum* is discussed and applied.

Material from previous chapters relevant to Chapter 8 includes:

1. All of the discussion of angular kinematics and torques from Chapter 4.
2. Translational laws of motion and principles of conservation of mechanical energy and linear momentum.
3. Definitions of work and impulse.

NEW DATA AND PHYSICAL CONSTANTS

1. The SI unit of moment of inertia is derived from its definition to be $kg\text{-}m^2$.
2. The SI unit of angular momentum is $kg\text{-}m^2/s$, which is equivalent to *joule-sec* (J-s).

NEW CONCEPTS AND PRINCIPLES

Understanding this section should enable you to do the following:

1. *Recognize the variables used to describe rotational motion and the analogous linear variables.*
2. *Identify the connection between rotational and linear descriptions of motion for an object which rolls without slipping.*
3. *Use the rotational equations of motion to solve problems involving constant angular acceleration about an axis fixed in direction.*

Rotational Variables and Equations of Motion.

In Chapter 4 angular variables were introduced in the description of circular motion, and the connection between them and tangential velocity and acceleration was derived. The results can be summarized as follows:

Angular Quantity	Connection with Linear Motion
Displacement: $\Delta\theta$ (radians)	$\Delta\theta = S/R$
Velocity: $\omega = \Delta\theta/\Delta t$ (rads/s)	$\omega = V/R$
Acceleration: $\alpha = \Delta\omega/\Delta t$ (rads/s^2)	$\alpha = a_p/R$

Important Observations

1. The linear variables v and a are the velocity and component of acceleration *tangential* or *parallel* to a circle of a radius R around the axis of rotation.
2. For such motion there is also a component of acceleration *normal* to the circle, the *centripetal* acceleration,

$$a_N = v^2/R ,$$

directed toward the center of rotation.

Example 8-1. A 100 kg cart having wheels of radius R = 15 cm rolls down a 45° ramp without the wheels slipping. Assume there is no friction. What is the angular acceleration of a wheel about its axle?

S: What is meant by "rolls without slipping"?

T: It means that every revolution of the wheels corresponds to a translation of the cart down the ramp by an amount $2\pi R$. Thus one rev = $2\pi(.15\text{ m}) = .942$ meters.

S: This same R also connects a_p with α, right? So the problem becomes one of finding a_p on an inclined plane.

T: Sound familiar? The component of gravitational acceleration down the ramp is $g \sin 45^o$, so

$$a_p = g \sin 45^o = .707\ g = 6.93\ \text{m/s}^2.$$

Then

$$\alpha = a_p/R = (6.93\ \text{m/s}^2)/(.15\ \text{m}) = 46.2\ \text{rads/s}^2 .$$

Angular Equations of Motion. If the *angular acceleration* is *constant*, the equations of angular motion are directly analogous to the corresponding linear ones. Thus

1. $\langle\omega\rangle = \frac{1}{2}(\omega_o + \omega)$
2. $\Delta\theta = \langle\omega\rangle t$
3. $\omega = \omega_o + \alpha t$
4. $\Delta\theta = \theta - \theta_o = \omega_o t + \frac{1}{2}\alpha t^2$
5. $\omega^2 = \omega_o^2 + 2\alpha(\Delta\theta)$

Important Observation

1. For a rotational axis *fixed in direction*, there can be only *two* possible directions of rotation, clockwise and counter-clockwise. You must remember to be *consistent* in assigning + and - signs to *all* angular variables in a problem accordingly.

Example 8-2. For the cart in the previous example,

(a.) how many revolutions have the wheels turned during the first 5 seconds, starting from rest, and
(b.) what is the wheel's angular velocity at that time?

S: I have $\alpha = 46.2\ \mathrm{rad/s^2}$ from before. I need the equation that relates $\Delta\theta$ to t for part (a.).

T: That equation is

$$\Delta\theta = \omega_o t + \tfrac{1}{2}\alpha t^2 = 0 + \tfrac{1}{2}(46.2\ \mathrm{rad/s^2})(5\ \mathrm{s})^2 = 577\ \text{radians}$$

S: To get revolutions we must divide by 2π, which gives part (a.):

$$577/2\pi = \underline{91.9\ \text{revs}}.$$

For part (b.), we just have

$$\omega = \omega_o + \alpha t = 0 + (46.2\ \mathrm{rad/s^2})(5\ \mathrm{s}) = \underline{231\ \mathrm{rad/s}}.$$

T: That's right. Try this: how much time would it take to turn the *next* 91.9 revs?

S: That would mean using the same equation with an initial velocity of 231 rad/s, I think.

T: That's possible, and will give you a *quadratic* equation for t:

$$\Delta\theta = 577\ \mathrm{rad} = (231\ \mathrm{rad/s})t + \tfrac{1}{2}(46.2\ \mathrm{rad/s^2})t^2.$$

Can you see another way?

S: You mean using another equation?

T: You've *got* the right equation. What if you found the time for 2 x 91.9 revs from the start? It's a simpler equation because the linear term in t is zero like before.

S: I get it. Then I subtract the first 5 seconds from the total time.

T: Exactly. Often there is more than one way to approach a problem. *It's best to turn the problem over in your mind and determine a strategy before getting enmeshed in calculations. If you rush headlong at the problem, you may miss a simpler way of solving it.*

S: So for $\Delta\theta$ = 184 revs = 1155 radians, we have

$$1155\ \text{rads} = 0 + \tfrac{1}{2}(46.2\ \mathrm{rad/s^2})t^2$$

Solving for t:

$$t^2 = 2(1155\ \text{rads})/(46.2\ \mathrm{rad/s^2}) = 50\ \text{sec; giving}\quad t = \underline{7.07\ \text{sec.}}$$

T: So the time for the second 91.9 revs is just 7.07 - 5 = 2.07 sec. You ought to verify that the full quadratic equation will give the same result.

Example 8-3. A wheel is spinning clockwise at 60 revs/min. It is then engaged by a frictional clutch to a drive shaft spinning at a constant 30 revs/min counter-clockwise. Suppose it takes 30 seconds for the wheel to acquire the same rotational velocity as the drive shaft.

(a.) What is the angular acceleration undergone by the wheel, assuming it to be constant?

(b.) How many revolutions did the wheel make in slowing to its momentary stop?

S: In the first part, the wheel is *decelerating*, and then accelerates after slowing to a stop. Don't I have to break the problem into two parts?

T: The direction of *change* of ω is the same in both parts. *Decreasing* clockwise speed is the same as *increasing* speed counter-clockwise. So you can treat the problem in one piece, with the same constant value of α throughout. Be careful of signs, though.

S: I can see that. The equation I need relates ω to t. If I take clockwise as negative and counter-clockwise as positive, I should have

$$\omega = \omega_o + \alpha t, \quad \text{or } +30 \text{ rev/min} = -60 \text{ rev/min} + \alpha t.$$

This will give a positive α. Is that right?

T: α *should* be positive with your choice of sign, because it represents a *counter-clockwise* change in ω. The units in the equation have to be straightened out, however.

$$30 \text{ rev/min} = \pi \text{ rad/s}; \quad 60 \text{ rev/min} = 2\pi \text{ rad/sec}.$$

S: Then

$$\alpha = (\omega - \omega_o)/t = \{\pi - (-2\pi)(\text{rad/s})\}/(30 \text{ s}) = \pi/10 \text{ rad/s}^2$$

For part (b.), I don't know how long it took to stop.

T: No, but we are assuming that α is constant, and there is an equation that relates ω to $\Delta\theta$ directly.

$$\omega^2 = \omega_o^2 + 2\alpha(\Delta\theta)$$

Putting $\omega = 0$ (the object momentarily stopped), and $\omega_o = 2\pi$ rad/s, we have

$$0 = 4\pi^2 \text{ rad}^2/\text{s}^2 + 2(\pi/10 \text{ rad/s}^2)\Delta\theta.$$

Thus

$$\Delta\theta = -20\pi \text{ radians} = -10 \text{ revolutions}.$$

S: Why is $\Delta\theta$ negative?

T: The wheel turns 20π radians *clockwise* while slowing to a stop. Only after that, when it starts moving counter-clockwise, does $\Delta\theta$ become positive.

Work and Power Delivered by Torques

Understanding this section should enable you to do the following:

1. *Calculate the work and power done by torques which cause rotation.*

Continuing the analogies between linear and rotational quantities, we replace force by torque and linear displacement by angular displacement to derive the expression for *rotational work.*

Rotational Work and Power--For an axis in a fixed direction, the product of a *torque* τ times an *angular displacement* $\Delta\theta$ constitutes the *work done by the torque.* Mathematically,

$$W = \tau(\Delta\theta)$$

For a constant torque, the *rate* of doing work, or the *power* delivered by the torque is

$$P = \Delta W/\Delta t = \tau(\Delta\theta/\Delta t) = \tau\omega .$$

Important Observations

1. Notice that, with units of N-m for torque, these expressions yield joules for work and watts for power.
2. As with the linear situation, rotational work and power can be positive or negative, depending on whether t and θ or are in the same rotational direction or opposite.
3. Keep in mind that the units of torque are N-m, not joules. See the comment about this in Chapter 4.

Example 8-4. It is found that a torque of .650 N-m is required to counter bearing friction and keep a flywheel moving at constant speed. If an electric motor rated at 2 kw output is hooked up to this wheel, how fast can it turn it?

S: Can I assume the bearing friction doesn't depend on speed so this motor will exert .650 N-m of torque at whatever speed?

T: That's implied by the statement, yes. The math is easy:

$$P = \tau\omega\text{, or } \omega = P/\tau = (2 \times 10^3 \text{ W})/(.65 \text{ N-m}) = 3.1 \times 10^3 \text{ rad/s} .$$

Rotational Kinctic Energy

Understanding this section should enable you to do the following:

1. *State the definition of moment of inertia and identify the moments of inertia for various symmetrical rigid bodies about their centers of mass.*
2. *Use the parallel axis theorem to calculate the moment of inertia of an object about any fixed axis, given its center-of-mass moment of inertia.*
3. *Calculate the rotational kinetic energy of an object from its moment of inertia and angular speed.*

4. *Calculate the total kinetic energy of an object which is translating while rotating about an axis fixed in direction.*
5. *Use the rotational and translational work-energy theorems to find linear and angular velocities.*

When a rigid body rotates, all parts of it share the *same angular velocity*, but *different linear velocities*, depending on how far a given part is from the axis of rotation. This should lead you to expect that the energy of rotation of an object must involve not only mass and ω, but also some description of how the mass is *distributed* relative to the axis of rotation.

Moment of Inertia--By considering an object to be made up of N small bits of mass m_i whose distances to a given axis of rotation are r_i, we can formally define the *moment of inertia* of an object *about that axis*:

$$I = \sum_{i=1}^{N} m_i r_i^2$$

By using calculus, this definition yields values of I for regularly shaped solids about axes through their centers, as contained in Table 8.2 in your text.

Important Observations

1. The SI units of moments of inertia are *kg-m*^{2}.
2. Mass which is relatively far from the axis contributes more to I than does mass nearer the axis. This means that for a given total mass, an object which has most of its mass near its periphery rather than near its center will have a considerably larger I about its center.
3. The definition requires a *choice of axis*, and a single object will have a different I about each chosen axis.
4. The axis need *not* pass through any material point of the object. It can be an *external* axis.
5. Calculations from the definition for regular shapes always yields I about their center of mass axis in the form $I = kM_{tot}L^2$, where k is some number or fraction, L is a radius or length of the object, and M_{tot} is its total mass. See the table below for examples.

Radius of Gyration--The "effective" radius k of *any* object which gives the moment of inertia about its center of mass the same value as a ring or cylindrical shell of the same total mass,

$$I = M_{tot}k^2$$

Table 8-1. Some Moments of Inertia and Radii of Gyration

Object	Axis	I	k
Cylindrical shell (hoop)	longitudinal through center axis	MR^2	R
Solid cylinder	longitudinal through center axis	$\frac{1}{2}MR^2$	$R/\sqrt{2}$

Solid sphere	any diameter	$(2/5)MR^2$	$(2/5)^{1/2}R$
Thin rod	normal to length through center	$(1/12)ML^2$	$L/\sqrt{12}$
Thin rod	normal to length through end	$(1/3)ML^2$	$L/\sqrt{3}$

For irregular solids, or about axes which lack symmetry, one must usually find I experimentally, then calculate k from

$$k = (I/M_{tot})^{1/2} .$$

If you know the moment of inertia about a center of mass axis, it is easy to calculate I about *any other axis parallel* to it, as follows:

Parallel Axis Theorem--The moment of inertia I through any axis *parallel to and a distance d from* a center of mass axis is related to the center of mass moment of inertia, I_{cm}, by

$$I = I_{cm} + M_{tot}d^2 .$$

Example 8-5. Find the center of mass moment of inertia of a square made of 4 uniform thin rods whose mass is m and whose length is L. Do this for the two cases shown in the figure.

(a.) Axis perpendicular to the plane of the square.
(b.) Axis in the plane of the square.

S: All I know is I for each rod through its center normal to its length, $I = (1/12)mL^2$.

T: You know somewhat more than that, as you'll see. It would be tedious even with calculus to calculate these from the basic definition of I, but the parallel axis theorem is easy -- it just requires a little geometry. Suppose you tell me what the moment of inertia is for a rod rotating around axis O in this diagram:

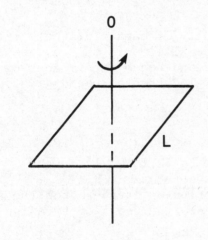

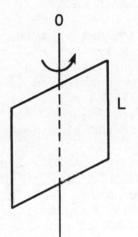

S: Well, axis O is a distance L/2 from the c.m. axis, so I think

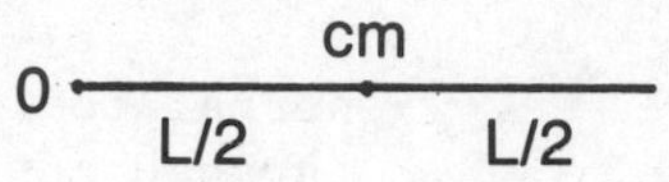

$$I_O = I_{cm} + m(L/2)^2$$

T: Right. With $I_{cm} = (1/12)mL^2$ this becomes

$$I_O = (1/12)mL^2 + 1/4mL^2 = (1/3)mL^2.$$

Can you take it from there?

S: In part (a.), each of the 4 rods has just this moment of inertia about the square's center. I think these contributions would add to give

$$(I_O)_{tot} = 4(1/3)mL^2 = (4/3)mL^2 .$$

T: Precisely. If you note that $M_{tot} = 4m$, this is $I_O = (1/3)M_{tot}L^2$. Now can you start (b.)?

S: Two of the rods are rotating about their centers, contributing $(1/12)mL^2$ each. But the others are just moving on end in a circle, parallel to the axis O. I don't know what to do with them.

T: I think you *do*. What is their mass distribution relative to the axis?

S: *All* their mass is a distance L/2 from the axis. So would their contribution be just $m(L/2)^2$ each?

T: Yes! *Any* distribution of mass which has *all* bits of mass a distance r from the axis has $I = M_{tot}r^2$. You could also use the parallel axis theorem, realizing that an *ideally* thin rod would have zero thickness and hence zero moment of inertia when rotated about its longitudinal axis. The theorem would give you $I_O = 0 + md^2$ with $d = L/2$, or $I_O = m(L/2)^2$.

S: So we'll add all these together and get

$$I_O = 2(1/12)mL^2 + 2m(L/2)^2 = (2/3)mL^2 = (1/6)M_{tot}L^2 .$$

T: Did you notice that in (a.), more mass is distributed farther from the axis than in (b.)? Thus (a.) has the larger I_O. What are the radii of gyration for the two cases?

S: The radius of gyration is given by $k = (I/M)^{1/2}$. So for (a.):

$$k = \{(1/3)ML^2/M\}^{1/2} = L/\sqrt{3} = .577\ L$$

and for (b.):

$$k = \{(1/6)ML^2/M\}^{1/2} = L/\sqrt{6} = .408\ L$$

T: Notice how the rotational characteristics of the *same object* depend on the axis of rotation.

<u>Rotational Kinetic Energy</u>--The product of *1/2* the *moment of inertia* of an object times the *square of its rotational velocity.* Both I and ω are taken with respect to the same given axis of rotation. Mathematically,

$$(KE)_{rot} = \tfrac{1}{2}I\omega^2$$

<u>Important Observations</u>

1. Notice the SI units are once again $(kg\text{-}m^2)(rad/s^2) = kg\text{-}m^2/s^2$ = *joules* of energy.
2. Most often the axis chosen will be an axis of symmetry, such as an axis passing through the center of mass of the object.
3. The analogy with translational KE is obvious. I replaces mass as the measure of *rotational inertia*, and of course replaces v.
4. As with translational KE, rotational KE is always *positive.*

<u>Rotational Work-Energy Theorems</u>--All the work-energy theorems developed previously can be extended to include rotational work and energy by noting the following generalizations:

A. The work done by forces is now

$$W = Fs\cos\Phi + \tau\theta,$$

where s is the displacement *of the c.m.* (center of mass) , Φ is the angle between **F** and **s**, and θ is the rotational variable about the c.m.

B. All potential energies involve the position of *the center of mass* of objects.

C. The translational KE of an object is the KE of its *center of mass,*

$$(KE)_{trans} = \tfrac{1}{2}MV_{cm}^2$$

D. The *total mechanical energy* is now:

$$ME = \tfrac{1}{2}MV^2_{cm} + \tfrac{1}{2}I_{cm}\,\omega_{cm}^2 + \text{all PE's of the c.m.}$$

This shows the great importance of the c.m., in that *simultaneous rotation and translation can be separated into translation of the c.m. and rotation about it.*

<u>Example 8-6</u>. Suppose a steamroller used in road construction consists of two solid cylindrical rollers each having $m = 1.2 \times 10^4$ kg and radius R = 60 cm., plus a body frame whose mass is $M = 2 \times 10^3$ kg. If the roller is moving at 10 mi/hr, how much kinetic energy does it have? What fraction of it is in the rotation of the rollers?

<u>S</u>: We know that $KE = KE_{trans} + KE_{rot} = \tfrac{1}{2}MV^2 + \tfrac{1}{2}I\omega^2$. Is M just the frame, or the *total* mass?

<u>T</u>: The frame *and* the centers of mass of the rollers are *both* translating at 10 mi/hr, so you must use *total* mass.

<u>S</u>: *Each* roller has $KE_{rot} = \tfrac{1}{2}I_{cm}\omega^2$. For cylinders I see that $I_{cm} = \tfrac{1}{2}mR^2$. So

$$I_{cm} = \tfrac{1}{2}(1.2 \times 10^4 \text{ kg})(.60 \text{ m})^2 = 2.16 \times 10^3 \text{ kg-m}^2.$$

What do we know about ω?

T: The rollers are not slipping, so $\omega = V_{cm}/R$, where you have to observe that the V_{cm} of the axle is the same as the V_{cm} of the whole machine. This is what we mean by a *rigid* body. Also,

$$V_{cm} = 10 \text{ mi/hr} \times 1600 \text{ m/mi} \times 1 \text{ hr}/3600 \text{ s} = 4.44 \text{ m/s}$$

S: Then

$$\omega = (4.44 \text{ m/s})/(.60 \text{ m}) = 7.4/\text{s}.$$

This is actually rad/s, right? Now for both rollers together we have

$$KE_{rot} = 2 \times \tfrac{1}{2} I_{cm}\omega^2 = 2(1/2)(2.16 \times 10^3 \text{ kg-m}^2)(7.4/\text{s})^2$$

$$= 1.18 \times 10^5 \text{ J}.$$

T: And

$$KE_{trans} = \tfrac{1}{2} M_{tot} V_{cm}^2 = (1/2)(2.6 \times 10^4 \text{ kg})(4.44 \text{ m/s})^2$$

$$= 2.56 \times 10^5 \text{ J}.$$

So

$$KE_{tot} = 3.74 \times 10^5 \text{ J}.$$

The fraction of this which is rotational is

$$(KE_{rotation})/KE_{tot} = (1.18 \times 10^5)/(3.74 \times 10^5) = .316 \text{ or } 31.6\% .$$

Example 8-7. A brick of uniform density is shown lying on its largest face. It has a mass of 10 kg. How much work is required to set it up so that it rests on its *smallest* face?

S: Sounds like a problem involving just $W = \Delta(PE)_{grav}$. How high do we lift the block?

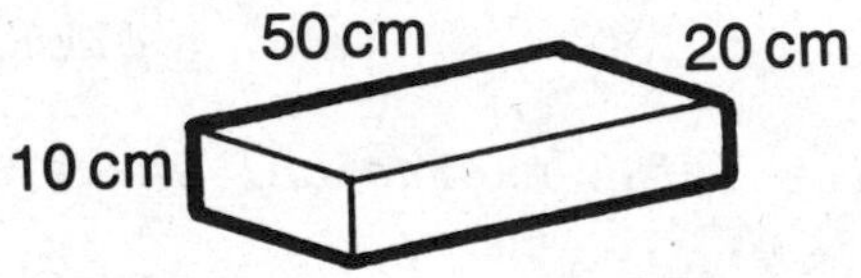

T: The *precise* question is: how far do we raise the c.m.?

S: The c.m. is at the geometric center of the block, so we raise it from 5 cm above the floor to 25 cm, half the longest side. That's a net rise of 20 cm.

T: Right. The PE of the block is Mgh_{cm}, so

$$W = \Delta PE = Mg(\Delta h)_{cm} = (10 \text{ kg})(9.8 \text{ m/s}^2)(.20 \text{ m}) = 19.6 \text{ J} .$$

Example 8-8. Derive a general expression for the *total* KE for an object of mass M and radius R rolling without slipping with a translational speed V_{cm}. Assume a general form for the moment of inertia of a regular solid, $I = AMR^2$.

S: I'm not sure where to start, except that rolling without slipping means $V_{cm} = R\omega$.

T: That's important. Start by writing $KE_{tot} = KE_{trans} + KE_{rot}$, and write each term out as a general expression.

S: You mean

$$KE_{rot} = \tfrac{1}{2}I\omega^2 = \tfrac{1}{2}(AMR^2)\omega^2$$

and

$$KE_{trans} = \tfrac{1}{2}MV^2{}_{cm}$$

What is that symbol A, again?

T: It represents in general the number that multiplies MR^2 for various regular shapes. For instance, A = 1/2 for cylinders, 2/5 for spheres, etc. Check Table 8.2 for other cases. Now connect ω with V_{cm}.

S: Alright.

$$KE_{rot} = \tfrac{1}{2}AMR^2(V_{cm}/R^2).$$

I see that the R's cancel in general. So

$$KE_{tot} = \tfrac{1}{2}AMV_{cm}{}^2 + \tfrac{1}{2}MV_{cm}{}^2.$$

T: This can be factored to give

$$KE_{tot} = \tfrac{1}{2}(A + 1)MV_{cm}{}^2.$$

The effect of geometric shape on the *rolling inertia* of an object is shown by the way A enters in the result.

Example 8-9. A cylindrical disc and a sphere are rolling without slipping at the same speed V_{cm}. They have the same mass and radius. Then they come to a hill, which will rise farther before stopping? How much farther?

S: This involves converting KE into gravitational PE. The last example showed us the general form of total KE in this kind of motion.

T: That's the key--the *total* KE, not just the translational, gets converted into PE as they reach their maximum height.

S: So we can write

$$KE_{tot} = Mgh_{max}, \quad \text{or} \quad \tfrac{1}{2}(A + 1)MV_{cm}{}^2 = Mgh_{max},$$

which gives

$$h_{max} = (A + 1)V_{cm}{}^2/2g\ .$$

Don't I need numbers now?

T: Examine the question and your result. You see, the h_{max} depends *only* on A and V. Mass and radius have disappeared. This general result demonstrates that the objects could be of entirely different masses and radii, but the maximum height wouldn't change. The only quantity left which is different for the two objects is the value of A.

S: For the disc, A = 1/2, and for the sphere, A = 2/5. So if we form a ratio for comparison,

$$h_{max}\ (\text{sphere})/h_{max}\ (\text{disc}) = (3/2)/(7/5) = 15/14 = 1.07$$

T: Again you can see how easy it is to work with ratios. This shows that the disc will rise about 7% higher than the sphere, due to the fact that it has more rotational energy when it is traveling at the same speed as the sphere.

Newton's Second Law for Rotational Motion

Understanding this section should enable you to do the following:

1. *Calculate the angular acceleration produced by a net torque about a given axis.*
2. *Calculate the linear and angular acceleration of an object on which a net force and net torque are acting.*

The rotational analog of the 2nd Law of Motion involves net external torque in place of net external force, moment of inertia instead of mass, and angular acceleration instead of linear acceleration. In mathematical form:

$$\tau_{net} = I\alpha$$

Important Observations

1. Notice that this equation is consistent with the units previously derived:

$$\text{N-m} = (\text{kg-m}^2)(1/\text{s}^2) = \text{N-m}.$$

2. You must again take care to be consistent with signs. Previously (Chapter 5) we agreed to take *counter-clockwise* torques to be *positive* and *clockwise* ones to be *negative*. Now we should extend that convention to *all* rotational variables.
3. It is necessary to calculate *all* the rotational quantities in a given problem about the *same* axis.
4. For an object free to translate as well as to rotate, *both forms of the 2nd Law can be applied simultaneously with respect to the center of mass.* That is,

$$\mathbf{F}_{net} = M\mathbf{a}_{cm}$$

and

$$(\tau_{net})_{cm} = I_{cm}\,\alpha_{cm}\,.$$

In this way, the total motion of a rigid body can be separated into the acceleration of the c. of m. plus the angular acceleration *about* the c. of m.

<u>Example 8-10</u>. In the following figure, a string is wound around the perimeter of a uniform disc of radius R = 10 cm and mass M = 2 kg. One end of the string is attached to the ceiling. The disc falls from rest. How long does it take to fall 6 meters, and how fast is the disc spinning at that time?

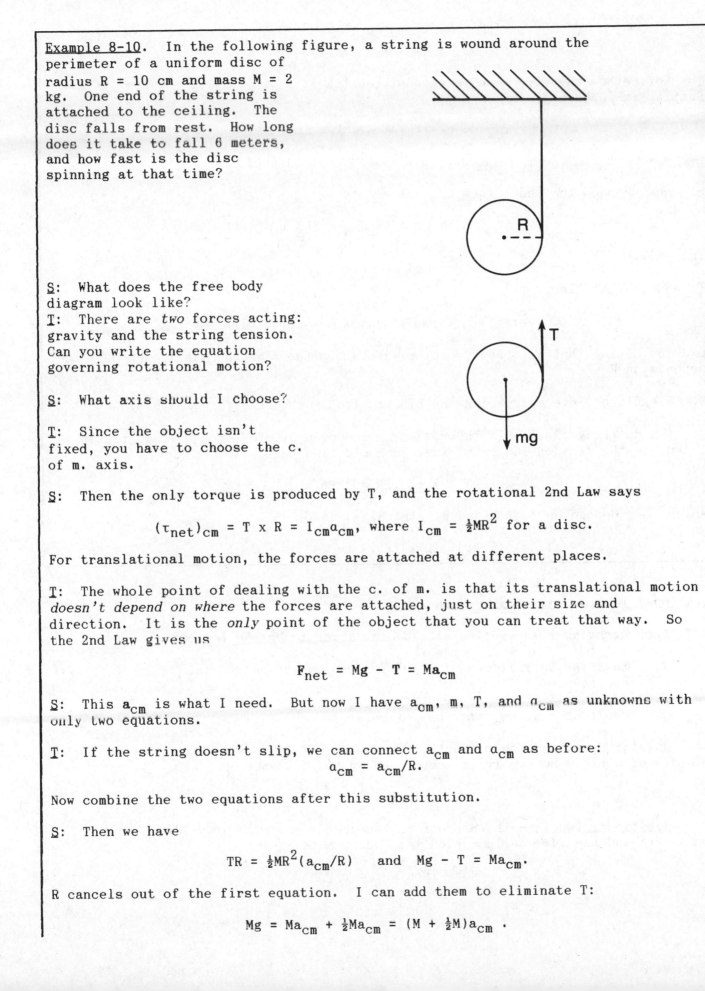

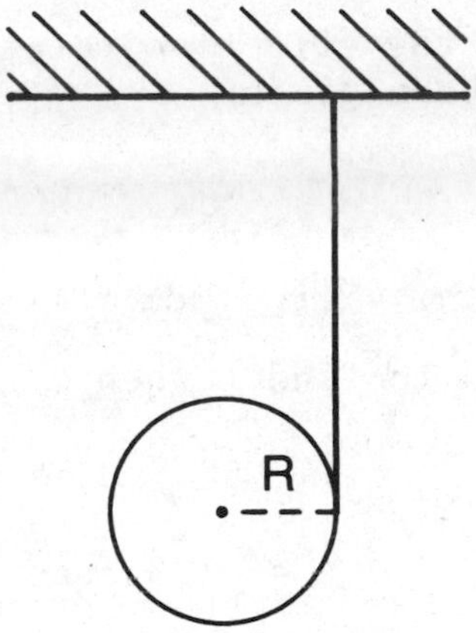

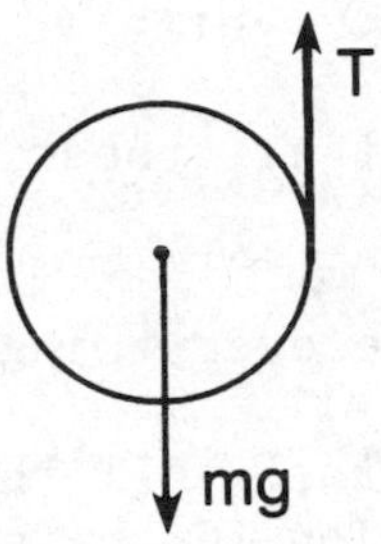

<u>S</u>: What does the free body diagram look like?
<u>T</u>: There are *two* forces acting: gravity and the string tension. Can you write the equation governing rotational motion?

<u>S</u>: What axis should I choose?

<u>T</u>: Since the object isn't fixed, you have to choose the c. of m. axis.

<u>S</u>: Then the only torque is produced by T, and the rotational 2nd Law says

$$(\tau_{net})_{cm} = T \times R = I_{cm}\alpha_{cm}, \text{ where } I_{cm} = \tfrac{1}{2}MR^2 \text{ for a disc.}$$

For translational motion, the forces are attached at different places.

<u>T</u>: The whole point of dealing with the c. of m. is that its translational motion *doesn't depend on where* the forces are attached, just on their size and direction. It is the *only* point of the object that you can treat that way. So the 2nd Law gives us

$$\mathbf{F_{net} = Mg - T = Ma_{cm}}$$

<u>S</u>: This $\mathbf{a_{cm}}$ is what I need. But now I have a_{cm}, m, T, and α_{cm} as unknowns with only two equations.

<u>T</u>: If the string doesn't slip, we can connect a_{cm} and α_{cm} as before:

$$\alpha_{cm} = a_{cm}/R.$$

Now combine the two equations after this substitution.

<u>S</u>: Then we have

$$TR = \tfrac{1}{2}MR^2(a_{cm}/R) \quad \text{and} \quad Mg - T = Ma_{cm}.$$

R cancels out of the first equation. I can add them to eliminate T:

$$Mg = Ma_{cm} + \tfrac{1}{2}Ma_{cm} = (M + \tfrac{1}{2}M)a_{cm}\ .$$

So

$$a_{cm} = Mg/(3M/2) = (2/3)g.$$

T: Notice how the effective rotational inertia $\frac{1}{2}M$ adds to M and makes $a_{cm} < g$. Putting in numbers gives

$$a_{cm} = (2/3)(9.8\ m/s^2) = 6.54\ m/s^2$$

and

$$\alpha_{cm} = a_{cm}/R = (6.54\ m/s^2)/(.10\ m) = 65.4/s^2$$

S: Now we need to find t when s_{cm} = 6 meters,

$$s_{cm} = \tfrac{1}{2}a_{cm}t^2 \ , \quad \text{so } t^2 = 2s/a_{cm} = 2(6\ m)/(6.54\ m/s^2) \ .$$

This gives $t = 1.35$ sec .

T: And in that time,

$$\omega = \alpha t = (65.4\ rad/s^2)(1.35\ s) = 88.3\ \ rad/s \ .$$

Did you notice that all these results were *independent* of the mass? And only ω depends on R.

S: I can't believe everything in this problem is the same for *any* mass.

T: I didn't say *that*! All the *motion* is independent of M, but we didn't stop to calculate the *tension* in the string. We would find that

$$T = I_{cm}\alpha_{cm}/R = \tfrac{1}{2}MR\alpha_{cm} \ ,$$

which *does* depend on *both* M and R. In this case

$$T = \tfrac{1}{2}(2\ kg)(.10\ m)(65.4\ rad/s^2) = 6.54\ N \ .$$

Rotational Impulse and Angular Momentum

Understanding this section should enable you to do the following:

1. *State the definitions of angular momentum and rotational impulse.*
2. *Calculate the angular velocity acquired by an object as the result of a given rotational impulse.*

Angular Momentum--The product of the *moment of inertia* and the *angular velocity* of a rigid body. It is symbolized by L. Mathematically,

$$L = I\omega.$$

Rotational Impulse--The product of the *torque* produced by a force about a given axis and the *time* during which the torque acts:

$$\text{Rot. Impulse} = \tau(\Delta t) \ .$$

Angular Momentum Form of Newton's Second Law--Notice that, with I constant, $\Delta L = I(\Delta\omega)$. So from the above definitions, it follows that

$$\tau_{net} = \Delta L/\Delta t.$$

or that

$$\text{Net Rotational Impulse} = \tau_{net}(\Delta t) = \Delta L = L_{final} - L_{initial}$$

Important Observations

1. The SI units of angular momentum are

$$(\text{kg-m}^2)(1/\text{s}) = \text{kg-m}^2/\text{s} = \underline{\text{joule-sec}} \ .$$

2. These results are very useful when applied to *systems* of mass as well as single objects, if you remember that the net torque is *external* to the system.
3. We are restricting these principles to axes whose *direction* is *fixed*. If the object or the system is translating, we must choose the axis passing through the c. of m.

Example 8-11. A flat disc is resting on a frictionless surface with three forces acting on it as shown in the figure. The wheel has a mass of 7 kg, a radius of 0.4 meters, and a radius of gyration of 0.3 meters. Describe the motion of the disc in both translation and rotation after the forces act for 3 seconds.

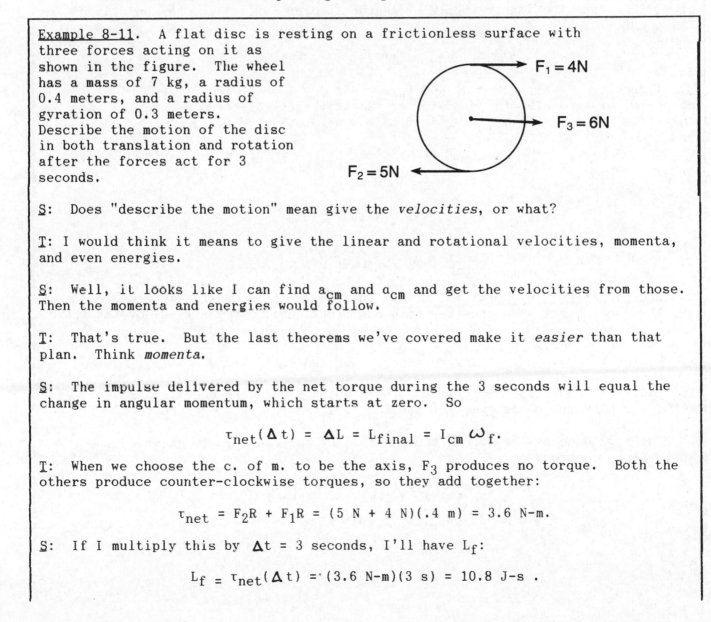

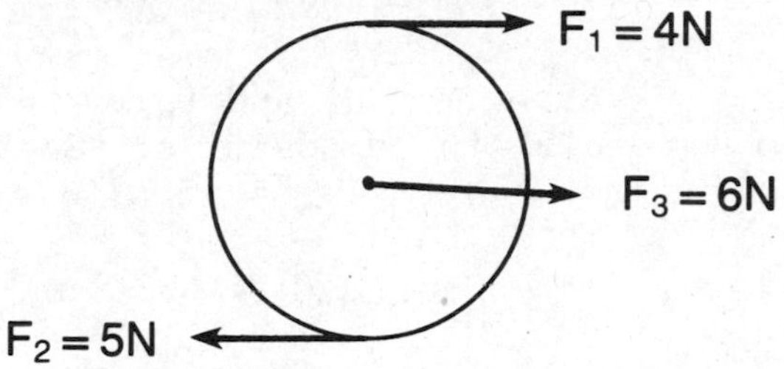

S: Does "describe the motion" mean give the *velocities*, or what?

T: I would think it means to give the linear and rotational velocities, momenta, and even energies.

S: Well, it looks like I can find a_{cm} and α_{cm} and get the velocities from those. Then the momenta and energies would follow.

T: That's true. But the last theorems we've covered make it *easier* than that plan. Think *momenta*.

S: The impulse delivered by the net torque during the 3 seconds will equal the change in angular momentum, which starts at zero. So

$$\tau_{net}(\Delta t) = \Delta L = L_{final} = I_{cm}\,\omega_f.$$

T: When we choose the c. of m. to be the axis, F_3 produces no torque. Both the others produce counter-clockwise torques, so they add together:

$$\tau_{net} = F_2R + F_1R = (5\text{ N} + 4\text{ N})(.4\text{ m}) = 3.6\text{ N-m}.$$

S: If I multiply this by $\Delta t = 3$ seconds, I'll have L_f:

$$L_f = \tau_{net}(\Delta t) = (3.6\text{ N-m})(3\text{ s}) = 10.8\text{ J-s}\ .$$

But I need the angular velocity.

T: Remember: $L = I\omega$, with I given by Mk^2, so

$$\omega_f = L_f/I = L_f/Mk^2 = (10.8 \text{ J-s})/(7 \text{ kg})(.3 \text{ m})^2$$
$$= 17.1 \text{ N-m-s}/(\text{kg m}^2) = 17.1 \text{ rad/s}.$$

Make sure that you see the units cancellation here.

S: Next,

$$KE_{rot} = \tfrac{1}{2}I\omega^2 = \tfrac{1}{2}Mk^2\omega^2 = \tfrac{1}{2}(7 \text{ kg})(.3 \text{ m})^2(17.1/\text{s})^2 = 92 \text{ J}.$$

Now what do we do about translation?

T: Remember, the motion of the c. of m. describes the translation, and a_{cm} is gotten from the 2nd Law, regardless of where the forces are attached:

$$\mathbf{F}_{net} = \mathbf{M}\mathbf{a}_{cm}$$

It follows that the linear impulse-momentum theorem applies to the momentum of the c. of m. :

$$F_{net}(\Delta t) = \Delta P_{cm} = M(V_f - V_i)_{cm}.$$

S: I get it. We have $V_i = 0$, and $F_{net} = 6 + 4 - 5 = 5\text{N}$, if we take the direction of F_1 and F_3 to be positive. So at the end of the 3 seconds,

$$P_{cm} = (5 \text{ N})(3 \text{ s}) = 15 \text{ N-s}$$

and

$$V_{cm} = P_{cm}/M = (15 \text{ N-s})/(7 \text{ kg}) = 2.14 \text{ m/s}$$

T: Then also we have

$$KE_{trans} = \tfrac{1}{2}MV_{cm}^2 = \tfrac{1}{2}(7 \text{ kg})(2.14 \text{ m/s})^2 = 16 \text{ J}.$$

Thus

$$KE_{tot} = 92 \text{ J} + 16 \text{ J} = 108 \text{ J} \quad .$$

You see how versatile energy is? It is the *one* quantity among all these for which you can simply add together both contributions. Can you find how far the wheel has moved and how much it has rotated in these 3 seconds?

S: Wouldn't I have to go back to finding a_{cm} and α_{cm} ?

T: You still don't fully appreciate the simplicity of work-energy language. Look, you know the energies--what are they related to? What caused them?

S: The *work* done by the forces and the torques, of course. I can write the work-energy theorems as

$$W_{trans} = F_{net}s_{cm} = KE_{trans}$$

and

$$W_{rot} = \tau_{net}\theta = KE_{rot}.$$

T: Exactly. So

$$s_{cm} = (16\ J)/(5\ N) = 3.2\ \text{meters}$$

and

$$\theta = (92\ J)/(3.6\ N\text{-}m) = 25.6\ \text{radians, or } 4.1\ \text{revs.}$$

Conservation of Angular Momentum

Understanding this section should enable you to do the following:

1. *Calculate the change in angular velocity which results from a change of the moment of inertia of a system or object.*

One of the most fundamental principles in all of physics is a corollary of the rotational impulse-momentum theorem. It is the *Principle of Conservation of Angular Momentum.*

Principle of Conservation of Angular Momentum--If no *net external torque* acts on a system, the *total angular momentum* of the system will remain *constant.*

Important Observations

1. There must be a *net external* torque to change the angular momentum of a system. The changes made in various parts of a system by *internal* torques just cancel out. This is the rotational counterpart of the Third Law of Motion.
2. Both the *direction* and *magnitude* of L remain constant.
3. Any axis may be chosen for the calculation of τ, I, and ω. An obvious choice wherever possible is that axis about which no net external torque exists. For problems involving gravity, this will usually mean taking an axis through the center of gravity.

Example 8-12. A uniform rod with mass of 1 kg and a length of 80 cm has two 500 gm weights of negligible size which are free to slide along its length. The rod is thrown into the air with an initial angular velocity of 3 revs/sec. The weights then slide out to opposite ends of the rod. When this happens, how is the rotational motion of the rod changed?

S: The rod and the weights comprise the system here, and since gravity acts through the center of mass, I'll be able to use the conservation principle if I choose that as my axis. So I can say that

$$L_f = I_f\omega_f = L_i = I_i\omega_i.$$

T: Very good. The whole problem comes down to the calculation of the change that occurs in the moment of inertia.

S: Since the weights are initially at the center of the rod, they don't contribute to I_i. It's just that of the rod taken about its center,

$$I_i = (1/12)M_{rod}l^2 = (1/12)(1\ kg)(.8\ m)^2 = 5.33 \times 10^{-2}\ kg\text{-}m^2.$$

The final value, I_f, is that of the rod plus two "point" masses a distance 1/2 from the c. of m.

$$I_f = 5.33 \times 10^{-2}\ \text{kg-m}^2 + 2(.5\ \text{kg})(.4\ \text{m})^2$$

$$= 2.13 \times 10^{-1}\ \text{kg-m}^2.$$

T: And so, very simply,

$$\omega_f = (I_i/I_f)\omega_i = (5.33 \times 10^{-2})(3\ \text{rev/s})/(2.13 \times 10^{-1}) = .751\ \text{rev/s}$$

Notice that we knew nothing about the forces acting, except that they were *internal*. Also, we didn't need to convert ω to rad/s, since we were doing a *ratio* calculation.

QUESTIONS ON FUNDAMENTALS

1. A solid cylinder of 5 kg mass has a radius of 8 cm. What is the moment of inertia about its c. of m. axis?

(a.) 40×10^{-4} kg-m^2 (b.) 80×10^{-4} kg-m^2 (c.) 160×10^{-4} kg-m^2
(d.) 20×10^{-4} kg-m^2 (e.) 320×10^{-4} kg-m^2

2. If the cylinder in #1 has a cord wound around its surface, and a force of 20 N is applied to the cord, what angular acceleration results?

(a.) 100 rad/s^2 (b.) 10 rad/s^2 (c.) 100 rev/s^2 (d.) 125 rad/s^2
(e.) 80 rad/s^2

3. If an object has an angular acceleration of 20 rad/s^2, what will its angular velocity be after 5 seconds, if it starts from rest?

(a.) 100 rad/s (b.) 100 rpm (c.) 4 rad/s (d.) 250 rad/s
(e.) 200 rpm

4. What is the expression for the moment of inertia of a sphere (mass M, radius R), about an axis tangent to a point on its surface?

(a.) MR^2 (b.) $(2/5)MR^2$ (c.) $(1/2)MR^2$ (d.) $(3/2)MR^2$
(e.) $(7/5)MR^2$

5. A solid sphere having mass 90 kg and radius 15 cm is spinning about a fixed axis at a rate of 3 revs/s. How much KE does the sphere have?

(a.) 288 J (b.) 7.6 J (c.) 360 J (d.) 144 J (e.) 3.65 J

6. If the sphere in #5 is rolling without slipping along a surface with a angular velocity of 3 revs/sec, what is the speed of its c. of m.?

(a.) 126 m/s (b.) 2.83 m/s (c.) 0.45 m/s (d.) 38 m/s
(e.) 4.5 m/s

7. What is the total KE of the sphere in #6?

(a.) 720 J (b.) 360 J (c.) 504 J (d.) 288 J (e.) 648 J

8. In #5, what is the angular momentum of the sphere?

(a.) 31 kg-m^2/s (b.) 7.63 kg-m^2/s (c.) 288 kg-m^2/s
(d.) 2.03 kg-m^2/s (e.) 15.3 kg-m^2/s

GENERAL PROBLEMS

Problem 8-1. A spherical bowling ball of radius R and mass M starts down the bowling alley *sliding without rolling* with initial velocity v_o. Show by direct application of Newton's 2nd Law that the ball will *roll without slipping* when its c. of m. velocity has become $(5/7)v_o$.

Q1: Let's start with the free body diagram.
Q2: What does the translational 2nd Law state?
Q3: What does the rotational 2nd Law state?
Q4: What is the connection between α and a_{cm}?
Q5: What are the equations of motion for linear and angular velocities?
Q6: What is the precise condition that marks the beginning of rolling without sliding?
Q7: How do I proceed to find V_{cm} when this happens?

Problem 8-2. A carousel of 500 kg mass has a moment of inertia I_{car} = 300 kg-m^2, and a radius of R = 1.5 m. A 60 kg boy is standing on its rim when it is rotating at a rate of 3 seconds per revolution. The boy proceeds to walk inward to a position at the radius of gyration of the carousel. After this has happened, what is the rotational speed of the carousel?

Q1: What is the original angular speed and angular momentum?
Q2: What is the effect of the boy moving inward?
Q3: Where does the boy move to? What is the radius of gyration?
Q4: What are the old and new values of the moment of inertia of the system of boy plus carousel?

Problem 8-3. Suppose you have a solid uniform cylinder of mass M = 1000 kg and radius R = 0.5 meters. As shown in the accompanying figure, two small identical rocket engines are positioned at opposite points on the rim of the cylinder so that their thrusts are tangential to the cylinder's sides. The rockets have exhaust velocities of 2000 m/s, and each has a mass of 10 kg. We want to design these engines so that they will give the cylinder an angular speed of 1 revolution/sec after burning for 2 seconds. What must be the rate of fuel consumption ($\Delta m/\Delta t$) be?

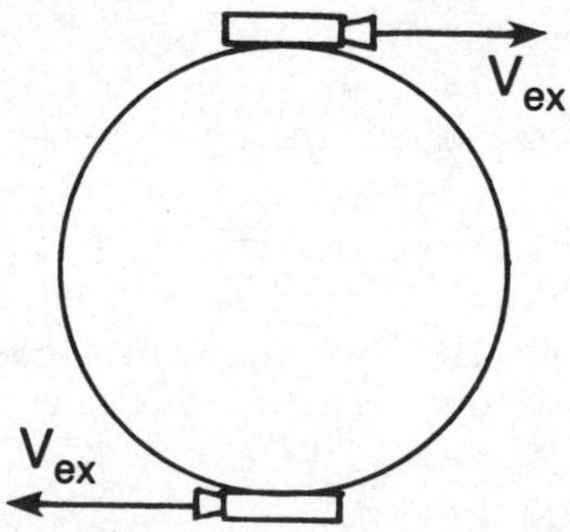

Q1: What is the simplest approach to this?
Q2: What is the moment of inertia of the cylinder plus rockets?
Q3: What is the angular momentum when ω = 1 rev/s?
Q4: This is equal to the rotational impulse given by both rockets. What must their individual thrust be?

Problem 8-4. A 1 kg wheel, R = 10 cm, is on a horizontal table

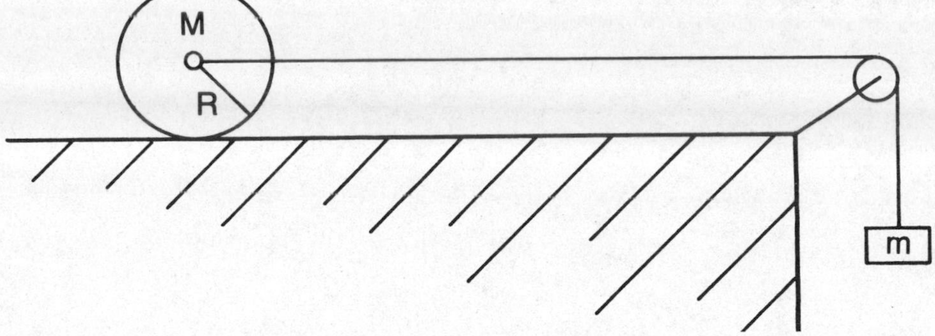

where it can roll without slipping as shown in the figure. A string is attached by looping it over small stubs which provide an axis through its center. The string passes over a massless pulley and has a 300 gram mass hanging on it. Neglect friction in the pulley or between the stubs and the string.

(a.) From energy considerations, what will be the angular velocity of the wheel when the 300 gram mass has descended 2 meters? Take the radius of gyration of the wheel to be 8 cm.

(b.) How long will this take?

Q1: Without friction, what does the work-energy theorem say in this situation?
Q2: What are ΔKE and ΔPE?
Q3: Is there a connection between ω and V?
Q4: For part (b.), what are the free body diagrams and the 2nd Law equations for the wheel and the hanging mass?
Q5: What time is asked for?

Problem 8-5. Consider a "dumbbell" made of two identical solid

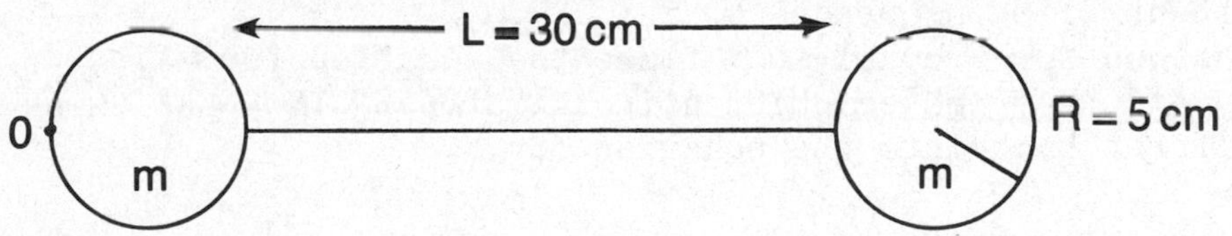

spheres separated by a thin rod of negligible mass, as shown in this figure. The mass of each sphere is 1500 grams, the radius is 5 cm, and the length of the rod is 30 cm. Starting from horizontal rest, the dumbbell is allowed to fall, pivoting around an axis 0 shown, which is fixed. Find rotational velocity of the dumbbell about 0 at the instant it has reached the vertical.

Q1: What is the relation which gives the rotational velocity?
Q2: What is the change of PE between horizontal and vertical positions?
Q3: What is the total moment of inertia of the dumbbell around point 0?
Q4: Where is the c.m. of the whole dumbell located?

ANSWERS TO QUESTIONS

FUNDAMENTALS:

1. (c.)
2. (a.)
3. (a.)
4. (e.)
5. (d.)
6. (b.)
7. (c.)
8. (c.)

GENERAL PROBLEMS:

Problem 8-1.

A1:

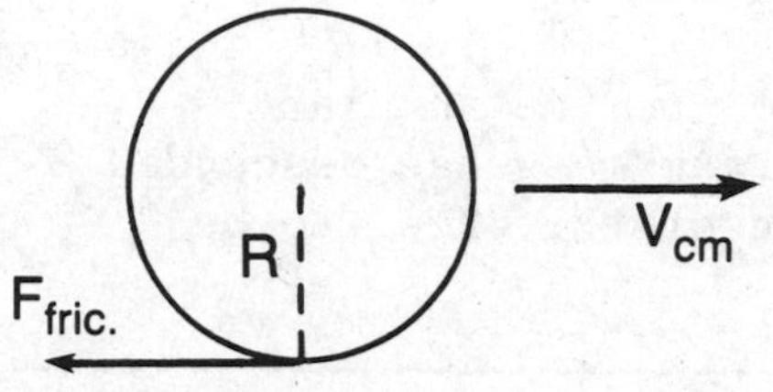

A2: $F_{fric} = Ma_{cm}$

A3: $\tau_{cm} = F_{fric}R = I_{cm}\alpha$, where $I_{cm} = (2/5)MR^2$

A4: a_{cm} is not equal to $R\alpha$, but that is not evident immediately, and you don't have to know this to solve the problem.

A5: $V_{cm} = V_o - a_{cm}t = V_o - (F_{fric}/M)t$;

$$\omega_{cm} = 0 + \alpha t = (F_{fric}R/I_{cm})t = (F_{fric}R)t/[(2/5)MR^2]$$

$$= (F_{fric})t/(2/5)MR$$

A6: Rolling without slipping means $\omega = V_{cm}/R$.

A7: Set the equation for V equal to R times the equation for ω. This will yield the time at which slipping stops. Then put this expression for t back into the equation for V.

Problem 8-2.

A1: 3 seconds/rev = 1/3 rev/s = 2.1 rad/sec.

$$L_o = I_{car}\,\omega_o + (M_{boy} + M_{pack})R^2\,\omega_o = 961 \text{ J-s.}$$

A2: He thus decreases the system's moment of inertia. Since angular momentum is conserved as he does this, angular speed must *increase*.

A3: $k = (I_{car})^{\frac{1}{2}}/M_{car}$ = .77 m.

A4: I_{old} = 300 kg-m^2 + (60 kg)(.15 m)^2 = 435 kg-m^2

I_{new} = 300 kg-m^2 + (60 kg)(.77 m)^2 = 336 kg-m^2

Problem 8-3.

A1: The rotational impulse-angular momentum theorem. Using torques, the 2nd Law, and rotational equations of motion is a more tedious alternative.

A2: $I_{tot} = \frac{1}{2}M_{cyl}R^2 + 2M_{rocket}R^2 = 130$ kg m^2

A3: $L = I_{tot}\omega$, with $\omega = 2\pi$ rad/s. Thus $L = 260\pi$ J-s.

A4: Rot. Impulse $= 260\pi$ J-s $= 2F_{rocket}Rt$.
$F_{rocket} = 130\pi$ N $= 408$ N.

Problem 8-4.

A1: $W_{ext} = 0$, and $\Delta KE + \Delta PE = 0$

A2: $\Delta PE = -mg(\Delta h) = -.588$ J
$\Delta KE = -\Delta PE = +.588 \text{ J} = \frac{1}{2}MV^2 + \frac{1}{2}Mk^2\omega^2 + \frac{1}{2}mV^2$

A3: Yes. $V = R\omega$.

A4:

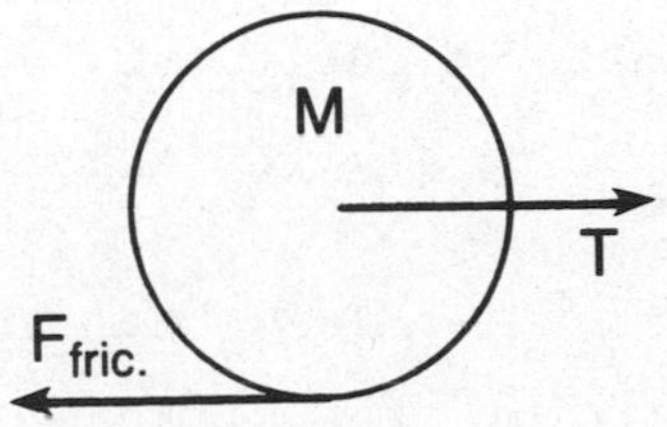

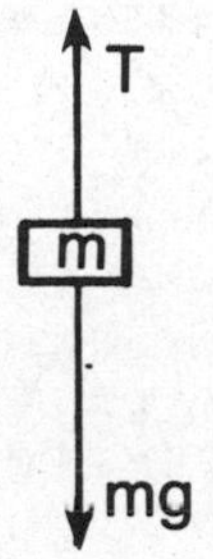

$T - F_{fric} = Ma$ $\qquad\qquad$ $mg - T = ma$

$F_{fric}R = I\alpha = Mk^2(a/R)$

These are three equations in the unknowns F_{fric}, T, and a.

A5: The value of t in the equation $s = \frac{1}{2}at^2$ when s = 2 meters. a is then found from the equations in A4 above.

Problem 8-5.

A1: $KE = \frac{1}{2}I_0\omega^2 = -\Delta PE$

A2: $\Delta PE = M_{tot}g(\Delta h_{cm})$, where $\Delta h_{cm} = -25$ cm.

A3: The nearer sphere's center is a distance R from O. The farther one is a distance 3R + L. The parallel axis theorem gives

$$I_0 = 2(2/5)mR^2 + mR^2 + m(3R + L)^2 = 3.11 \times 10^{-1} \text{ kg m}^2$$

A4: By symmetry, at the center of the rod, a distance 25 cm from the axis O.

CHAPTER 9. THE STRUCTURE OF ATOMS, MOLECULES, AND GASES

BRIEF SUMMARY AND RESUME

The fundamental *structure* of *atoms* and *molecules* is emphasized. Properties of *electrons*, *protons*, and *neutrons* are discussed, along with the way they combine to form the various chemical *elements* and their *isotopes*. Basic ways of measuring amounts of matter, such as the *unified mass unit* and *Avogadro's number* are defined. Properties of the *gaseous* state of matter are examined, and the *kinetic theory of gases* is derived from mechanical principles. *Temperature* is introduced as an additional fundamental dimension of measurement of physical properties. The *ideal gas law* is introduced and relationships between the *microscopic* and *macroscopic* descriptions of gases are discussed.

Material from previous chapters relevant to Chapter 9 includes:

1. Kinetic and potential energy.
2. Linear momentum and impulse.

NEW DATA AND PHYSICAL CONSTANTS

1. Fundamental masses.

   a. Unified Mass Unit (u).

   $1\ u = 1.661 \times 10^{-27}$ kg.

   b. Electron mass (m_e).

   $m_e = 9.11 \times 10^{-31}$ kg $= 5.49 \times 10^{-4}$ u.

   c. Proton mass (m_p).

   $m_p = 1.67265 \times 10^{-27}$ kg $= 1.00728$ u.

   d. Neutron mass (m_n).

   $m_n = 1.67495 \times 10^{-27}$ kg $= 1.00866$ u.

2. Avogadro's Number (N_A).

   $N_A = 6.022 \times 10^{23} = 1$ mole.

3. Universal Gas Constant (R).

   R = 8.314 J/mole-K.

4. Boltzmann's Constant (k) .

   $k = R/N_A = 1.380 \times 10^{-23}$ J/molecule-K.

5. Standard Temperature and Pressure (STP).

$$P_{std} = 1.0135 \times 10^5 \text{ N/m}^2 = 1 \text{ standard atmosphere}$$

$$T_{std} = 273 \text{ K.}$$

NEW CONCEPTS AND PRINCIPLES

Atomic Structure

Understanding this section should enable you to do the following:

1. *Identify the mass number and atomic number of an isotope from its symbolic notation.*
2. *Calculate the natural abundance of two isotopes of an element, given the masses of the pure isotopes.*

In the search for the way in which all matter is fundamentally structured, it has been found that three basic, or *elementary*, particles in various combinations form *atoms* of all the chemical elements. These elements in turn can combine to form all the varieties of molecular compounds. The three elementary particles are the *electron*, *proton*, and *neutron*, whose basic physical properties are summarized in the previous section and in Table 9.1 of your text. The following are definitions used in the description of various atomic structures:

Atomic Number (Z)--The number of protons in an atom's nucleus.

Mass Number (A)--The number of protons plus neutrons in an atomic nucleus.

Ion--An atom which has a net electric charge because of having more or fewer electrons then protons.

Isotope--Species of atoms having the same proton number but differing in neutron number.

Natural Isotopic Abundance--The percentage mixture of various isotopes in a sample of any element as that element is found in its natural state.

Unified Mass Unit (u)--Exactly 1/12 of the mass of one atom of the carbon-12 isotope, including its electrons. The conversion to SI units is:

$$1 \text{ u} = 1.661 \times 10^{-27} \text{ kg.}$$

Important Observations

1. All atoms having the same proton number, Z, belong to the same element. Thus isotopes are varieties of the same element, with differing nuclear structure due to the different number of neutrons.
2. Each chemical element has a name and an alphabetical symbol, such as C for carbon and He for helium. This symbol is *redundant* with the value of Z.
3. Each isotope is specified by a symbolic notation ${}^{A}_{Z}X$, where X indicates the alphabetical element symbol corresponding to a particular Z. A family of isotopes of the same element differ in A but have the same Z (and X) .

4. Ions can be either *positive* (electron number *less* than Z) or *negative* (electron number *greater* than Z) .
5. Electrons do *not* exist *within* the nucleus, but in orbital shells or clouds external to it. The nucleus is composed of protons and neutrons. Since the proton and neutron masses are so much greater than the electron mass, the great majority of an atom's mass is in its nucleus. Hence the name *mass number* for the total number of protons and neutrons.
6. With the exception of ${}^{12}_{6}C$, the masses of isotopes in u are not exactly integers equal to the mass number A. Very often in calculations involving atomic masses, we find it necessary to use all seven or eight significant figures.

Example 9-1. In nature carbon has an atomic mass of 12.01114 u. We know that the ${}^{12}_{6}C$ isotope has by definition an atomic mass of *exactly* 12.00000 u. The other naturally occurring carbon isotope is found to have a mass of 13.00335 u. What is this isotope's symbol, and what is the natural isotopic abundance for carbon?

S: With a mass so close to 13 u, can we guess that A = 13?

T: That's a safe statement for a sample of a *pure* isotope. So the symbol is ${}^{13}_{6}C$.

S: We can say that the percentage of ${}^{12}_{6}C$ plus that of ${}^{13}_{6}C$ must equal 100% if they are the only contributing isotopes.

T: Very true. Let's make a mathematical statement out of that. Let x = % of ${}^{12}_{6}C$, and y = % of ${}^{13}_{6}C$. Then x + y = 100.

S: We need another relation. Something like Example 9.3 in the text:

$$x(12.00000) + y(13.00335) = 12.01114$$

T: Good! Substituting y = 1 - x into the second equation gives

$$12.00000x + 13.00335 - 13.00335x = 12.01114$$

Collecting terms in x gives

$$x = 98.89\% \text{ and then } y = 1.11\%.$$

Notice how important it is to keep the large number of significant figures.

Example 9-2. In the isotope ${}^{23}_{11}Na$, approximately what percentage of the atom's mass resides in the electrons, the neutrons, and the protons?

S: Do we need to look up the very precise mass of this isotope?

T: Not for this approximate calculation. The actual mass happens to be 22.989771 u. But we can just use 23 u for this case.

S: Well, we have 11 protons, and therefore 11 electrons. The number of neutrons is (23-11) = 12. Can we use the masses of these particles as stated earlier?

T: Strictly speaking, that would not suffice if we need great precision, for reasons that will become clear later in the course. For our rough calculations here, we can use 1 u for each proton or neutron.

S: Then the mass of the neutrons would be 12 u, and that of the protons would be 11 u. The electrons would have only 11(.0005 u), or 0.0055 u.

T: So the *fraction* of the mass of the ${}^{23}_{11}Na$ atom due to electrons is only about

$$(.0055\ u)/(23\ u) = .00024 = .024\%\ !$$

For the neutrons, we have 12/23 = 52.2%

and for the protons, 11/23 = 47.8% .

Amounts of Matter

Understanding this section should enable you to do the following:

1. *Convert between the number of moles of a substance and the mass of the sample in grams and atomic mass units.*

There are various ways to describe quantities of matter. Previously we have designated the *mass* of macroscopic objects. When we realize that macroscopic objects are composed of "standardized" building blocks or particles having fundamental amounts of mass, we can use *numbers* of fundamental units to specify the quantity of matter as an alternate to mass.

The Mole and Avogadro's Number--The number of ${}^{12}_{6}C$ grams of the pure isotope is 6.022×10^{23}, and is called *Avogadro's number* (N_A). This amount of any substance is called one *mole*.

Corollary--The mass of one mole (molar mass) of any pure substance in grams equal to the mass of the elementary unit of the substance expressed in u.

Important Observations

1. The concept of the *mole* is the same as any other pure counting number, such as dozen, score, etc. It designates a specific (very large!) number of *anything* you wish to apply it to.
2. In the corollary, "pure substance" means a collection of *identical* fundamental units, like atoms of a certain isotope, electrons, bricks, etc. Obviously, since the mass of one u is so very small, the *practical* application of molar mass is usually restricted to atoms and molecules.

Example 9-3. Find the following masses:

(a.) 2.3 moles of ${}^{235}_{92}U$ in grams.
(b.) 1 mole of electrons in grams.
(c.) 10^{-6} moles of marbles, each having 1 gram mass, in u.

S: In (a.), 1 mole of ${}^{235}_{92}U$ has a mass of 235 u .

T: No! 1 mole of a substance has a mass *in grams numerically equal to the mass of the fundamental unit, expressed in u.* Here *one atom* of $^{235}_{92}U$ has a mass of 235 u. One *mole* (N_A of them) has a mass of 235 *grams.* 2.3 moles has a mass of 2.3(235) = 540.5 grams.

S: Let me try again. In (b.), *each* electron has a mass of *.00055 u.* So a *mole* of electrons would have a mass of .00055 x N_A = *3.31 x* 10^{20} *u*, which is 3.31 x 10^{20} u x 1.661 x 10^{-27} kg/u = .00055 grams.
I get it now. If you can express the *individual unit mass in u*, that same number is the mass of one mole in *grams.*

T: There you go. Now how about (c.)?

S: Here I have the unit of substance expressed in grams. The mass of *one mole* would be N_A times this, or 6.022 x 10^{17} gm. Furthermore, going in reverse, the mass of a single marble in *u* should be the same number as the mass of one mole in grams, or

$$M_{marble} = 6.022 \times 10^{23} \text{ u.}$$

T: Nice going!

Basic Description of Gases

The following definition of a gas is given in your text.

Gas--A substance which expands to fill uniformly the volume of any container in which it is placed.

The macroscopic properties of a gas are measured by the following:

Volume (V)--The space occupied by a gas.

Mass (M)--The total mass of all the atoms or molecules in a given sample of gas.

Density (d)--The mass per unit volume of a substance,

$$d = M/V$$

Pressure (P)--The force per unit area normal to any boundary surface of the gas.

$$P = F_N/A.$$

Temperature (T)--A measure of the "hotness" or "coldness" of a substance.

Important Observations

1. As discussed in the previous section, the mass of a sample can be related to the *number* of standard units (molecules or atoms in this case) making up the sample. If M is the atomic or molecular mass and n the number of moles, it follows that M = nM.
2. The SI units of density are kg/m^3

3. The SI units of pressure are N/m^2.
4. The above definition of temperature is very vague, and will become more precise and meaningful as we progress. Notice the comments in Section 9.6 of your text about temperature being a *new physical dimension*, apart from mass, length and time. The SI unit of temperature is the *kelvin* (K).

Kinetic Theory of an Ideal Gas

Understanding this section should enable you to do the following:

1. Calculate the total KE of an ideal gas given its pressure and volume.

The properties defining an ideal gas are given in Section 9.7 of your text. Let's briefly summarize the two points of basic importance:

1. The molecules or atoms of the gas are taken to be *point* masses, requiring negligible volume compared to the container's volume.

2. No forces act between the atoms or molecules except when they collide, and the collisions with themselves and with boundary surfaces are *perfectly elastic*.

Kinetic Energy of a Gas--The total kinetic energy of the atoms or molecules of an ideal gas is equal to 3/2 of the product of gas *pressure* times *volume*. Mathematically,

$$KE_{tot} = (3/2)PV$$

Important Observations

1. The detailed derivation of this principle is given in Section 9.7 in your text.
2. This equation relates microscopic properties (KE of the atoms or molecules) to the bulk, or macroscopic, properties (pressure and volume).
3. The kinetic energies of the atoms are those of *random direction* of motion, and do not represent the bulk translation of the sample as a whole. They represent energy *internal* to the sample.

The Ideal Gas Law

Understanding this section should enable you to do the following:

1. Calculate the density of a given ideal gas at standard temperature and pressure.
2. Use the ideal gas law to calculate tbe relationship between changes in pressure, volume, temperature, and amount of gas.

A number of empirical laws of ideal gases have been observed over a period of time beginning with the 17th century.

Boyle's Law--If *temperature* is held constant, changes in pressure and volume can occur only in such a way that *the product of gas pressure and volume remains constant.* Mathematically,

$$PV = \text{constant}, \quad \text{or} \quad P_1V_1 = P_2V_2 = \ldots \text{etc.}$$

Charles' Law--If *pressure* is constant, changes in volume and temperature can occur only in such a way that *the ratio of volume occupied by a gas to its absolute temperature remains constant.* Another way of stating this is that V is linearly proportional to T. Mathematically,

$$V/T = \text{constant}, \quad \text{or } V_1/T_1 = V_2/T_2 = \ldots \text{ etc.}$$

Gay-Lussac's Law--If the *volume* is constant, changes in the pressure and temperature can occur only in such a way that *the ratio of the pressure of a gas to its absolute temperature remains constant.* Or, P is linearly proportional to T. Mathematically,

$$P/T = \text{constant}, \quad \text{or } P_1/T_1 = P_2/T_2 = \ldots \text{etc.}$$

Ideal Gas Law--By combining the above laws and experimentally determining the constant of proportionality, we find the general statement of the *ideal gas law*:

$$PV = nRT, \quad \text{where } n = \text{the number of moles and}$$

$$R = \text{the universal gas constant} = 8.314 \text{ J/mol-K.}$$

Important Observations

1. This law can be written as **PV/nT = constant = R,** from which any of the more restrictive laws can be found easily.
2. The law often lends itself to comparative or ratio calculations, before and after changes in the variables. In such calculations, where R *cancels* in the ratios, you can use any convenient units for P and V as long as you are *consistent* throughout a problem.
3. In *all* calculations, *even those involving ratios*, you must use *absolute*, or *kelvin*, temperatures.
4. In any explicit calculation where the value of R enters, remember that *its units dictate* the units you have to use for P and v.
5. P is the *absolute* pressure, relative to vacuum. Occasionally we refer to a *gauge* pressure, which is the difference between absolute and ambient pressure outside the container, the latter being often approximately one atmosphere.

Example 9-4. What are the mass densities of helium and air at STP?

S: We need to know mass and volume to calculate density.

T: An STP volume of *any* ideal gas is 22.4 liters per mole.

S: So we need the masses of one mole for each gas. For He, it is approximately 4 grams. But air is a *mixture*. Does it obey the ideal gas law?

T: Yes. We can assume that the nitrogen and oxygen molecules don't take up any room. So we can treat air as an ideal gas with an *average* molecular mass given by the approximate mixture of 80% N_2 and 20% O_2. Thus

$$M_{air} = (.80)(28) + (.20)(32) = 28.8 \text{ gm/mole.}$$

S: Then

$$d_{air} = (28.8 \text{ gm/mole})/(22.4 \times 10^{-3} \text{ m}^3\text{/mole}) = 1.29 \text{ kg/m}^3.$$

T: And

$$d_{He} = (4 \text{ gm/mole})/(22.4 \times 10^{-3} \text{ m}^3\text{/mole}) = 0.18 \text{ kg/m}^3$$

Example 9-5. At constant atmospheric pressure, a container of air is observed to expand from 21.0 liters to 55.8 liters as the temperature is increased from 273 K to 720 K. Is there a leak in the container?

S: Now wait a minute. What do you mean, *leak*?

T: Can air get in or out? If so, what property of the gas would that change?

S: It would change n. By knowing P, V, and T before and after, we could find n each time and see if it changes.

T: You could, but it would be simpler to check on whether or not Charles' law holds. It *assumes* a constant sample of gas.

S: O.K. Charles' law says $(V_2/V_1) = (T_2/T_1)$. So it would predict

$$T_2 = (55.8/21.0)(273 \text{ K}) = 725 \text{ K}$$

The temperature is actually *lower* than this, *so n must have changed.* Can I tell whether n has increased or decreased without actually calculating n_1 and n_2?

T: That's a good instinct--to try to sniff out the results qualitatively before getting wrapped up in detailed calculations. In the ideal gas law, T is proportional to (1/n), so to get a *lower* T from a change in n, you would expect n to *increase*, i.e., air is leaking *into* the container. Let's find out if we're right:

$$n = PV/RT \;, \quad \text{so} \quad n_2/n_1 = (V_2/V_1)(T_1/T_2) \;, \text{ since P and R cancel.}$$

This gives

$$n_2/n_1 = (55.8/21.0)(273/720) = 1.0075$$

Can you tell what n_1 is with a quick calculation? Look at the original condition of the gas.

S: I could use the whole gas law directly, but I can see that at first we have STP, but V is *less* than 1 mole would occupy, so n_1 must be *less* than 1 mole by the ratio of 21.0 to 22.4.

T: Good! That makes $n_1 = (21.0/22.4) = 0.938$ moles, and then $n_2 = (.938)(1.0075) = 0.945$ moles.

Relationship Between Microscopic and Macroscopic Models

Understanding this section should enable you to do the following:

1. *Calculate the RMS speed of the molecules of a known ideal gas at a given temperature.*
2. *Calculate the total KE of a sample of ideal gas at a given temperature.*

The macroscopic description of a gas (ideal gas law) can be combined with the microscopic description (kinetic theory) to give a clearer interpretation of *what temperature describes*. If N is the number of atoms (molecules) in the sample, then we have

$$KE_{tot} = N\langle KE\rangle = (3/2)PV = (3/2)nRT.$$

Temperature of an Ideal Gas--A measure of the average translational kinetic energy of the atoms (molecules) of an ideal gas. Mathematically,

$$\langle KE\rangle = (3/2)kT,$$

where

$$k = \text{Boltzmann's constant} = 1.38 \times 10^{-23} \text{ J/molecule-K} = (n/N)R = R/N_A.$$

Root-Mean-Square Speed (V_{rms})--The *square root* of the *mean* of the *square* of the atomic (molecular) speeds in a gas. Mathematically,

$$V_{rms} = \langle v^2\rangle^{\frac{1}{2}} = (3RT/M)^{\frac{1}{2}} \quad \text{for an ideal gas}$$

Important Observations

1. The physical interpretation of absolute zero temperature becomes obvious from the above. At absolute zero the atoms (molecules) of an ideal gas would have *zero kinetic energy* and *zero gas pressure* would result. Since *real* gases condense into liquids or solids before absolute zero is reached, this must be *extrapolated* from the behavior of gases at higher temperatures. *All* gases show extrapolated zero pressures at -273 ^{o}C, or 0 K.
2. V_{rms} is *not* the same as the average or mean speed. In fact, Vrms = 1.088 $\langle v\rangle$ for the distribution of speeds in an ideal gas at uniform temperature.

Example 9-6. If you have 66 grams of CO_2 gas at STP, what is the average kinetic energy per molecule? What is V_{rms}?

S: If I can find KE, I would have to know the number of molecules to find $\langle KE\rangle$ per molecule

T: Right. You *are* given the amount of the gas.

S: As *mass*, though. Let's see, CO_2 has a molecular mass of (12 + 2x16) = 44. That means 44 grams of CO_2 would be 1 mole. Our 66 grams then is 1.5 moles.

T: And 1.5 moles = $(1.5)N_A = N = 9.033 \times 10^{23}$ molecules. The kinetic theory says

$$KE_{tot} = N<KE> = (3/2)PV.$$

S: The volume and pressure now need to be known.

T: At STP, pressure is 1.01×10^5 N/m^2, and the molar volume of any ideal gas is 22.4 liters/mole.

S: So our sample would occupy 1.5 times this, or 33.6 liters.

T: We now have everything needed for the answers:

$$<KE> = \frac{(3/2)PV}{N} = \frac{(3/2)(1.01 \times 10^5 \text{ N/m}^2)(33.6 \times 10^{-3} \text{ m}^3)}{9.033 \times 10^{23} \text{ molecules}} = 5.64 \times 10^{-21} \text{ J}$$

S: For V_{rms}, we need T and M. Which mass, per molecule or mole?

T: Be sure you use m, the mass of one molecule if you use k, or M, the molecular mass, if you use R.

$$V_{rms} = \sqrt{3kT/m} = \sqrt{3RT/M} = \sqrt{\frac{3(8.314 \text{ J/mole-K})(273 \text{ K})}{.044 \text{ kg/mole}}} = 393 \text{ m/s}$$

This is 884 mi/hr! Notice that the mass has to be in kg, as always in the SI.

Example 9-7. How much kinetic energy is contained in the air of an average classroom at 300 K? Take the classroom to be 10 feet high, 30 feet wide, and 30 feet long.

S: We have to know the amount of air in the room. The volume is (10 ft x 30 ft x 30 ft) = 9000 cubic feet, or $9000/(3.28 \text{ ft/m})^3 = 255 \text{ m}^3$. We can't assume STP. How do we find n?

T: In the absence of other information, we will assume that P = 1 atmosphere. Find out how much volume 1 mole needs at 1 atm and 300 K.

S: If V = 22.4 liters for one mole at T = 273 K, what would it be at T = 300 K?

T: At constant P, V is proportional to T, so (V/22.4) = (300 K)(273 K), giving V = 24.6 liters for one mole at 300 K.

S: Then the room contains

$$n = (255 \text{ m}^3)/(24.6 \times 10^{-3} \text{m}^3) = 1.04 \times 10^4 \text{ moles of air.}$$

T: The total KE is given by either (3/2)mRT or (3/2)NkT. Take your pick.

S: Since we have moles,

$$KE_{tot} = (3/2)(1.04 \times 10^4 \text{ moles})(8.314 \text{ J/mole})(273 \text{ K}) = 3.54 \times 10^7 \text{ J}.$$

T: If all this energy could be given to a 1000 kg sports car, how fast would it be going?

S: We would set $\frac{1}{2}mv^2$ for the car equal to 3.54×10^7 J:
So

$$v = \{2(3.54 \times 10^7 \text{ J})/(1000 \text{ kg})\}^{1/2} = 266 \text{ m/s, almost 600 mi/hr !}$$

Can this be right?

T: It's right. Rather ordinary temperatures can represent large amounts of energy. We'll see that there are restrictions on how much conversion from thermal to mechanical energy can occur. It isn't as simple as this example makes it seem.

QUESTIONS ON FUNDAMENTALS

1. The mass number of gold is 197. How many atoms of gold are in a one gram sample?

(a.) 6.023×10^{23} (b.) 3.06×10^{21} (c.) 1.35×10^{21}
(d.) 4.18×10^{21} (e.) need more information

2. How many neutrons does the isotope $^{232}_{90}Th$ have?

(a.) 232 (b.) 90 (c.) 142 (d.) 322

3. How much total KE do the atoms possess in a volume of 1 m^3 of gas at standard atmospheric pressure?

(a.) not enough information given (b.) 1.01×10^5 J (c.) 1.5 J
(d.) 1.5×10^5 J (e.) $.667 \times 10^5$ J

4. How much total KE is in a 1 mole sample of gas at room temperature (300K)?

(a.) 3.74×10^3 J (b.) 1.66×10^3 J (c.) 2.49×10^3 J
(d.) $.621 \times 10^3$ J

5. What is the rms speed of an ideal gas of neon atoms at T = 300 K? The atomic mass of neon is 20.2.

(a.) 3.7×10^5 m/s (b.) 4.8×10^3 m/s (c.) 609 m/s
(d.) 1.2×10^3 m/s (e.) 393 m/s

6. What is the temperature of 2 moles of an ideal gas when it occupies a volume of 25 liters at standard atmospheric pressure?

(a.) 152 K (b.) 273 K (c.) 373 K (d.) 1500 K (e.) 300 K

GENERAL PROBLEMS

<u>Problem 9-1.</u> Suppose you have a rectangular box 1 meter in length, having cross-sectional area A. The box is divided by a gas tight partition as shown in the figure below. In the left compartment you have 120 grams of argon gas at 310 K. In the right compartment you have 8 grams of helium gas at 255 K.

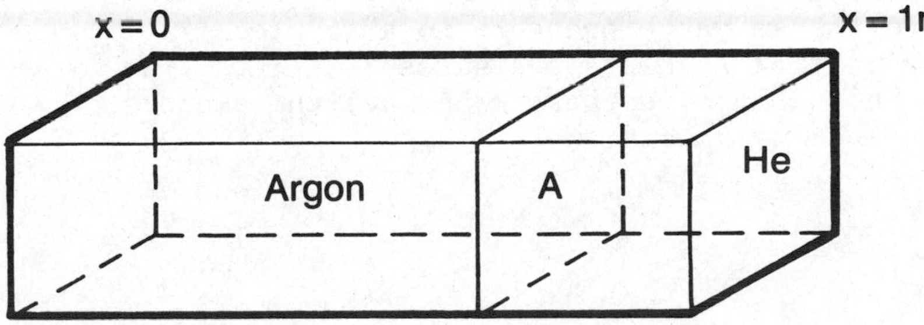

(a.) If the partition is held fixed at the 50 cm mark, what will be the net force on it? In what direction?

(b.) If the partition could move freely while remaining gas tight, what position would it assume, if the temperatures in both sides are held constant?

Q1: The volumes are equal in (a.). How many moles of each gas do we have?
Q2: How can I find the forces on the partition when I don't know A?
Q3: For (b.) what is the condition that determines the position of the partition?

Problem 9-2. A 1000 cm^3 tank contains 30 grams of CO_2 gas. The temperature in the warehouse is 260 K when the tank is filled during the night. By next afternoon the temperature has risen to 310 K. If the tank cannot withstand more than 15 atmospheres pressure, how much CO_2 must be vented by the relief valve during this period?

Q1: To find the initial pressure, we need to know n.
Q2: What is the initial pressure?
Q3: What would the pressure be at 310 K if all the gas remained in the tank?
Q4: How many moles at 310 K would create a 15 atmosphere pressure?

Problem 9-3. The atmosphere of Venus is practically all CO_2. The temperature is approximately 750 K and the pressure is 90 earth atmospheres. Find the mass density of CO_2 and the v_{rms} for CO_2 molecules on Venus.

Q1: What is a mole on Venus? Does R have the same value?
Q2: What is the number of moles per volume?
Q3: Doesn't pressure affect v_{rms}?
Q4: What is the mass per molecule for CO_2?

Problem 9-4. There are two containers at standard temperature, one containing 10 liters of O_2 at 1800 torr and the other holding 7 liters of N_2 at 5 atmospheres, both *gauge* pressures, in a room where air pressure is 0.95×10^5 Pa. The two tanks are connected and the contents allowed to mix without changing temperature. What is the final *gauge* pressure and the total mass of gas in the combined container?

Q1: Qualitatively, what does connecting the tanks do?
Q2: Doesn't the presence of one gas affect the other?
Q3: What determines the final pressure?
Q4: What remains constant upon connecting the tanks?
Q5: Do we have to change gauge pressure to absolute?
Q6: What are the initial absolute pressures?
Q7: The mass doesn't change, either. What are the amounts of each gas from the initial conditions?

ANSWERS TO QUESTIONS

FUNDAMENTALS:

1. (b.)
2. (c.)
3. (d.)
4. (a.)
5. (c.)
6. (a.)

GENERAL PROBLEMS:

Problem 9-1.

A1: The atomic masses for argon and helium are 120 and 4, respectively. Thus $n_{Ar} = 3$ moles and $n_{He} = 2$ moles.

A2: $V = AL$, $F = PA$, and $P = nRT/V$. Combine these and notice that $L = .5$ m for both compartments.

A3: That it is in equilibrium. Solve for the value of L_{Ar}, that equalizes the forces. Note that $L_{He} + L_{Ar} = 1$.

Problem 9-2.

A1: The molecular mass of CO_2 is 44 u. Thus n = .682 moles.

A2: You don't need to know. It is 14.6 atm (absolute).

A3: You don't need to know! It is 17.4 atm.

A4: That's the right question! n = .588 moles.

Problem 9-3.

A1: Both are the same. Mole is a pure number, and R is the *universal* gas constant.

A2: $n/V = P/RT = 1.46 \times 10^3$ moles/m^3.

A3: No. It depends only on mass and temperature.

A4: You can work with the mass per mole. But $m = M/N_A = 7.31 \times 10^{-23}$ gm.

Problem 9-4.

A1: It allows additional volume at constant temperature for each gas.

A2: Remember the basic assumptions of an ideal gas: point masses taking up no volume in the container. Each gas can be treated as if it was alone under these assumptions.

A3: It is the sum of the pressures of the independent gases, $P_{tot} = P(O_2) + P(N_2)$.

A4: n of each gas, R, and T. So PV is constant for each, where 17 liters is the final volume for both.

A5: Absolutely! *Always* in using the ideal gas law, use absolute pressure and absolute temperature.

A6: The surrounding pressure is $.95 \times 10^5$ Pa = .94 atm = 715 torr. This has to be *added* to both gauge pressures. Then it can be subtracted from the final absolute pressure.

A7: At standard temperature (273 K), $n(O_2) = 1.47$ moles, $n(N_2) = 1.85$ moles.

CHAPTER 10. LIQUIDS AND SOLIDS

BRIEF SUMMARY AND RESUME

A brief description of the *structure of liquids* begins this chapter, which includes a discussion of *surface tension* and *evaporation*. The *densities* of materials are compared to that of water in the concept of *specific gravity*. *Pascal's Principle* governing the undiminished transmission of pressure throughout the volume of a confined liquid provides the basis for discussing the operation of *barometers* and *manometers*. Pressure created by the *weight of columns of liquids* is introduced in this connection as well. *Hydraulic devices* are examined as examples of simple machines. These ideas are further developed in *Archimedes' Principle of buoyant forces* on an object immersed in a liquid.

The velocities and pressures of a fluid in motion are described by the *equation of continuity* and *Bernoulli's Equation*, which are statements of fluid incompressibility and conservation of mechanical energy. The concept of fluid *viscosity* is introduced and the *coefficient of viscosity* is defined. The volume rate of flow through pipes is described by *Poiseuille's Law*.

Diffusion and *osmosis* in solutions are discussed, and the equation for *osmotic pressure* is developed.

Solids are contrasted to fluids, and the process of *melting* is discussed. *Elastic properties* of solids are defined, and described by the *elastic moduli* relating *stress* on a solid to the *strain* produced. An example is *Hooke's Law*. The *elastic limit* and *ultimate strength* of a solid are discussed.

Material from previous chapters relevant to Chapter 10 includes:

1. The concepts of pressure and density.
2. Kinetic and potential energy.
3. The conservation of mechanical energy.
4. The conditions for static equilibrium.

NEW DATA AND PHYSICAL CONSTANTS

1. Pressure Units.

   SI: $1\ N/m^2 = 1$ pascal (Pa)

   Mercury column: 1 mm Hg = 1 torr

   Standard atmosphere: $1\ atm = 1.0135 \times 10^5\ N/m^2$

2. Coefficient of Viscosity (η).

   $\eta = (F/A)/(v/L)$

   SI units of η are $N\text{-}s/m^2$.

   $1\ N\text{-}s/m^2 = 10^3$ centipoise (cP).

3. Elastic Moduli.

   a. Young's Modulus. $Y = (F_N/A)/(\ \ L/L_o)$

b. Shear Modulus. $S = (F_t/A)/\Phi$

c. Bulk Modulus. $B = (-P)/(\Delta V/V_o)$

All the moduli have SI units of N/m^2.

NEW CONCEPTS AND PRINCIPLES

Hydrostatic Pressure

Understanding this section should enable you to do the following:

1. *Calculate the pressure at any depth in a liquid due to the weight of the liquid above that point.*
2. *Calculate the difference between gauge pressure and absolute pressure.*
3. *Convert between various common units used for pressure.*

In this section it is assumed that the liquids are of *uniform density* and essentially *incompressible*.

Pascal's Principle--Any pressure applied to a confined liquid is *transmitted undiminished to every part* of the liquid and is *normal to any boundary* of the liquid.

Pressure of the Weight of a Liquid Column--At a depth h below the top surface of a liquid, the pressure produced by the weight of liquid above the depth is given by

$$P = dgh, \quad \text{where } d = \text{mass density of the liquid.}$$

Important Observations

1. The pressure at *all* points at a depth h in a liquid is the same, dgh, regardless of the shape or size of the container. In the figure below for instance, $P_a = P_b = P_c$.

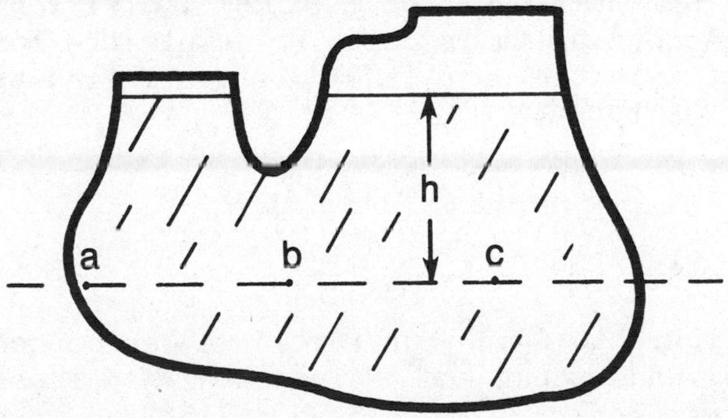

2. If the surface of the liquid is subjected to a gas pressure P_o, as when open to the atmosphere for example, this pressure is added equally to every point in the fluid, according to Pascal's Principle. Thus the total or *absolute* pressure at point a would be $P_{tot} = P_o + dgh$, whereas the *gauge* pressure would be just dgh.

Example 10-1. What would be the total pressure under a depth of 760 mm of mercury on Mars, when the mercury is exposed to the Martian atmosphere, which is 0.7% of earth's standard atmosphere?

S: 760 mm Hg is 1 standard atmosphere, so the total pressure at that depth would be 1.007 atmospheres.

T: *Nope.* You missed an *underlying assumption* in the conditions under which we can equate 760 mm Hg to 1 atmosphere. It *is not defined* as equal.

S: Let's see. P = dgh, and d should be the same on Mars, since mass is. But g is not *earth's* g, is it?

T: That's right. g(Mars) is about 36% of g(earth), so 760 mm Hg would produce only 0.36 of a standard (earth) atmosphere. That makes

$$P_{tot} = .007 + .36 \text{ atmospheres} = .367 \text{ atm}$$

$$= .367(1.01 \times 10^5 \text{ Pa}) = 3.71 \times 10^4 \text{ Pa.}$$

The atmosphere of Mars is so thin that it hardly makes much of a contribution.

Hydraulic Devices

Understanding this section should enable you to do the following:

1. *Calculate the performance of simple hydraulic devices such as hydraulic levers, barometers, and manometers.*

The previous principles of liquid or hydraulic pressure can be easily applied to many devices, among which are the *hydraulic lever*, the *barometer*, and the *manometer*.

Hydraulic Lever--The basis for all hydraulic pistons, jacks, etc., is the *multiplication of force* by applying a small force to a piston of small area pushing on a confined liquid. According to Pascal's Principle the force transmitted to a larger piston in contact with the same liquid is multiplied by the *ratio* of the *piston areas.* Mathematically,

$$F_2 = F_1(A_2/A_1).$$

See Figure 10.4 in your text.

Barometer--(See Figure 10.5 in your text). A device consisting of a vertical tube, closed at the top, whose open end is immersed in a reservoir of liquid whose surface is open to the atmosphere. The closed end of the tube above the liquid contains negligible air pressure. The liquid will rise in the tube to a height above the reservoir level at which the pressure due to the weight of the liquid column balances the surrounding air pressure. Thus

$$P_{air} = dgh, \quad \text{where } d = \text{density of the liquid (usually Hg).}$$

<u>Manometer</u>--(Figure 10.6 in your text). A U-shaped tube partially filled with liquid (usually Hg). If one end of the tube is connected to a container with gas at pressure P_2 and the other is open to a gas at pressure P_1, the liquid level in the two arms will differ by a height h given by

$$P_2 = P_1 + dg(h_1 - h_2) = P_1 + dgh.$$

If P_1 is the surrounding air pressure, the manometer directly measures the *gauge* pressure of the gas in the container,

$$P_{gauge} = P_2 - P_{air} = dgh$$

<u>Important Observations</u>

1. In both the barometer and the manometer, if the liquid used is mercury and pressure is expressed in *torr*, these equations become $P = h$ and $P_{gauge} = h$, respectively.
2. In general, P_{air} is *not* 1 standard atmosphere. Both P and T may not be standard in given ambient conditions.

<u>Example 10-2</u>. An unusual manometer, using alcohol in the U-tube, shows a difference in alcohol level of $h_1 - h_2 = 85$ cm when end #1 is open to the room and end #2 is connected to a container of gas. In the same room, the mercury barometer shows a height of 746 mm. What are the gauge and absolute pressures of the gas in the container in torr and in SI units?

<u>S</u>: Since the manometer fluid is not mercury, I can't just directly express the pressures in torr, can I?

<u>T</u>: That's right. You can immediately call the *barometer* reading is $P_{air} = 746$ torr, but the manometer reading is *not* 850 torr.

<u>S</u>: What is the density of alcohol?

<u>T</u>: To two significant figures, its *specific gravity* is 0.79.

<u>S</u>: Then the density is 790 kg/m^3, and the manometer reading is

$$P_{gauge} = dgh = (790 \text{ kg/m}^3)(9.8 \text{ m/s}^2)(.85 \text{ m}) = 6.6 \times 10^3 \text{ Pa}.$$

I now have to convert this to torr and the barometer reading to SI to complete the problem. I see that pressure units can be tricky.

<u>T</u>: Yes, you have a lot of options, many of which we haven't even mentioned. We'll stick to the torr, the standard atmosphere, and the Pascal. Remember

$$1 \text{ atm} = 760 \text{ torr} = 1.01 \times 10^5 \text{ Pa}$$

S: We have

$$P_{gauge} = (760\ \text{torr/atm})(6.6 \times 10^3\ \text{Pa})/(1.01 \times 10^5\ \text{Pa/atm}) = 50\ \text{torr}$$

So

$$P_{abs} = P_{gauge} + P_{air} = 50\ \text{torr} + 746\ \text{torr} = 796\ \text{torr}$$

$$= (796/760) \times 1.01 \times 10^5\ \text{Pa} = 1.06 \times 10^5\ \text{Pa}.$$

Archimedes' Principle

Understanding this section should enable you to do the following:

1. *Calculate the buoyant force on objects floating or submerged in a fluid.*
2. *Calculate the density of an object from the values of its net weight in air and submerged.*
3. *Calculate the percentage of a floating object which is submerged.*

This ancient and ingenious principle involves the concept of the *buoyant force* produced on an object partially or fully submerged in a fluid. The derivation is given in your text.

Buoyant Force--An object partially or fully submerged in a fluid (liquid or gas) experiences a *buoyant force* upward equal to the *weight of the fluid the object displaces.* Mathematically,

$$F_B = (d_{fluid})g\ V_{displaced}$$

Important Observations

1. $V_{displaced}$ is the *whole* volume of the object if it is *submerged* or some fraction of it if it is floating.
2. Since the *weight* of the object is

   $$w = (d_{object})gV_{object},$$

   the question of whether the object will float or sink depends on whether F_B can become as large as w. With no other forces acting, on a fully submerged object the net vertical force is

   $$F_{net} = F_B - w = gV_{obj}(d_{fluid} - d_{obj}).$$

   Sinking will occur if $d_{fluid} < d_{obj}$. Otherwise it will float.
3. If the fluid used is water, the specific gravity of an object is just the ratio of its weight in air, w, to the difference between that weight and its submerged net weight, w_w. Mathematically,

   $$\text{sp. grav.} = (d_{obj}/d_{water}) = w/(w - w_w)$$

Example 10-3. The bag of a balloon when filled with helium is a sphere 50 m in diameter. What *total* weight, including bag, gondola, and contents can it lift in pure air at STP? If you wanted to lift more than this, would you wait for a cooler or a warmer day?

S: It seems we found the STP densities of He and air before.

T: It was Example 9-4. We have $d_{He} = .18\ kg/m^3$ and $d_{air} = 1.29\ kg/m^3$.

S: We need to find the buoyant force exerted on the balloon by the surrounding air. Any weight more than that can't be lifted.

T: You have that

$$F_B = (d_{air})gV_{He}, \quad \text{and} \quad V_{sphere} = (4/3)\pi R^3.$$

S: So

$$F_B = (1.29\ kg/m^3)(9.8\ m/s^2)(4/3)\pi(25\ m)^3 = 8.27 \times 10^5\ N.$$

This is the maximum weight the balloon can lift.

T: But this includes the helium.

S: Isn't the helium weightless? It's lighter than air.

T: Come on, now! The helium has a weight $d_{He}gV_{He}$. The fact that it floats in a denser gas doesn't mean it isn't attracted by gravity.

S: Then

$$w_{He} = (.18\ kg/m^3)(9.8\ m/s^2)(4/3)\pi(25\ m)^3 = 1.15 \times 10^5\ N.$$

It sure *isn't* weightless!

T: So the *net* weight that can be lifted by the helium is

$$F_{net} = F_B - w_{He} = (8.27 - 1.15) \times 10^5\ N = 7.12 \times 10^5\ N$$

What about temperature effects?

S: The temperature shouldn't have any effect, since it would change the densities of air and He by the same amount.

T: Sounds like you're *guessing*. The densities would change by the same *fraction*, but that doesn't mean the same *absolute* amount. Let's take a more careful look. The *amount* of He in the bag remains the same, so w_{He} doesn't change. The buoyant force depends on d_{air} and V_{He}. At constant pressure, we know that

$$V_2 = V_1(T_2/T_1) \text{ and } d_2 = d_1(T_2/T_1),$$

so the *product* $d_{air}gV_{He}$ remains *constant*. Thus at higher T the bag expands but displaces less dense air, and vice versa at lower temperatures. Your guess was right, but you had no reasoning to support it.

Fluids in Motion--Hydrodynamics

Understanding this section should enable you to do the following:

1. *Calculate the variation of fluid flow velocity which accompanies a variation in the cross-sectional area of a pipe.*
2. *Relate changes in fluid pressure and flow speed and calculate the effect of fluid height on both for non-viscous flow.*

To decribe fluid flow, we make the following assumptions:

1. The flow is what we call "*laminar*," or along *streamlines*. There is no turbulence or mixing, and hence no energy is dissipated as heat.
2. The fluid is of *uniform density* and *incompressible*. This assumption leads directly to the *equation of continuity*.

Equation of Continuity--For fluid flow through a pipe of varying cross-sectional area, the product of *area* times *flow velocity* is *constant* at all points in the flow. Mathematically,

$$v_1A_1 = v_2A_2 = \ldots \text{ etc.}$$

The application of the work-energy theorem with no heat production to fluid flow leads to *Bernoulli's Equation*.

Bernoulli's Equation--For incompressible flow *without viscosity*, the energy per unit volume of the fluid remains constant at all points in the flow. Mathematically,

$$P + \tfrac{1}{2}dv_1^2 + dgy = \text{constant},$$

where P = fluid pressure
d = fluid density
v = flow velocity
y = vertical height of the fluid above some reference.

Important Observations

1. Each term, including pressure, has SI units of J/m^3. You should verify this.
2. Since so many pressure units are encountered, be careful to use Pa for pressure in this equation.

Special Cases

1. No variation in height. Then at any two points,

$$P_1 + \tfrac{1}{2}dv_1^2 = P_2 + \tfrac{1}{2}dv_2^2.$$

This means the pressure is *least* where flow velocity is *greatest*.

2. Effluent Velocity from Tank. Assume a small hole at vertical position y_2 below the fluid level, y_1, in a large tank. Both the hole and the fluid surface are exposed to the same external pressure, so $P_1 = P_2$. Furthermore, the area of the fluid surface is much greater than the area of the hole, so the speed of fall of the fluid surface, v_1, is negligible ($v_1 = 0$). Then the *speed of effluent* through the hole is

$$v_2 = [2g(y_2 - y_1)]^{\frac{1}{2}} = (2gh)^{\frac{1}{2}} .$$

This is known as Torricelli's Theorem.

Example 10-4. In the accompanying figure, an open tank filled with water 3 meters deep is sitting on the ground. A spigot 2 meters below the water level is connected to a pipe 5 cm in radius. Just before the spigot there is a small vertical glass tube extending upward from a small hole in this pipe. Further on, the exit pipe is connected to a hose of 3 cm radius lying on the ground, with its end open to the atmosphere.

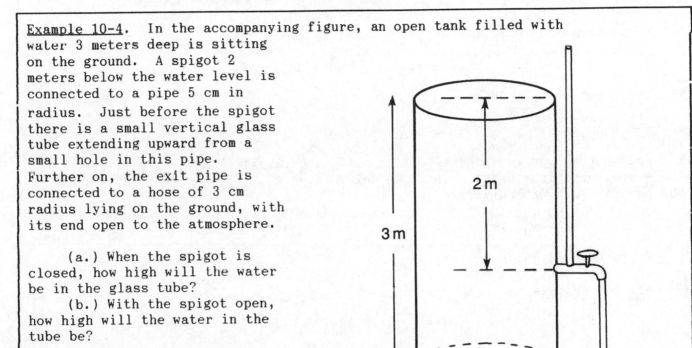

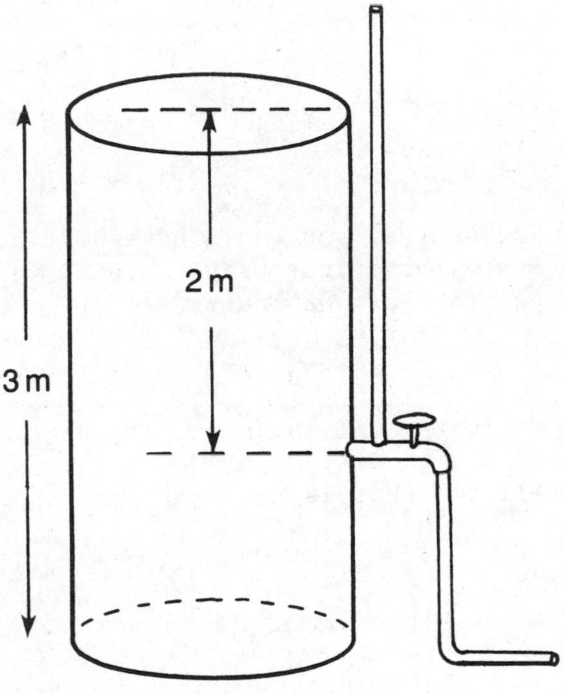

(a.) When the spigot is closed, how high will the water be in the glass tube?

(b.) With the spigot open, how high will the water in the tube be?

S: When the spigot is closed, we have a *hydrostatic* problem. The pressure at the bottom of the glass tube is just that of the weight of two meters of water, so it would support a two meter column of water in the tube.

T: That's right. The glass tube is a manometer of sorts, measuring *gauge* pressure at the spigot. Now when water starts *flowing*, what happens?

S: Pressure *diminishes* as flow speed *increases*, so the water level in the tube should lower? What does Bernoulli's Equation look like when applied to this?

T: The important thing is to identify two points of the flow which are crucial to the problen. Certainly the spigot and the end of the hose are examples. If we label the spigot #1 and the end of the hose #2, the equation of continuity gives

$$v_1 \pi r_1^2 = v_2 \pi r_2^2,$$

and Bernoulli's equation gives

$$P_1 + \tfrac{1}{2}dv_1^2 + dgy_1 = P_2 + \tfrac{1}{2}dv_2^2 + dgy_2.$$

S: If we use the top of the tank as the reference level, we have $y_1 = -2$ m and $y_2 = -3$ m. But we still have too many unknowns!

T: Let's count. Actually, we know $P_2 = P_{air}$ since the hose is open to the atmosphere. The unknowns are then P_1, v_1 and v_2. Do you see one more equation anywhere?

S: We know the water surface at the top of the tank is at P_{air}, and I think we can say that the flow speed of the water surface is zero.

T: Exactly. Can you make an equation between these conditions and those at the end of the hose?

S: We've taken y = 0 at the water surface, so

$$P_{air} = P_{air} + \tfrac{1}{2}dv_2^{\,2} + dg(-3\text{ m}).$$

Thus P_{air} cancels out, and

$$v_2 = \{2(9.8\text{ m/s}^2)(-3\text{ m})\}^{\frac{1}{2}} = 7.67\text{ m/s}.$$

T: Notice that you could have written this immediately from Torricelli's Theorem. The spigot plays no part in determining v_2. Now we can find v_1 from the equation of continuity.

$$v_1 = v_2(r_2/r_1)^2 = 7.67\text{ m/s }(3/5)^2 = 2.76\text{ m/s}.$$

S: P_1 is the only unknown left.

T: All you want is the *gauge* pressure, $P_1 - P_{air}$.

$$P_1 - P_{air} = -\tfrac{1}{2}dv_1^{\,2} - dgy_1$$

$$= -\tfrac{1}{2}(10^3\text{ kg/m}^3)(2.76\text{ m/s})^2 - (10^3\text{ kg/m}^3)(9.8\text{ m/s}^2) = 15.8 \times 10^3\text{ Pa}$$

S: How do I translate that into water level in the tube?

T: Easy.

$$P_{gauge} = dgh = 15.8 \times 10^3\text{ Pa},$$

giving

$$h = (15.8 \times 10^3\text{ N/m}^2)/(10^3\text{ kg/m}^3)(9.8\text{ m/s}^2) = 1.61\text{ meters}.$$

The *static* column height was 2 meters, so it has lowered by 0.39 m.

Viscous Flow

Understanding this section should enable you to do the following:

1. *Identify the meaning of the coefficient of viscosity of a fluid and the role it plays in determining the fluid flow through pipes.*

2. *Calculate the volume rate of viscous flow through pipes of various cross-sections and pressure gradients, using Poiseuille's Law.*

In the previous section, we assumed there was no resistance to flow, i.e., no viscosity. Actually, the volume rate of flow through pipes is very dependent on the fluid viscosity, which is measured by a quantity called the *coefficient of viscosity.*

Coefficient of Viscosity (η)--When two plates of area A are separated by a thickness of fluid L and subjected to a *shear stress* F/A (see Figure 10.21 in your text), the resulting constant speed v of one plate relative to the other is given by

$$v = (F/A)L/\eta ,$$

where η is the *coefficient of viscosity* of the fluid. Values of η are given in Table 10.2 in your text.

Shear Stress--If two parallel plates of area A have equal and opposite forces F applied *in the plane* of the plates, the *shear stress* on the plates is *F/A.*

Important Observations

1. The SI units of shear stress are derived to be N/m^2. This seems to be the same as for pressure, but it is quite a different physical situation, since shear forces are *parallel* to the areas whereas forces creating pressure are *normal* to the areas. This is a good reason to reserve the name *pascal* as the SI unit of *pressure* specifically.
2. The SI unit for coefficient of viscosity is derived from its defining equation as:

$$\eta = (F/A)/(v/L) = (N/m^2)/(m/s)/(m) = (N/m^2)s$$

   Another commonly used unit for η is the *centipoise* (cP), where $1\ cP = 10^{-3}\ Ns/m^2$.

Poiseuille's Law--The volume rate of flow through pipes of ciruclar cross-section, including the effects of viscosity, is given by *Poiseuille's Law.* Mathematically,

$$Volume/t = \pi R^4(P_1 - P_2)/8\eta L$$

Here R = radius of the pipe
P_1 = pressure at the *upstream* end of the pipe section
P_2 = pressure at the *downstream* end of the pipe section
L = length of the pipe section

Important Observations

1. Notice the very strong dependence on R, as discussed in your text.
2. $(P_1 - P_2)/L$ is called the *pressure gradient* between the two ends of the pipe.
3. Be careful with the units of η. When they are in cP, you must remember to furnish the factor of 10^{-3} to match the SI units of the rest of the expression.

Example 10-5. Suppose you have one section of a pipe with $R_1 = 3$ mm

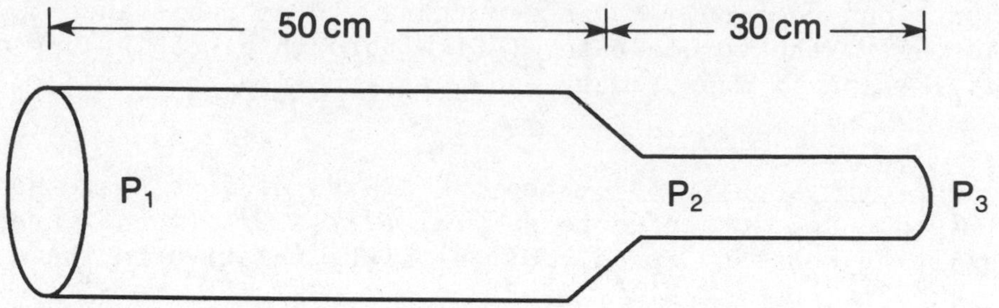

connected to a section with $R_2 = 2$ mm as shown in this figure. The large section is 50 cm long, and the shorter section is 30 cm long. If the pressures at the ends of the whole pipe are $P_1 = 2$ atmospheres and $P_3 = 1$ atmosphere, how much benzene will flow through the pipe per second? What will the pressure P_2 be at the connection between sections? What flow *velocity* will the benzene have as it exits the pipe?

S: Poiseuille's Law applies only to pipes of constant radius. How can I use it here?

T: It applies to any *section* with constant radius.

S: This means I can write

$$\pi R_1^{\,4}(P_1 - P_2)/\eta L_1 = V/t \qquad \text{for the 1st section and}$$

$$\pi R_2^{\,4}(P_2 - P_3)/\eta L_2 = V/t \qquad \text{for the 2nd section.}$$

Can I set these rates of flow equal?

T: You've got it. Viscous or not, the fluid is still *incompressible*, so the equation of continuity still applies.

S: Thus we can set

$$\pi R_1^{\,4}(P_1 - P_2)/\eta L_1 = \pi R_2^{\,4}(P_2 - P_3)/\eta L_2$$

and solve for P_2. It looks like P_2 doesn't depend on η.

T: Right, but flow rate and velocity *will*. The algebra gets a little tedious here, so watch out for correct exponents, especially in calculating the R^4 terms. Collecting terms in P_2, you get

$$P_2 = \frac{\left(\frac{R_1^4 P_1}{L_1}\right) + \left(\frac{R_2^4 P_3}{L_2}\right)}{\left(\frac{R_1^4}{L_1}\right) + \left(\frac{R_2^4}{L_2}\right)} = \frac{(3.27 \times 10^{-5}) + (5.39 \times 10^{-6})}{(1.62 \times 10^{-10}) + (5.33 \times 10^{-10})}\ \text{Pa}$$

$$= 1.77 \times 10^5 \text{ Pa} = 1.75 \text{ atm}.$$

S: Now to get flow rate, we can use

$$V/t = \frac{\pi R_1^4(P_1 - P_2)}{\eta L_1} = \frac{\pi(3 \times 10^{-3}\,\text{m})^4 (2.02 - 1.77) \times 10^5 \text{ Pa}}{(0.65 \times 10^{-3}\,\text{N-s/m}^2)(0.5\,\text{m})}$$

$$= 1.96 \times 10^{-2} \text{ m}^3/\text{s}.$$

T: Good. Then the flow *velocity* is again obtained from vA = V/t. Thus

$$v_2 = (1.96 \times 10^{-2} \text{ m}^3/\text{s})/\pi(2 \times 10^{-3} \text{ m})^2 = 1.56 \times 10^3 \text{ m/s}.$$

Elastic Properties of Solids

Understanding this section should enable you to do the following:

1. *Calculate the changes in length, volume, or shape which result from various stresses applied to solids.*

Although we often treat solids as rigid and incompressible, we know that small changes in length, volume, or shape accompany the action of forces on solids. The important quantity involving these forces is measured by the *stress*, or force per unit area applied to the object. The effect of these stresses is to produce a *strain*, or dimensional change, in the object.

Stress--If a force F is applied to a surface area A of an object, the *stress* produced is the ratio F/A. There are three categories of stress, depending on how the force is applied.

a. Longitudinal Stress--Here F is *normal* to A. This produces *tensile* stress if F pulls *outward*, and *compressional* stress if it pushes *inward* on the object. These tend to change the *length* of the object.

b. Shear Stress--When F is *tangential*, or parallel to the plane of A, it produces a *shear stress* which tends to *deform the shape* of the object.

c. Pressure Stress--A uniform compressional stress which is *everywhere* normal to the entire boundary of an object tends to *decrease its volume.* This is the case of an object subjected to a uniform pressure P.

Strain--The three categories of strain caused by the above stresses are:

a. Longitudinal Strain--The ratio of the change in length to original length, $\Delta L/L_o$. This can be either *lengthening* or *shortening* depending on whether the stress is tensile or compressional.

b. Shear Strain--As an object distorts from a shear stress, the angular displacement Φ of a line in the object originally perpendicular to the applied force.

c. Volume Strain--The ratio of volume change to initial volume, $\Delta V/V_o$, which results from the compression of an object subjected to a surrounding pressure stress.

Elastic Moduli--If when the stress is removed the object returns to its original configuration, the strain is said to have been *elastic*. In such cases, each stress will be directly proportional to the strain it produces. The constants of proportionality relating them for a given material are called that material's *elastic moduli*. The three moduli are:

a. Length (Young's) Modulus.

$$Y = (F_n/A)/(\Delta L/L_o)$$

b. Shear (Torsional) Modulus.

$$S = (F_t/A)/\Phi$$

c. Bulk (Volume) Modulus.

$$B = (-P)/(\Delta V/V_o)$$

B is just the reciprocal of the *compressibility* of a material.

All the moduli have SI units of N/m^2. Values for some materials are given in Table 10.4 of your text.

Example 10-6. We have so far treated liquids as incompressible. To see how good an approximation this is, calculate the density of sea water at the bottom of an oceanic trench 10 km deep, and compare this to its standard density of 1.025×10^3 kg/m^3. The bulk modulus of sea water is $.21 \times 10^{10}$ N/m^2.

S: I sense that I'm supposed to consider the pressure created by that depth of sea water. But what water sample should I consider being compressed?

T: *Any* representative sample small enough to have essentially uniform pressure surrounding it. Why not take a *unit* mass of 1 kg? Then its volume would be

$$V_o = m/d = (1\ \text{kg})/(1.025 \times 10^3\ \text{kg/m}^3) = 9.76 \times 10^{-4}\ \text{m}^3.$$

S: I can find ΔV from the bulk modulus if I know P. Should it be absolute or gauge pressure?

T: Strictly speaking, gauge pressure, since the standard density assumes standard pressure. It won't make substantial difference in this case, though.

S: Since we're admitting that density changes with depth, what value should I use in calculating $P_{gauge} = dgh$?

T: That's a perceptive question. Here we expect the change in d to be very small, so as a first approximation we can treat it as constant to calculate P. If we get a result showing a *large* change in d, this procedure would be incorrect.

S: Then

$$P_{gauge} = dgh = (1.025 \times 10^3 \text{ kg/m}^3)(9.8 \text{ m/s}^2)(10^4 \text{ m})$$
$$= 1.005 \times 10^8 \text{ N/m}^2 = 1000 \text{ atmospheres!}$$

Then

$$\Delta V/V_o = (-P)/B = (-1.005 \times 10^8 \text{ N/m}^2)/(.21 \times 10^{10} \text{ N/m}^2)$$
$$= -4.8 \times 10^{-2}.$$

T: Thus the change in volume is only a *4.8% change* under all that pressure. The new volume is

$$V = V_o(1 - .048) = 9.3 \times 10^{-4} \text{ m}^3.$$

We then have

$$d = (1 \text{ kg})/(9.3 \times 10^{-4} \text{ m}^3) = 1.076 \times 10^3 \text{ kg/m}^3.$$

Notice that $\Delta d/d_0 = -\Delta V/V_o$. This is true for any quantities which are inversely related: the fractional change in one is the negative of the fractional change in the other.

Example 10-7. A copper wire of radius 1 mm has reference marks on it every cm which are precise to the nearest mm. It is attached to a 100 kg mass whose volume is 1.5×10^{-2} m^3. When the wire shows the mass is 2.250 meters under water (pure), how far under water is it, *really*?

S: The mass is going to stretch the wire, so that the depth will be *greater* than the reading shows, by an amount equal to the stretch experienced by a length of 2.250 meters.

T: Very good summary. What determines the amount of stretch?

S: The stress on the wire. We have its radius and hence A. What *force* is being exerted?

T: The *weight* of the mass *submerged.*

S: We need to know the density of the object. Then the submerged weight is

$$w_w = gV(d_{obj} - d_w).$$

T: The object's density is

$$d = m/V = (100 \text{ kg})/(1.50 \times 10^{-2} \text{ m}^3) = 6.667 \times 10^3 \text{ kg/m}^3.$$

So

$$w_w = (9.8 \text{ m/s}^2)(1.5 \times 10^{-2} \text{ m}^3)(5.667 \times 10^3 \text{ kg/m}^3) = 8.33 \times 10^2 \text{ N}.$$

S: The stress on the wire is then

$$F/A = F/\pi r^2 = (8.33 \times 10{=}2 \text{ N})/(1 \times 10^{-3} \text{ m})^2 = 8.33 \times 10^8 \text{ N/m}^2$$

T: And the strain is just

$$\Delta L/L_o = \text{Stress}/Y = (8.33 \times 10^8\ N/m^2)/(11.0 \times 10^{10}\ N/m^2)$$

$$= 7.57 \times 10^{-3}$$

With L_o = 2.250 m, we have that the new length is

$$L = L_o(1.0076) = 2.267 \text{ meters, the } \textit{real} \text{ depth of the mass.}$$

QUESTIONS ON FUNDAMENTALS

1. What pressure is exerted bya 1.5 m column of gasoline (density = 680 kg/m^3)?

   (a.) 10^4 N/m^2 (b.) 10^5 N/m^2 (c.) 1020 N/m^2
   (d.) 6.66 x 10^3 N/m^2 (e.) 10^3 N/m^2

2. What is the buoyant force exerted on a 1000 cm^3 object submerged in water (density 10^3 kg/m^3)?

   (a.) 10^3 N (b.) 1 N (c.) 9.8 N (d.) 4.9 N
   (e.) not enough information given

3. In one section (A) of a horizontal pipe carrying a fluid, the fluid velocity is twice as much as in another section (B). How does the pressure at A compare with that at B?

   (a.) 2 x (b.) 4 x (c.) 1/2 as much (d.) 1/4 as much (e.) same

4. Assuming the pipe in #3 to have circular cross-section, how does the radius at A compare with that at B?

   (a.) 1/2 as much (b.) $1/\sqrt{2}$ as much (c.) 2 x
   (d.) $\sqrt{2}$ times as much (e.) the same

5. What rate of water flow results from a pressure difference of 0.1 atmospheres between the ends of a 10 meter section of pipe whose radius is 5 cm? The coefficient of viscosity of water is 0.8 cP.

   (a.) 1.23 x 10^3 m^3/s (b.) 3 x 10^{-3} m^3/s (c.) 45 m^3/s
   (d.) 0.67 m^3/s (e.) 3.1 m^3/s

6. By what fraction does an aluminum wire 1 mm in cross-sectional area extend when a 10^3 N weight is hung on it?

   (a.) 1/70 (b.) 1/7 (c.) 1/220 (d.) 1/10 (e.) 1/100

GENERAL PROBLEMS

Problem 10-1. A 4 cm diameter pipe is emptying water into the top of a large open tank. The tank has one hole in it 3.5 meters below its top. When a pressure gradient of .01 atm/meter is maintained along the pipe, the tank remains full to the brim without spilling over. What is the hole's radius? If you have a pump that can create 4 x 10^5 Pa gauge pressure, what is the maximum length of this pipe that you can have between the pump and the tank and maintain the tank full?

Q1: What principle determines the rate of inflow?
Q2: What condition enables the tank to stay full?
Q3: What is the leak rate?
Q4: Do we know v_{out}?
Q5: Is the pump pressure, P_1, the *gauge* pressure?
Q6: What condition determines the maximum pipe length?

Problem 10-2. Devise a way to calculate the specific gravity of an unknown fluid from the results of the following measurements: the weight of an object in air, its weight submerged in water, and its weight submerged in the unknown fluid.

Q1: What are the expressions for these three weighings?
Q2: What is the definition of the specific gravity?
Q3: Are there enough equations? How many unknowns do we have?
Q4: What's the most convenient way to start the algebraic solution?

Problem 10-3. A carelessly made barometer has some air trapped above the mercury column. The glass tube has a total length of 1000 mm *above the reservoir* level of mercury. On a day when an accurate barometer reads 745 torr, this defective one reads 730 torr. What will it read when actual air pressure is 760 torr?

Q1: How does the trapped air affect the reading? What is the pressure equation for the defective barometer?
Q2: Will the gas pressure change for the new reading?
Q3: How much volume of air is trapped at the initial reading?
Q4: At the new reading, what is the expression for the new volume?
Q5: What will determine the new pressure of the trapped air?
Q6: What is the barometer equation when the true pressure is 760 torr?

Problem 10-4. To demonstrate that a metal boat can float, assume a steel cube of side length L and wall thickness t. Steel's specific gravity is 7.8. Find out in terms of L what the *maximum* wall thickness can be for the cube to float.

Q1: What is the general requirement for floating?
Q2: Under what condition will F be maximum? What will it be then?
Q3: What is the weight of the box?
Q4: What is the mathematical condition that will then determine t_{max}?

Problem 10-5. Water flows through the horizontal pipe shown in the following figure. Open vertical pipes are fitted into the pipe, and the height of water in them measures the gauge pressure of the fluid. In Section 1, the water height is 2 meters, and the flow speed is 2 m/s. If the area of Section 1 is 400 cm^2, how small a radius can a downstream section of the pipe have and still not suck air in through its tube? (This is an effect known as *cavitation*, and can cause serious damage in many flow situations.) A barometer shows the air pressure to be 780 torr.

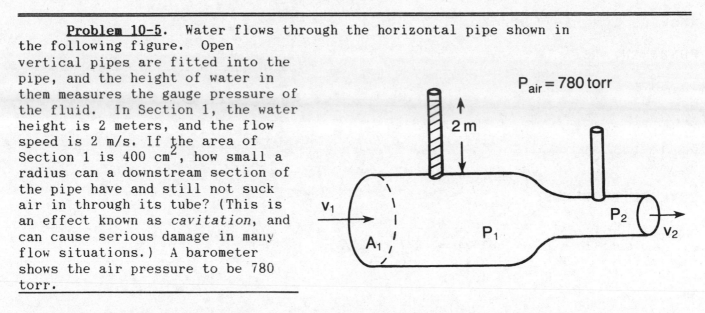

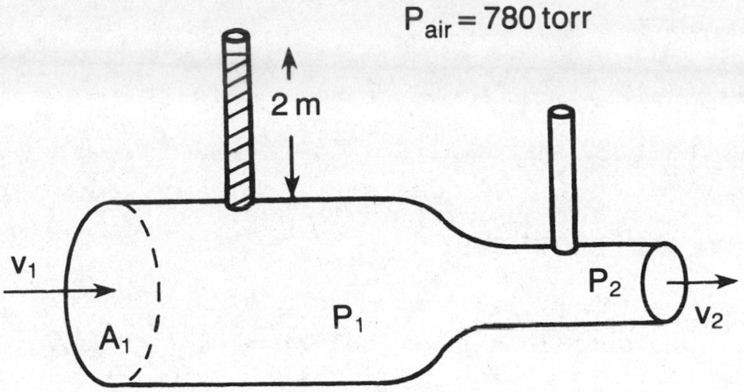

Q1: What specific condition will cause cavitation in section 2?
Q2: What determines P_2?
Q3: Do we know P_1?
Q4: Then v_2 can be found by setting $P_2 = P_{air} = 780$ torr. How do we find the corresponding pipe radius?

ANSWERS TO QUESTIONS

FUNDAMENTALS:

1. (a.)
2. (c.)
3. (d.)
4. (b.)
5. (e.)
6. (a.)

GENERAL PROBLEMS:

Problem 10-1.

A1: Poiseuille's Law: $(Vol)/t = \pi R^4(\Delta P/L)/\eta$,

with $P/L = .01 \times 1.01 \times 10^3$ Pa/m

$R = 2 \times 10^{-2}$ m

and

$\eta = 10^{-3}$ N-s/m^2.

A2: That the leak rate from the hole equal this input flow rate.

A3: $(Vol/t)_{out} = v_{out}A_{hole} = v_{out}\pi r_{hole}^2$

A4: From Torricelli's Theorem, $v_{out} = (2gh)^{\frac{1}{2}}$, where h = 3.5 meters.

A5: Yes, since the pipe opens to the atmosphere.

A6: $(P_{gauge})_{max}/L_{max} = .01$ atm/meter, or

$L_{max} = (4 \times 10^5 \text{ Pa})/(.01 \times 1.01 \times 10^5 \text{ Pa/m})$

Problem 10-2.

A1: in air: $w = dgV$

in water: $w_w = (d - d_w)gV$

in fluid: $w_{fl} = (d - d_{fl})gV$

A2: $(sp.\ grav)_{fl} = d_{fl}/d_w$

A3: We don't know d, d_{fl} and V. There are three unknowns and three pieces of data.

A4: Calculate the weight *differences*, $(w - w_w)$ and $(w - w_{fl})$.

Problem 10-3.

A1: The trapped gas *adds* its pressure to the weight of the mercury column. The barometer equation is $P = P_{true} - P_{gas}$. At the initial reading, $P_{gas} = 15$ torr.

A2: Yes, because as the mercury column rises the volume of the trapped gas changes.

A3: The total glass tube length is 1000 mm above the reservoir reference. Since the mercury column height is 730 mm, the air volume is (270 mm)A, where A is the tube's bore area.

A4: If we call the new mercury height h', the new volume is (1000 mm - h')A.

A5: The temperature is constant, so PV is constant. Thus $P'_{gas}(1000 - h') = (15 \text{ torr})(270)$. Notice that A cancels.

A6:

$h' = 760 \text{ torr} = P'_{gas} = 760 \text{ torr} - (15 \text{ torr})270/(1000 - h')$.

This is a *quadratic* equation for h'.

Problem 10-4.

A1: That $F_B \geq$ weight of the cube.

A2: When all the volume is just submerged, displacing the maximum possible amount of water. Then $F_B = d_w g L^3$.

A3: $w = d_{steel} g(6L^2t)$, since the 6 sides each have volume L^2t. Also, $d_{steel} = 7.8\ d_w$.

A4: That $7.8(6L^2t) \leq L^3$.

Problem 10-5.

A1: If the gauge pressure in section 2 becomes *negative*, i.e., $P_2 < P_{air}$.

A2: Bernoulli's Equation is the relation between P_1, v_1 and v_2. $P_1 + \frac{1}{2}d_w v_1^2 + P_2 + \frac{1}{2}d v_2^2$. Remember to use *absolute SI* units for pressure.

A3: $(P_1)\text{gauge} = d_w g(2\ \text{m}) = P_1 - P_{air}$.

A4: From the equation of continuity, $v_1A_1 = v_2\pi r_2^2$.

CHAPTER 11. PERIODIC MOTION AND WAVES

BRIEF SUMMARY AND RESUME

The concept of *periodic motion* is introduced. As an example, *simple harmonic motion (SHM)* and the physical conditions of *elastic forces* which produce it are examined. The motions of a *mass on a spring* and *simple pendulums* are discussed and expressions for their *SHM frequencies* are developed. *Forced* and *damped* SHM are mentioned, as well as the concept of *resonance* in an oscillating system.

The concept of *transverse* and *longitudinal waves* is introduced and mathematically described. *Superposition* of waves and *interference* resulting from *phase differences* are discussed. The conditions for *standing wave resonance* for waves on strings are examined.

Material from previous chapters directly relevant to Chapter 11 includes the following:

1. Forces which produce elastic strains.
2. Frequency, period, and angular frequency of circular motion.

NEW CONCEPTS AND PRINCIPLES

Periodic Motion

Understanding this section should enable you to do the following:

1. *Describe a given periodic motion in terms of its frequency, period, and amplitude.*
2. *State the defining characteristic of simple harmonic motion and describe the characteristics of the force which causes a system to perform SHM.*
3. *Calculate the period of the SHM of a mass on a spring and of a simple pendulum.*

An extremely important type of motion in physics is *periodic* motion. We use the following vocabulary for its description.

Periodic Motion--Motion which repeats itself regularly at a fixed time interval.

Period (T)--The fixed time interval in which a full cycle of periodic motion takes place.

Frequency (f)--The number of cycles of periodic motion per second. f is the *reciprocal* of T.

Displacement--The distance of any part of a system from its equilibrium position at any instant of time. This can be either a linear or angular displacement.

Amplitude--The *maximum* value of the displacement.

Obviously, a mathematical description of periodic motion must use functions which are periodic in position and time. The *simplest* such functions are the sine and cosine, which repeat themselves every 2π radians.

Simple Harmonic Motion (SHM)--Periodic motion which can be described by a *single* sine or cosine function.

Mathematical Description of SHM

a. Position. Here x designates any general position coordinate of a mass m undergoing SHM. The function

$$x(t) = x_0 \cos 2\pi(t/T) = x_0 \cos 2\pi ft = x_0 \cos \omega t$$

is *periodic*, i.e., it repeats its values of x every time interval T. The sine function has this same property.

b. Velocity. The velocity of the mass at any time t is the *slope* of the x(t) curve, which turns out to be

$$v(t) = -(\omega x_0)\sin \omega t = -v_0 \sin \omega t$$

c. Acceleration. In turn, the acceleration of the mass is the *slope* of the v(t) curve, which gives

$$a(t) = -(\omega^2 x_0)\cos\omega t = -a_0 \cos \omega t = -\omega^2 x(t)$$

Important Observations

1. In Chapter 4 the meaning of f and ω in describing circular motion was introduced. Your text derives the connection between circular and linear periodic motion in the present chapter. Remember that f is expressed in hertz and ω in radians/second.
2. SHM is only *one special kind* of periodic motion.
3. As the sine or cosine has maximum and minimum values of ±1, the displacement x(t) goes between extreme values of $\pm x_0$, the *amplitude* of the motion. Similarly, the extreme values of velocity and acceleration are v_0 and a_0 respectively. Their relation to x_0 can be seen from the above equations.
4. The velocity is 1/4 cycle out of synchronization with displacement. This amounts to a *phase difference* of 90°, or $\pi/2$ radians. The acceleration is 180°, or π radians out of phase with displacement.

Hooke's Law and SHM--Any system for which the *restoring force* created by a displacement from equilibrium is *proportional to the displacement* and *opposite in direction* obeys *Hooke's Law*. Such a system will perform SHM.

Physical Determination of SHM Frequency--By combining Hooke's Law ($F = -kx$), Newton's 2nd Law ($F = ma$), and the SHM description ($a = -\ {}^2x$), the angular frequency is clearly given by $\omega = \sqrt{(k/m)}$. Thus

$$f = \omega/2\pi = (1/2\pi)(k/m)^{\frac{1}{2}} \quad \text{and} \quad T = 2\pi(m/k)^{\frac{1}{2}}\ .$$

Important Observations

1. m, f, and T are *not* dependent on *amplitude*, but only on the "stiffness", k, of the restoring force and the mass responding to that force.

2. If you determine the specific expression for the restoring force in a given problem, you can immediately identify the SHM frequency of the system.

Example 11-1. Suppose a wooden cube whose mass is 8 kg and whose density is 700 kg/m^3 is floating in water. Show that if the cube is pushed down slightly below its equilibrium floating depth and then released, it will undergo SHM, neglecting viscosity effects. Also determine the frequency and period of the SHM.

S: What do I have to find to demonstrate SHM?

T: It is sufficient to show that when pushed or pulled below or above equilibrium by a small distance h, that the net restoring force is proportional to h and directed toward the equilibrium position.

S: What is the equilibrium condition?

T: That the cube's *weight*, mg, is equal to the *buoyant force*, $d_w g V_{sub}$. Here V_{sub} is the volume which is *submerged*. If A is the area of a cube face and h_o is the depth to which it is submerged, then $V_{sub} = Ah_o$. So equilibrium is where

$$mg = d_w g A h_o.$$

S: What will happen when the cube is pushed further down by an amount h?

T: The *volume* of water displaced will then be increased to $A(h_o + h)$, so F_B will increase to $d_w A(h_o + h)g$. If you *lift* the block by an amount h, F_B will decrease to $d_w A(h_o - h)g$.

S: I see. Then the *net* restoring force either way will be

$$F_{net} = F_B - w = (d_w g A)h,$$

opposite to the direction of the displacement. In vector form, I could write

$$\mathbf{F}_{net} = -(d_w g A)\mathbf{h}.$$

This is an example of Hooke's Law, so SHM will result when the system is released. How do I find the frequency?

T: The general form of the frequency is $f = (1/2\pi)/(k/m)^{\frac{1}{2}}$, where k is the constant of proportionality between the restoring force and the displacement. You can find k by inspecting the form of F_{net} you derived above.

S: It looks like $k = d_w g A$, so the frequency should be

$$f = (1/2\pi)\ (d_w g A/m)^{\frac{1}{2}}.$$

We are given m. What is A?

T: If the cube sides are length L, given by $m = dL^3$, then $A = L^2$. Putting all the numbers in,

$$L = \sqrt[3]{m/d} = \sqrt[3]{\frac{8\,\text{kg}}{700\,\text{kg/m}^3}} = 0.225\,\text{m}.$$

So

$$f = (1/2\pi)\sqrt{\frac{(10^3\,\text{kg/m}^3)(9.8\,\text{m/s}^2)(.225\,\text{m})^2}{8\,\text{kg}}} = 1.25\,\text{Hz}$$

The period is $T = 1/f = 0.8$ seconds.

Example 11-2. A perfectly elastic glass marble is dropped onto a marble slab from a height of 10 cm.

(a.) Does the marble undergo periodic motion?
(b.) Where is the acceleration of the marble the greatest?
(c.) Does the marble undergo SHM?
(d.) What is the period of the marble's motion?

S: As long as the marble bounces *elastically*, it will repeat the *same* bounce in the same period of time. Therefore the motion *is* periodic.

T: Right. Notice that if the bounces were to get *smaller*, they would take *less* time, which would be a situation where the period *would* depend on the *amplitude* of the motion.

S: The acceleration is greatest at the top of the bounce.

T: I think you know better. The marble, as in all our previous trajectory problems, always has a *constant* acceleration of g downward. Only when it collides with the slab does another force occur, and this is probably its greatest acceleration, but we don't know the time duration of the collision.

S: That's right, because the force of gravity is constant throughout its motion. Doesn't that mean that the motion is *not* SHM, since the restoring force on the marble is not proportional to height?

T: Exactly. This is an example of *periodic* motion, but the position, velocity, and acceleration are *not* describable by single sine or cosine functions.

S: Then we cannot obtain the period by direct inspection of the force acting. How do we find it?

T: You have to work out the time of fall. By symmetry, the rise time will be the same. The period will then be *twice* the time of fall.

$$T = 2t_{fall} = 2(2s/g)^{\frac{1}{2}} = 2\{2(.10\text{ m})/(9.8\text{ m/s}^2)\}^{\frac{1}{2}} = .286\text{ s.}$$

Simple Pendulum--A point mass m at the end of a massless string of length L will *approximately* undergo SHM. The angle the string makes with the vertical follows a sine or cosine variation in time *when the amplitude is small* (generally less than 15° for a 1% departure from pure SHM). The period, derived in your text, is given by

$$T = 2\pi(L/g)^{\frac{1}{2}} .$$

Notice that T is *independent* of mass!

Example 11-3. At sea level at the north pole, a .992 m long simple pendulum is timed for 1000 seconds and makes 1996 complete swings. At sea level at the equator, it makes 2001 swings in 1000 seconds.

(a.) Calculate the values of "local" g at the pole and at the equator.
(b.) Can the centripetal effect (Chapter 4) at the equator account for all the difference in g thus observed?

S: The two periods are

$$T_{pole} = 1.996 \text{ sec} \quad \text{and} \quad T_{eq} = 2.001 \text{ sec}.$$

Since $g = 4\pi^2 L/T^2$, we find

$$g_{pole} = 9.832 \text{ m/s}^2 \quad \text{and} \quad g_{eq} = 9.7804 \text{ m/s}^2 .$$

How do we calculate the centripetal effect?

T: The pendulum will be in circular motion at the equator, and hence its apparent weight will be reduced by an amount $m\omega^2 R$, where R is the earth's radius. This is equivalent to saying that "g" is reduced by an amount $\omega^2 R$:

$$g_{apparent} = g - \omega^2 R$$

We know that $\omega = 2\pi$ rad/24 hr, and $R = 6.4 \times 10^6$ m. Let's see if we get g_{eq} when we subtract $\omega^2 R$ from g_{pole}.

S: O.K. We have

$$\omega^2 R = 4\pi^2(6.4 \times 10^6 \text{ m})/(24 \text{ hr} \times 3600 \text{ s/hr})^2 = 3.4 \times 10^{-2} \text{ m/s}^2.$$

Then

$$g_{pole} - \omega^2 R = (9.832 - 0.34) \text{ m/s}^2 = 9.798 \text{ m/s}^2.$$

This is still .0176 m/s^2 *larger* than g_{eq} from the data. What's the explanation?

T: The spin of the earth also causes the *equatorial radius* to be slightly larger than the *polar radius*, so a mass at the equator is slightly farther from the earth's center, and hence experiences weaker gravity.

Waves

Understanding this section should enable you to do the following:

1. *State the relation between wavelength, wave speed, and frequency.*
2. *Identify frequency and wavelength of a wave from sinusoidal wave form.*
3. *Identify the phase of a given wave at any point and time.*
4. *Calculate the speed of waves on a string under given tension.*

A body undergoing SHM can cause a sinusoidal variation in displacement to be propopagated through a medium to which it is coupled. This is called *wave propagation*. The properties of the medium will determine the *speed* of the wave. Along with the SHM frequency of the source, this determines the *wavelength* of the wave.

<u>Wavelength (λ)</u>--The distance between any two consecutive equivalent points on a wave pattern.

$\lambda = v/f$, where v = speed of wave propagation.

<u>Transverse Waves</u>--Waves in which the vibrations in the medium are *perpendicular* to the direction of energy transport. Mathematically, such a wave traveling *toward increasing x* has the form

$$y(x,t) = y_0 \cos 2\pi(x/\lambda - t/T),$$

where y = the transverse displacement of the medium
y_0 = the maximum displacement, or *amplitude* of the wave
= wavelength
T = period of the wave.

<u>Important Observations</u>

1. At any point x, the medium's displacement has SHM, as shown for example at x = 0:

$$y(0,t) = y_0 \cos 2\pi(t/T).$$

2. At any instant of time, the medium has a displacement profile which is periodic in x, as shown for example at t = 0:

$$y(x,0) = y_0 \cos 2\pi(x/\lambda).$$

3. To see that this describes a wave traveling toward +x, we can examine how fast we would have to move to "ride with the crest," i.e., to keep up with the maximum displacement, where $2\pi(x/\lambda - t/T) = 0$. This would require $x/t = \lambda/T$. But x/t is our required speed, and $\lambda/T = \lambda f$ is the wave speed. These are both positive, and so they represent motion toward +x. It follows that a wave traveling toward *-x* would be described by

$$y(x,t) = y_0 \cos 2\pi(x/\lambda + t/T)$$

<u>Phase</u>--The angle represented by $2\pi(x/\lambda \pm t/T)$ at any given values of x and t is called the *phase* of the wave at that point and at that instant. It represents the *part of the cycle* the wave is in. As a wave goes through a complete cycle, its phase goes through 2π radians.

<u>Speed of Waves on a String</u>--The speed of transverse waves propagating along a string having *linear mass density* (mass per length) p and under longitudinal tension F is given by

$$v = (F/\mu)^{\frac{1}{2}} .$$

Important Observations

1. Notice that the units above are correct. When F is in N and μ is in kg/m, v is in m/s.
2. A large F means a taut string, with a large restoring force acting on transverse displacement. This will cause the wave to propagate rapidly. A large μ means the string has a large inertia per length and will respond sluggishly to a restoring force, slowing the wave propagation. Thus the formula qualitatively agrees with "intuitive" expectations, although the actual square root dependence requires a complete derivation.

Example 11-4. A wave on a string is of the form

$$y = .12 \sin(1.26x - 440t) ,$$

with all quantities in SI units.

(a.) What points on the string have zero displacement? What points have maximum displacement? Find the general expressions as a function of time.

(b.) Where are the first two zero displacement positions on the +x axis at t = .02 seconds?

(c.) What are the wavelength and frequency of this wave?

(d.) How fast and in what direction will these zeros of displacement move along the string?

(e.) If there is a tension of 200 N in this string, how much total mass would the string have if it was 8 meters long?

S: For zero displacement the sine has to be zero. What is the general condition for this?

T: The sine has zeros wherever its *phase is zero or any integral multiple of* π *radians*. So

$$y = 0 \quad \text{where } 1.26\ x - 4405 = n\pi,$$

where n = 0, ±1, ±2, ... Similarly, the maxima ($\pm\ y_o$) occur where the sine is ±1, or

$$y = \pm y_o \quad \text{where } 1.26x = 440t = n(\pi/2),$$

with n = ±1, ±3, ±5, ...

S: Now where are the first two zeros for +x at t = .02 sec.?

T: We would have

$$y = 0 \quad \text{at} \quad x = \{440(.02)-n\pi\}/1.26$$

You can see that the first two zeros are for n = 2 and n = 1.
For n = 2,

$$x = (8.8 - 2\pi)/1.26 \;=\; 2.00 \text{ meters.}$$

For n = 1,

$$x = (8.8 - \pi)/1.26 \quad = \quad 4.49 \text{ meters.}$$

S: How can I recognize the wavelength and frequency from the wave form?

T: The "standard" wave form shows that the number multiplying x represents $2\pi/\lambda$, and that multiplying t is $2\pi/T = 2\pi f$.

S: So then $2\pi/\lambda = 1.26$ and $2\pi f = 440$. This gives

$$\lambda = 4.99 \text{ m} \quad \text{and } f = 70 \text{ Hz.}$$

T: You could also infer wavelength from the spacing of the zeros. Adjacent zeros on a sine or cosine curve are $\lambda/2$ apart. Our previous result showed zeros at 2.00 m and 4.49 m when t = .02 s. This would give $\lambda = 4.98$ m.

S: Now that we have λ and f, the wave speed is $v = f\lambda$, or

$$v = (70 \text{ Hz})(4.99 \text{ m}) = 349 \text{ m/s.}$$

Since the wave form involves a *minus* sign in (x - t), it represents a wave moving *toward +x.*

T: That's right. Also notice that at <u>t = 0</u>, the n = 1 zero displacement is at

$$x = (0 - \pi)/1.26 \quad = \quad -2.49 \text{ m.}$$

At t = .02 s, we found this zero at x = +4.49 m. This means the zero moved 6.98 m *to the right* in .02 sec, giving a speed for the wave of

$$v = 6.98 \text{ m}/.02 \text{ s} = 349 \text{ m/s.}$$

S: Since we know v and F, we can find the linear mass density from

$$v = \sqrt{(F/\mu)} \text{ , or } \mu = F/v^2.$$

So

$$\mu = (200 \text{ N})/(349 \text{ m/s}) \quad = \quad 1.64 \times 10^{-3} \text{ kg/m.}$$

T: Finally, since the string is 8 meters long, its total mass is

$$M = \mu L = (1.64 \times 10^{-3} \text{ kg/m})(8 \text{ m}) = 13.1 \text{ grams.}$$

Superposition and Interference

Understanding this section should enable you to do the following:

1. *Identify the amplitude and the positions of nodes and antinodes in a standing wave pattern.*
2. *State the resonance condition for standing waves on a clamped string, and identify the harmonic of a given wave pattern.*
3. *Calculate the tension required to produce a given harmonic pattern on a string of given length and mass density.*

Superposition Principle--If two or more waves exist simultaneously in the same space, the total wave form they produce is described by the *linear addition* or *superposition* of their individual wave forms.

Interference--When two waves which differ in phase by any *odd number of half cycles* (π, 3π, 5π, etc.) superpose, their *amplitudes will subtract*, and we have *destructive interference*. If the waves are in phase, or have a phase difference 2π. 4π, 6π, etc., their amplitudes will *add*, and we have *constructive interference*. If the amplitudes are *equal*, destructive interference results in *total cancellation* of the wave, and constructive interference results in *doubling* the individual amplitudes.

Special Case: Standing Waves on a String--The superposition (derived in your text) of two waves of *equal* amplitude, wavelength, and frequency, traveling in opposite directions along a string gives the resultant waveform

$$y(x,t) = [2y_o \cos(2\pi x/\lambda + \Phi)]\cos 2\pi t/T$$

Important Observations

1. The arbitrary phase constant Φ is the wave's phase at $x = 0$ in the general case. It depends on the choice of origin relative to the wave, and the condition of the string at $x = 0$.

   a. String clamped at $x = 0$. Then $y(0,t)$ must be zero, so $y(x)$ must have the form $y_o \sin 2\pi x/\lambda$, which means $\Phi = -\pi/2$, since $\cos(\theta - \pi/2) = \sin\theta$.

   b. String with maximum amplitude at $x = 0$. Here $y(x)$ must have the form of $y_o \cos 2\pi x/\lambda$, so $\Phi = 0$.

2. This wave form describes *standing* waves, since each position x oscillates in SHM with an amplitude depending sinusoidally on x. The pattern of this varying amplitude "stands" *fixed* in position on the string, as in Figure 11.27 (text).

Node--A position in a standing wave pattern with *zero amplitude*. This occurs at values of x given by $\cos(2\pi x/\lambda + \Phi) = 0$, or

$$2\pi x/\lambda + \Phi = +\pi, +3\pi, +5\pi, \text{etc.}$$

Antinode--A position in a standing wave pattern with *maximum amplitude*. This occurs at values of x given by $\cos(2\pi x/\lambda + \Phi) = \pm 1$, or

$$2\pi x/\lambda + \Phi = 0, +2\pi, +4\pi, +6\pi, \text{etc.}$$

Important Observation

1. The distance between successive nodes or successive antinodes is 1/2 wavelength, $\lambda/2$.

<u>Example 11-5</u>. A meter stick is held next to a standing wave pattern on a string. Nodes are observed at x = 10 cm, 42 cm, and 74 cm. Find the phase constant Φ and the amplitude of the wave at x = 0 if the antinode amplitude is 6 mm.

<u>S</u>: The distance between nodes is 32 cm, so x = 2(32 cm) = 64 cm. But I still don't quite understand what Φ means.

<u>T</u>: It is the phase of the wave, the part of its cycle, at x = 0. Here we have the first node, where y = 0, at x = .10 m. So

$$2y_0 \cos[2\pi(.10)/(.64) + \Phi] = 0, \text{ giving}$$

$$2\pi(.1/.64) + \Phi = \pi/2, \qquad \text{or } \Phi = \pi/2 - .3125 = .1875\pi \text{ radians}$$

Now you should be able to find y at x = 0.

<u>S</u>: The wave form at x = 0 looks like this:

$$y(0,t) = 2y_0 \cos(0 + \Phi)\cos 2\pi t/T \text{ , with } 2y_0 = 6 \text{ mm.}$$

Now $\cos \Phi = \cos(.1875\pi) = .831$, so

$$y(0,t) = (6 \text{ mm})(.831) \cos 2\pi t/T = (4.99 \text{ mm}) \cos 2\pi t/T.$$

What does the time dependence mean?

<u>T</u>: The amplitude at x = 0 is 4.99 mm. This point on the string is doing SHM between ±4.99 mm with a period T.

<u>Resonant Waves on a String</u>--Resonance occurs when a standing wave pattern matches the length of the string. If both ends of the string are *clamped*, there must be *nodes* at both ends, so an *integral number of half wavelengths* must equal L. Mathematically,

$$n(\lambda_n/2) = L, \text{ where } n = 1, 2, 3, \text{ etc.}$$

<u>Resonant Frequencies</u>--Since the wavelengths are physically determined by the wave speed on the string and the frequency of the source of oscillation, the conditions for resonance will be satisfied for only specific *resonant frequencies* where a standing wave pattern can exist. Mathematically,

$$f_n = v/\lambda_n = (F/\mu)^{\frac{1}{2}} = (n/2L)(F/\mu)^{\frac{1}{2}},$$

where n = 1, 2, 3, etc.

<u>Important Observations</u>

1. As shown in Figure 11.29 (text), you can immediately determine n by counting the number of loops in the standing wave pattern.
2. The various resonant frequencies are called the *harmonics* of the string. Notice that successive harmonics have simple whole number ratios, i.e.,

$$f_2/f_1 = 2/1; \quad f_3/f_2 = 3/2; \quad \ldots \quad f_{n+1}/f_n = (n + 1)/n$$

<u>Example 11-6</u>. Suppose you have a wire 3 meters long with linear mass density 10 grams/m. It is clamped at both ends and placed under a tension F. Two successive resonances are observed at 45.67 Hz and 54.8 Hz. What is the *lowest* resonant frequency that this tension will produce on this wire, and what is the value of F?

<u>S</u>: What is the significance of knowing two resonances?

<u>T</u>: Two *successive* resonances. The ratio of these ought to be the ratio of two identifiable whole numbers, which in turn identify the harmonic numbers of the resonances.

<u>S</u>: Well their ratio is 54.8/45.67 = 1.20. We can set this equal to $f_{n+1}/f_n = (n + 1)/n$. Thus

$$(n + 1)/n = 1.20, \quad \text{so } (n + 1) = 1.2n.$$

This gives n = 5 and n + 1 = 6. These are the 5th and 6th harmonics!

<u>T</u>: Now it ought to be easy to find the fundamental, f_1.

<u>S</u>: Can I put $f_1/f_5 = 1/5$?

<u>T</u>: Sure. The frequencies are generally related by $f_m/f_n = m/n$. Simple, huh?

<u>S</u>: Yeah. So $f_1 = f_5/5 = 45.67/5 = 9.13$ Hz. For the value of F, do we have determine wave speed?

<u>T</u>: Not once we have derived the resonant frequency conditions. We have $f_n = (n/2L)(F/\mu)^{\frac{1}{2}}$, which we can use with any of the given resonant frequencies. For example,

$$F = (2L/n)^2 f_n^{\,2} \mu = 4(3\text{ m})^2\ (9.13\text{ Hz})^2 (10^{-2}\text{ kg/m}) = 30\text{ N}.$$

Can you find the wavelengths of these resonances?

<u>S</u>: I could find $v = (F/\mu)^{\frac{1}{2}}$ and then $\lambda_n = v/f_n$.

<u>T</u>: Yes, but that's too much work. Notice that f_1 has 1 loop in 3 meters. Each loop is 1/2 wavelength, so $\lambda_1 = 6$ m. Similarly, f_5 has 5 loops., each worth 1/2 $_5$, so $\lambda_5 = 1.2$ m, and so on.

QUESTIONS ON FUNDAMENTALS

1. What is the period of a system which undergoes periodic motion with a frequency of 100 Hz?

(a.) 0.1 sec (b.) 0.01 sec (c.) 100 sec (d.) 1 sec
(e.) 10 sec

2. What is the frequency of oscillation of a 10 kg mass on a spring whose force constant is 300 N/m?

(a.) 5.5 Hz (b.) .18 Hz (c.) .03 Hz (d.) .87 Hz (e.) 34.4 Hz

3. How long will it take for a simple pendulum of length 150 cm to make one oscillation of small amplitude? Assume standard gravity.

(a.) 2.46 s (b.) .41 s (c.) 4.92 s (d.) 50.3 s (e.) 8.0 s

4. If a transverse wave has the form:
$y(x,t) = y_0 \cos 2\pi(x/.025 - t/.01)$, what is the speed of the wave?

(a.) 0.4 m/s (b.) depends on the amplitude (c.) 2.5 cm/s
(d.) 2.5 m/s (e.) 4 m/s

5. A tension of 50 N is applied to a string, and the speed of transverse waves on the string is then observed to be 70.7 m/s. What would be the mass of a 1 meter length of this string?

(a.) 10 kg/m (b.) .707 kg/m (c.) 0.5 kg/m (d.) 0.1 kg/m
(e.) .01 kg/m

6. If the length of the string in #5 is 1 meter, what is the frequency of the fundamental mode of standing waves?

(a.) 35.4 Hz (b.) 70.7 Hz (c.) 17.7 Hz (d.) 60 Hz (e.) 5 Hz

7. If the 1 meter string in #6 is vibrating with 5 loops along its length, what is the wavelength of the standing wave?

(a.) 0.2 m (b.) 0.1 m (c.) 0.4 m (d.) 0.8 m (e.) 1 m

GENERAL PROBLEMS

Problem 11-1. When the law of gravitation is applied to a point mass m *inside* the surface of a uniform spherical mass of density d, it is found that, at a distance r from the sphere's center, only the mass M(r) *inside* r contributes to the gravitational force on m. This mass is $M(r) = (4/3)\pi r^3 d$ (see the figure to the right). Show that if you had a tunnel straight through the center of the earth, m would undergo SHM in the tunnel, and find the period of its motion.

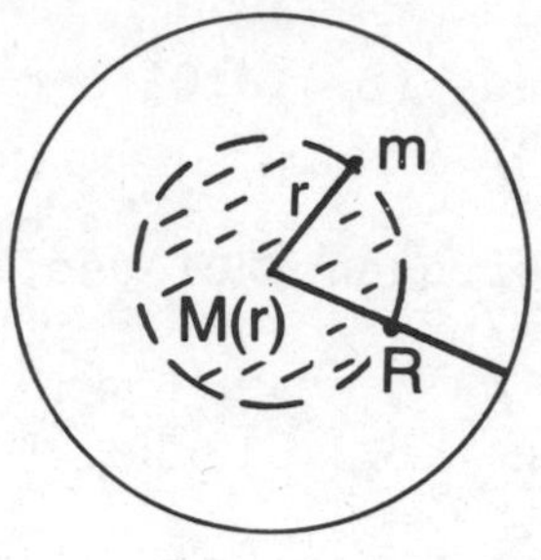

Q1: What is the gravitational force on m placed a distance r from Earth's center?
Q2: Where is m in equilibrium?
Q3: Is this an SHM force?
Q4: What is the force constant, and the expression for the period?
Q5: What is d?

Problem 11-2. A *physical pendulum* consists of any rigid object of mass m free to swing about an axis O. If its center of mass is a distance L from O and its moment of inertia about O is I_o then its period is given by

$$T = 2\pi(I_o/mgL)^{\frac{1}{2}} \quad .$$

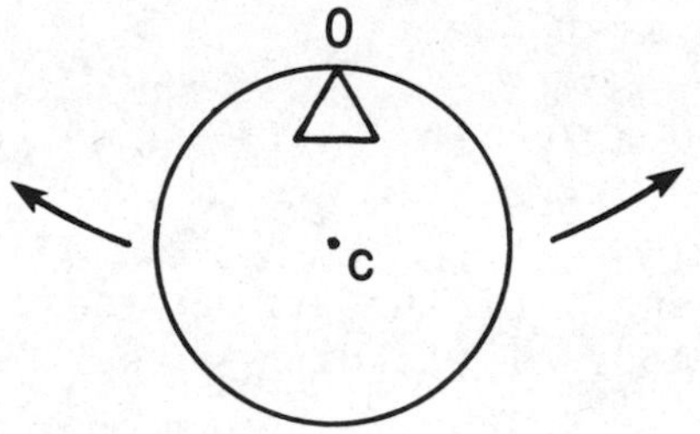

(a.) Show that this becomes $2\pi(L/g)^{\frac{1}{2}}$ for a simple pendulum.

(b.) Find the period of a uniform stick 2 meters long pivoting about its end.

(c.) Find the period of a solid disc of radius 2 meters pivoting about a point on its rim. (See the figure below.)

(d.) For (b.) and (c.), find the length of an equivalent simple pendulum.

Q1: For a point mass on the end of a string of length L, what is I_o?
Q2: What is the moment of inertia about its end?
Q3: Is the length of the rod the same as L in the expression for T?
Q4: What is the moment of inertia of a disc about a point on its perimeter?
Q5: What does "equivalent" simple pendulum mean?

Problem 11-3. An aluminum wire 10 m long has a cross section of 0.5 mm^2. Young's modulus for aluminum is 7.0×10^{10} N/m^2. If a 5 kg mass is hung on this wire, pulled down a short distance below equilibrium, and then released, with what frequency would it perform SHM?

Q1: What determines f?
Q2: What determines k?
Q3: How does k relatc to Young's modulus?
Q4: What is the expression for f in terms of Young's modulus?

Problem 11-4. A point on a vibrating string is moving vertically in SHM with a frequency of 256 Hz. The amplitude of motion is 1.2 mm.

(a.) What is the maximum velocity of the point?
(b.) What is the maximum acceleration of the point?

Q1: How is maximum velocity related to frequency and amplitude of SHM?
Q2: How is maximum acceleration related to frequency and amplitude?

ANSWERS TO QUESTIONS

FUNDAMENTALS

1. (b.)
2. (d.)
3. (a.)
4. (d.)
5. (e.)
6. (a.)
7. (c.)

GENERAL PROBLEMS:

Problem 11-1.

A1: From the statement of the problem, M(r) is the mass which enters the gravitational force expression. The effect of mass at a greater distance from the center than r just cancels out, as can be shown with calculus. So

$$F_g = GmM(r)/r = G(4\pi/3)mdr, \text{ radially } \textit{inward}.$$

A2: Where F_g is zero, i.e., at r = 0, the earth's center.

A3: Yes. It is proportional to displacement from equilibrium, and directed back toward equilibrium.

A4: $k = G(4\pi/3)md$. The SHM period is $T = 2\pi(3/4\pi Gd)^{1/2}$.

A5: The mass of the earth is 6×10^{24} kg, and its radius is 6.4×10^6 m. d = M/Vol.

Problem 11-2.

A1: mL^2.

A2: $I = (1/3)m(\text{length})^2$.

A3: L is the distance between O and the c of m of the object.

A4: Use the parallel axis theorem, $I_o = I_{cm} + mL^2$.

A5: One having the *same period* as in (b.) and (c.).

Problem 11-3.

A1: $f = (1/2\pi)\sqrt{(k/m)}$.

A2: k is the ratio of force applied to the resulting amount of stretch, i.e., $k = F/\Delta L$.

A3: $Y = (F/A)/(\Delta L/L) = k(L/A)$

A4: $f = (1/2\pi)(YA/mL)^{1/2}$

Problem 11-4.

A1: $|v_{max}| = \omega A = 2\pi fA$

A2: $|a_{max}| = \omega^2 A = (2\pi)^2 f^2 A$.

CHAPTER 12. SOUND WAVES

BRIEF SUMMARY AND RESUME

This chapter begins with the description of *longitudinal compression waves* which travel through all media and which we call *sound*. The *speed of sound* and factors which determine it in various media are investigated. *Resonance* of sound waves in pipes is discussed and the mathematical conditions for *standing sound waves* in open and closed organ pipes are derived. The formation of *beat frequencies* from interference between waves of different frequencies is described. Concepts of *intensity* and *loudness* of sound are introduced and defined in measurable units. The changes in observed frequency produced by the motion of a sound source or an observer, called the *Doppler effect*, are derived. The description of an object moving through air at *supersonic speed* is introduced.

Material from previous chapters relevant to Chapter 12 includes:

1. The description of wave phenomena, including traveling wave forms, interference, standing waves, wave speed, frequency, and wavelength.
2. Elastic moduli of materials.
3. Units of power.

NEW DATA AND PHYSICAL CONSTANTS

1. SI Units of wave intensity (I).

$$I = \text{Power/Area} = \text{watts/m}^2$$

2. Decibel scale of intensity.

$$\beta(\text{dB}) = 10 \log_{10} I/I_o ,$$

where I_o is some reference intensity. β is a *dimensionless* quantity.

NEW CONCEPTS AND PRINCIPLES

Understanding this section should enable you to do the following:

1. *Identify the amplitude, wavelength, frequency, and speed of a traveling sound wave from its mathematical form.*
2. *Calculate the speed of sound from the elastic properties and density of the medium carrying the wave.*

Longitudinal Wave--A wave in which the vibrations of the medium are *parallel* to the direction of propagation of the wave.

Sound Wave--A longitudinal wave consisting of periodic variations in the displacement of molecules of the medium in which it is traveling. Mathematically, the wave form traveling along the x axis is

$$X(x,t) = X_{max} \cos[2\pi(x/\lambda \pm t/T)].$$

Important Observations

1. The negative sign in the cosine function again describes a wave traveling in the +x direction, and the positive sign describes wave motion toward -x.
2. Note that X describes a small displacement of a molecule *along* the x axis from its equilibrium position. x is the position coordinate along the entire length of the wave.
3. A *transverse* wave, as in Chapter 11, is easier to visualize than a longitudinal one, because a graph of transverse displacement, y(x), at a given time shows an *actual picture* of the wave. A graph of a sound wave is more abstract, representing displacements in the *x* direction along the y axis of the graph.

Speed of Sound Waves--The speed of sound in gases and liquids depends on the square root of the ratio of bulk modulus to the equilibrium density of the material. For solids, Young's modulus replaces the bulk modulus. Mathematically,

$$v = (B/d_o)^{\frac{1}{2}} \quad \text{(gases and fluids)}; \quad v = (Y/d_o)^{\frac{1}{2}} \quad \text{(solids)}$$

Important Observations

1. As was the case for wave speed on a string, notice the dependence of speed directly on a term describing the *elastic stiffness* (B or Y) and inversely on an *inertial* term (d_o).
2. Since solids are more rigid than liquids, which in turn are much more rigid than gases (see Table 12.1, text), it follows that the speed of sound is fastest in solids and slowest in gases. The lower densities of gases do not compensate for their much lower values of bulk modulus.
3. Notice the correctness of units in these formulas:

$$[(N/m^2)/(kg/m^3)]^{1/2} = m/s.$$

Temperature Dependence of Speed of Sound--Only in *gases* is the speed of sound significantly affected by temperature. The bulk modulus for a gas is directly proportional to absolute (kelvin) temperature, so the relationship can be stated as follows:

$$v_1/v_2 = \sqrt{T_1}/\sqrt{T_2}$$

Example 12-1. A tuning fork is vibrating under water at a frequency of 1000 Hz. If the tines of the fork move a total distance of 10^{-6} meters as they vibrate,

(a.) Write the mathematical form of this wave in specific numerical form, using SI units.

(b.) Some of the sound energy is transmitted into STP air above the water surface. Write down the mathematical form of the sound wave in air, making it as quantitative as you can.

S: To write the mathematical form, we need the period, wavelength, and amplitude. The amplitude should be *half* the total travel of the tines, so X_{max} = $.5 \times 10^{-6}$ m. Also, we are given f = 1000 Hz, so T = 1/f = 10^{-3} sec. But how do we know the wavelength?

T: It's determined by the *speed* of the waves, i.e., how far each compression has moved by the time the next compression is formed. From Table 12.1 (text) we have for water:

$$v = 1.4 \times 10^3 \text{ m/s} = f\lambda .$$

So

$$\lambda = (1.4 \times 10^3 \text{ m/s})/(10^3 \text{ Hz}) = 1.4 \text{ meters.}$$

S: Then we can write down the wave form:

$$X = (.5 \times 10^{-6})\cos[2\pi(x/1.4 - t/.001)],$$

or

$$X = (.5 \times 10^{-6})\cos(4.49x - 6.28 \times 10^3 t) .$$

We haven't talked about what happens when sound passes from one medium to another. What can I say?

T: There 's a lot you can deduce by simple reasoning. Is the *speed* going to change?

S: Yes.

T: That means the product $f\lambda$ will change. Will f change?

S: I don't see how it can, since 1000 compressions per second are being produced by the fork, and that many per second will be launched into the air as they arrive at the water surface.

T: Very good! You're exactly right. Thus *speed changes* and *frequency doesn't.* This must mean that *wavelength changes*, since $\lambda = v/f$.

S: What about amplitude?

T: That *will* change, since some of the wave energy at the surface is reflected back into the water. But that calculation isn't obvious at this point, so we'll just observe that the wave will have some new amplitude in the air. Now calculate the air wavelength.

S: In air at STP, v = 330 m/s, so

$$\lambda_{air} = (330 \text{ m/s})/(1000 \text{ Hz}) = .330 \text{ m.}$$

I see that the air waves become more "bunched up" when the speed is slowed.

T: Right. The wave form in air is then

$$X = (X_{max})_{air} \cos[2\pi(x/.330 - t/10^{-3})]$$

$$= (x_{max})_{air} \cos(19.0x - 6.28 \times 10^3 t)$$

Example 12-2. A hammer strikes a long brass railing. Some distance away, a person hears the sound with her ear pressed against the railing. Half a second later she hears the hammer blow in the air. How far is she from where the hammer struck? The air is at STP and brass has the following characteristics: bulk modulus = 8.3×10^{10} N/m^2, and specific gravity = 8.44.

$\underline{S}$: The speed of sound in STP air is 332 m/s, and in brass it is

$$v_{brass} = (B/d_o)^{\frac{1}{2}} = \{(8.3 \times 10^{10} \text{ N/m}^2)/(8.44 \times 10^3 \text{ kg/m}^3)\}^{\frac{1}{2}}$$

$$= 3.14 \times 10^3 \text{ m/s.}$$

I can't find the distance from either of these.

$\underline{T}$: No, but using them both gives you the following:
In brass, $s = v_b t_b$, where t_b = time of wave travel in brass.
In air, $s = v_a t_a$, where s is the *same* , and t_a = time of wave travel in air.

$\underline{S}$: There are *3* unknowns (s, t_b, and t_a) in these *2* equations.

$\underline{T}$: But you also have a *third* equation, namely that $t_a = t_b + .5$ sec.

$\underline{S}$: I forgot that. Then setting the expressions for s equal, we have

$$v_b t_b = v_a t_a = v_a(t_b + .5).$$

This yields

$$t_b = (.5v_a)/(v_b - v_a) = (.5 \text{ s})(332 \text{ m/s})/(3140 - 332) \text{ m/s}$$

$$= 5.91 \times 10^{-2} \text{ sec.}$$

$\underline{T}$: And so t = .559 sec., and

$$s = v_b t_b = (3140 \text{ m/s})(5.91 \times 10^{-2} \text{ s}) = 1.86 \times 10^2 \text{ meters.}$$

Standing Sound Waves; Resonance

Understanding this section should enable you to do the following:

1. *Calculate the resonant frequencies and wavelengths of pipes of given length, both with open and with closed ends.*
2. *Calculate the dependence of these resonances on the type of gas filling the pipes and on the gas temperature.*
3. *Identify the difference in harmonic content of pipes open on both ends compared to those with one closed end.*

Just as was the case for transverse waves on a string, two longitudinal sound waves of equal amplitude and frequency traveling in opposite directions along a pipe or tube will interfere to produce *standing waves* having the form

$$X = [2X_{max} \cos(2\pi x/\lambda)]\cos 2\pi t/T .$$

Important Observations

1. As you move along the x axis, the longitudinal displacement amplitude varies sinusoidally, going through *maxima* of $2X_{max}$ wherever $\cos 2\pi(x/\lambda) = \pm 1$ and *minima* of 0 wherever $\cos 2\pi(x/\lambda) = 0$. The maxima are called *antinodes* of the standing wave pattern, while the minima are called *nodes*.

<u>Resonance Conditions at the Ends of a Pipe</u>--For *resonant* standing waves to be set up in the column of gas in a pipe, the following displacement conditions must occur:

(a.) At a *closed* end, *no displacement can occur*, so there must be a displacement *node* there.
(b.) At the *open* ends, a *maximum* displacement, or *antinode*, must occur.

<u>Important Observations</u>

1. As it was with resonances on a vibrating string, the distance between successive nodes or successive antinodes represents *half a wavelength*.
2. The distance between a node and an adjacent antinode is a *quarter wavelength*.
3. As discussed in your text, a *node* of *displacement* is at an *antinode* or *maximum* of *pressure* variation, while zero pressure variation or *pressure node* takes place at a *displacement antinode*. This means that pressure and displacement oscillations are 1/4 cycle ($\pi/2$ radians) *out of phase* with each other.

<u>Mathematical Conditions for Resonance</u>--

(a.) <u>Pipe open at both ends</u>. Using either a displacement or a pressure standing wave description gives the same result: *the pipe length must equal an integral number of half wavelengths*, or

$$L = n(\lambda_n/2), \text{ where } n = 1, 2, 3, \ldots$$

The resulting harmonic frequencies are

$$f_n = n(v/2L), \quad \text{where } n = 1, 2, 3, \ldots$$

(b.) <u>Pipe closed at one end</u>. Here the pipe must be one quarter wavelength longer than an integral number of half wavelengths. The extra length is required so that the open end, a displacement antinode, is 1/4 wavelength beyond the last node in the pipe, while the closed end is at a displacement node. Thus

$$L = (n/2 + 1/4)\lambda_n, \quad \text{where } n = 0, 1, 2, 3, \ldots$$

The harmonic frequencies are

$$f_n = (2n + 1)(v/4L), \quad \text{where } n = 0, 1, 2, 3, \ldots$$

<u>Important Observations</u>

1. In case (a.), notice that the *fundamental* frequency is $f_1 = v/(2L)$, and *all* harmonics, *even and odd multiples* of f_1, resonate in such a pipe. This is the same condition on harmonic content as for a string clamped at both ends.

2. In case (b.), notice that the fundamental frequency is $f_1 = v/(4L)$ (for $n = 0$). Furthermore, $(2n + 1)$ generates *only odd integers*, so *only odd harmonics* resonate in a pipe closed at one end, i.e., f_3, f_5, etc.

Example 12-3. Suppose you have an organ pipe closed at one end which is 2 meters long.

(a.) What is its fundamental resonant frequency when filled with STP air?
(b.) What would be its fundamental frequency if filled with STP helium?
(Helium has a bulk modulus of 1.67×10^5 N/m^2 at STP, and a density of 1.80×10^{-1} kg/m^3).

S: The fundamental for a pipe like this is $f_1 = v/4L$. For STP air, $v = 330$ m/s, and so

$$f_1 = (330 \text{ m/s})/4(2 \text{ m}) = 41.3 \text{ Hz}.$$

What changes when you change the gas in the pipe?

T: The speed of sound. In this case, both B *and* d_o change.

$$v_{He} = (B/d_{He})^{\frac{1}{2}} = (1.67 \times 10^5 \text{ N/m}^2)/(1.80 \times 10^{-1} \text{ kg/m})$$

$$= 963 \text{ m/s},$$

So

$$(f_1)_{He} = (963 \text{ m/s})/(8 \text{ m}) = 120 \text{ Hz}.$$

Example 12-4. A wire .4 m long with linear mass density of 2 gm/m is under 315 N tension. It resonates with 5 loops between its clamped ends. This creates a resonance in a nearby organ pipe which is closed on one end, is 50 cm long, and contains STP air. What harmonic of the pipe is resonating with the wire, and what is its frequency?

S: We can find the speed of waves on the string with what's given, but how do we find the frequency?

T: From the previous chapter, we have the conditions for resonant frequencies on a string, $f_n = n(v/2L)$. The *number of loops* tells us the harmonic number n. Here $n = 5$.

S: O.K. Then

$$f_5 = \frac{5}{2(.4 \text{ m})}\sqrt{(315 \text{ N})/(2 \times 10^{-3} \text{ kg/m})} = 2.48 \times 10^3 \text{ Hz}$$

Is this the same harmonic that is resonating in the pipe?

T: It is the *same frequency*, but the conditions for resonance in the pipe are different, so it probably is a different *harmonic* of the pipe.

S: What is the expression for harmonic frequencies in this pipe?

T: That $f_n = (2n + 1)(v/4L)$, where v = sound speed and L = length of the pipe.

S: So we want to set this $f = 2.48 \times 10^3$ Hz and find n? But n can only be a whole number.

T: n should indeed come out to be at least *close* to a whole number. However, the quantities in the problem are not exact and resonant conditions experimentally have some frequency *width* to them, so we have no reason to expect an exactly whole number for n from the problem.

S: Well, the numbers give us

$$(2n + 1)(332 \text{ m/s})/4(.5\text{m}) = 2.48 \times 10^3 \text{ Hz, which means}$$

$$2n + 1 = 14.9, \text{ or } n = 6.97.$$

T: Thus the resonance must be the 7th harmonic of the pipe.

Interference of Different Sound Frequencies; Beats

Understanding this section should enable you to do the following:

1. *Calculate the beat frequency resulting from the interference of two sound waves of different frequencies, or conversely.*
2. *Compare two frequencies, given the beat frequency resulting from their interference.*

The application of the principle of superposition to the interference of two sound waves of *equal amplitude* but *different frequencies* (f_1 and f_2) results in the combined wave form

$$X(t) = \left[2X_{max} \cos \frac{2\pi(f_1 - f_2)t}{2} \right] \cos \frac{2\pi(f_1 + f_2)t}{2}$$

This shows how the displacements oscillate at any point.

Important Observations

1. The factor in brackets above is oscillating sinusoidally with a relatively *low* frequency, $f_1 - f_2$, and represents a slowly oscillating *amplitude* of the other factor.
2. The other cosine factor is varying at a frequency $\frac{1}{2}(f_1 - f_2)$, which is the *average* of the individual frequencies.
3. As we shall see, the loudness of sound depends on the wave amplitude. In each cycle the slowly varying factor produces a maximum sound level *twice*, a phenomenon which we call *beats*.

Beats--The periodic variation in the absolute value of the amplitude of a wave caused by the interference of two waves of slightly different frequencies superimposing in the same medium.

Beat Frequency--The above amplitude variation occurs at a frequency equal to the *difference* in the frequencies of the individual waves, $|f_1 - f_2|$.

Example 12-5. An organ pipe is tuned so that its third harmonic is at 1320 Hz at the start of a concert. At intermission, the organist hears 5 beats per second when this same harmonic is compared with a 1320 Hz tuning frequency. If the concert hall was 23° C at the start, what has its temperature become? (Assume the pipe length has remained constant.)

S: What critical parameter is the temperature affecting?

T: If we neglect slight length changes (which we'll consider in coming chapters), the temperature affects the *speed of sound* in air. We are aware that

$$v_2/v_1 = \sqrt{T_2}/\sqrt{T_1}$$, where temperatures are in *kelvins*.

S: A given harmonic frequency is $f_n = n(v/2L)$ or $(2n + 1)(v/4L)$, depending on the type of pipe. Which type do we have?

T: I don't know. Do you *need* to know?

S: I guess not. Either way the speed ratio is obtained by the ratio of the same harmonic at intermission to that at the start,

$$f_3(T_2)/f_3(T_1) = v_2/v_1 = \sqrt{T_2}/\sqrt{T_1} .$$

T: Very good! Most often a difficult sounding problem is quite easy when you get to the essentials. The beat frequency tells us that $f_3(T_2)$ is 1325 Hz, and we know that $T_1 = 23 + 273 = 296$ K. So we get

$$1325 \text{ Hz}/1320 \text{ Hz} = \sqrt{T_2}/\sqrt{(296 \text{ K})} ,$$

which gives $T_2 = 298$ K.

This shows how *sensitive* the beat frequency method of comparing frequencies is, and why musicians must retune during a concert. We can easily perceive one beat per second, *no matter how high* the frequencies we're comparing.

Sound Intensity and Loudness

Understanding this section should enable you to do the following:

1. *To compare intensities of sound waves based on their frequencies and amplitudes.*
2. *To compare sound intensities from a point source at various distances.*
3. *To relate decibel differences in sound intensity level to the actual intensity ratios.*

Wave Intensity (I)--The *rate* of energy transport *per area* normal to the direction of the wave. Mathematically,

$$I = \text{Power/Area} = (\text{Energy/time})/\text{Area}$$

Important Observations

1. The SI units of intensity are *watts/m*^{2}.
2. The factor of A in the denominator means that sound radiating outward from a *point source* becomes *less intense*, since the radiated power is diluted over greater areas at greater distances from the source. At a distance R, the power is spread over an area $4\pi R^2$. Thus the intensity of sound from a point source diminishes as $1/R^2$. This is known as the *inverse square law of intensity*.

Dependence of Intensity on Wave Parameters--Your text derives the dependence of intensity on wave parameters to be

$$I = 2\pi^2 f^2 X^2_{max} vd,$$

where

f = frequency
X_{max} = amplitude
v = wave speed
d = mass density of the medium.

Important Observations

1. The most important dependencies to note are the proportionality of I to the *squares* of *frequency* and *amplitude*.
2. Since the displacement amplitude is proportional to the *pressure* amplitude of the wave, we can conclude that I is proportional to the square of the pressure amplitude as well.

Decibel Scale (ß)--A *logarithmic* scale for comparing intensity levels of sound is the *decibel (dB) scale*. It is defined by the following mathematical relation:

$$\beta \text{ (in dB)} = 10 \log_{10}(I/I_o) .$$

where I_o is a reference intensity. If I_o is taken to be the "threshold of hearing," i.e., $I_o = 10^{-12}$ watts/m^2, then this expression gives the *absolute* intensity value of I.

Important Observations

1. The decibel is a *dimensionless* quantity.
2. The same scale can be used for comparing *any two* intensities with each other. For instance the decibel difference between I_1 and I_2 would be

$$\beta = 10 \log_{10}(I_2/I_1) .$$

Notice that if $I_1 = I_2$, $\beta = 0$, since log 1 = 0.

3. Since $\log_{10}2 = .30103$, an intensity *twice* as great as another is about *+3 dB* relative to it. An intensity 2^n as great as I_1 would be *+3n dB* relative to it.

Example 12-6. Suppose the sound from a single trumpet reaches you with an intensity I_1. A second trumpet producing the same amount of sound energy is only half as far from you as the first one.

(a.) What is the *total* intensity of sound reaching you, in terms of I_1?
(b.) What intensity difference in dB does this represent?

S: We don't know the actual power being emitted by the trumpets. But if they were the *same* distance from us, we should measure a total intensity $2I_1$.

T: That would be right. However, considering each trumpet to be essentially a point source of sound, the one at a distance R/2 would give 4 times the intensity as the one at R, according to the inverse square law. So total intensity would be

$$I_{tot} = I_1 + 4I_1 = 5I_1.$$

S: The dB difference would then be

$$\beta = 10 \log(5I_1/I_1) = 10 \log 5 = +7 \text{ dB}.$$

T: Right.

Example 12-7. The OSHA (Occupational Safety and Health Administration) uses a standard of 80 dB for the maximum sound levels which can be endured for sustained periods of time without causing hearing impairment.

(a.) What intensity of sound is this?
(b.) A rock concert often produces a 115 dB sound level in the audience. By what factor does this intensity exceed OSHA standard?

S: Do these levels refer to a 10^{-12} W/m^2 threshold?

T: That's assumed unless you're directly comparing two levels as in the last example.

S: The definition of decibels gives 80 dB = 10 $\log(I/I_0)$ for the OSHA standard. How do I find I?

T: You have to take the *inverse* log, or *antilog*,

$$(I/I_0) = \log^{-1} (80/10) = \log^{-1} 8 = 10^8.$$

So

$$I = 10^8 I_0 = (10^8)(10^{-12} \text{ w/m}^2) = 10^{-4} \text{ w/m}^2.$$

S: We can do the same thing for part (b.).

T: Yes, but it would be more direct to find I_2/I_1 corresponding to the 35 dB difference in intensity level between the OSHA standard and the rock concert.

S: You mean

$$35 \text{ dB} = 10 \log(I_2/I_1) \text{ , which gives } I_2/I_1 = \log^{-1}(3.5)\text{, or}$$

$$I_2 = 3.16 \times 10^3 I_1 \text{ !!}$$

Ouch!!

T: This reveals that the rock concert sound intensity level reaches more than 3,000 times the occupationally safe level. They're *your* ears!

The Doppler Effect

Understanding this section should enable you to do the following:

1. *Calculate the change in frequency of a wave produced by either the motion of the observer or of the wave source, or both.*
2. *Calculate the speed of an observer or wave source from an observed change of frequency.*

Doppler Effect--A change in the *observed frequency* of a wave which is produced by *motion* of the observer and/or the source of the wave. The derivation in your text results in the following mathematical expressions:

I. Observer moving:

$$f' = f(1 \pm v_o/v) \quad (+ = \text{approaching}, - = \text{receding})$$

II. Source moving:

$$f' = f(1 \pm v_s/v)^{-1} \quad (+ = \text{receding}, - = \text{approaching})$$

Here f is the true source frequency, v_o is the observer velocity, and v_s is the source velocity.

Important Observations

1. As mentioned in the text, there is no change in wavelength when the *observer* moves. She or he simply runs into more or fewer waves per second due to her or his velocity relative to the source.
2. When the *source* moves, the wavelength actually shortens in the direction of motion and lengthens in the opposite direction, but the speed with which the waves hit the observer is the same as in still air.
3. In *both* cases, frequency increases with approaching motion and decreases with receding motion, but *not by the same amount*, even when $v_o = v_s$.
4. When *both* source and observer are moving, the observed frequency is given by combining the two effects,

$$f' = f\frac{(1 \pm v_o/v)}{(1 \mp v_s/v)}$$

Example 12-8. An ambulance siren changes pitch from 850 Hz to 770 Hz as it rushes past where you are standing by the curb. Assume STP air. How fast is the ambulance going?

S: It doesn't seem that we have enough information. We don't know v_s or f.

T: Well, let's write down what we know.

When approaching, $(f')_{app} = f(1 - v_s/v)^{-1}$

and when receding, $(f')_{rec} = f(1 + v_s/v)^{-1}$.

We also are given both f'_{app} and f'_{rec}, and know that v = 330 m/s.

S: That gives us enough equations, all right. What's the easiest way to proceed?

T: If you take the *ratio* of the two expressions, the unknown f will cancel.

S: O.K. Then

$$(f')_{app}/(f')_{rec} = 850 \text{ Hz}/770 \text{ Hz} = 1.10 = (1 + v_s/v)/(1 - v_s/v)$$

T: Solving for v_s/v,

$$1.10(1 - v_s/v) = 1 + v_s/v, \quad \text{so}$$

$$v_s/v = (.1)/(2.1) = .0476.$$

Thus

$$v_s = .0476(330 \text{ m/s}) = 15.7 \text{ m/s, or about 40 mi/hr} .$$

Supersonic Speed

Understanding this section should enable you to do the following:

1. *Calculate an object's speed, knowing the speed of sound and the object's Mach number.*
2. *Calculate the angle of the shock wave produced at a given Mach number.*

The speed of sound varies with pressure, density, and temperature of the air. It thus changes with altitude above the earth's surface. If all aircraft's speed through the air is *faster than the speed of sound at its altitude*, we say that it has *supersonic* speed. In this case a conical surface of abrupt pressure change is set up, with the leading point of the aircraft at the apex of the cone. (See Figure 12.22, text.) This is called the *Mach cone*, and has an angle, θ_m, called the *Mach angle*. As this Mach cone moves along the ground at the speed of the aircraft, a "sonic boom" is experienced by observers on the ground. The *Mach number* of the aircraft describes these phenomena.

Mach Angle (θ_m)--The angle of the shock wave cone relative to the direction of motion of an aircraft at supersonic speed. Mathematically,

$$\sin\theta_m = (\text{Mach Number})^{-1}$$

Important Observations

1. For *subsonic* flight, the Mach number is *less than 1*. For *supersonic* flight, it is *greater* than 1.
2. *Only* when the Mach number is greater than 1 is a shock wave formed and a Mach angle defined.

Example 12-9. An aircraft is flying at Mach 1.8 over the flat desert. A sonic boom is heard on the ground 8 seconds after the aircraft has passed directly overhead. Assume the aircraft is flying where the speed of sound is v_s = 300 m/s. At what altitude is the aircraft flying?

S: I think I need a diagram showing what the shock wave looks like.

T: O.K. Here's a rough sketch (not to scale).

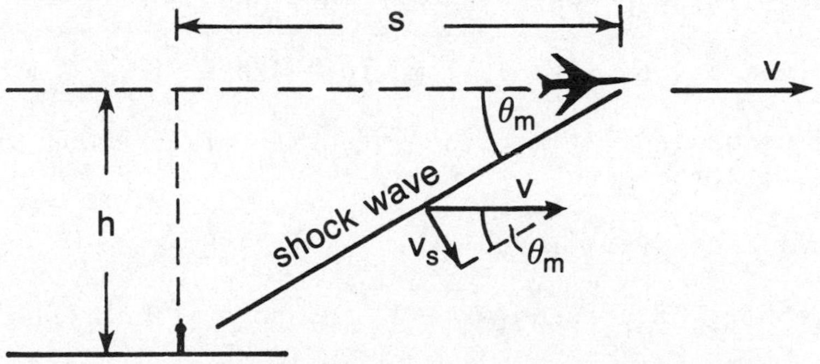

S: I see! We have to calculate how far the plane has gone in the 8 seconds to find the side of the triangle labelled s. That should be easy. $v = 1.8v_s$ = 1.8(300 m/s) = 540 m/s, so

$$s = vt = (540 \text{ m/s})(8 \text{ s}) = 4320 \text{ meters}.$$

Now the altitude h can be found from $h/s = \tan\theta_m$.

T: And θ_m is given by

$$\theta_m = \sin^{-1}(\text{Mach number})^{-1} = \sin^{-1}(1/1.8).$$

Thus $\theta_m = 33.75^\circ$, giving $h = s \tan 33.75^\circ$ = 2.89 km, or about 1.75 mi.

In reality, the shock wave would not be the completely straight line depicted above, because of the variation in the speed of sound with altitude.

QUESTIONS ON FUNDAMENTALS

1. If the temperature in a room goes from 20 ^{o}C to 40 ^{o}C, what happens to the speed of sound in the room's air?

(a.) doubles (b.) increases by a factor of $\sqrt{2}$
(c.) increases by 6.8% (d.) increases by 3.3%
(e.) stays the same

2. If a sound wave has the mathematical form (in SI units) of cos(3.5x - 2000t), what is the distance between successive displacements of air molecules?

(a.) 1.8 m (b.) 3.5 m (c.) 0.56 m (d.) 571 m (e.) 1.75 x 10^{-3} m

3. What would the frequency of the wave described in #2 be?

(a.) 2000 Hz (b.) 318 Hz (c.) 571 Hz (d.) 7000 Hz (e.) 1000 Hz

4. What would be the wavelength of the third harmonic standing sound waves in a pipe 3 meters long if it was open at both ends?

(a.) 3 m (b.) 2/3 m (c.) 2 m (d.) 6 m (e.) 4 m

5. Answer #4 above if the pipe was closed at one end.

(a.) 7/12 m (b.) 5/4 m (c.) 12/7 m (d.) 4/5 m (e.) 1 m

6. If one siren is producing 7 times the intensity of a second siren, how many more decibels of sound is it producing?

(a.) 5 db (b.) 3 db (c.) 10 db (d.) 7 (e.) 8.5

7. What is the intensity of a sound wave 33 db above the threshold of hearing?

(a.) 2 x 10^{-9} W/m^2 (b.) 33 x 10^{-12} W/m^2 (c.) 3.3 x 10^{-12} W/m^2

(d.) 2 x 10^{-10} W/m^2 (e.) 2000 W/m^2

8. If a source of sound is approaching at half the speed of sound, how does the frequency you are hearing compare with the frequency actually being emitted?

(a.) 4 times higher (b.) not enough information to tell
(c.) 50 % higher (d.) twice the frequency (e.) 10 times

GENERAL PROBLEMS

Problem 12-1. A girl is equidistant from 2 loudspeakers which are producing the same single frequency tone. She moves laterally until she hears the sound intensity fade to a minimum. At this position she is 10 ft from one speaker and 8 ft from the other. Assume the speed of sound in the room to be 1100 ft/s.

(a.) What is the frequency of the tone?
(b.) If she moved to a position where she was 20 ft from one speaker and 16 ft from the other, would she hear a maximum or a minimum sound intensity?

<u>Q1</u>: Is she directly between the speakers?
<u>Q2</u>: Why does she experience a minimum intensity as she moves?
<u>Q3</u>: In part (b.), she has merely doubled both distances. Doesn't this maintain the condition of destructive interference?

Problem 12-2. An organ pipe is observed to have three adjacent resonances at 225, 315, and 405 Hz in STP air. What is its length?

<u>Q1</u>: What type of pipe is it, open or closed?
<u>Q2</u>: Is 225 Hz the fundamental?
<u>Q3</u>: What are the ratios of these adjacent frequencies?
<u>Q4</u>: Which harmonics are these?
<u>Q5</u>: How is this related to length?

Problem 12-3. In the figure below, A and B are stationary sources of sound waves, both emitting the same frequency f. An observer is in a car, traveling toward A and away from B at 100 km/hr. He measures a beat frequency of 20 Hz in the interference of waves from the two sources. If the speed of sound in air is 330 m/s, calculate what frequency f is being emitted.

<u>Q1</u>: What is the significance of the 20 Hz beat frequency?
<u>Q2</u>: Which frequency would be higher?
<u>Q3</u>: What is the expression for $(f')_A$?
<u>Q4</u>: What is the expression for $(f')_B$?

ANSWERS TO QUESTIONS

FUNDAMENTALS:

1. (d.)
2. (a.)
3. (b.)
4. (c.)
5. (c.)
6. (e.)
7. (a.)
8. (d.)

GENERAL PROBLEMS:

Problem 12-1.

A1: It doesn't matter.

A2: At the first position, the two waves travel equal distances to reach her, and so they are *in phase*, interfering *constructively*. At the latter position, they are interfering *destructively*, meaning that the waves are arriving one-half cycle out of phase. The *difference* in distance thus corresponds to one-half the wavelength of the tone.

A3: The *difference* in distance is the important quantity. It is now *twice* what it was.

Problem 12-2.

A1: Precisely what you have to find out.

A2: If it was, would the next frequency be 315 Hz in *either* kind of pipe?

A3: Now we're getting somewhere. The ratios are (315/225) = 1.4 and (405/315) = 1.286. Also, (405/225) = 1.8.

A4: f_5, f_7, f_9.

A5: $f_n = n(v/4L)$.

Problem 12-3.

A1: That the sound waves received by the moving observer *differ* by 20 Hz in frequency.

A2: The frequency from A, $f'_A = f'_B + 20$ Hz.

A3: $f'_A = f(1 + v_o/v)$, where $v = 330$ m/s and $v_o = 100$ km/hr.

A4: $f'_B = f(1 - v_o/v)$.

CHAPTER 13. TEMPERATURE AND HEAT

BRIEF SUMMARY AND RESUME

The concept of *temperature* is recalled from the kinetic theory of gases, and discussion of a number of types of *thermometers* follows. Definitions for various *temperature scales* are given. The *contraction* and *expansion* of solids and liquids which accompany *temperature changes* are described, and *coefficients of thermal expansion* are introduced.

Heat is introduced as the *transfer* of *thermal energy*, and is related to temperature changes of an object by the concept of *specific heat capacity*. Non-SI units of heat, the *kilocalorie* and *BTU*, are defined. The *molar heat capacities* of gases for constant pressure and constant volume processes are derived and connected with the *ideal gas law*. The *heats of fusion* and of *vaporization* required to change the state, or phase, of a substance are introduced, and vaporization, or *boiling*, is discussed in detail. The *vapor pressure*, *boiling point*, and *triple point* of a substance are defined.

Material from previous chapters relevant to Chapter 13 includes:

1. The ideal gas law and Kelvin temperature.
2. Thermal energy.

NEW DATA AND PHYSICAL CONSTANTS

1. Absolute zero temperature.

$$0 \text{ K} = -273.15\ ^\circ\text{C} = -459.67\ ^\circ\text{F}.$$

2. Temperature degree sizes.

$$1 \text{ K} = 1 \text{ C}^\circ = (9/5) \text{ F}^\circ.$$

3. SI units of coefficients of thermal expansion.

$$(\text{C}^\circ)^{-1} = (\text{K})^{-1}.$$

4. Units of specific heat capacity.

<u>SI</u>: J/kg-K.

<u>Common</u>: kcal/kg-C°.

5. Units of molar heat capacity.

<u>SI</u>: J/mole-K.

<u>Common</u>: kcal/mole-C°.

6. Mechanical Equivalent of Heat.

$$1 \text{ kcal} = 4.186 \times 10^3 \text{ J}.$$

7. Units of latent heats of fusion and vaporization.

SI: J/kg.

Common: kcal/kg.

8. Triple point of water.

$$P_3 = 4.58 \text{ torr}; \quad T_3 = .01\ ^\circ\text{C}.$$

NEW CONCEPTS AND PRINCIPLES

Temperature Scales and Thermometers

Understanding this section should enable you to do the following:

1. *State the definition of absolute zero.*
2. *Locate various reference temperatures on the Kelvin, Celsius, and Fahrenheit temperature scales, and be able to convert from one scale to the others.*

Although many devices can be used to measure temperature, the device used to establish temperature *standards* is the *constant volume gas thermometer.* A main reason is that, at low pressure, *all* gases show the *same* change of pressure for a given temperature change, assuming the same number of moles and the same volume. Thus the temperature scale does *not* depend on the *type* of gas used. Since absolute (Kelvin) temperature is directly proportional to the average translational kinetic energy of the gas molecules, and this kinetic energy is responsible for creating the pressure of the gas, the following statement *defines absolute zero.*

Absolute Zero--That temperature at which the pressure of an ideal gas would become zero.

Important Observations

1. This temperature cannot actually be attained, but it is the *limiting* lowest temperature that *any* substance can approach.
2. Actual gases change state well above absolute zero, so this definition requires an *extrapolation* of the P vs. T behavior of ideal gases at considerably higher temperatures.

Temperature Scales--Three temperature scales are used in the text. They are the *Kelvin, Celsius,* and *Fahrenheit* scales. The following chart summarizes and compares these scales, to three significant figures.

	Absolute Zero	Water Freezes	Human Body	Water Boils (P = 1 atm)	Relation to Other Scales
Kelvin	0	273	310	373	T(K)=T(C)+273
Celsius	-273	0	37	100	T(C)=(5/9)[T(F)-32]
Fahrenheit	-460	32	98.6	212	T(F)=(9/5)T(C)+32

Important Observations

1. As mentioned before, the *Kelvin* scale is one with a zero which is defined *fundamentally* rather than arbitrarily. Another example is the *Rankine* Scale, based on the size of the Fahrenheit degree. These scales must be used in *thermodynamic* calculations such as the ideal gas law and the kinetic theory of gases. The *kelvin (K)* is the SI unit of temperature.
2. The *size* of a degree is the same in the Kelvin and Celsius scales, and is 9/5 times the size of a Fahrenheit degree. Note there is a subtle difference between talking about a *definite temperature T* in ˚C or ˚F and about a *temperature range* ΔT in C˚ or F˚. In the first case we say "degrees Celsius or degrees Fahrenheit," while in the latter we say "Celsius degrees or Fahrenheit degrees."

Example 13-1. Express the following temperatures and temperature changes in the other two temperature scales.

(a.) 120 ˚F (b.) -100 ˚C (c.) 20 K (d.) ΔT = 3.6 C˚
(e.) ΔT = -27 F˚ (f.) ΔT = 240 K

S: (a.) T(˚C) = (5/9)[120 ˚F - 32] = 48.9 ˚C; T(K) = 273 +T(˚C) = 322K.

(b.) T(K) = 273 +T(˚C) = 173 K; T(˚F) = (9/5)(-100) + 32 = -148 ˚F.

(c.) T(˚C) = T(K) =273 = -253 ˚C; T(˚F) - (9/5)(-253) + 32 = -423˚F.

(d.) ΔT(F^{O}) = (9/5)(3.6 C˚) = 6.5 F˚; ΔT(K) = ΔT(C˚) = 3.6 K.

(e.) ΔT(C˚) = (5/9)(-27 F˚) = -15 C˚ = -15 K.

(f.) ΔT(C˚) = ΔT(K) = 240 C˚; ΔT(F˚) = (9/5)(240 C˚) = 432 F˚.

Thermal Expansion of Solids and Liquids

Understanding this section should enable you to do the following:

1. *Calculate the changes in length, area, and volume of liquids and solids which result from temperature changes.*

Coefficients of Thermal Expansion for Solids--The *fractional* change in the dimensions of a solid due to temperature change are found to be directly proportional to that temperature change. The constant of proportionality is called the *coefficient of thermal expansion*. Values for various materials are found in Table 13.2 (text). Mathematically,

$$\text{Length:} \quad \alpha = (\Delta L/L_o)/\Delta T$$

$$\text{Area:} \quad 2\alpha = (\Delta A/A_o)/\Delta T$$

$$\text{Volume:} \quad 3\alpha = \beta = (\Delta V/V_o)/\Delta T$$

Coefficient of Thermal Expansion for Liquids--The only dimensional change which applies to liquids is that of *volume*, since they cannot maintain their own shape. The *coefficient of volume expansion* is given by

$$\beta = (\Delta V/V_o)/\Delta T$$

Important Observations

1. Since the expansion coefficients involve *changes* in temperature, $(\Delta T)^{-1}$, their units are $(K)^{-1}$ or, equivalently, $(C^o)^{-1}$.
2. Thermal expansion is like a photographic enlargement, in that *all* dimensions undergo the *same ratio* of change per degree. This includes holes, cavities, and interior volumes, as well as the material itself.
3. The coefficients of volume expansion for liquids are considerably larger than those for solids, in general.
4. Near the freezing point, water is an exception to the general rule of expansion, actually contracting with a temperature increase.

Example 13-2. Steel rails are laid on a winter day when the temperature is -10 ^oF. How much gap must be left between the ends of adjacent rails to allow for expansion so that the rails just touch on a summer day when the temperature is +40 ^oC?

S: We can't find ΔT until we have the temperature scales sorted out.

T: Since the expansion coefficient for steel is given in units of $(C^o)^{-1}$, we should convert ^oF to ^oC. Thus -10^oF is $(5/9)(-10 - 32) = -23.3\ ^oC$.

S: Then $T = 40^oC - (-23.3^oC) = +63.3\ C^o$. The fractional change in length will be α times this:

$$\Delta L/L_o = \alpha \Delta T = 11 \times 10^{-6} (C^o)^{-1} \times 63.3\ C^o = 6.96 \times 10^{-4}.$$

Thus the change in length is

$$\Delta L = (6.96 \times 10^{-4})\ L_o = 2.79 \times 10^{-2} \text{ feet.}$$

Is this the gap that should be left?

T: Yes. Imagine the center of each rod to remain fixed. Each end will then expand into the gap by *half* the above amount, from *each* side of the gap. The expansion amounts to a little more than 1/3 of an inch.

Example 13-3. Suppose you buy 1500 gallons of bulk carbon tetrachloride (a cleaning fluid) in winter when the warehouse temperature is -15 ^oC. You pay $5 a gallon. Next summer, you retail the fluid at $10 a gallon when the temperature is 35 ^oC. What fraction of your profit is due to thermal expansion? The coefficient of volume expansion for CCl_4 is $\beta = 58.1 \times 10^{-5} (C^o)^{-1}$.

S: Without a temperature change, the profit would be \$5 on each gallon, or a total of \$5 x 1500 gallons = \$7500. We have to find out how many more gallons we have in summer, when the temperature is 50 C^o higher than when we bought it.

T: The increase in volume (gallon is a unit of volume) is given by

$$V/V_o = \beta \ T = 58.1 \times 10^{-5} (C^o)^{-1} \times 50 \ C^o = 2.91 \times 10^{-2}$$

So

$$V = (1500 \text{ gallons})(2.91 \times 10^{-2}) = 43.6 \text{ gallons.}$$

S: That's quite a bit. You make a profit of \$5 on each gallon, so this amounts to an extra profit of \$5 x 43.6 gallons = \$218.

T: *No*, that's *not* right. These extra gallons didn't cost you *anything*, so you get \$10 pure profit on them, an extra profit of \$436. The *total* profit is \$7500 + \$436 = \$7936, so the fraction that is due to expansion is

$$(\text{extra profit})/(\text{total profit}) = \$436/\$7936 = .055, \text{ or } 5.5\%.$$

Heat and Thermal Energy of Solids and Liquids

Understanding this section should enable you to do the following:

1. *Define the difference between heat and thermal energy.*
2. *Make calculations involving the relationship between heat transferred to or from a system, its temperature change, and the amount and kind of material in the system, specified by its specific heat capacity.*
3. *Convert between the SI unit of energy, the joule, and the common thermal units, the kilocalorie and BTU.*
4. *Perform calculations of equilibrium temperature resulting from the mixing of hot and cold substances, based on the conservation of thermal energy.*

Thermal Energy--The sum of the *random mechanical energies* of the atoms and molecules of a substance.

Heat (Q)--The thermal energy *transferred* between a system and its surroundings as a result of *temperature differences*.

Specific Heat Capacity (c)--As long as a substance doesn't *change state*, i.e., melt, vaporize, etc., the temperature change resulting from the transfer of heat Q into or out of the system is *directly* proportional to Q and *inversely* proportional to the amount of material in the system. The constant of proportionality depends on the *type* of material, and is called the *specific heat capacity* of the substance. Mathematically,

$$T = Q/mc$$

where m = the mass of the system
c = specific heat capacity of the material of the system.

Important Observations

1. The relation shows the difference between temperature and heat, and the relation between them. Since the dimensions of ΔT, Q, and m are already defined, the specific heat capacity has dimensions of energy/mass-degree. *In SI units, c is measured in J/kg-K.*
2. Substances with large specific heat capacities undergo relatively small temperature changes for a given Q, while small values of c result in larger swings of T for the same Q. You might think of c as being a kind of "temperature inertia" of a system in responding to heat transfers.
3. If the amount of substance is measured in *moles*, n, rather than mass, we have the *molar specific heat* (C) of the substance, defined by

$$\Delta T = Q/nC$$

   C then has the dimensions of energy/*mole*-degree.
4. The algebraic sign of this equation shows that *positive* Q represents heat going into the system, resulting in an *increase* in temperature, and vice versa for *negative* Q.

Example 13-4. How many joules would be required to increase the temperature of 2 kg of water from 0 ^oC to 20 ^oC? If the same heat was added to 2 kg of methyl alcohol at 20 ^oC, what would its final temperature be? The specific heat capacities are: for H_2O, $c = 4.186 \times 10^3$ J/kg-C^o; for CH_3OH, $c = 2.512 \times 10^3$ J/kg-C^o.

S: We know $Q = mc\Delta T$. For the water, this gives

$$Q = (2\ \text{kg})(4.186 \times 10^3\ \text{J/kg-C}^o)(+20\ \text{C}^o) = 1.674 \times 10^5\ \text{J}.$$

T: And for the alcohol,

$$\Delta T = Q/mc = (1.674 \times 10^5\ \text{J})/(2\ \text{kg})(2.512 \times 10^3\ \text{J/kg-C}^o) = 33.3\ \text{C}^o$$

Since the starting temperature was 20 ^oC,

$$T_f = 20\ ^oC + 33.3\ C^o = 53.3\ ^oC.$$

Notice that it takes a considerable energy to effect much temperature change.

Common Units of Heat--Traditionally, two units of heat and thermal energy are used more commonly than the SI units. These are the *kilocalorie* and the *BTU* (British Thermal Unit), defined as follows:

kilocalorie (kcal)--The amount of heat required to raise the temperature of *1 kg* of water by *1 C^o*.

BTU--The amount of heat required to raise the temperature of *1 pound* of water by *1 F^o*.

Important Observations

1. The specific heat of water is thus *defined* to be *unity* in both systems. The usual units of specific heat capacities

are thus kcal/kg-C^o or BTU/lb-F^o. The numerical values of c are the same in both systems of units.

2. The conversion factor between kilocalories and joules is called the *mechanical equivalent of heat*:

$$1 \text{ kcal} = 4.186 \times 10^3 \text{ J}.$$

3. The kilocalorie is the same amount of energy as the Calorie (notice the capital letter) used in nutritional calculations. The calorie (or gram-calorie) is 1/1000 of the kilocalorie.
4. The remaining conversion factor follows from the above:

$$1 \text{ kcal} = 3.96 \text{ BTU}$$

Example 13-5. A 50 gram copper cup contains 125 grams of water at 20 ^{o}C, and is insulated from its surroundings. 300 grams of a mixture of brass and gold filings at 150 ^{o}C is placed into the water. It is found that an equilibrium temperature of 40 ^{o}C results. What percentage of the filings is gold? In units of kcal/kg-C^o, the specific heats are as follows: c_{gold} = .031, c_{brass} = .090, and c_{copper} = .093.

S: Our equation will come from the principle that the heat lost by the filings must equal the heat gained by the cup plus the water.

T: In absolute value, yes. If you take into account the algebraic signs of Q, you would have

$$Q_{lost} = - Q_{gain}$$

so that both sides of the equation have the same sign.

S: The heat gained is

$$Q_{gain} = (m_w c_w - m_c c_c)(T_f - T_i)$$

$$= [(.125 \text{ kg})(1 \text{ kcal/kg-C}^o)+(.050 \text{ kg})(.093 \text{ kcal/kg-C}^o)][40\ ^oC - 20\ ^oC]$$

$$= 2.59 \text{ kcal}.$$

Since I don't know the mass of either the brass or the gold, how can I calculate Q_{lost}?

T: You know that the *total* mass is 300 gm. So let X = mass of gold and (.300 - X)kg = mass of brass. Then

$$Q_{lost} = [X(.031 \text{ kcal/kg-C}^o)+(.300 - X)(.090 \text{ kcal/kg-C}^o)][(150 - 40)^oC]$$

S: Now we have to set these equal and solve for X. It looks rather messy.

T: Just be very methodical in collecting terms and watch your algebraic signs carefully. Let's omit the units for the moment.

$$(.031X = .090X)(110) + (.300)(.090)(110) = 2.59$$

Thus

$$-6.49X = 2.59 - 2.97.$$

So there is X = .0586 kg of gold, and (.300 - X) = .242 kg brass.

The percentage of gold is

% gold = (.0586 kg/.300 kg) x 100 = 19.5%

Heat and Temperature Changes of Gases

Understanding this section should enable you to do the following:

1. *Calculate the total thermal energy in a sample of an ideal monatomic gas at temperature T.*
2. *Calculate the temperature change of an ideal monatomic gas when heat is added or subtracted in constant volume or constant pressure processes.*
3. *Calculate the fraction of heat which does work in a constant pressure process in an ideal monatomic gas.*

The temperature change of a sample of gas resulting from a transfer of heat into or out of the gas depends on the *process* of heat transfer, i.e., whether it happens while volume is held constant, or at constant pressure, or some other process. Thus the specific heat capacity of a gas is different for different processes also. When dealing with gases it is customary to use *molar* heat capacities.

Thermal Energy (U) of a Monatomic Ideal Gas--The only form of mechanical energy of the gas atoms here is *translational kinetic* energy. From the kinetic theory of gases we conclude that the total thermal energy is

$$U = (3/2)nRT,$$

where n = number of moles
R = universal gas constant

Constant Volume Molar Specific Heat (C_v)--If the volume of the gas is held constant as heat is added, no work is done by the gas pressure, so all the heat must go into the increase of internal energy:

$$Q = \Delta U = (3/2)nR\Delta T.$$

From the definition of specific heat, we have

$$C_v = Q/n\Delta T = (3/2)R$$

Constant Pressure Molar Specific Heat (C_p)--When pressure is constant, the volume *must* change as heat is added. The gas then does an amount of mechanical work $P\Delta V$. Thus the heat added accomplishes *two* things: it does work, and it increases thermal energy. Mathematically,

$$Q = P\Delta V + \Delta U = nR\Delta T + (3/2)nR\Delta T = (5/2)nR\Delta T.$$

This defines the specific heat capacity at constant pressure:

$$C_p = Q/n\Delta T = (5/2)R = C_v + R.$$

Important Observations

1. For a given number of moles, the thermal energy of a gas depends *only* on the absolute temperature T.
2. The values of $C_v = (3/2)R$ and $C_p = (5/2)R$ apply only to *monatomic* ideal gases. Molecular gases have higher values due to additional internal mechanical energies of rotation and vibration. Notice, however, that *all monatomic gases have the same molar specific heats.*
3. Since at constant pressure the amount of work done in expansion of an ideal gas is always $P\,\Delta V = nR\,\Delta T$, with the result that $C_p = C_v + R$ is *not* restricted to monatomic gases.
4. Since $C_p > C_v T$, the temperature change during a constant pressure heat transfer must be *less* than that for an equal heat transfer at constant volume.
5. The *ratio* of specific heats, C_p/C_v, is symbolized by γ , and offers a convenient test of the kinetic theory of gases. For monatomic gases the theory predicts $\gamma = 1.67$, which agrees well with experiment.

Example 13-6. Suppose you have 20 grams of argon, whose mass number is 40.

(a.) How many *moles* of argon do you have?

(b.) How much heat must be added to this sample to raise its temperature from $-10^{o}C$ to $100^{o}C$ at *constant pressure*?

(c.) If the same amount of heat was added at *constant volume*, what would the final temperature be, starting at $-10^{o}C$?

(d.) In part (b.) what percentage of the heat went for doing the *work* of expansion?

S: 40 grams of Ar would be one mole, so I have .5 moles. Part (b.) looks straightforward. We have

$$Q = nC_p\,\Delta T = n(5/2)R\,\Delta T.$$

What value do I use for R? And do I have to convert to Kelvin temperatures?

T: Your choice of R will determine the units of Q. For instance, R = 8.314 J/mole-K will give you joules, while R = 1.99 cal/mole-K will give you calories (*not* kilocalories!). As for temperature, ΔT is the same in K or C^{o} since they represent the same *change* in temperature.

S: O.K. Then, since $\Delta T = 100\ ^{o}C - (-10\ ^{o}C) = 110\ C^{o}$,

$$Q = (.5 \text{ mole})(5/2)(1.99 \text{ cal/mole-K})(110\ C^{o}) = 274 \text{ calories}.$$

In part (c.), the final temperature is the unknown.

T: Yes, and you use $C_v = (3/2)R$ instead of C_p. You have

$$274 \text{ cal} = Q = nC_V\,\Delta T = n(3/2)R(T_f - (-10^{\circ}C)).$$

Don't forget the double negative in the last factor.

S: Solving for T_f,

$$274 \text{ cal} = (.5 \text{ mole})(3/2)(1.99 \text{ cal/mole-K})(T_f+10\ ^\circ\text{C}) = 1.49T_f + 14.9$$

So $\quad T_f = 174\ ^\circ\text{C}$.

That's a big difference from the 100 °C final temperature in the constant pressure case.

T: Uh-huh, when none of the heat has to do work, the temperature rise is much higher. Part (d.) should demonstrate the difference even more.

S: I don't see how to separate the work out of the total Q.

T: The work done is $P\,\Delta V = nR\,\Delta T$. Thus you have

$$Q_{tot} = Q_{work} + Q_{int.\ energy} = nR\,\Delta T + nC_v\,\Delta T = nC_p\,\Delta T.$$

S: So

$$Q_{work} = (.5 \text{ mole})(1.99 \text{ cal/mole-K})(110\ ^\circ\text{C}) = 109 \text{ cal}$$

and

$$Q_{tot} = 274 \text{ cal}.$$

So the percentage of Q which did work is

$$100 \times Q_{work}/Q_{tot} = 100 \times 109 \text{ cal}/274 \text{ cal} = 39.9\%.$$

T: It's even more revealing when you do it algebraically:

$$Q_{work}/Q_{tot} = nR\,\Delta T/nC_p\,\Delta T = R/C_p = 2/5 = .40, \quad \text{or } 40\%$$

for monatomic gases. Can you generalize this in order to calculate this ratio for a gas whose value of $\gamma = C_p/C_v = 1.4$?

S: We've seen that Cp = Cv + R even for gases that aren't monatomic. So the generalization should be that

$$Q_{work} = nR\,\Delta T = n(C_p - C_v)\,\Delta T$$

Then

$$Q_{work}/Q_{tot} = n(C_p - C_v)\,\Delta T/nC_p\,\Delta T = (C_p - C_v)/C_p$$

T: Excellent. Thus for *constant pressure* processes,

$$Q_{work}/Q_{tot} = 1 - 1/\gamma = (\gamma - 1)/\gamma \quad .$$

For monatomic gases, $\gamma = 1.67$, and this ratio is .40 . For diatomic gases, $\gamma = 1.4$, and this ratio is .286 .

Change of Phase: Heats of Fusion and Vaporization

Understanding this section should enable you to do the following:

1. *Calculate the heat transferred between a substance and its surroundings when it changes state.*

2. *Identify the pressure and temperature conditions for changes of state of a substance from its phase diagram.*
3. *Identify the triple point and critical point of a substance and its fusion, vaporization, and sublimation curves from its phase diagram.*

State or Phase of a Substance--Whether the substance has the physical properties of a *solid*, *liquid*, or *gas*.

Phase Change--This occurs when a substance undergoes a change of state, which happens at various combinations of pressure and temperature for different substances. This is accompanied by a heat transfer between the substance and its surroundings. The following is a summary of phase changes:

Table 13-1. Summary of Phase Changes

Phase Change	Common Name	Direction of Heat Flow
solid → liquid	melting, or fusion	$Q>0$, heat input
liquid → solid	freezing	$Q<0$, heat outflow
liquid → gas	boiling, or vaporization	$Q>0$, heat input
gas → liquid	condensation	$Q<0$, heat outflow
solid → gas	sublimation	$Q>0$, heat input
gas → solid	deposition	$Q<0$, heat outflow

Important Observation

1. Phase changes represent structural changes in the atomic or molecular bonds of a substance, and hence involve *energy* changes. Thus the heat transfers accompanying a phase change *do not result in a change of temperature*, but only of *structure*. Only when the substance is *entirely* in a single phase do temperature changes result from heat transfer according to the equations involving specific heat capacities.

Phase Diagram--A graph showing the pressure-temperature dependence of phase changes for a substance. The pairs of pressure and temperature values at which phase changes take place form *curves* on a graph of P vs. T which *separate* the phases of a substance, as shown in the accompanying figure. (see also Figure 13.16, text.)

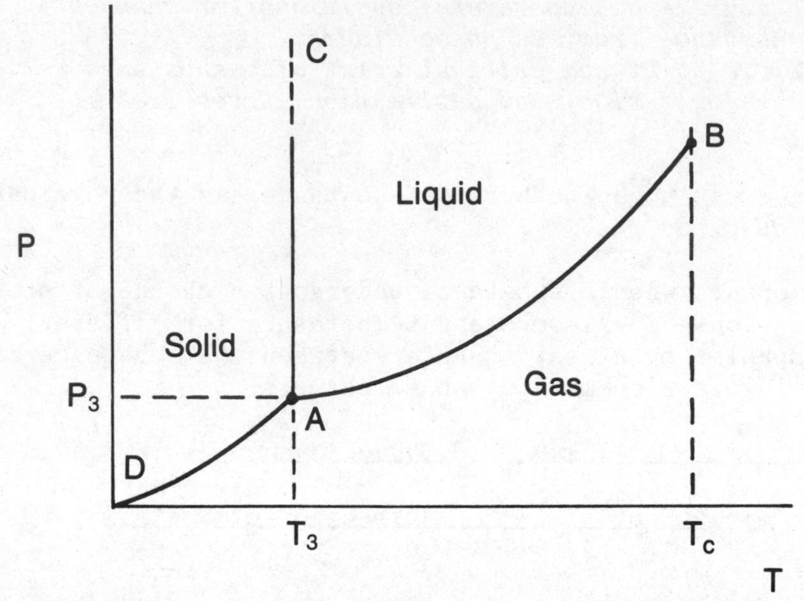

Important Observations

1. Point A in the figure is called the *triple point* of the substance, where all three states coexist in equilibrium. It occurs at only *one* set of values of pressure and temperature, P_3 and T_3. *Below* P_3, no increase of temperature will melt the substance. It will *sublime* directly from solid to gas as you cross the curve AD. *Below* T_3, no amount of pressure will liquefy the substance. It will undergo *deposition* directly from gas to solid. For water, T = .01°C, P_3 = 4.58 torr; for carbon dioxide, T_3 = -56.6°C, P_3 = 3884 torr = 5.11 atm.
2. Curve AD is called the *sublimation* curve, AB is the *vaporization* curve, and AC is the *fusion* curve.
3. The steepness of the fusion curve shows that the melting temperature is only *slightly* dependent on pressure, whereas the vaporization curve shows a wide variation in boiling temperature with changing pressure.
4. Water is unusual in that its fusion curve has a *negative* slope. This means that, at temperatures just below the freezing point, the application of a large pressure (moving vertically upward on the graph) will cause ice to melt.
5. There is a *maximum* temperature for which a liquid can be formed, *no matter how high the pressure.* This is called the *critical temperature*, T_c. The vaporization curve *ends* at this temperature, at point B in the figure, called the *critical point* of the substance. For water, T_c = 374°C, and P_c = 218.4 atm; for carbon dioxide, T_c = 31.1°C, P_c = 72.9 atm. Most other gases have T_c below room temperature, and so cannot be liquified by pressure alone, without first cooling them below their T_c.

Example 13-7. From U.S. and Soviet space probes we have the following data:

(a.) The surface temperature of Venus is about 900 ^{o}F.
(b.) The atmospheric pressure at the surface of Mars is approximately .006 atmospheres. What can you imply about the possible state of water on these planets?

S: Without calculating, it would seem that water is gaseous on Venus and solid on Mars.

T: How can you be sure? What if the Venusian pressure is very great, such as the measured 90 atmospheres?

S: I'd have to see the vaporization curve of water in detail.

T: No you wouldn't. Think *critically*!

S: No puns, please. I see what you mean. The critical temperature of water is 374 ^{o}C, or *705* ^{o}*F*. So Venus' temperature is well above T_c, and no state of water can exist except *gas*!

T: Precisely. Now how about Mars?

S: Well, .006 atm = 4.56 torr. This is *barely* below the *triple point* pressure for water, which would *prohibit liquid from forming even at high summer temperatures*.

T: How sublime! This is a close one to call, and our data needs to be more precise. Also, the sublimation of a lot of water vapor into the Martian atmosphere would *create* a higher pressure, and could move it above P_3. The fate of water on Mars remains a puzzle at present.

Heat of Fusion (L_f)--The amount of heat required to melt 1 kg of a substance from a solid to a liquid. Mathematically,

$$L_f = Q_f/m \ .$$

Heat of Vaporization (L_v)--The amount of heat required to change 1 kg of a substance from liquid to gas(vapor). Mathematically,

$$L_v = Q_v/m \ .$$

Important Observations

1. These are also known as *latent* heats, hence the symbol L. The units are obvious:

   SI: J/kg ; Common: kcal/kg .

2. There is also a latent heat of sublimation, L_s, defined analogously.
3. Refer to Table 13-1 in this guide for a reminder of the direction of heat flow in these phase changes. Table 13.5 (text) gives values of latent heats for various substances.

Example 13-8. You have 6 kg of water at 20 °C. How many kcal of heat are required to turn it all into steam at 1 atm pressure and to further heat the steam to 150 °C at constant pressure? Use Tables 13.4 and 13.5 (text) for data.

S: Since there is a phase change, I can't take the whole ΔT = 130 C° at once, can I?

T: That's right, for two reasons. First, some latent heat is required to vaporize the water at 100 °C and second, steam has quite a different specific heat than the liquid. Take it one step at a time.

S: First I should find the heat required to take the water *to* the boiling point.

$$Q_1 = mc\Delta T = (6\text{ kg})(1.00\text{ kcal/kg-C}^\circ)(80\text{ C}^\circ) = 480\text{ kcal.}$$

Next we have to boil the water, requiring heat $Q_2 = mL_v$.

T: The heat of vaporization is 539.6 kcal/kg. Thus

$$Q_2 = (6\text{ kg})(539.6\text{ kcal/kg}) = 3238\text{ kcal.}$$

S: For heating the steam, I know that C_p = 8.20 cal/mole-K, and $Q_3 = mC_p\Delta T$.

T: Not quite right. Since C_p is a molar specific heat, you need to know how many *moles* there are in 6 kg of water. H_2O has a mass number of 2(1) + 16 = 18, so 6 kg is 1/3 Ikilomoles, or n = 333 moles. Then $Q_3 = nC_p\Delta T$.

S: O.K.

$$Q_3 = (333\text{ moles})(8.20\text{ cal/mole-K})(50\text{ C}^\circ) = 1.37 \times 10^5\text{ cal!}$$

Isn't this a lot?

T: It's only 137 kilocalories, which corresponds to your previous units. Thus

$$Q_{tot} = 480\text{ kcal} + 3238\text{ kcal} + 137\text{ kcal} = 3855\text{ kcal.}$$

Notice how much of this was required just for the phase change.

QUESTIONS ON FUNDAMENTALS

1. What is the Celsius temperature which is equivalent to 0 ^{o}F?

   (a.) 0 ^{o}C (b.) -32 ^{o}C (c.) -18 ^{o}C (d.) -40 ^{o}C (e.) 256 ^{o}C

2. How much of a temperature rise (in Celsius) would increase the length of a steel meter stick by 1 mm?

   (a.) 110 ^{o} (b.) 91 ^{o} (c.) 9 ^{o} (d.) 1100 ^{o} (e.) 180 ^{o}

3. How much temperature rise will 2 kg of iron undergo with the addition of 5000 J of heat?

   (a.) 22.7 x 10^{3} C^{o} (b.) .55 C^{o} (c.) 273 C^{o} (d.) 5.4 C^{o}
   (e.) 180 C^{o}

4. For 3 moles of a monatomic gas, how much heat must be added at constant pressure to raise the temperature by 10 C^{o}?

   (a.) 150 cal (b.) 90 cal (c.) 50 cal (d.) 624 cal
   (e.) 208 cal

5. What percentage of the heat added in #4 went into an increase in the internal energy of the gas?

   (a.) 20% (b.) 100% (c.) 40% (d.) 50% (e.) 60%

6. The point on a phase diagram at which gas, solid, and liquid coexist in equilibrium is called

   (a.) the critical point (b.) the fusion point (c.) the sublimation point (d.) the triple point (e.) the node of phases

7. If it takes 30 kcal of heat to melt 4 kg of a substance, what is its heat of fusion?

   (a.) 120 kcal/kg (b.) 0.133 kcal/kg (c.) 7.5 kcal/kg
   (d.) 75 kcal/kg (e.)

8. How much internal energy is there in 2.5 moles of Helium at T = 0 ^{o}C?

   (a.) 2050 J (b.) 0 J (c.) 8500 J (d.) 14200 J (e.) 3400 J

GENERAL PROBLEMS

Problem 13-1. How much heat in joules must be added to 3 kg of mercury initially at 25 °C in order to increase its volume by 1%? Use available data from Tables 13.2 and 13.3 (text).

Q1: What ΔT is necessary to accomplish a 1% volume expansion?
Q2: What determines the heat necessary to accomplish this ΔT?

Problem 13-2. At -10 °C suppose you have a steel container with 500 cm capacity. In it you have a brass sphere of radius 3 cm. The container is filled to the top with turpentine. By the time the container and its contents have warmed to room temperature of 27 °C, how much turpentine has spilled? Use Table 13.2 (text).

Q1: How much volume of turpentine is there originally?
Q2: What is the volume of the ball?
Q3: What determines how much turpentine spills?
Q4: What is the net container capacity?
Q5: Will this increase or decrease on heating?
Q6: What is the mathematical form of the changes in volume?

Problem 13-3. Suppose you have 1000 cm of steel at one atm pressure and at 95 °F. The steel sinks to the bottom of an oceanic trench 11 x 10^3 meters deep, where the water temperature is 35 °F. Considering both thermal and pressure effects, calculate the new volume of the steel. What percentage of the total volume decrease is due to pressure effects? Use Tables 10.4 and 13.2 for data.

Q1: What is the equation describing the thermal compression?
Q2: Since β is given in $(C^o)^{-1}$, what is ΔT in those units?
Q3: What is the equation describing mechanical compression?
Q4: What is ΔP?

Problem 13-4. 30 grams of steam at 200 °C and 1 atm pressure, and 150 grams of ice at -20 °C are added to an insulated sample of 100 grams of water at 20 °C. The steam cools to its vaporization point at constant pressure. What is the final equilibrium temperature? Use data from Tables 13.3, 13.4, and 13.5 (text).

Q1: How can I tell on which side of the phase changes the final temperature will lie?
Q2: Is enough heat released by the cooling and condensing steam to heat and melt all the ice?
Q3: I still don't know if all the steam will condense. Will warming the melted ice plus original water up to 100 ^oC require all the rest of the steam's energy?
Q4: Shall I assume T_f to be above or below 20 °C?
Q5: What is the expression for $|Q_{lost}|$?
Q6: What is the expression for $|Q_{gain}|$?

Problem 13-5. A steel ring is to be placed around a shaft whose outside diameter (o.d.) is 12.0094 cm at 25 ^{o}C. The ring is in the form of a band whose rectangular cross-sectional area is 2.2 cm^2, and it has an inside diameter (i.d.) of 11.9030 cm at 25 ^{o}C.

(a.) To what temperature must the ring be heated so that it will just slip onto the shaft?

(b.) What is the longitudinal tension in the ring after it has cooled back down to 25 ^{o}C?

Q1: Should I apply the expression for *linear* or *area* expansion?
Q2: What causes a longitudinal tension when the ring cools?
Q3: What relation involves the tension produced when a substance is stretched?
Q4: Is L_0 the circumference or the radius?
Q5: Are there any units problems in using this expression?

ANSWERS TO QUESTIONS

FUNDAMENTALS

1. (c.)
2. (b.)
3. (d.)
4. (a.)
5. (e.)
6. (d.)
7. (c.)
8. (c.)

GENERAL PROBLEMS:

Problem 13-1.

A1: An increase in temperature according to $\Delta V/V_o = \beta\,\Delta T$, with $\Delta V/V_o = .01$.

A2: Mercury's specific heat. $Q = mc\,\Delta T$.

Problem 13-2.

A1: 500 cm^3, minus the original volume of the brass ball.

A2: It is given by $(4\pi/3)R^3$.

A3: Its increase in volume minus the increase (or plus the decrease) in net container capacity.

A4: The volume of the steel container, minus the volume of the ball.

A5: Brass has a higher coefficient of expansion than steel, but you'll have to work out the *actual* ΔV for each, not just the *fractional* change. You should find in this case that the net capacity will increase. But notice that if the ball was bigger, it could decrease.

A6: $\Delta V_{cont} = (500\ cm^3)\beta_{steel}\,\Delta T$

$\Delta V_{ball} = (4\pi/3)R^3\beta_{brass}\,\Delta T$

$\Delta V_{turp} = [500 - (4\pi/3)R^3]\beta_{turp}\,\Delta T$, with $\Delta T = 37$ C°.

Problem 13-3.

A1: $\Delta V/V_o = \beta_{steel}\,\Delta T$

A2: $\Delta T(C^\circ) = (5/9)\Delta T(F^\circ) = -33.3$ C°. Thus ΔV will be *negative*.

A3: $\Delta V/V_o = -\Delta P/B_{steel}$, where B = bulk modulus.

A4: ΔP is the *gauge* pressure due to the weight of 11×10^3 m of water. $\Delta P = d_w gh = 1.08 \times 10^8$ Pa.

Problem 13-4.

A1: A crucial question, and one that has to be estimated before the final solution. Ask a more specific question.

A2: Cooling the steam to 100 °C will release $Q = nC_p\,\Delta T = -1370$ cal. (n = 30/18 moles). Condensing the steam will release $Q = mL_v = -16{,}200$ cal. Heating the ice to 0 °C requires $Q = mc\,\Delta T = 1500$ cal. ($c_{ice} = .5$ cal/gm-C°). Melting the ice requires $Q = mL_f = 12{,}000$ cal. So enough energy is available in the steam alone to melt *all* the ice, and the final temperature will be above 0 °C.

A3: From A2, the steam still has a surplus of 4070 cal after melting all the ice. Heating the melted ice plus the original water to 100 °C would require 15,000 cal + 8000 cal = 23,000 cal. This would require that *all* the steam condense, and T_f must lie below 100 °C.

A4: It makes no difference now. Just be consistent in identifying gains and losses of energy in the mixing.

A5: $|Q_{lost}|$ = 1370 cal + 16,200 cal + (30 gm)(1 cal/gm-C°)(100 - T_f), assuming T_f > 20 °C.

A6: Q_{gain} = 1500 cal + 12,000 cal + (150 gm)(1 cal/gm-C°)(T_f - 0) + (100 gm)(1 cal/gm-C°)(T_f - 20).

Problem 13-5.

A1: Either one. The linear expansion applies to the diameter (or circumference), while the area expansion would apply to changes in the *square* of the diameter. So the easiest choice is to use linear expansion.

A2: The shaft is holding the ring in a *stretched* condition compared to its normal length at 25 °C.

A3: (F/A) = Y($\Delta L/L_o$), where Y = Young's modulus.

A4: The ratio ($\Delta L/L_o$) is the same for either, namely 8.94 x 10^{-3}.

A5: The units of Y determine the units of F/A.

CHAPTER 14. THE TRANSFER OF HEAT AND THE FIRST LAW OF THERMODYNAMICS

BRIEF SUMMARY AND RESUME

The three ways of transferring heat, *conduction, convection*, and *radiation*, are discussed and described mathematically. *Stefan's Radiation Law* is introduced and the concepts of *emissivity* and *radiation balance* between an object and its surroundings are discussed. Applications of heat transfer are treated.

The equivalence between *mechanical work* and *heat* is discussed. Work done by an expanding gas is reviewed and the concepts of heat, work, and internal energy are combined in the expanded statement of energy conservation known as the *1st Law of Thermodynamics*. The 1st Law is applied to various processes involving ideal gases.

Material from previous chapters relevant to Chapter 14 includes the following:

1. Definition of absolute temperature, power, and heat.
2. Conservation of energy.
3. Work done by a gas at constant pressure.
4. Molar heat capacities at constant volume and constant pressure.

NEW DATA AND PHYSICAL CONSTANTS

1. Thermal Conductivity (k).

k = (rate of heat transfer per area of surface) ÷(temperature gradient through surface)

<u>SI units</u>: $(\text{watts/m}^2)/(\text{K/m}) = (\text{watts})/(\text{K-m})$

<u>Common Units</u>: $(\text{kcal})/(\text{s-m-C}^\circ)$

2. Stefan-Boltzmann Constant (σ).

For a perfect radiator,

$$P = \sigma A T^4, \text{ where } \sigma = 5.67 \times 10^{-8}\ \text{W/m-K}^4.$$

3. Emissivity (e).

For an imperfect radiator, Stefan's Law becomes

$$P = e\sigma A T^4$$

where e is the object's *emissivity*, a dimensionless number between 0 and 1.

4. Wien's constant.

For a radiating object the product of its absolute temperature and the wavelength at which maximum radiation intensity occurs is constant,

$$\lambda_{max} T = 2.898 \times 10^{-3}\ \text{m-K}.$$

5. Mechanical Equivalent of Heat. (Joule's Equivalent).

$$1 \text{ kcal} = 4.186 \times 10^3 \text{ J}.$$

NEW CONCEPTS AND PRINCIPLES

Heat Transfer

Understanding this section should enable you to do the following:

1. *Calculate the rate of heat flow through various thicknesses of various materials for a given temperature difference through the material.*
2. *Calculate the radiated power per unit area of surface for an object of given emissivity and temperature.*
3. *Find the temperature of an object of known emissivity, given its radiated power per area.*
4. *Calculate the wavelength at which the intensity of radiated power is maximum for objects at various temperatures.*

Heat Conduction--The transfer of heat through a material by vibration and collision of molecules which remain relatively fixed in position. The *rate* of heat flow by conduction is proportional to the cross-sectional *area* of material normal to the heat flow and to the *temperature gradient* across the thickness of the material. Mathematically,

$$Q/t = kA(T_2 - T_1)/L$$

where k = coefficient of thermal conductivity
L = thickness of the material.

Important Observations

1. Q/t represents a *rate* of energy transfer, or *power* transfer. SI units would be J/s or watts, while common thermal units would be kcal/s.
2. You can easily see that the units of k are either W/m-K (SI) or kcal/(s-m-C^o) in common units. Unfortunately, in the U.S. Q/t is usually measured in BTU/hr, L in inches, and A in ft^2. Thus sometimes you see k expressed in the awkward units of (BTU-in.)/(hour-ft^2-F^o)!
3. Values of k for various materials are given in Table 14.1 (text).

Example 14-1. In a test of an ice chest, it is found that 10 pounds of ice will melt in 3 hours when the chest is placed in an oven at 450 ^oF. How long would you expect the ice to last on a day when the temperature stays at 90 ^oF? Assume that the interior chest temperature remains at the melting point of ice.

S: We're not given the dimensions or the material of the chest. It doesn't seem that enough information is given.

T: k, A, and L all remain the same, so kA/L can be regarded as *one* unknown. Time is the other. Thus the two conditions given are enough as long as we don't want to know k, A, or L *separately*.

S: In both cases, the Q involved seems to be that required to melt 10 pounds of ice at 0 °C.

T: That's right. 10 pounds is 4.55 kg, so

$$Q = mL_f = (4.55\ \text{kg})(80\ \text{kcal/kg}) = 364\ \text{kcal}.$$

S: Changing to the Celsius scale, $T_2 = (450\ ^oF - 32\ ^oF)(5/9) = 232\ ^oC$ in the oven, and $T_2 = (90\ ^oF - 32\ ^oF)(5/9) = 32\ ^oC$ in the second case. Am I right in thinking a ratio of the two cases is the easiest approach?

T: Quite right. We have

$$(Q/t)_{oven}/(Q/t) = [kA(232\ ^oC - 0\ ^oC)/L]/[kA(32\ ^oC - 0\ ^oC)/L].$$

You can see that you didn't even need Q, since it cancels. Thus

$$t = t_{oven} \times (232\ ^oC)/(32\ ^oC) = (3\ \text{hours})(7.25) = 21.75\ \text{hours}.$$

Suppose the chest had walls 2 cm thick and its total surface area was 1.2 m^2. What would the material's conductivity have to be?

S: We could use either case, with $A/L = (1.2\ m^2)/(2 \times 10^{-2} m) = 0.6 \times 10^2$ m. Thus

$$k = \frac{(\Delta Q/\Delta t)_{oven}}{(232\ ^\circ C)(A/L)} = \frac{(364\ \text{kcal})/(1.08 \times 10^4\ \text{s})}{(232\ ^\circ C)(0.6 \times 10^2\ \text{m})}$$

$$= 2.42 \times 10^{-6}\ \text{kcal/(s-m-C}^o)$$

T: This is the value of k for styrofoam.

<u>Heat Convection</u>--A process in which molecules of a fluid carry thermal energy with them as they move physically into regions of different temperature. It is not possible to derive a general mathematical formula to describe all convective situations, but an empirical description is

$$Q/t = hA(T_2 - T_1),$$

where h is different for each individual set of conditions. T_2 is the temperature of the surface area A with which the convective fluid is in contact, and T_1 is the fluid temperature. If $T_2 < T_1$, we have convective *heating*, and if $T_2 > T_1$, there is convective *cooling*.

Radiative Heat Transfer--The vibrations of the atoms and electrons of a heated substance create *electromagnetic waves* which carry energy away from the substance. These waves are treated in detail in Chapter 22. A great range of wavelengths can occur from various physical processes, but those commonly produced by heated objects include mostly ultraviolet (.1 μm $< \lambda <$ .4 μm), visible light (.4 μm $< \lambda <$ .7 μm), and infrared (.7 μm $< \lambda <$ 1 μm). This wave energy is emitted or absorbed witiout any transfer of material and without any material medium required to propagate the waves.

Thermal Radiation--The electromagnetic radiation emitted by hot objects because of their temperature. Two laws express this mathematically.

Stefan's Radiation Law--The rate of energy radiated from an object at absolute temperature T is proportional to the *fourth* power of T and also proportional to the object's area A. Mathematically,

$$P = Q/t = e\sigma AT^4.$$

Wien's Displacement Law--The wavelength at which the radiated power has maximum intensity is inversely proportional to T. Mathematically,

$$\lambda_{max} = (2.898 \times 10^{-3} \text{ m-K})/T$$

Important Observations

1. The symbol e is the *emissivity* of the object, and is dependent on the material and the surface conditions. It is a *dimensionless* number between 0 and 1. Objects which are good emitters are equivalently good absorbers. A perfect absorber (e = 1) is referred to as a blackbody. Perfect reflectors have e = 0, and can neither absorb or emit radiation.
2. The Stefan-Boltzmann constant, σ, has an SI value of

$$\sigma = 5.67 \times 10^{-8} \text{ W/m}^2\text{K}^4 .$$

3. An object radiates at *any* temperature above absolute zero. At very low temperatures, however, the emitted power per area becomes very small due to the very strong dependence of P on T^4, and the wavelength for maximum intensity becomes very long (far infrared or even microwave wavelengths).
4. Stefan's Law includes radiation energy summed up over *all* emitted wavelengths.

Example 14-2. The melting point of tungsten is about 3400 °C. At what wavelength would the radiation intensity from a tungsten filament this hot be a maximum? How large an area would it take to radiate 1 kilowatt if the tungsten's emissivity is .35?

S: Wien's Law states that

$$\lambda_{max} = 2.898 \times 10^{-3} \text{ m-K}/T.$$

Do I use T = 3673 K?

T: Yes, in both Wien's Law and Stefan's Law, Kelvin temperatures are required.
S: Then

$$\lambda_{max} = 2.898 \times 10^{-3} \text{ m-K}/3673 \text{ K} = 7.89 \times 10^{-7} \text{ m}.$$

T: This is .789 μm, in what we call the near "infrared," just beyond the long wavelength visible limit of .7 μm. Even at its melting temperature, a tungsten filament isn't hot enough to reproduce the color content of sunlight.

S: If e = .35, the radiated power *per area* would be

$$P/A = e\sigma T^4 = (.35)(5.67 \times 10^{-8} \text{ W/m}^2\text{K}^4)(3673 \text{ K})^4 = 3.6 \times 10^6 \text{ W/m}^2.$$

T: So it would take only

$$(1000 \text{ W})/(3.6 \times 10^6 \text{ W/m}^2) = 2.77 \times 10^{-4} \text{ m}^2 = 2.77 \text{ cm}^2$$

to radiate 1000 watts.

Radiation Balance--The *net* power radiated or absorbed by an object is the difference between the amount it radiates into its surroundings and the amount it absorbs from its surroundings. Mathematically,

$$P_{net} = P_{rad} - P_{abs}.$$

If the object is at a Kelvin temperature T_2 and the surroundings are at temperature T_1, this means

$$P_{net} = e\sigma A(T_2{}^4 - T_1{}^4),$$

where we assume the usual case that e is the same for emission and absorption, and that the emitting and absorbing areas are the same. There are some exceptions to this rule, as shown in some of the examples to follow. When $P_{net} = 0$, the object's radiated power is in *balance* with the power being absorbed from its environment.

Example 14-3. Suppose a black plate (e = .9) is in direct sunlight whose intensity is 800 W/m^2. What final temperature does the plate reach when radiation balance is achieved with the sunlight?

S: What do we use for the sunlight temperature, T_1?

T: Here you are directly given the power incident on the plate, 90% of which is absorbed:

$$P_{abs}/A = (.90)(800 \text{ W/m}^2) = 720 \text{ W/m}^2.$$

S: Don't we need to know the plate area?

T: The final temperature doesn't depend on it. What counts is P/A. When the plate is at temperature T, it radiates power according to

$$P_{rad}/A = e\sigma T^4.$$

What is the condition for radiation balance?

S: That $P_{abs}/A = P_{rad}/A$. This means $e(800\ W/m^2) = e\sigma T^4$. It seems that e cancels out, too.

T: That's right, as long as absorptivity is equal to emissivity. Also, we have assumed that the absorbing area and emitting areas are the same. This would mean, for instance, that the "back side" of the plate was insulated from the surroundings.

S: Then

$$T^4 = (800\ W/m^2)/(5.67 \times 10^{-8}\ W/m^2K^4) = 1.41 \times 10^{10}\ K^4.$$

This gives

$$T = 345\ K = 72\ ^oC.$$

Somehow this seems somewhat low. Shouldn't the plate be able to boil water?

T: The plate is probably in an ambient air temperature of around $T_1 = 300$ K, so it will absorb more heat than just the sunlight. Our radiation balance should include this:

$$P_{abs}/A = e(800\ W/m^2) + e\sigma T_1^4 = e\sigma T_2^4.$$

This gives

$$T_2^4 = 800\ W/m^2 + T_1^4 = 2.22 \times 10^{10}\ K^4.$$

So

$$T_2 = 386\ K = 113\ ^oC.$$

Example 14-4. Since sunlight is roughly half visible light and half infrared, and the radiated energy from a solar collector is essentially all infrared, the collector's performance can be improved by coating it with a substance which has e = .9 at visible wavelengths and e = .1 at infrared wavelengths. Such a coating is called a *selective coating*. Calculate the final temperature of the plate in Example 14-3 if it had such a coating.

S: Do you mean half of the sunlight is 90% absorbed (visible wavelengths), and half of it is 10% absorbed (infrared)?

T: Right. Furthermore, *all* the emission from the plate occurs at 10% emissivity. The surroundings at 300 K will also be entirely in the infrared, so the plate will absorb only 10% of that radiation. Can you collect all this in mathematical form?

S: I think so. We have

$$P_{abs}/A = .9(400\ W/m^2) + .1(400\ W/m^2) + 0.1\sigma T_1^4,$$

and

$$P_{rad}/A = 0.1\sigma T_2^4.$$

We need to set these equal and solve for T_2.

T: Very good. This leads to

$$T_2^4 = (1/\sigma)4000\ W/m^2 + (300\ K)^4$$

which gives

$$T_2 = 529\ K = 256\ ^oC.$$

As you can see, this is a remarkable improvement in performance.

Heat and Work

Understanding this section should enable you to do the following:

1. *Convert between mechanical and thermal units of energy.*
2. *Calculate the work of expansion done by a gas as it changes pressure, volume and temperature in certain simple processes.*
3. *Use the 1st Law of Thermodynamics to relate changes in internal energy of a system to the exchange of work and heat with its surroundings.*

The concepts of mechanics and heat are unified through the *mechanical equivalent of heat*, which reconciles the units of work and thermal energy.

Mechanical Equivalent of Heat--

$$1\ kcal = 4.186 \times 10^3\ J.$$

Work Done by a Gas--This depends on the particular process by which the gas changes its pressure, temperature, and volume. For an ideal gas, it is sufficient to specify any two of the above quantities, since the ideal gas law then determines the third. It is convenient to describe changes in a gas by tracing its pressure and volume changes on a graph, which gives a "P-V curve" for any given process. *The work done by the gas can then be found by measuring the area under its P-V curve.* For a constant volume process, the P-V curve is a vertical line, which has no area under it. This means no work is done unless the volume changes. For a process at constant pressure, the P-V curve is a horizontal line with area $P(V_2 - V_1)$ or $W = P\ V$.

Important Observations

1. When a gas changes P, V, or T, it is said to change its *thermodynamic state.* The work done in such a change depends on the *method* of change, not just on the initial and final states.
2. If ΔV is *positive*, W is positive, meaning the gas *did* work when expanding. If ΔV is *negative*, the gas was *compressed*, and *had work done on it.* This is the meaning of negative W.

Example 14-5. The P-V diagram below shows two processes by which a gas is changed from state A to state B. Calculate the work done by the gas in the processes AB and ACB.

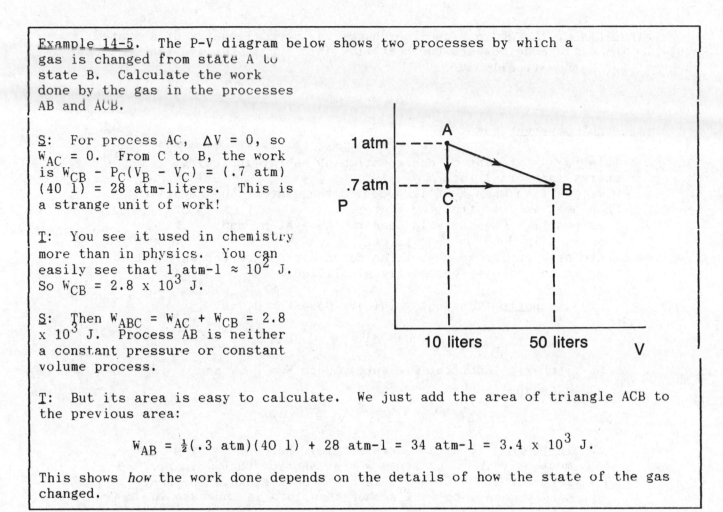

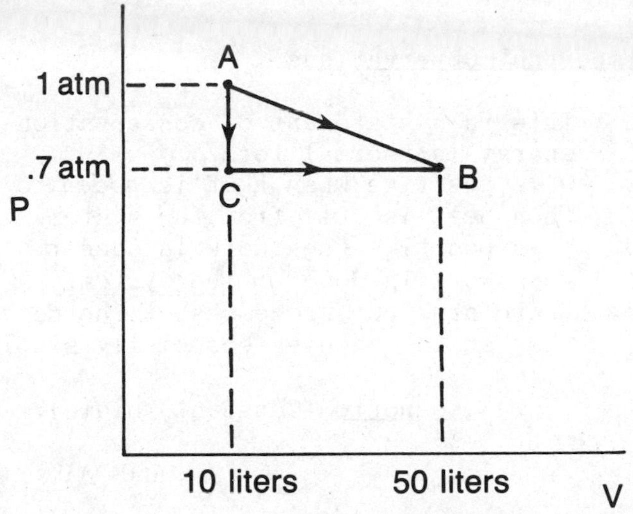

S: For process AC, $\Delta V = 0$, so $W_{AC} = 0$. From C to B, the work is $W_{CB} = P_C(V_B - V_C) = (.7\text{ atm})(40\text{ l}) = 28$ atm-liters. This is a strange unit of work!

T: You see it used in chemistry more than in physics. You can easily see that 1 atm-l $\approx 10^2$ J. So $W_{CB} = 2.8 \times 10^3$ J.

S: Then $W_{ABC} = W_{AC} + W_{CB} = 2.8 \times 10^3$ J. Process AB is neither a constant pressure or constant volume process.

T: But its area is easy to calculate. We just add the area of triangle ACB to the previous area:

$$W_{AB} = \tfrac{1}{2}(.3\text{ atm})(40\text{ l}) + 28\text{ atm-l} = 34\text{ atm-l} = 3.4 \times 10^3\text{ J}.$$

This shows *how* the work done depends on the details of how the state of the gas changed.

Internal Energy (U)--The sum of the kinetic and potential energies of all the individual particles making up the system. This can change for two reasons: *heat* can flow into or out of the system, and the system can do *work* or have work done on it.

State Functions--Quantities which depend *only* on the values of the *thermodynamic state variables*, P, V, and T, and do *not* depend on the *process* by which these values were achieved.

Important Observations

1. From the preceding discussion, *work is not* a state function. In Chapter 13 we saw that the heat exchanged with a gas depends on the type of process, so Q cannot be a state function, either.
2. Internal energy depends *only* on temperature, regardless of how the temperature was reached. Therefore U *is* a state function.
3. In a thermodynamic *cycle*, where the system returns to its original thermodynamic state, there can be *no net change* of a state function, even though Q and W may be non-zero for the cycle.

First Law of Thermodynamics--The change in internal energy of a system is equal to the difference between the heat flow *into* the system and the work *done* by the system. Mathematically,

$$\Delta U = Q - W$$

Important Observations

1. This is a statement of conservation of energy *including* energy in thermal form.
2. Q is positive when heat is *added* to the system, and negative when heat is *lost* from the system.
3. W is positive when work is done *by* the system, and negative when work is done *on* the system.
4. There are four processes of thermodynamic change for which the 1st Law becomes epsecially simplified:

   a. Isochoric (Constant Volume). Here V = 0, so

   $$W = 0 \quad \text{and} \quad \Delta U = Q = nC_v \Delta T$$

   b. Isobaric (Constant Pressure). Here $W = P\Delta V$ and $Q = nC_p \Delta T$, so

   $$\Delta U = nC_p \Delta T - P\Delta V = nC_v \Delta T.$$

   c. Isothermal (Constant Temperature). Here $\Delta T = 0$, so we must have $\Delta U = 0$, hence $Q = W$. Notice that Q is *not* zero for $\Delta T = 0$, which is a common mistake. The heat merely goes into work rather than into an increase in internal energy.

   d. Adiabatic (No Heat Transfer). Here $Q = 0$, so $\Delta U = -W$.

5. Notice that in *all* cases, we can find ΔU from the expression $\Delta U = nC_v \Delta T$. If a system undergoes a temperature change ΔT, U must change by a certain amount, *regardless* of the process.
6. For *any* complete thermodynamic *cycle*, the system returns to its original T, so $\Delta U = 0$ and $Q = W$ for the cycle.

Example 14-6. Return to Example 14-5, and calculate ΔU and Q for each of the processes AC, AB, and CB. Assume one mole of an ideal monatomic gas.

S: We've seen that $W_{AC} = 0$, which leaves $\Delta U_{AC} = Q_{AC} = nC_V \Delta T$. It seems as though we need to find the temperatures T_A, T_B, and T_C.

T: Those come from the ideal gas law. Remember to choose R so that the units are consistent.

$$T_A = P_A V_A/nR = (1\text{ atm})(10\text{ l})/(1\text{ mole})(.0821\text{ atm-l/mole-K}) = 122\text{ K}.$$

The other temperatures are easily found by ratios:

$$T_C = (P_C/P_A)T_A = (.7)(122\ K) = 85\ K$$

$$T_B = (V_B/V_C)T_C = (5)(85\ K) = 425\ K$$

S: Then

$$\Delta U_{AC} = Q_{AC} = (1\ mole)(3/2)(.0821\ atm\text{-}l/mole\text{-}K)(85 - 122)\ K$$

$$= -4.56\ atm\text{-}l = -4.56 \times 10^2\ J.$$

T: Notice that this means the system gave up this much internal energy as heat loss.

S: For process CB, we have W_{CB} = 28 atm-l. Is it easier to calculate ΔU or Q next?

T: It doesn't make much difference. You know that $\Delta U = nC_V \Delta T$ for *all* processes, so you may as well use this again.

S: This gives

$$\Delta U_{CB} = (1\ mole)(3/2)(.0821\ atm\text{-}l/mole\text{-}K)(425 - 85)\ K$$

$$= +42\ atm\text{-}l = 42 \times 10^2\ J.$$

T: Then $Q_{CB} - \Delta U_{CB} + W_{CB} = 70 \times 10^2$ J. Of this, 42×10^2 J went into internal energy, and 28×10^2 J did the work of expansion.

S: Finally, W_{AB} - 34 atm-l = 34×10^2 J. But now I don't know how to write down either ΔU or Q for the process AB.

T: Ah, but you do! Again, $\Delta U = nC_V \Delta T$ for *all* processes, and you know ΔT.

S: You mean

$$\Delta U_{AB} = (1\ mole)(3/2)(.0821\ atm\text{-}l/mole\text{-}K)(425 - 122)\ K$$

$$= 37.4\ atm\text{-}l = 37.4 \times 10^2\ J?$$

T: That's right. Or, you could say that

$$\Delta U_{AB} = \Delta U_{AC} + \Delta U_{AB} = -4.6 \times 10^2 + 42 \times 10^2\ J = 37.4 \times 10^2\ J.$$

You *cannot* make similar statements about Q or W, which depend on the particular processes involved. It follows that

$$Q_{AB} = \Delta U_{AB} + W_{AB} = 37.4 \times 10^2 + 34 \times 10^2\ J = 71.4 \times 10^2\ J.$$

Notice that a larger fraction of this heat went into work than in the process CB.

Example 14-7. How much heat has to be added to 36 grams of water at 100 °C and 1 atmosphere pressure to boil it and raise the steam temperature to 120 °C, maintaining the same pressure? How much work has the water done, and what is the change in its internal energy?

S: The heat needed will be $Q = mL_v + nC_P \Delta T$. What are n and C_P?

T: Since the molar mass of water is 18, we have n = 2 moles. The value of $\gamma = C_P/C_V = 1.3$ for steam. Along with $C_P = C_V + R$, this means that

$$C_P = 1.3C_V = 1.3(C_P - R), \text{ giving } C_P = 4.33\ R$$

S: Then

$$Q = (36\ \text{gm})(540\ \text{cal/gm}) + (2\ \text{moles})(4.33)(1.99\ \text{cal/mole-K})(20\ \text{C}^o)$$

$$= 19440 + 345\ \text{cal} = 19800\ \text{cal}.$$

The work done at constant pressure is $P\Delta V$. How do we find ΔV?

T: $\Delta V = V_{steam} - V_{liquid}$, and $V_{liq} = m/d = 36\ \text{gm}/1(\text{gm/cm}^3) = 36\ \text{cm}^3$. V_{steam} can be found from the ideal gas law.

S: O.K. That gives

$$V_{steam} = nRT/P = (2\ \text{moles})(.0821\ \text{atm-l/mole-K})(393\ \text{K})/(1\ \text{atm})$$

$$= 64.5\ \text{l} = 64.5 \times 10^3\ \text{cm}^3.$$

T: Good. You remembered to use kelvins for T. Thus

$$W = (1.01 \times 10^5\ \text{N/m}^2)(\ 64.5 \times 10^{-3}\ \text{m}^3 - 36 \times 10^{-6}\ \text{m}^3)$$

$$= 6.5 \times 10^3\ \text{J} = 1.55 \times 10^3\ \text{cal}.$$

S: You really have to keep concentrating with so many choices of units.

T: That's right, but don't let them intimidate you. The more you work with conversion factors, the easier it becomes. Now we can easily find ΔU:

$$\Delta U = Q - W = 1.98 \times 10^4\ \text{cal} - 1.55 \times 10^3\ \text{cal} = 1.82 \times 10^3\ \text{cal}.$$

QUESTIONS ON FUNDAMENTALS

1. How much heat is lost per hour by conduction through the wall of a house which is 8 feet high and 20 feet long, if the wall is essentially made of 6 inches thick fiberglass (k = 0.27 (BTU inches)/(hr ft^2 F°) and the outside and inside temperatures are -20 °C and +20 °C, respectively?

(a.) 1024 BTU/hr (b.) 518 BTU/hr (c.) 288 BTU/hr
(d.) 160 BTU/hr (e.) 86.5 BTU/hr

2. How much heat per hour would an area equal to the wall in #1 lose by radiation to the outside if the emissivity was 0.5? Take the wall temperature to be +20 °C and the environmental temperature to be -20 °C. (1 Watt = 3.4 BTU/hr; 1 ft^2 = .093 m^2)

(a.) 4690 BTU/hr (b.) 9380 BTU/hr (c.) 1380 BTU/hr
(d.) 10570 BTU/hr (e.) 5160 BTU/hr

3. At what wavelength would a +20 °C wall radiate at maximum intensity?

(a.) 9.89×10^{-3} m (b.) 1.45×10^{-4} m
(c.) 5.8×10^{-2} m (d.) 9.89×10^{-6} m (e.) 1.45×10^{-6} m

4. An ideal monatomic gas at a constant pressure of 3 atmospheres expands its volume by 15 liters. While this is happening, 1500 J of heat are added to the gas. What is the resulting change in the gas' internal energy?

(a.) +3000 J (b.) +1455 J (c.) -3000 J (d.) -1455 J (e.) +1545 J

5. What would be the temperature change of one mole of the gas undergoing the process in #4?

(a.) +144 K (b.) +70 K (c.) -86 K (d.) -70 K
(e.) -144 K

6. How far would a 200 pound (890 N) man have to climb vertically in order to "work off" 1000 kcalories of food energy?

(a.) 480 m (b.) 1150 m (c.) 4700 m (d.) 108 m
(e.) 654 m

GENERAL PROBLEMS

Problem 14-1. The sun's radiant power is 3.9×10^{26} watts. At the earth, what is the intensity of solar radiation? If 70% of the solar radiation falling on the side of the earth facing the sun is absorbed, and the entire surface of the earth radiates at the temperature determined by radiation balance, what would this temperature be?

Q1: How far from the sun is the earth?
Q2: Over how many square meters is the solar radiation distributed this far from the sun?
Q3: What is the effective absorbing area of the earth?
Q4: What is the effective radiating area of the earth?
Q5: What is the equation determining T_{rad}?

Problem 14-2. Suppose you are stirring 2 kg of a liquid using a hand cranked paddle wheel. You exert a tangential force of 5 N at the end of a 25 cm handle, which causes the paddles to turn at a rate of 2 revs/sec. If the liquid has a specific heat of 0.6 cal/gm-C^{o}, and is in a well insulated container, how much temperature rise would result after 2 minutes of stirring?

Q1: Why does the temperature rise?
Q2: How much work has been done during 2 minutes?
Q3: What expression connects the work to ΔT?
Q4: What is the connection between Q and W?

Problem 14-3. An insulated beverage container is filled with liquid at 50 ^{o}C. A 1 cm thick layer of insulation having thermal conductivity $k = 2.4 \times 10^{-6}$ kcal/s-m-C^{o} fills the space between the walls. The inner wall has an emissivity e = .10. Assume the insulation is transparent to the radiated wavelengths. Compare the rates of heat loss per area by conduction and by radiation when the outer wall is at 20 ^{o}C.

Q1: What is the rate of conduction of heat per area?
Q2: What is the net rate of radiation of heat per area?
Q3: Are there problems with units?

Problem 14-4. 2 moles of an ideal gas with $\gamma = 1.4$ is taken through the thermodynamic process ABC shown in the figure below. Find the work done, the change of internal energy, and the heat exchanged for this process.

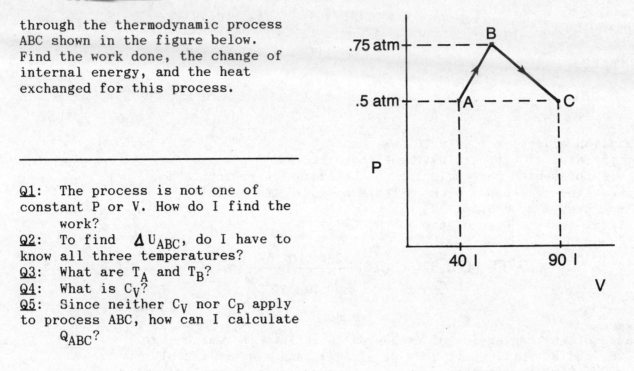

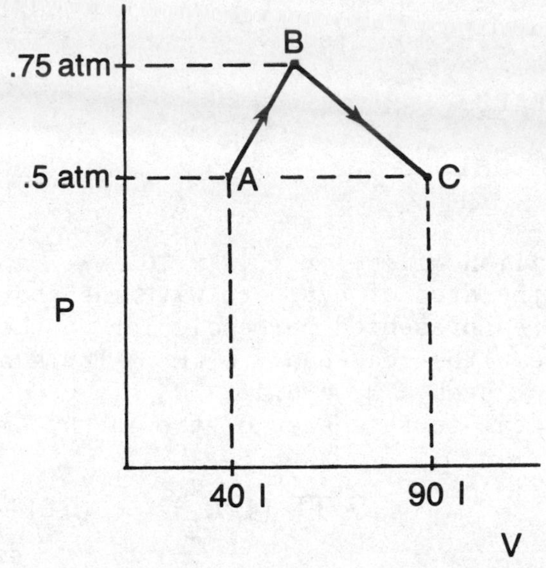

Q1: The process is not one of constant P or V. How do I find the work?

Q2: To find ΔU_{ABC}, do I have to know all three temperatures?

Q3: What are T_A and T_B?

Q4: What is C_V?

Q5: Since neither C_V nor C_P apply to process ABC, how can I calculate Q_{ABC}?

ANSWERS TO QUESTIONS

1. (b.) 2. (a.) 3. (d.) 4. (c.)
5. (e.) 6. (a.)

FUNDAMENTALS:

GENERAL PROBLEMS:

Problem 14-1.

A1: 93 million miles, or 1.49×10^{11}m.

A2: Over the area of a sphere with the above radius.

A3: The area presented perpendicular to the radiation, which is a circle with area equal to πR^2. It is *not* half the earth's surface area. $R_E = 6.4 \times 10^6$ m.

A4: The total surface area of the earth, $4\pi R_E^2$.

A5: That

$$\sigma T_{rad}^4 (4\pi R_E^2) = (.70)\frac{3.9 \times 10^{26}\,\text{W}}{4\pi(1.49 \times 10^{11})^2}(\pi R_E^2)$$

Problem 14-2.

A1: Because of the conversion of mechanical work into thermal energy.

A2: $W = \tau\theta$, and $\theta = \omega t$, where $\omega = 2(2\pi)$rad/s, and $\tau = (5\text{N})(.25\text{m})$.

A3: $\Delta T = Q/mc$, where Q and c are usually expressed in common thermal units.

A4: The mechanical equivalent of heat, 1 cal = 4.186 J.

Problem 14-3.

A1: $(1/A)(Q/t)_{cond} = k\,\Delta T/L$, where L = 1 cm

A2: $(1/A)(Q/t)_{rad} = e\sigma(T_{in}^4 - T_{out}^4)$

A3: In the conduction formula you'll get units of kcal/s-m^2. In the radiation formula, you *must* use T in *kelvins*, and σ is in SI units. Thus you'll get watts/m^2. You have to use the mechanical equivalent of heat to reconcile these answers.

Problem 14-4.

A1: The process lends itself to easy calculation of area under the P-V curve. Thus $W_{ABC} = (.5\text{ atm})(90\text{ l} - 40\text{ l})$ + area of triangle ABC

A2: No. Since U is a state function, $\Delta U_{ABC} = \Delta U_{AC} = nC_V(T_C - T_A)$. Only the end points matter.

A3: The ideal gas law gives $T_A = (P_AV_A/nR)$. Then $T_B = (V_B/V_A)T_A$.

A4: From $\gamma = 1.4$ and $C_P = C_V + R$, you should find $C_V = (5/2)R$.

A5: You have both W_{ABC} and U_{ABC}. Now use the 1st Law of Thermodynamics, $Q_{ABC} = \Delta U_{ABC} + W_{ABC}$.

CHAPTER 15. THE SECOND LAW OF THERMODYNAMICS

BRIEF SUMMARY AND RESUME

The difference between *reversible* and *irreversible* processes is discussed. Pressure-volume relations in *isothermal* and *reversible adiabatic* processes are described and expressions are given for the *work* done in these processes. The concept of order and disorder in systems is introduced and *entropy* is defined as a quantitative measure of disorder. The tendency of any closed system to become more disordered is linked to the principle of *entropy increase* in irreversible processes. *Heat engines* are introduced, and the concept of *1st Law efficiency* is applied to various examples. The *Carnot cycle* is derived as the heat engine of *maximum* efficiency for a given temperature difference. The *Second Law of Thermodynamics* is discussed through various interpretations of the foregoing principles. Finally, *heat pumps* and *refrigeration cycles* are discussed and their *coefficients of performance* are defined with the help of the 2nd Law.

Material from previous chapters which is relevant to Chapter 15 includes:

1. The 1st Law of Thermodynamics.
2. Molar specific heats of ideal gases.
3. P-V description of thermodynamic processes.

NEW DATA AND PHYSICAL CONSTANTS

1. Entropy ($\Delta S = Q/T$).

   SI units: J/K

   Common units: kcal/K

NEW CONCEPTS AND PRINCIPLES

Reversibility and Irreversibility

Understanding this section should enable you to do the following:

1. *Distinguish between reversible and irreversible thermodynamic processes.*
2. *Calculate the P-V behavior and work done in isothermal and reversible adiabatic processes.*

Reversible Process--A change in thermodynamic state in which all intermediate states between the initial and final states are well defined. This means the process must be slow enough to allow the system to always be in thermal equilibrium. Such a process is an *ideal* which can be closely approximated but never *exactly* attained.

Irreversible Process--A sudden change of state in which the system cannot pass through a series of intermediate equilibrium states. Also a process in which heat is irreversibly lost to the system's surroundings.

Isothermal Process--A change of state in which the system's *temperature* is held constant. Boyle's Law for ideal gases states that under this condition,

$$PV = \text{constant, or } P_1V_1 = P_2V_2 = \ldots \text{ etc.}$$

Also, calculus yields the following expression for the work done in an isothermal process:

$$W_{isotherm} = P_oV_o \ln(V_f/V_o)$$

Adiabatic Process--A change of state in which *no heat is exchanged* between the system and its surroundings. This can occur either because of complete insulation or because the process happens so suddenly that heat transfer doesn't have time to occur. The methods of calculus give us the following behavior of adiabatic processes:

$$PV^{\gamma} = \text{constant, or } TV^{\gamma-1} = \text{constant,}$$

where γ is the ratio for molar specific heats, C_p/C_V. Also, the work done in an adiabatic process is given by

$$W_{adiabatic} = (P_oV_o - P_fV_f)/(\gamma - 1)\,.$$

Important Observations

1. Isothermal and slow adiabatic processes can be considered reversible.
2. Isothermal processes are represented by *hyperbolas* in the P-V plane, since this is the curve described by PV = const. Examples are showm in Figure 15.3 (text).
3. Adiabatic processes are *steeper* curves in the P-V plane than isotherms, since PV^{γ} = const., and $\gamma > 1$. Examples are shown in Figure 15.6 (text).
4. Even though calculus is required to derive the expressions for work in these processes, notice that the results are very easy to apply.
5. Review Chapter 14 in this guide for the way the 1st Law of Thermodynamics applies to these processes. To summarize:

   Isothermal: $\Delta T = 0$, so $\Delta U = 0$. Then Q = W. (Note that Q is *not* necessarily 0!)

   Adiabatic: Q = 0, so $\Delta U = -W$.

Example 15-1. A sample of air ($\gamma = 1.4$) is slowly compressed from 1 atm pressure to 2 atm. The original volume and temperature are V_1 = 20 liters and T_1 = 300 K. The temperature remains at 300 K during this process. Next the air is suddenly (adiabatically) expanded back to a pressure of 1 atm.

(a.) Find the final volume and temperature.
(b.) Find ΔU, Q, and W for each of the two processes.

S: Do I have to find the intermediate volume?

T: Yes, because the processes involve different relationships between their initial and final state variables.

S: The isothermal case is easy, since $P_1V_1 = P_2V_2$. Thus doubling the pressure causes the volume to be cut in half, to $V_2 = 10$ l.

T: This is now the initial volume for the adiabatic process.

S: Now we have $P_2V_2^{\gamma} = P_3V_3^{\gamma}$, with $P_2 = 2$ atm, $P_3 = 1$ atm, and $V_2 = 10$ l. So

$$(2 \text{ atm})(10 \text{ l})^{\gamma} = (1 \text{ atm})(V_3)^{\gamma}.$$

T: To get V_3, you have to raise the equation to the $1/\gamma$ power:

$$V_3 = (2)^{1/\gamma} (10 \text{ l}) = (2)^{(1/1.4)} (10 \text{ l}) = 16.4 \text{ l}.$$

S: For the temperature, we have $TV^{\gamma-1}$ = const. So with $T_2 = 300$ K, $V_2 = 10$ l, and $V_3 = 16.4$ l, this gives

$$T_3 = T_2(V_2/V_3)^{\gamma-1} = (300 \text{ K})(10 \text{ l}/16.4 \text{ l})^{(.4)} = (300 \text{ K})(.82) = 246 \text{ K}.$$

T: Very good. Another way of finding this is from the ideal gas law. Since n is constant, PV/T is constant, so we can write

$$P_2V_2/T_2 = P_3V_3/T_3 \text{ , which gives } T_3 = (P_3/P_2)(V_3/V_2)T_2 = 246 \text{ K}.$$

S: For the isotherm, $\Delta T = 0$, so $\Delta U = 0$. I don't see how I can calculate Q when $\Delta T = 0$.

T: Remember, Q doesn't need to increase internal energy. It can go into work too. Here *all* of it converts to work. For an isotherm

$$W = P_1V_1 \ln(V_2V_1) = (1 \text{ atm})(20 \text{ l}) \ln(1/2)$$

$$= (20 \text{ atm-l})(-.693) = -13.9 \text{ atm-l} = -13.9 \times 10^2 \text{ J}.$$

Also, Q = W. This is the amount of work done *on* the gas to compress it. To remain at constant temperature, this amount of heat must flow *out* of the gas.

S: Now, for the adiabatic process Q = 0 by definition. We're given the expression for adiabatic work:

$$W = (P_2V_2 - P_3V_3)/(\gamma - 1) = [(2 \text{ atm})(10 \text{ l}) - (1 \text{ atm})(16.4 \text{ l})]/(0.4)$$

$$= +9 \text{ atm-l} = +9 \times 10^2 \text{ J}.$$

Is this right?

T: Sure, it's an easy expression with which to calculate. This is the amount of work done by the gas in expanding, and since Q = 0, $\Delta U = -W$ means the internal energy has *decreased* by this amount in order to do the work.

Entropy and Disorder

Understanding this section should enable you to do the following:

1. *Calculate the entropy change in reversible thermodynamic processes.*
2. *Identify the connection between entropy increase in a system and the degree of disorder in the system.*

One of the most profound accomplishments in physics is the understanding of why thermal energy "flows" from high temperature to low temperature and never the reverse, even though the 1st Law of Thermodynamics would be satisfied either way. This means, for instance, that mixing hot and cold objects always results in them coming to some common equilibrium temperature rather than the energy flow resulting in further heating the hot object and cooling the cold object. The answer lies in observing that the equilibrium condition is one of *maximum randomness* or *disorder*. It seems that any isolated system will tend to a condition of increased randomness because this is a condition having *greater probability* than a highly ordered state. In order to quantify this amount of disorder, the concept of *entropy* is introduced.

Entropy (S)--A thermodynamic state function related to the amount of disorder in system. The *change* in entropy of a system during an *isothermal* process is the ratio of the heat flow into the system to the system's temperature during the process. Mathematically,

$$(\Delta S)_{isotherm} = Q/T$$

Important Observations

1. The units of entropy are:

   SI: J/K

   Common: kcal/K

2. Since Q = 0 for an *adiabatic* process, $\Delta S = 0$ also. Thus an adiabatic process can be called *isoentropic*.
3. Entropy is a state function for reversible processes. That means that ΔS will be the *same* for *any* choice of reversible processes between the same two states, A and B. We can always choose to go from A to B via a combination of an adiabatic plus an isothermal process. Since $\Delta S = 0$ for an adiabatic, ΔS_{AB} will simply be that of the isotherm involved, Q/T. Thus we have the general result that for *all* reversible processes, $\Delta S = Q/T$.
4. For any reversible *cycle*, we must have $\Delta S_{cycle} = 0$.
5. Remember, for isothermal processes, Q = W, so

$$\Delta S_{isotherm} = W_{isotherm}/T .$$

Example 15-2. In Example 14-5, what is the entropy change ΔS for the isobaric process CB? Assume one mole of an ideal monatomic gas.

S: We have $Q = nC_p\Delta T$ for isobaric processes, and so

$$\Delta S_{CB} = Q/T = (nC_P \Delta T)/T .$$

T: That is *not* a correct approach. For instance, what T will you use, since CB is not an isotherm? This expression doesn't have any meaning.

S: I thought that expression $\Delta S = Q/T$ was good for *any* reversible process. We don't have a general expression for ΔS_{isobar}.

T: The following will show in what sense $Q/T = \Delta S$. It is possible to go from C to B by taking an *isotherm* to an intermediate state A which in turn is connected to B by an *adiabatic*. This is shown in the figure to the right. Then $\Delta S_{CA} = Q_{CA}/T_C$ and $\Delta S_{AB} = 0$.

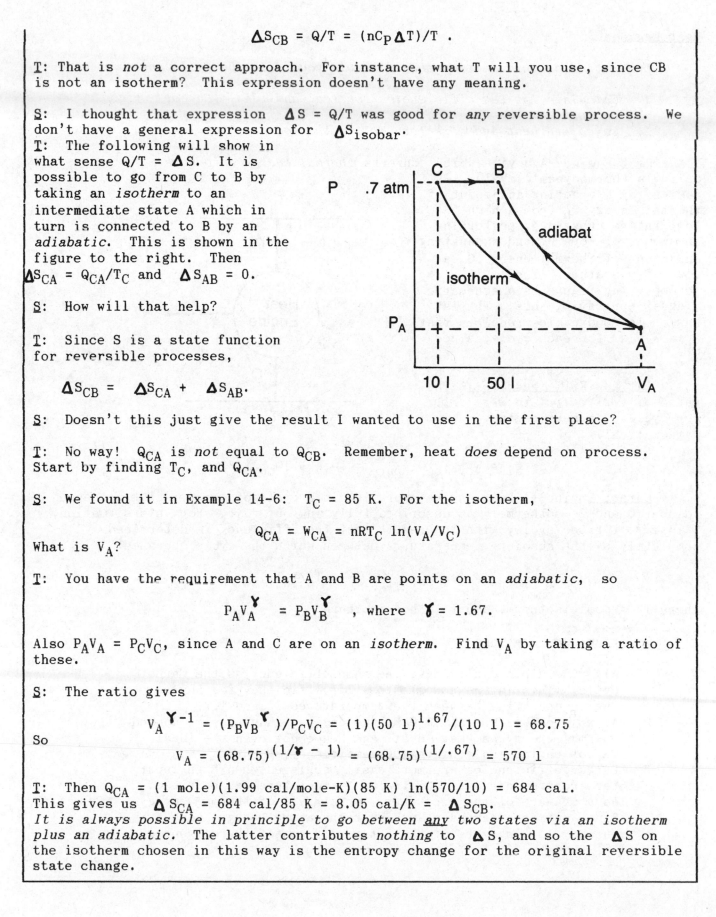

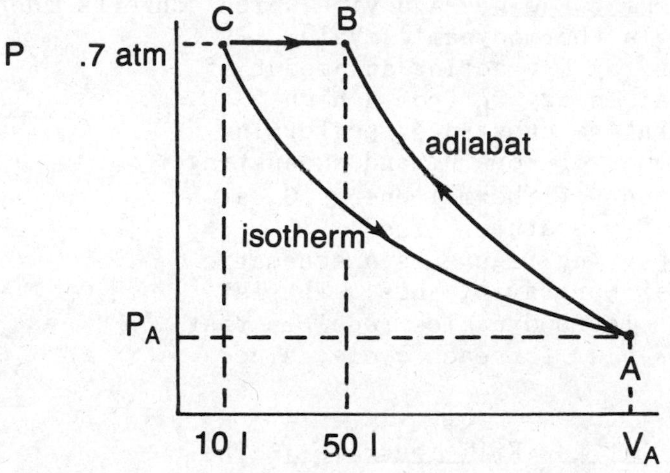

S: How will that help?

T: Since S is a state function for reversible processes,

$$\Delta S_{CB} = \Delta S_{CA} + \Delta S_{AB}.$$

S: Doesn't this just give the result I wanted to use in the first place?

T: No way! Q_{CA} is *not* equal to Q_{CB}. Remember, heat *does* depend on process. Start by finding T_C, and Q_{CA}.

S: We found it in Example 14-6: $T_C = 85$ K. For the isotherm,

$$Q_{CA} = W_{CA} = nRT_C \ln(V_A/V_C)$$

What is V_A?

T: You have the requirement that A and B are points on an *adiabatic*, so

$$P_A V_A^{\gamma} = P_B V_B^{\gamma} \text{ , where } \gamma = 1.67.$$

Also $P_A V_A = P_C V_C$, since A and C are on an *isotherm*. Find V_A by taking a ratio of these.

S: The ratio gives

$$V_A^{\gamma - 1} = (P_B V_B^{\gamma})/P_C V_C = (1)(50\ l)^{1.67}/(10\ l) = 68.75$$

So

$$V_A = (68.75)^{(1/\gamma - 1)} = (68.75)^{(1/.67)} = 570\ l$$

T: Then $Q_{CA} = (1 \text{ mole})(1.99 \text{ cal/mole-K})(85 \text{ K}) \ln(570/10) = 684$ cal. This gives us $\Delta S_{CA} = 684 \text{ cal}/85 \text{ K} = 8.05 \text{ cal/K} = \Delta S_{CB}$. *It is always possible in principle to go between* <u>*any*</u> *two states via an isotherm plus an adiabatic.* The latter contributes *nothing* to ΔS, and so the ΔS on the isotherm chosen in this way is the entropy change for the original reversible state change.

Heat Engines

Understanding this section should enable you to do the following:

1. *Calculate 1st Law efficiencies for real and ideal (Carnot) engines.*
2. *Calculate heat and work for various thermodynamic cycles.*

Heat Engine--A device which converts *thermal energy into work* during a thermodynamic cycle. It does so by extracting an amount of thermal energy Q_h from a high temperature substance, performing an amount of work W, and exhausting an amount of thermal energy Q_c at a lower temperature. The accompanying figure is a schematic diagram portraying this. The 1st Law of Thermodynamics requires that $Q_h = W + Q_c$ for each cycle, since $\Delta U = 0$.

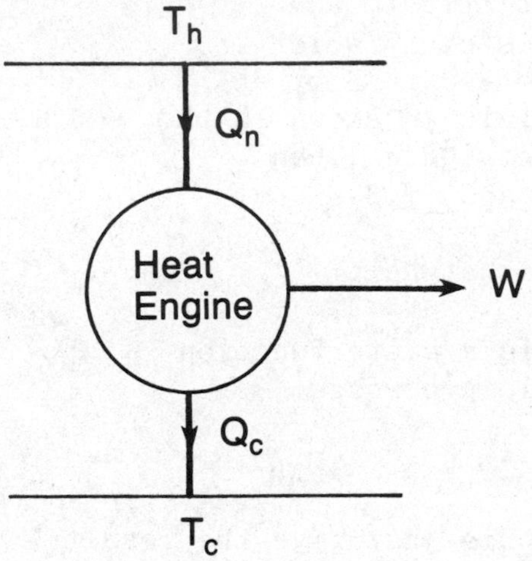

1st Law Efficiency (e_1)--The ratio of *work output* to *heat input* per cycle of a heat engine. Mathematically,

$$e_1 = W_{cycle}/Q_{input} = [Q_{in} - |Q_{out}|]/Q_{in}$$

Carnot Engine--A heat engine operating in a cycle of 4 *reversible* processes, 2 adiabatic and 2 isothermal, as described fully in your text. For this situation, S is *zero* for the cycle, with the result that the efficiency is determined completely by the absolute temperatures between which the cycle operates:

$$(e_1)_{Carnot} = 1 - (T_c/T_h) ,$$

where T_c = cool isotherm, and T_h = hot isotherm.

Important Observations

1. All reversible cycles have the same efficiency as the Carnot cycle. This is the *maximum possible* efficiency for any engine operating between the temperatures T_c and T_h.
2. At *any* temperature above 0 K, the engine exhausts some unused thermal energy as waste heat, and hence not even the ideal Carnot engine can achieve 100% efficiency.
3. Larger efficiencies are more easily achieved when high input temperatures are available. Thus temperature can be viewed as a measure of the *usefulness*, or *quality* of thermal energy. Since entropy increases as high temperature energy cools, entropy is a measure of the *non-usefulness* of thermal energy.

Example 15-3. An engine, operating between temperature extremes of 30 °C and 300 °C at 70% of maximum theoretical efficiency, exhausts 2 kcal of heat per cycle. How much work does it perform per cycle?

S. The work done is

$$W = e_1 \times Q_{in}.$$

We don't know e_1 or Q_{in} but $e_1 = .70(e_1)_{Carnot}$.

T: Right. And

$$(e_1)_{Carnot} = 1 - (T_c/T_h) = 1 - (303\text{ K}/573\text{ K}) = .471 \quad \text{(or 47.1\%)}.$$

Then

$$e_1 = .70\ (.471) = .330.$$

S: I still don't know Q_{in}.

T: Write the 1st Law for the whole cycle.

S: $\Delta U_{cycle} = 0$, so $W = Q_{in} - |Q_{out}|$. I see! Since also $W = (e_1)Q_{in}$, we have

$$e_1 Q_{in} = Q_{in} - |Q_{out}|$$

giving

$$Q_{in} = -|Q_{out}|/(e_1 - 1) = (2\text{ kcal})/(.67) = 2.99\text{ kcal}.$$

T: Right. The two statements you used were the definition of e_1 and the 1st Law of Thermodynamics. So

$$W_{cycle} = 2.99 - 2\text{ kcal} = .99\text{ cal}.$$

Example 15-4. Suppose a gas turbine has a combustion temperature of 2400 °C and an exhaust temperature of 400 °C. It operates at 1/3 of ideal efficiency. The exhaust is used to turn a lower temperature steam turbine which exhausts heat at 70 °C and achieves 70% of Carnot efficiency. This is an example of a *combined cycle* power plant.

(a.) What are the individual efficiencies of each cycle?
(b.) What is the total efficiency of the combined cycle operation?

S: The efficiency of the gas turbine would be

$$(e_1)_{gas} = .333(1 - T_c/T_h) = .333[(1 - (673\text{ K}/2673\text{ K})] = .247$$

That of the steam turbine would be

$$(e_1)_{steam} = .70[1 - (343\text{ K}/673\text{ K})] = .343\ .$$

How do we combine these to get *total* efficiency?

T: Just follow what happens to a given amount of input energy, say 100 J. In the gas turbine we have:

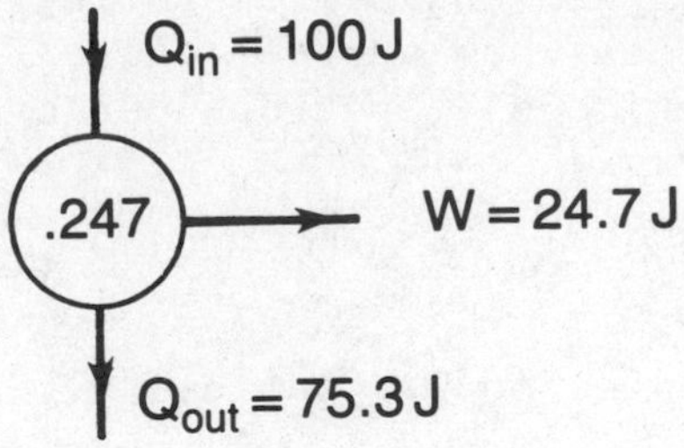

For the steam turbine, we have

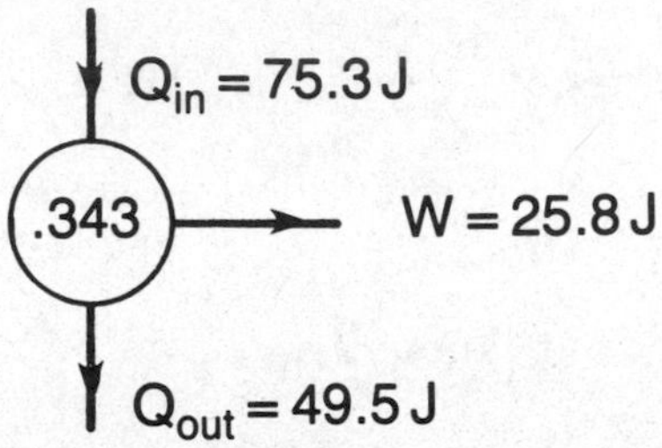

Taken together, we have the following picture:

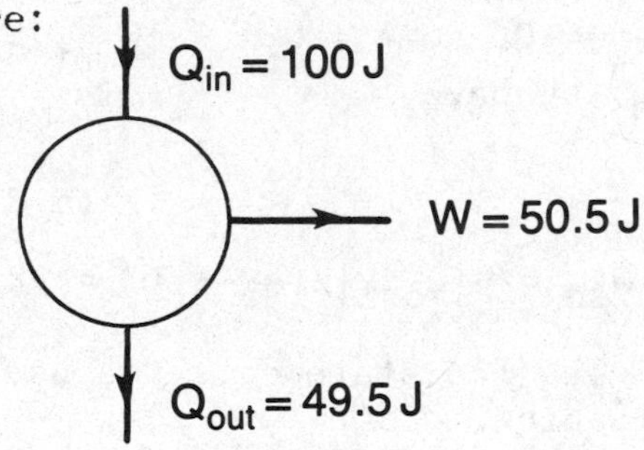

So $(e_1)_{total} = .505$.

This more than *doubles* the useful work output, or cuts the input fuel bill by more than half for a given output requirement. It shows that the exhaust of the gas turbine is still high quality thermal energy.

Example 15-5. How much farther can the combined cycle concept go? Suppose the lowest exhaust temperature *freely* available to you is 20 ^{o}C, and you have a *perfect* heat engine as a 3rd cycle, operating off the exhaust of the steam turbine. How much would the total efficiency improve?

S: Well, this third engine would have

$$e_1 = 293\text{ K}/343\text{ K} = .146.$$

So its energy diagram would look like this:

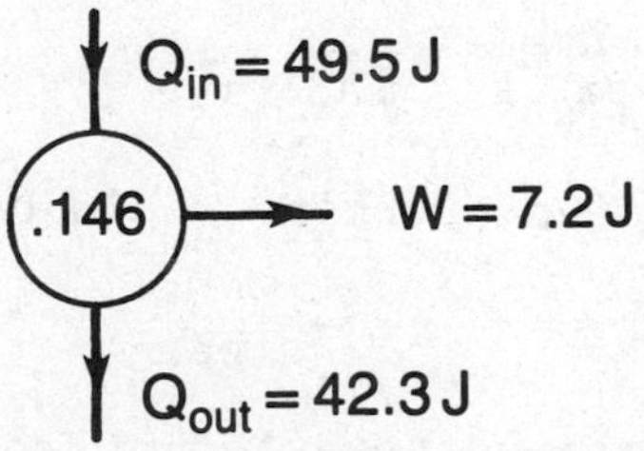

T: Right. And so $(e_1)_{total}$ = .577. You can see that we're running out of high quality thermal energy. The exhaust of this 3rd engine isn't useful at all unless we could find some environment with even higher entropy (lower temperature). It doesn't solve the problem to *create* such an environment, since that *consumes* energy.

Second Law of Thermodynamics

The 1st Law of Thermodynamics is a statement about energy conservation which includes thermal energy on an equivalent basis with mechanical work and energy. The 2nd Law of Thermodynamics summarizes the special properties of the way heat and thermal energy behave, which has its ultimate explanation in the statistical behavior of the vast numbers of atoms and molecules that make up macroscopic samples of matter. There are a number of equivalent ways of stating the second law:

1. Entropy Statement. The entropy of an isolated system increases in irreversible processes, and remains constant in reversible processes.

2. Kelvin Statement. It is impossible to construct a heat engine whose sole function is to convert a given amount of thermal energy entirely into mechanical work.

3. Clausius Statements.
   1st Law. The energy of the universe is constant.
   2nd Law. The entropy of the universe is continually increasing.

Heat Pumps and Refrigerators

Understanding this section should enable you to do the following:

1. *Calculate the coefficients of performance of real and ideal heat pumps and refrigerators.*

Heat Pump--A device which utilizes an input of work to extract thermal energy Q_c from a low temperature reservoir at T_c and exhaust thermal energy Q_h to a high temperature reservoir at T_h. For each cycle, the 1st Law requires $W + Q_c = |Q_h|$, since again $\Delta U_{cycle} = 0$.
The figure below shows that schematically a heat pump is a heat engine with the energy flows reversed.

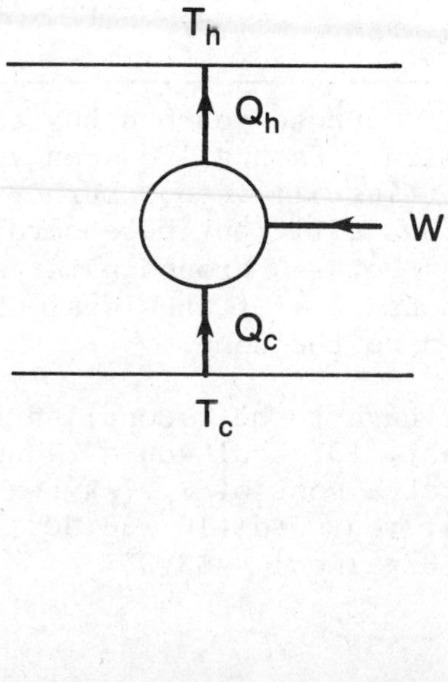

Coefficient of Performance(CP)--In the same way that *efficiency* is a figure of merit for a heat engine, we define a *coefficient of performance* as a figure of merit for a heat pump. It is the ratio of the desired output of the pump, Q_h, to the required input of work, W. Mathematically,

$$CP = Q_h/W = |Qh|/\{|Q_h| - Q_c\}$$

Carnot Heat Pump--A Carnot engine operating in *reverse* represents an *ideal* heat pump, having the *maximum possible* CP for *any* heat pump operating between a given temperature difference. Mathematically,

$$(CP)_{Carnot} = T_h/(T_h - T_c)$$

Important Observations

1. From the definition of CP, you can see that CP > 1 for *all* heat pumps. Thus *more heat* can be delivered at T_h than the work required for operation. This does *not* mean that energy conservation is being violated. Instead, it assumes that low temperature thermal energy is *freely* available in the environment, and that the only energy you have to *pay* for to get Q_h is the work W, supplied by the electricity or natural gas required to run a compressor.
2. In practical terms, the heat pump consists of two heat exchangers, a compressor, and an expansion valve. As the working fluid passes through the expansion valve, it cools *below* the temperature T_c, and so absorbs heat Q_c from the environment. The fluid is then heated by the compressor (which requires an input W) to a temperature *above* T_h. Thus heat Q_h flows out into the space where heating is required.

Refrigerators and Air Conditioners--The basic mechanism of the heat pump can be designed so that the desired function is the *removal* of heat Q_c from an enclosed volume. This application is a *refrigerator* or *air conditioner*, where the low temperature heat exchanger is in the interior of the refrigerator or the house. The coefficient of performance is slightly different than for the heat pump, since its function is different. We have

$$(CP)_{refrig} = Q_c/W = Q_c/(|Q_h| - Q_c)$$

and

$$(CP)_{Carnot\ refrig.} = T_c/(T_h - T_c)$$

Example 15-6. Suppose you can buy electricity for \$.07 per kwh and fuel oil for \$1.20 per gallon. Each gallon can yield 36,000 kcal of heat. To heat your home, you can either install an oil burner which is 75% efficient, use electricity directly in 100% efficient baseboard heaters, or use electricity to run a heat pump operating between ground water at 40 ^{o}F and a high temperature of 100 ^{o}F. The heat pump has 50% of the ideal CP. Compare the costs per 100,000 kcal of heat delivered to the home.

S: The units have to be reconciled first. 1 kwh = 860 kcal. So for the direct electrical heat, 10^5 kcal would require $(10^5/860) = 1.16 \times 10^2$ kwh of electricity, at a cost of \$.07/kwh x 1.16×10^2 kwh = \$8.14. To get 10^5 kcal from fuel oil, you need $(10^5/36{,}000) = 2.78$ gallons, costing (2.78 gallons)(\$1.20/gallon) = \$3.33.

T: But only 75% of the heat from the oil is delivered to heating the home. 25% is lost through the chimney and incomplete combustion.

S: Well, then you would need (4/3)(2.78 gallons) = 3.71 gallons, costing $4.45. For the heat pump, we have to find its CP. Those Fahrenheit temperatures don't help.

T: No problem. 40 $^{o}F = 4.4\ ^{o}C = 277$ K, and 100 $^{o}F = 38\ ^{o}C = 311$ K. So

$$(CP)_{ideal} = 311\ K/(311\ K - 277\ K) = 9.15$$

S: Our actual heat pump has 1/2 of this, or $(CP)_{actual} = 4.6$. How is this related to the electrical energy required to produce 10 kcal of heat?

T: We require $Q_h = 10^5$ kcal, delivered into the house. The energy you need to do this, W, is the electricity needed to run the compressor.

S: So from $CP = 4.6 = Q_h/W$, we find $W = (10^5\ kcal)/4.6 = 2.17 \times 10^4$ kcal. This is 25.3 kwh, costing only $1.77!

T: You can see the amazing advantage of a heat pump, which utilizes the free low temperature energy in the environment.

QUESTIONS ON FUNDAMENTALS

1. 2 moles of an ideal monatomic gas has an initial pressure of 5 atm. It undergoes an isothermal expansion from a volume of 30 l to 50 l. How much work does the gas do?

(a.) 7740 J (b.) 12,900 J (c.) 4640 J (d.) 10,100 J
(e.) 15,480 J

2. 2 moles of an ideal monatomic gas at an initial pressure of 5 atm undergoes an adiabatic expansion from 30 l to 50 l. What is the final pressure of the gas?

(a.) Impossible to tell (b.) 11.8 atm (c.) 2.45 atm
(d.) 10.3 atm (e.) 2.13 atm

3. How much work did the gas in question #2 do?

(a.) 7740 J (b.) 6570 J (c.) 11,000 J (d.) 15,700 J
(e.) 2640 J

4. What was the entropy change of the gas in question #1?

(a.) 4.25 J/K (b.) 12.1 J/°C (c.) 0 (d.) 8.5 J/K
(e.) Impossible to calculate

5. If an ideal (Carnot) heat engine has an efficiency of 66.7%, and it exhausts heat at room temperature (20 °C), at what high temperature must it be receiving an input of heat?

(a.) 60°C (b.) 333 K (c.) 98 K (d.) 879 K (e.) 440 K

6. If the engine in question #5 is to produce 5 kW of power, at what rate must it exhaust energy to its surroundings?

(a.) 2.5 kW (b.) 7.5 KW (c.) 5 kW (d.) 10 kW
(e.) 3.33 kW

7. If an air conditioner has a CP of 3 and extracts 10,000 BTU of heat per hour from a room, how much input of electrical power will it need to consume?

(a.) 3333 W (b.) 732 W (c.) 976 W (d.) 1460 W
(e.) 650 W

GENERAL PROBLEMS

Problem 15-1. Show from the 1st Law of Thermodynamics that, for an adiabatic process,

$$W_{adiab.} = (P_1V_1 - P_2V_2)/(\gamma - 1) .$$

Q1: What is the 1st Law for an adiabatic process?
Q2: How can I express $\Delta U_{adiab.}$?
Q3: Can C_V be expressed in terms of γ ?
Q4: How do I replace T_1 and T_2?

Problem 15-2. A heat engine containing 2 moles of an ideal monatomic gas operates in the thermodynamic cycle showN in the figure below. Process AB is isochoric, BC is adiabatic, and CA is isobaric. Calculate Q, W, and ΔU for each of the three processes.

Q1: Process AB is isochoric. What does that tell me about Q_{AB}, W_{AB}, and ΔU_{AB}?
Q2: For the adiabatic BC, $Q_{BC} = 0$. What do I need to know for W_{BC} and ΔU_{BC}?
Q3: Can we summarize all the information we know?
Q4: What is V_A?
Q5: Can we now find T_c or V_c?
Q6: For CA, $W_{CA} = P_A(V_A - V_c)$, but I need T_c to find Q_{CA} or ΔU_{CA}.

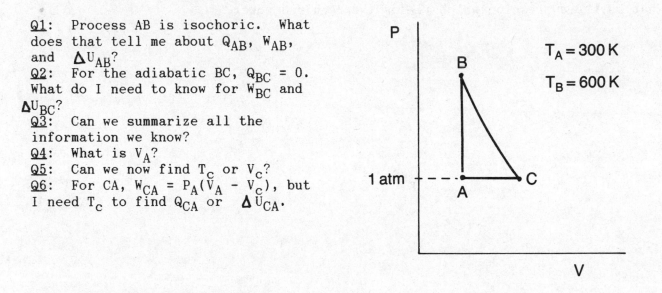

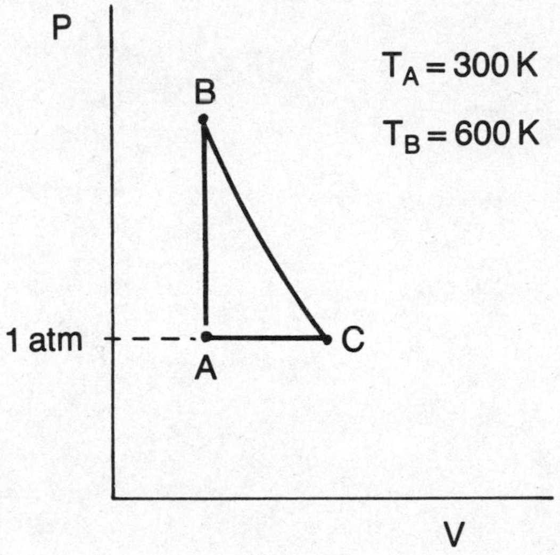

Problem 15-3. For the heat engine in Problem 15-2, calculate the efficiency and the ratio of its efficiency to the maximum possible efficiency for an engine operating between these temperature extremes.

Q1: What defines the engine's efficiency?
Q2: What is Q_{out} and Q_{in}?
Q3: What is W_{out}?
Q4: What is the maximum possible efficiency?

Problem 15-4. Return once more to the two previous problems. If process BC was an *isotherm* instead of an adiabatic, would the engine be more or less efficient? Calculate the efficiency for this case.

Q1: What would change if BC was isothermal?
Q2: Which of these changes would effect the efficiency?
Q3: What is Q_{BC}?
Q4: What is V_c?
Q5: What is Q_{out}?
Q6: Has $(e_1)_{ideal}$ changed?

Problem 15-5. Your home freezer has a wall area of 1.5 m^2, and a wall thickness of 3 cm, which is filled with styrofoam insulation. It is designed to maintain a compartment temperature of -25 ^oC. The refrigeration cycle is designed with a CP of 3.5 when the room temperature is +25 ^oC. How much average power must be delivered to the compressor to achieve this?

Q1: How much thermal power is lost by conduction under these conditions?
Q2: What is the connection between this and required power for the refrigerator?
Q3: What is the expression which yields the required power?

ANSWERS TO QUESTIONS

FUNDAMENTALS:

1. (a.)
2. (e.)
3. (b.)
4. (d.)
5. (d.)
6. (a.)
7. (c.)

GENERAL PROBLEMS:

Problem 15-1.

A1: $Q_{adiab.} = 0$, so $\Delta U_{adiab.} = -W_{adiab.}$

A2: ΔU *always* can be written as $nC_V(T_2 - T_1)$, where now these temperatures lie on the same adiabatic curve.

A3: Yes. $C_V = C_P/\gamma = (C_V + R)/\gamma$. Solving for C_V gives

$$C_V = R/(\gamma - 1)$$

A4: The idea gas law would say that $P_1V_1 = nRT_1$ and $P_2V_2 = nRT_2$.

Problem 15-2.

A1: $W_{AB} = 0$, and $Q_{AB} = nC_V(T_B - T_A) = \Delta U_{AB}$.

A2: First, since $Q_{BC} = 0$, $\Delta U_{BC} = -W_{BC}$. You can use either

$$W_{BC} = (P_BV_B - P_CV_C)/(\gamma - 1), \quad \text{with } \gamma = 1.67, \text{ or}$$

$$\Delta U_{BC} = nC_V(T_C - T_B).$$

With one you have to know P_B, V_B, and V_C. With the other you have to know T_C.

A3: Sure. We know $P_A = 1$ atm $= P_C$, $T_A = 300$ K, $T_B = 600$ K. This immediately tells us that $P_B = 2P_A = 2$ atms. Also, $V_A = V_B$, although you're not given any of the volumes directly.

A4: $V_A = (nRT_A/P_A) = (2 \text{ moles})(.0821 \text{atm-l/mole-K})(300 \text{ K})/(1 \text{ atm}) = 49.3 \text{ l} = V_B$.

A5: For the adiabatic, $P_BV_B^{\gamma} = P_CV_C^{\gamma}$. So $V_C = (P_B/P_C)^{1/\gamma} V_B = 74.7$ l. Now you have all you need to find W_{BC}, and hence ΔU_{BC}.

A6: $T_C = (V_C/V_A)T_A = 454$ K, and $Q_{CA} = nC_P(T_A - T_C)$, $\Delta U_{CA} = nC_V(T_A - T_C) = Q_{CA} - W_{CA}$. You should notice that $\Delta U_{cycle} = 0$.

Problem 15-3.

A1: $e_1 = W_{out}/Q_{in} = \{Q_{in} - |Q_{out}|\}/Q_{in} = 1 - |Qout|/Q_{in}$. Take your pick.

A2: Q_{out} is the sum of negative Q's for the processes in the cycle. Q_{in} is the sum of all positive Q's. In this case, $Q_{in} = Q_{AB}$ and $Q_{out} = Q_{CA}$.

A3: The algebraic sum of $W_{AB} + W_{BC} + W_{CA}$. *This is the same as the area enclosed by the entire cycle.* In this case, the P-V area is *not* easy to calculate.

A4: The temperature extremes are $T_{hot} = 600$ K and $T_{cold} = 300$K. So

$$(e_1)_{ideal} = 1 - T_{cold}/T_{hot} = .50$$

Problem 15-4.

A1: $P_A(= P_C)$, P_B, and V_A $(= V_B)$ would stay the same. T_C would now equal $T_B = 600$ K. V_C, Q_{BC}, Q_{CA}, W_{BC}, W_{CA} would all change.

A2: Q_{in} will now include Q_{BC}, and $Q_{out} = Q_{CA}$ will be different.

A3: On an isotherm, $\Delta U_{BC} = 0$, so $Q_{BC} = W_{BC} = nRT_B \ln(V_C/V_B)$.

A4: Now we have $P_BV_B = P_CV_C$, so $V_C = 2V_B = 98.6$ l.

A5: $Qout = Q_{CA} = W_{CA} + \Delta U_{CA}$, where $W_{CA} = P_A(V_A - V_C)$, and $\Delta U_{CA} = nC_V(T_A - T_C)$.

A6: No, since the temperature extremes are still 300 K and 600 K.

Problem 15-5.

A1: $(Q/t)_{cond} = kA(\Delta T/L)$.

A2: This is the rate at which heat must be removed to maintain the freezer temperature, Q_C/t. This is related to the required input power by the CP, $CP = Q_c/W$.

A3: $P = W/t = (Q_c/t)/CP = (1/CP)(kA\,\Delta T/L)$.

CHAPTER 16. ELECTROSTATICS

BRIEF SUMMARY AND RESUME

A new physical property of matter, the existence of *electrical charge*, is introduced. *Coulomb's Law*, describing the fundamental force of interaction between charges, is examined, and *amounts* of charge are defined. The concept of *electric field* is introduced, and graphical and analytical methods of its calculation for various charge distributions are described, including *Gauss' Law*. The differences between metals and insulators are discussed.

Material from previous chapters relevant to Chapter 16 includes:

1. Newton's Laws of Mechanics.

NEW DATA AND PHYSICAL CONSTANTS

1. Quantities of Electric Charge.

   Coulomb (C): The basic SI unit of charge.

   Electron Charge (e): The amount of charge on an electron or proton. $1\ e = 1.6 \times 10^{-19}$ C.

2. Coulomb Constant (k_e).

   $k_e = 8.988 \times 10^9$ N-m^2/C^2.

3. Permittivity of Free Space (ϵ_0).

   $\epsilon_0 = 1/4\pi k_e = 8.85 \times 10^{-12}$ C^2/N-m^2.

4. Electric Field Units.

   $E = F_e/q$ gives SI units of N/C, equivalent to V/m.

NEW CONCEPTS AND PRINCIPLES

Basic Concepts of Electric Charge

1. There exist two kinds of electric charge, designated positive (+) and negative (-). Objects charged with *like* charges exhibit a *repulsive* interaction; those charged with *opposite* charges exhibit an *attractive* interaction.
2. Atoms contain charged fundamental particles. The proton carries a definite amount of (+) charge, and the electron carries the *same* amount of (-) charge.
3. In any closed system, the net amount of charge remains constant, or *conserved*. Equal amounts of (+) and (-) charge can physically *neutralize* each other, but neither kind of charge can be created or destroyed *alone*.

4. Materials are distinguished as *conductors* or *insulators* by whether or not some of their electrons are free to move throughout the volume of the material. In conductors, notably *metals*, electrons *can* move and thus conduct electricity. In insulators, the electrons are bound in place to the atoms of the material and cannot move over macroscopic distances.

Electric Force Between Charges

Understanding this section should enable you to do the following:

1. *Calculate the electric force between two charges separated by a given distance.*
2. *Use the superposition principle to calculate the electric force on a charge due to various other charges in its vicinity.*

Coulomb's Law of Electric Force--The force between two point charges at rest is directly proportional to the product of the charges and inversely proportional to the square of the distance between them. If the charges are of *like* sign, the forces are *repulsive*, i.e., directed *away* from the charges along the line joining them. If the charges are of *opposite* sign, the force is *attractive*, i.e., directed toward the charges along the line joining them. Mathematically, the *magnitude* of the electric force is given by Coulomb's Law:

$$F_e = k_e(q_1q_2)/r^2 \text{ , where } k_e = \text{Coulomb's constant} = 9.0 \times 10^9 \text{ N-m}^2/\text{C}^2.$$

Corollary: Definition of the Coulomb (C) of Charge--A *coulomb* of charge, when placed *one meter* away from an equal charge in vacuum, experiences 9.0×10^9 N of force.

Important Observations

1. The quantity of charge can be considered a *fourth* fundamental physical dimension, since it represents a new property of matter which cannot be derived from the previous dimensions of mass, length, and time.
2. The coulomb is an extremely large amount of static charge, since two such charges one meter apart in the laboratory would experience 9 *billion* N of force! Charges usually encountered in the lab would be in the range of 10^{-6} C (μC) to 10^{-12} C (pC).
3. The smallest, or *elementary*, charge in nature is that carried by individual electrons or protons. This is given the symbol e and *always* has the value of $e = 1.6 \times 10^{-19}$ C, + for protons, and - for electrons. Thus it would take 1/e, or 6.25×10^{18} electrons to equal -1 C of charge!
4. The Coulomb force obeys Newton's 3rd Law, in that in the interaction between two charges,

$$(\mathbf{F}_e)_{1 \text{ on } 2} = -(\mathbf{F}_e)_{2 \text{ on } 1}.$$

The Principle of Superposition--If two or more charges are acting on charge A, the contributions of force on A due to *each* charge are calculated by the Coulomb Law *as if no other charges were present*, and these individual contributions are added *vectorially* to find the *total* net force on A. Mathematically,

$$(F_{on\ A})_{net} = F_{1\ on\ A} + F_{2\ on\ A} + F_{3\ on\ A} + \cdots$$

Example 16-1. A positive charge of 1.0×10^{-8} C is fixed 3 cm directly above a charged object having a mass of .05 gm. The object is free to move, but is neither rising nor falling toward tne earth. What charge must it be carrying?

S: Why doesn't it move if it is free to do so?

T: The electrical force must be balanced by another force in the opposite direction, which in this case is gravity.

S: Gravity would be downward, so the electrical force would have to be *upward*, *toward* the positive charge.

T: This tells you that the force is *attractive*, so the object must carry a *negative* charge.

S: We should be able to find out how *much* charge by setting the electrical force equal to mg:

$$k_e q_1 q_2/r^2 = m_1 g,$$

where q_1 = unknown, $q_2 = 1.0 \times 10^{-8}$ C, $r = 3 \times 10^{-2}$ m, and $m_1 = 5 \times 10^{-5}$ kg

T: Good. You've put everything into SI units. The calculation gives

$$q_1 = mg(r^2/k_e q_2) = \frac{(5 \times 10^{-5}\,\text{kg})(9.8\,\text{m/s}^2)(3 \times 10^{-2}\,\text{m})^2}{(9 \times 10^9\,\text{N-m}^2/\text{C}^2)(10^{-8}\,\text{C})}$$

$$= 4.9 \times 10^{-9}\ \text{C, or } 4.9\ \text{nC}.$$

Example 16-2. Two charges of 5 μC are located at opposite corners of a square 0.5 m on a side as shown in the figure below. A -5 μC charge is at another corner. If the object in Example 16-1 was placed at the fourth corner and released, what will be its initial acceleration? (Neglect earth's gravity in this case.)

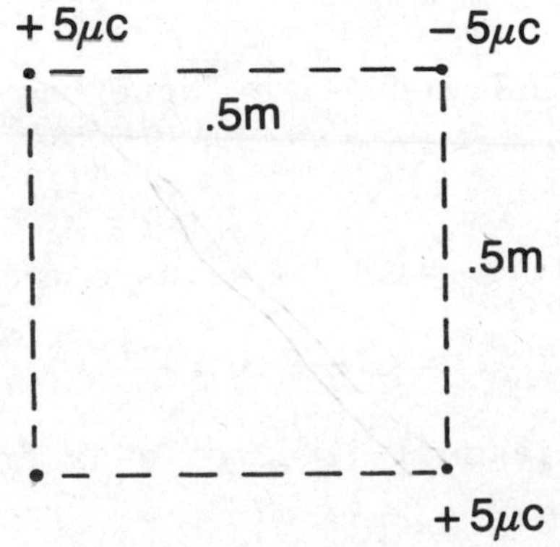

S: The positive charges will exert *repulsive* forces on q of magnitude

$$F_e = k_e(4.9\ \text{nC})(5\ \mu\text{C})/(.5\ \text{m})^2$$

each.

T: Remember to replace the

prefixes nano- and micro- by their numerical value.

S: Then

$$F_e = (9 \times 10^9\ N\text{-}m^2/C^2)(4.9 \times 10^{-9}\ C)(5 \times 10^{-6}\ C)/(.5\ m)^2$$

$$= 8.8 \times 10^{-4}\ N.$$

T: As shown in the second figure, one of these will be a vector to the *left* and one will be *downward*, both at the location of q. How about the effect of the -5 μC charge?

S: It will *attract* q along the square's diagonal. We have to find the diagonal's length.

T: The diagonal of a square is √2s, where s is the side length. Here the diagonal is √2(.5 m) = .707 m long.

S: The attractive force is then

$$F_e = (9 \times 10^9\ N\text{-}m^2/C^2) \times (4.9 \times 10^{-9}\ C)(-5 \times 10^{-6}\ C)/(.707\ m)^2$$

$$= 4.4 \times 10^{-4}\ N.$$

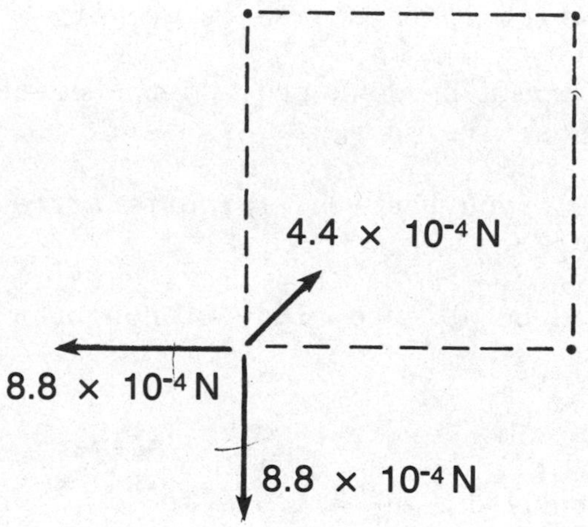

What do I do with the minus sign?

T: It simply reminds you that the force is *attractive* along the line between those two charges. Since you've already identified that direction, the sign is redundant. The force diagram on q is shown on the second figure.

S: This can be solved by resolving the vectors into x and y components.

T: You could, but there is a more convenient choice of axes. Notice the symmetry relative to the diagonal. The components normal to the diagonal will cancel.

S: O.K. Each repulsive force has a component of

$$(8.8 \times 10^{-4}\ N) \cos 45^\circ$$

along the diagonal. So we have a *net force* of

$$F_{net} = 2(8.8 \times 10^{-4}\ N) \cos 45^\circ - 4.4 \times 10^{-4}\ N = 8.0 \times 10^{-4}\ N$$

along the diagonal directed away from the center of the square.

T: So the object will have an *initial* acceleration in this direction of

$$a = F_{net}/m = (8.0 \times 10^{-4}\ N)/(5.0 \times 10^{-5}\ kg) = 16\ m/s^2.$$

This will not remain constant, however, because as the object moves, the forces on it will change.

Electric Field

Understanding this section should enable you to do the following:

1. *Use the definition of electric field to calculate the field from simple distributions of point charges.*
2. *Use Gauss' Law to calculate electric fields due to continuous distributions of charge possessing simple symmetry.*
3. *Identify qualitative aspects of electric fields from mappings or diagrams of the fields.*
4. *Calculate the electric force on a charge placed in a given field.*

Electric Field Strength (E)--The electric field strength at a point in space is the electric force experienced at that point by a small positive "test" charge q, divided by q. Mathematically,

$$\mathbf{E} = (\mathbf{F}_e)_{on\ q}\ .$$

Important Observations

1. The field is a *vector* quantity.
2. The SI units of E are N/C.
3. **E** is in the *same* direction as the force experienced by a + charge, and *opposite* in direction to the force on a - charge.
4. A corollary to the above definition is that, if the field due to some distribution of charge is known as a function of positive r, the force on a charge q at r is

$$\mathbf{F}_e(r) = q\mathbf{E}(r).$$

This again clearly shows that $\mathbf{F}_e$ on + charges is *parallel* to **E**, while on - charges it is *anti-parallel* to **E**.

Electric Field Due to a Point Charge--The electric field due to a point charge Q is

$$\mathbf{E}(r) = (k_eQ/r^2)\ \mathbf{r},$$

where r is the distance measured from Q, and **r** is a unit vector radially outward from Q.

Important Observations

1. A unit vector like **r** is a device whose sole purpose is to display *direction*. It is a dimensionless quantity whose magnitude is one.

2. Notice that when Q is +, **E** is in the +**r** direction, that is, *radially outward* from Q. When Q is -, the field is *directed inward toward* Q (i.e., the -**r** direction).
3. If we know this expression for a point charge, we can *in principle* use the superposition principle to find **E** for *any* distribution of charges. In practice this is very difficult except for distributions having simple symmetry.

<u>Electric Field Mapping</u>--We can visualize qualitative aspects of the electric field by drawing a series of lines with arrows in the direction of the field. Thus the field maps of point charges are shown in the figures below.

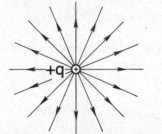

For more complicated distributions of charge, these patterns become more complicated as shown in the following figures:

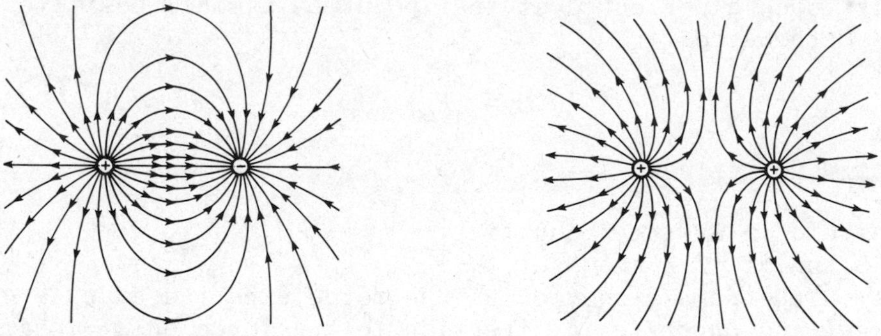

<u>Important Observations</u>

1. The direction of the field lines at any point tell you the direction of the force a + charge would experience there. A - charge would experience a force opposite to the field line direction.
2. The relative *density* of lines in the map tells you the relative *strength* or *magnitude* of the electric field.
3. Notice that field lines *originate* on (+) charges and *terminate* on (-) charges.

<u>Electric Flux Through a Surface</u>--If we imagine any small surface area ΔA in a region where there is an electric field, the product of ΔA times the component of **E** *normal* to the plane of ΔA is called the *electric flux through* ΔA. Mathematically, the electric flux through ΔA is written $E_\perp \Delta A$, shown schematically in the following figure.

In a qualitative sense, when looking at a mapping of a field, the electric flux is the number of field lines which pass through the area ΔA.

Gauss' Law--For any closed surface, the net electric flux through the entire surface is proportional to the net amount of charge in the volume enclosed by the surface. Mathematically,

$$\sum_{\text{surface}} E_{\perp} \Delta A = 4\pi k_e (Q_{net})_{enclosed}$$

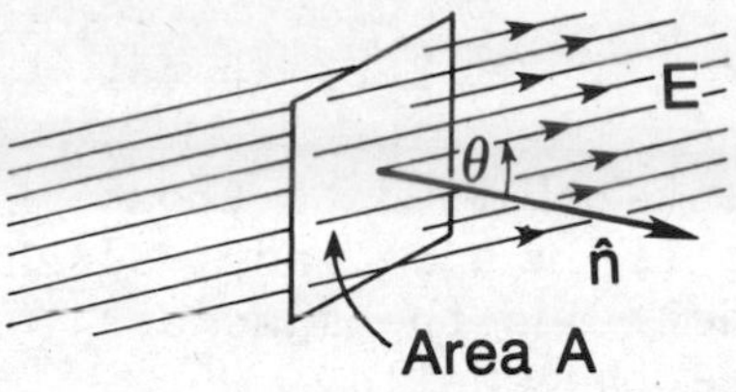

Flux = EA cos θ

Important Observations

1. Gauss' Law provides a powerful way to determine the electric field from regions of charge distributions having simple geometric symmetry. The key to using Gauss' Law lies in selecting a surface which simplifies the calculation of flux. This usually means using the symmetry of the charge distribution so that the field is constant on those parts of the surface to which it is normal and parallel to the rest of the surface, in which case $E_{\perp} \Delta A = 0$.
2. If no net charge is contained within the surface, there are no origins or terminations of electric field lines within. Any field line which crosses the surface must do so *twice*, once *entering* (negative flux) and once leaving (positive flux), so that its flux contributions add up to zero.

Illustrations of Gauss' Law

A. Point Charge. The field is everywhere radial. If we select a sphere of radius r, E will be constant and everywhere normal on the sphere. The sphere contains the charge Q, so

$$\sum_{\text{surface}} E_{\perp} \Delta A = E_r(4\pi r^2) = 4\pi k_e Q, \text{ giving } E_r = k_e(Q/r^2)$$

B. Spherical Shell Charge. Charge Q is homogeneously spread over a spherical shell of radius a. The same arguments of spherical symmetry as for the point charge hold here. Thus

$$E_{\perp} \Delta A = E_r(4\pi r^2) = 4\pi k_e Q \qquad \text{if } r > a$$

$$= 4\pi k_e(0) \qquad \text{if } r < a.$$

Thus $E_r = k_e Q/r^2$ if $r > a$, just as if the charge was a point charge at the center of the sphere, and $E_r = 0$ if $r < a$. *Inside a spherically symmetric charge, no field exists.*

C. Infinite Line of Charge. Charge is uniformly distributed along an infinite line. By symmetry, components of **E** parallel to the line cancel, leaving a field that is purely in the direction radial to the line. A cylindrical surface coaxial with the line will have **E** normal to it and constant in magnitude. On the end caps of the cylinder, **E** will be parallel to ΔA, contributing no flux. The area of the side of the cylinder is $2\pi rL$, so we have $\Sigma(E_{\perp}\Delta A) = E_r(2\pi rL)$. If λ is the linear charge density on the line, then the charge enclosed in the cylinder is $Q = \lambda L$. Gauss' Law then gives

$$E_r(2\pi rL) = 4\pi k_e \lambda L, \text{ or } \quad E_r = 2k_e(\lambda/r).$$

Arguments similar to those in B. above show that if the charge is on a cylindrical shell, the field is zero inside the shell and the same as for a line charge if the point is outside the shell.

D. Infinite Sheet of Charge. Here the charge is uniformly distributed over a sheet with area charge density σ. The electric field lines are everywhere normal to the sheet, so for the appropriate Gaussian surface we can select a right cylinder with axis normal to the sheet, having cross -sectional area A. E is parallel to the sides of the cylinder, contributing no flux except at the ends, where the flux is E_xA at *each* end. Thus $\Sigma(E_{\perp}\Delta A) = 2E_xA$. The surface encloses a charge $Q = \sigma A$. Thus $2E_xA = 4\pi k_e\sigma A$, or $E_x = 2\pi k_e\sigma$, *a constant value which does not vary with distance from the sheet of charge.*

Example 16-3. Point charges are placed as follows: $q_1 = 2\ \mu C$ at $x = -3$ cm, $q_2 = -5\ \mu C$ at $y = -4$ cm, and $q_3 = 3\ \mu C$ at $x = 2$ cm. What is the electric field at the origin? How much force in what direction would a -5 nC charge experience if placed there?

S: We have to calculate the contributions to the field one at a time. That due to q_1 is

$$E_1 = (9 \times 10^9\ N\text{-}m^2/C^2)(2 \times 10^{-6}\ C)/(3 \times 10^{-2}\ m)^2 = 2 \times 10^7\ N/C.$$

How do I tell the direction?

T: Field lines from a + charge are radially outward from the charge. For q_1 this mean that E_1 is toward +x.

S: Now from q_2,

$$E_2 = (9 \times 10^9\ N\text{-}m^2/C^2)(-5 \times 10^{-6}\ C)/(-4 \times 10^{-2}\ m)^2 = -2.8 \times 10^7\ N/C,$$

directed toward negative y, since q_2 is a - charge.

T: Right. The third charge contributes

$$E_3 = (9 \times 10^9)(3 \times 10^{-6})/(2 \times 10^{-2})^2 = 6.8 \times 10^7\ N/C.$$

Can you tell me the direction?

S: q_3 is +, so the field will be directed *away* from it. This means toward -x for the point at the origin. Now we have to add these as vectors?

T: Yep, and since each contribution is already in component form, it should be easy.

S: O.K., the magnitude is

$$(E_{tot})_x = (E_1)_x + (E_2)_x + (E_3)_x = 2 \times 10^7 - 6.8 \times 10^7 \text{ N/C}$$
$$= -4.8 \times 10^7 \text{ N/C}.$$

And

$$(E_{tot})_y = 0 - 2.8 \times 10^7 \text{ N/C} + 0 = -2.8 \times 10^7 \text{ N/C}.$$

Then

$$E_{tot} = (E_x{}^2 + E_y{}^2)^{\frac{1}{2}} = (30.9)^{\frac{1}{2}} \times 10^7 \text{ N/C} = 5.6 \times 10^7 \text{ N/C}.$$

What is the direction?

T: With both components negative, you know it's in the third quadrant. The angle below the negative x axis is given by

$$\theta = \tan^{-1} (E_y/E_x) = \tan^{-1}(2.8 \times 10^7)/(4.8 \times 10^7) = 30.3^o.$$

S: The force on -5 nC there would be

$$F_e = qE = (-5 \times 10^{-9} \text{ C})(5.6 \times 10^7 \text{ N/C}) = -.28 \text{ N}.$$

T: The minus sign indicates the direction is opposite to **E**, i.e., 30.3^o *above the +x axis.*

Example 16-4. Suppose there is a region in which charge is somehow distributed *homogeneously* within a very long cylinder of radius R = 10 cm. The charge density within this cylinder is $\rho = +10^{-3}$ C/m^3. Use Gauss' Law to find the electric field:

(a.) 7 cm from the axis.
(b.) 15 cm from the axis.

S: Our Gaussian surface will be a cylinder of radius 7 cm for part (a.). What length should we choose?

T: That is completely arbitrary, and can't affect the result. Just call it L.

S: How do we find the charge enclosed by the surface?

T: Since the charge is distributed with uniforn density ρ, the charge in any volume V is just Q = ρV. The volume of our cylinder is $V = \pi r^2 L$, so $Q_{encl} = \rho \pi r^2 L$.

S: How can we use the symmetry to our advantage?

T: We can assume **E** is purely radial, E_r, since components parallel to the axis cancel out. Furthermore, E_r has constant magnitude over the side of the cylinder. Thus **E** is everywhere perpendicular to the side of the cylinder, and parallel to the area of the end caps, so $E_\perp \Delta A = 0$ there. All in all, the total flux through the cylinder is $E_r(2\pi rL)$. Now put these results together in Gauss' Law and solve for E_r.

S: We get

$$E_r(2\pi rL) = 4\pi k_e Q_{encl} = 4\pi k_e \rho \pi r^2 L.$$

Thus

$$E_r = 2k_e \pi r^2 \rho / r = (2\pi k_e \rho) r.$$

T: Notice E_r increases in direct proportion to r. Putting in numbers, at r = 7 cm :

$$E_r = 2\pi(9 \times 10^9 \text{ N-m}^2/\text{C}^2)(10^{-3} \text{ C})(7 \times 10^{-2} \text{ m})$$

$$= 4.0 \times 10^6 \text{ N/C} .$$

S: At r = 15 cm we should be able to multiply this by 15/7, since E_r is proportional to r.

T: *Not true!* At r = 15 cm, we're *outside* the charge distribution, and the terms in Gauss' Law take on different values. Let's go back and see. What is the flux through the cylinder at r = 15 cm?

S: The *same*, namely $E_r(2\pi rL)$.

T: What is the charge enclosed?

S: $Q_{encl} = \rho \pi r^2 L$. Oh no! The charge only extends to R = 10 cm, so $Q_{encl} = \rho \times \pi R^2 L$.

T: Now what does Gauss' Law tell you?

S: That

$$E_r(2\pi rL) = 4\pi k_e \rho \pi R^2 L,$$

So

$$E_r = 2k_e \pi R^2 \rho / r .$$

Since the r's no longer cancel, this field falls off as 1/r, just like the field for a *line* charge.

T: Also, $\pi R^2 \rho$ is just the charge per length in the cylinder, i.e., the linear charge density λ. So $E_r = 2k_e \lambda / r$. Thus when you are *outside* the charge distribution, it acts like it is concentrated in a line charge on the axis. But inside, where r < R, E_r has a different dependence going to zero at r = 0. *Only the charge inside the chosen Gaussian surface contributes to the field.* This is also true of spherical charge distributions. (See problem D8.) Graphically, we can summarize this;

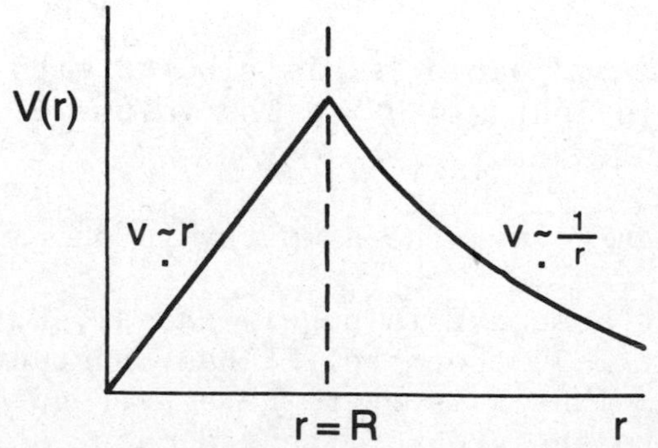

The field at r = 15 cm is

$$E_r = 2\pi(9 \times 10^9)(10^{-1})^2(10^{-3})/(15 \times 10^{-2}) = 1.2 \times 10^6 \text{ N/C}.$$

QUESTIONS ON FUNDAMENTALS

1. What is the magnitude of the force exerted by a 3 μC charge on a 20 nC charge if they are separated by 2 cm?

(a.) .027 N (b.) 27 N (c.) 1.35 N (d.) 135 N

2. If the 20 nC charge is at the origin of a coordinate axis and one 3 μC charge is at y = 2 cm while a second 3 μC charge is at x = -2 cm, what is the direction and magnitude of the force on the 20 nC charge?

(a.) 1.91 N, 45° above the -x axis
(b.) 2.7 N, 45° below the +x axis
(c.) 1.91 N, 45° below the +x axis
(d.) 2.7 N, 45° above the -x axis
(e.) zero

3. If a 5 μC charge is at the origin of coordinates, what magnitude of electric field does it produce at the coordinates x = 3 m and y = -4 m?

(a.) 9000 N/C (b.) 1.8 N/C (c.) 1800 N/C
(d.) 7810 N/C

4. If an infinite plane is covered with a uniform surface charge density of 50 $\mu C/m^2$, what electric field strength is produced at a distance of 2 meters from the sheet?

(a.) 2.83×10^6 N/C (b.) 4.5×10^5 N/C
(c.) 1.42×10^6 N/C (d.) 7.08×10^5 N/C
(e.) 9.01×10^5 N/C

5. What electric field strength is produced at a distance of 5 meters from the sheet of charge in question #4?

(a.) 1.13×10^6 N/C (b.) 1.44×10^5 N/C
(c.) 1.8×10^5 N/C (d.) 1.42×10^6 N/C
(e.) 2.83×10^6 N/C

6. What linear charge density must be on an infinitely long straight string to produce an electric field strength of 1.20×10^4 N/C at a distance of 3 cm from the line?

(a.) 40 nC/m (b.) 20 nC/m (c.) 0.6 nC/m
(d.) 6.37 nC/m (e.) 3.69 nC/m

GENERAL PROBLEMS

Problem 16-1. Suppose there is an infinite line charge along the y axis with linear charge density $\lambda = 2 \times 10^{-5}$ C/m. There is also a uniform sphere of charge centered on the point x = 10 cm, y = 10 cm. This sphere has a radius R = 3 cm and a volume charge density $\rho = 10^{-4}$ C/m^3. Find the electric field at the point x = 10 cm, y = -8 cm.

Q1: What is the contribution of the line charge?
Q2: In what direction is this?
Q3: What is the contribution of the sphere?
Q4: In what direction is this?
Q5: What is the sphere's charge, Q?

Problem 16-2. An uncharged spherical conducting shell with inner radius R_i = 2 cm and outer radius R_o = 4 cm contains a point charge q at its center. Using Gauss' Law, find the expression for the electric field:

(a.) inside R_i.
(b.) between R_i and R_o.
(c.) outside R_o.

Q1: What Gaussian surfaces should I choose?
Q2: What charge does the sphere at r_1 enclose?
Q3: What does Gauss' Law state about this sphere?
Q4: What can we say about the field on the sphere at r_2?
Q5: What charge must be enclosed by this sphere?
Q6: How much charge does the sphere at r_3 enclose?
Q7: What is Gauss' Law applied to this sphere?

ANSWERS TO QUESTIONS

FUNDAMENTALS:

1. (c.)
2. (c.)
3. (c.)
4. (a.)
5. (e.)
6. (b.)

GENERAL PROBLEMS:

Problem 16-1.

A1: It produces an electric field, $E_{line} = k_e(\lambda/r)$.

A2: The direction is radially outward from the line. At x = 10, y = -8, this is in the +x direction.

A3: Since the point (10,-8) is outside the charge distribution, the sphere will act as though all its charge Q was at a point at its center.

A4: Since the sphere is a negative charge, the field will be radially inward toward it. At the point (10,-8), this means it will be in the +y direction.

A5: $Q = \rho$ Volume $= \rho(4/3)\pi R^3$, R = radius of the sphere.

Problem 16-2.

A1: There is complete spherical symmetry, so the surfaces would be spheres concentric with q, of radii $r_1 < R_i$, $R_i < r_2 < R_o$, and $r_3 > R_o$.

A2: q.

A3: $E_r(4\pi r_1^2) = 4\pi k_e q$.

A4: This region is in conducting material, so E = 0.

A5: Since E = 0, and $4\pi r^2 E = 4\pi k_e Q_{encl}$, Q_{encl} must be *zero*. This means a charge -q has been induced on the inner surface of the conductor.

A6: The conductor has *no* net charge, so the sphere at r^2 must enclose only the charge q. When the -q is induced on the inner surface of the conductor, it leaves the outer surface with a +q charge. These add up to zero.

A7: $4\pi r_3^2 E_r = 4\pi k_e q$.

CHAPTER 17. ELECTRIC POTENTIAL ENERGY, POTENTIAL DIFFERENCE, AND CAPACITANCE

BRIEF SUMMARY AND RESUME

Electric potential is defined and its relation to electric field and potential energy is analyzed. The storage of charge in *capacitors* is described, and the concept of *capacitance* is defined. The effect of *dielectrics* on capacitance is described by the definition of *dielectric constant*.

Material from previous chapters relevant to Chapter 17 includes:

1. Concepts of Kinetic and Potential Energy.

NEW DATA AND PHYSICAL CONSTANTS

1. Electric Potential Units.

   $V = W/q$ gives units of J/C = volts (V).

2. The Electron-Volt of Energy.

   $1 \text{ eV} = 1.6 \times 10^{-19}$ J.

3. Units of Capacitance.

   $C = q/V$ gives units of coul/volt = farad (F).

NEW CONCEPTS AND PRINCIPLES

Electric Potential Difference: Voltage

Understanding this section should enable you to do the following:

1. *Identify the relation between electrostatic potential energy and electrostatic potential, and calculate one from the other.*
2. *Identify the relation between potential difference and electric field.*
3. *Calculate the electrostatic potential due to point charges.*

The electric force is another example of a *conservative* force. Therefore, in a region where an electric field exists, the work done on a charge q by the field as q changes position can be described by the change in the charge's *electric potential energy*. To describe this as a property of the charges producing the field rather than of q, we define the concept of *electric potential*.

Electric Potential (V)--The electric potential energy *per unit of charge* at a point in an electric field is called the *electric potential* at that point. Mathematically,

$$V = (PE)_{el}/q.$$

Voltage--The *potential difference* $V_{ab} = V_a - V_b$ between two points a and b is called the *voltage* between those points.

Important Observations

1. The unit of potential (and potential difference) is J/C, defined to be the *volt*. 1 J/C = 1 volt (V).
2. If you know the potential at point a, V_a, the potential *energy* of charge q at a is simply qV_a.
3. As in mechanics, the work done by the conservative force is equal to the loss in potential energy. For example, if a charge q moves from a to b, the work done by the field is $W_{ab} = qV_{ab} = q(V_a - V_b) = -\Delta V$.
4. Positive charges will tend to "fall downhill" to lower values of potential, whereas negative charges will "fall uphill" to regions of higher potential. This just reflects the opposite directions of the forces on them from Coulomb's Law, except that we are now dealing with a *scalar* quantity.

Absolute Potential of a Point Charge--A single isolated point charge Q produces a potential given by the expression

$$V(r) = k_e(Q/r),$$

where r is measured from the position of Q.

Important Observations

1. In this expression V is zero only at r = infinity. This choice of zero is what *defines* the *absolute* potential.
2. The potential is everywhere *positive* for a + charge, and everywhere *negative* for a - charge.
3. As you approach a + charge, the potential *increases* to larger + values. As you approach a - charge, the potential *decreases* to larger negative values.
4. If a number of point charges are distributed in a region, the principle of superposition of their individual contributions can be used to find the total potential at a given point. Here the calculation is a *scalar* one, with only algebraic signs to be mindful of.

Example 17-1. Using the charge distribution q_1, q_2, and q_3 described in Example 16-3, calculate the electric potential at x = 0, y = 0. What is the voltage between this point and the point x = 4 cm, y = 4 cm?

S: At the origin, the three contributions to the potential are

$$V_1 = k_e(q_1/r_1) = (9 \times 10^9 \text{ N-m}^2/\text{C}^2)(2 \times 10^{-6} \text{ C})/(3 \times 10^{-2} \text{ m})$$

$$= 6 \times 10^5 \text{ volts.}$$

$$V_2 = k_e(q_2/r_2) = (9 \times 10^9)(-5 \times 10^{-6})/(4 \times 10^{-2}) = -1.13 \times 10^6 \text{ V.}$$

$$V_3 = k_e(q_3/r_3) = (9 \times 10^9)(3 \times 10^{-6})/(2 \times 10^{-2}) = +1.35 \times 10^6 \text{ V.}$$

So I just add these together?

T: Yes. Scalar problems are much simpler than vector problems.

$$V(0,0) = (6 \times 10^5 - 1.13 \times 10^6 + 1.35 \times 10^6) \text{ V} = +8.2 \times 10^5 \text{ V.}$$

S: Next we have to find V at x = 4, y = 4. This means we have to calculate the distances from each charge to this point.

T: Right. This can simply be done using the x and y distances and Pythagoras' theorem.

$$r_1 = (7^2 + 4^2)^{1/2} \text{ cm} = 8.06 \text{ cm}$$

$$r_2 = (4^2 + 8^2)^{1/2} \text{ cm} = 8.94 \text{ cm}$$

$$r^3 = (2^2 + 4^2)^{1/2} \text{ cm} = 4.47 \text{ cm}$$

S: Then

$$V_1 = (9 \times 10^9)(2 \times 10^{-6})/(8.06 \times 10^{-2}) = 2.23 \times 10^5 \text{ V.}$$

$$V_2 = (9 \times 10^9)(-5 \times 10^{-6})/(8.94 \times 10^{-2}) = 5.03 \times 10^5 \text{ V.}$$

$$V_3 = (9 \times 10^9)(3 \times 10^{-6})/(4.47 \times 10^{-2}) = 6.04 \times 10^5 \text{ V.}$$

So

$$V(4,4) = (2.23 - 5.03 + 6.04) \times 10^5 \text{ V} = +3.24 \times 10^5 \text{ V.}$$

T: The voltage between the two points is then

$$V(0,0) - V(4,4) = (8.2 - 3.24) \times 10^5 \text{ V} = 4.96 \times 10^5 \text{ V.}$$

The origin is at the higher potential.

Equipotential Surface--A surface on which all points are at the same potential.

Important Observations

1. Equipotential surfaces must be everywhere perpendicular to the electric field, as explained in the text. It is easy to visualize the shapes of these surfaces for the simple charge distributions:

Charge Distribution	Electric Field Geometry	Equipotential Surfaces
Point or sphere	radial lines	concentric spheres
Line or cylinder	cylindrically radial lines	coaxial cylinders
Infinite plane	straight lines normal to plane	planes parallel to charged plane

2. A mapping of equipotentials is analogous to a contour map of topographic elevations, since lines of equal elevation are *gravitational* equipotentials.
3. No work is done on a moving charge if that charge stays on an equipotential surface.
4. The surfaces of *conductors* are equipotentials, as shown in the text. This means the following must be true:

    (a.) Electric fields must be normal to the surface of a conductor.
    (b.) There can be no static electric field within the material of a conductor.
    (c.) Any charge placed on a conductor must reside on its *exterior surface*.
    (d.) Fields can exist within a closed conducting shell only if the interior cavity contains charges. The cavity is totally *shielded* from fields caused by external charges.

<u>Example 17-2</u>. How much work must you do to move a -2 nC charge from a position 5 cm away from a -1 μC charge to a position 2 cm away?

<u>S</u>: Work = Force x displacement.

<u>T</u>: This is true only when the force is constant over the displacement, which *isn't* the case here. Use energy concepts.

<u>S</u>: O.K. We can write down the potential energy of the -2 nC charge at the two positions. Would the change in PE equal work?

<u>T</u>: Sure. You can use the same work-energy theorems that we developed in mechanics. We're not considering anything but a change in position here (no friction, no kinetic energy, etc.). So work done by you equals change in PE.

<u>S</u>: Since $V(r) = k_eQ/r$, where Q = -1 μC, the PE of q = -2 nC is PE = qV(r) = k_eqQ/r. Then

$$\text{Work} = k_eqQ[1/(2\text{ cm}) - 1/(5\text{ cm})]$$

$$= (9 \times 10^9\text{ N-m}^2/\text{C}^2)(-2 \times 10^{-9}\text{ C})(-1 \times 10^{-6}\text{ C})/(3.3 \times 10^{-2}\text{m})$$

$$= 5.4 \times 10^{-4}\text{ J.}$$

T: The work done *by the field* is the *negative* of this, since F_e is *outward* on the -2 nC charge, while its displacement is *inward*.

Example 17-3. How much work would it take to assemble 3 protons at the corners of an equilateral triangle whose sides are 1 μm long?

S: I don't know where the charges were originally.

T: Assume they were so far apart that their original interaction was negligible. This is the *practical* meaning of "infinitely separated."

S: Then originally the protons would have PE = 0. Looking at the last example, the work of assembly ought to be just the PE of the final arrangement. Is it that easy?

T: Sure is! With conservative forces you don't have to worry about the *route* taken, but just the initial and final conditions. Here you can superpose the three pair interactions.

S: The PE for each pair is $k_e e^2/r$, where $r = 10^{-6}$ m. This gives

$$(PE)_{pair} = (9 \times 10^9)(1.6 \times 10^{-19}\ C)^2/(10^{-6}\ m) = 2.3 \times 10^{-22}\ J.$$

T: So $(PE)_{tot} = 3(PE)_{pair}$, since e and r is the same for each interaction.

$$(PE)_{tot} = 6.9 \times 10^{-22}\ J$$

$$= (6.9 \times 10^{-22}\ J)(1.6 \times 10^{-19}\ J/eV) = 4.31 \times 10^{-3}\ eV$$

The Storage of Electric Charge

Understanding this section should enable you to do the following:

1. *Identify the definition of capacitance and calculate its value for simple capacitors.*
2. *Calculate the effect of dielectric materials on electric field, potential, and capacitance.*

When positive and negative charges are separated and stored on conductors, a potential difference or voltage is developed between them which is proportional to the stored charge. The proportionality constant depends on the specific size and geometrical arrangement of the conductors.

Capacitor--A device for storing separated electric charge.

Capacitance (C)--The ratio of charge stored in a capacitor to the voltage between the charges. Mathematically,

$$C = Q/V.$$

Parallel Plate Capacitor--An arrangement consisting of two parallel conducting plates whose area A is much larger than the plate separation s. The capacitance of this device is derived in your text:

$$C = \epsilon_o(A/s) ,$$

where $\epsilon_o = 1/(4\pi k_e) = 8.85 \times 10^{-12}\ C^2/N\text{-}m^2$.

Important Observations

1. The unit of capacitance is coul/volt, which is *defined* to be a *farad* (F). (1 Coulomb/Volt = 1 F).
2. Since V is always proportional to Q, Q cancels in every calculation of capacitance, leaving a result which depends only on ϵ_o, and critical dimensions of the device.
3. As the coulomb represents an enormous charge in practical terms, so the farad is an impractically large amount of capacitance. Practical values usually range from a μF to a pF (pico = 10^{-12}).

Example 17-4. Two parallel plates are separated by 2 mm become charged with 6 μC when they are connected to a 120 volt battery.

(a.) What area do the plates have?
(b.) If the plates are disconnected from the battery and then allowed to move until separated by only .5 mm, find the new values of Q, C, and V.
(c.) What is the electric field strength between the plates? Does it change?

S: The information given tells us the capacitance directly.

$$C = Q/V = (6 \times 10^{-6}\ C)/(120\ V) = 5 \times 10^{-8}\ F = 50\ nF.$$

What's the connection between this and plate area?

T: For parallel plates, $C = \epsilon_o A/s$, so

$$A = Cs/\epsilon_o = (50 \times 10^{-9}\ F)(2 \times 10\text{-}3\ m)/(8.85 \times 10^{-12}\ C^2/N\text{-}m^2) = 11.3\ m^2$$

S: How did you get m^2 from that mess of units?

T: Well, of course you have to break farads up into more fundamental units. You have:

$$(F)(m)/(C^2/N\text{-}m^2) = (C/V)(m)(N\text{-}m^2)/C^2 ,\ \text{with } V= J/C = N\text{-}m/C.$$

Then algebraically all that survives cancellation is m^2.

S: When the separation s changes, C is going to change:

$$C_{new} = \epsilon_o A/(0.5\ \text{mm}) = 4C_{orig} = 200\ \text{nF}.$$

Does anything else change?

T: The stored charge is trapped on the plates once they are disconnected, so Q can't change. Since $V = Q/C$, that means V must *decrease* by the same factor as C increased. Thus

$$V_{new} = V_{orig}/4 = 30\ \text{volts}.$$

S: How can this be?

T: The charge on the plates determines the electric field between the plates, so *E remains the same.* Remember the discussion in the text showing that the field is a measure of voltage per meter along the field lines. Thus the *voltage per meter* is the same in both cases, but since the new separation is 1/4 the original, it allows only 1/4 the total plate to plate voltage of the original arrangement.

S: This idea gives us $E = (30\ \text{V})/(.5 \times 10^{-2}\ \text{m}) = 6 \times 10^4\ \text{V/m}$. Is there a more basic way to determine the field?

T: Yes. The positively charged plate and the negatively charged one both contribute an amount $2\pi k_e \sigma$ in the same direction, as we've seen from Gauss' Law. This makes the total field

$$E_{tot} = 4\pi k_e (Q/A) = 4\pi(9 \times 10^9)(6 \times 10^{-6})/(11.3) = 6 \times 10^4\ \text{N/C}.$$

Dielectrics--Non-conducting materials which have the effect of partially cancelling the electric field which would exist in a vacuum.

Dielectric Constant (K)--The ratio of capacitance C *with* a dielectric filling the capacitor to the capacitance C_o with vacuum filling the capacitor. Mathematically, $K = C/C_o$.

Important Observations

1. The dielectric constant is a dimensionless number equal to or greater than 1 ($K_{vacuum} = 1$). Some values are given in Table 17.1 (text).
2. In the presence of a dielectric, all the expressions for F_e, E, and V developed previously for vacuum are correct if you make one change. That is to replace k_e by k_e/K, or equivalently, replace ϵ_o by $K\epsilon_o$. The latter, given the symbol $\epsilon = K\epsilon_o$, is called the *permittivity* of the dielectric.
3. Thus E and V are *reduced* by a factor of K, whereas C is *increased* by the same factor.

Example 17-5. Consider the original capacitor of Example 17-4 again. Now suppose that, when the plate separation was diminished to 15 mm, the plates were immersed in oil, with dielectric constant K = 2.2. What are the final values of E, V, Q, and C under this condition?

S: You say that E and V are *reduced* by K, so where we had $E = 6 \times 10^4$ V/m and V = 30 volts, we would now have $E_{in\ oil} = 6 \times 10^4/2.2 = 2.73 \times 10^4$ V/m and V = 30/2.2 = 13.6 V. What happens to Q? It must change too, since E changes.

T: Good question. E is reduced *between the plates* by the behavior of the molecules of the dielectric. But this has no physical effect on the charge on the plates, so Q remains unchanged. This means that since C = Q/V, we have as much charge stored as before, but with a voltage 2.2 times smaller. Thus $C_{in\ oil} = 2.2C_{vacuum}$ = (2.2)(200 nF) = 440 nF.

Example 17-6. Compare the values of E and V at r = 10 cm from a 1 μC point charge in vacuum and under water.

S: In vacuum,

$$E_r = k_e(Q/r^2) = (9 \times 10^9)(10^{-6})/(10^{-1})^2 = 9 \times 10^5 \text{ N/C}.$$

and

$$V = k_e(Q/r) = 9 \times 10^4 \text{ volts}.$$

What happens under water?

T: Water has a very large dielectric constant, K = 80. Both of the above results are reduced by this factor under water.

$$E_{water} = 9 \times 10^5/80 = 1.13 \times 10^4 \text{ N/C}$$

$$V_{water} = 9 \times 10^4/80 \quad 1.13 \times 10^3 \text{ volts}$$

The Coulomb force between charges under water is reduced by the same factor as well.

QUESTIONS ON FUNDAMENTALS

1. What is the electric potential produced at the origin of coordinates by two charges each of 3 μC placed at x = 2 cm and y = -2 cm?

(a.) 13,500 v (b.) zero (c.) 6750 v (d.) 27,000 v
(e.) 54,000 v

2. Answer question #1 if the charges are located at x = -2 cm and y = -2 cm.

(a.) 27,000 v (b.) zero (c.) 6750 v (d.) 13,500 v
(e.) 54,000 v

3. What would be the electric potential energy of a 5 μC charge if it was placed at the origin of coordinates in the situation described in question #1?

(a.) 67.5 mJ (b.) 33.75 mJ (c.) 135 mJ (d.) zero
(e.) 270 mJ

4. What is the capacitance in air of a pair of parallel plates having area A = 100 cm^2 separated by a gap of 2 mm?

(a.) 442.5 pF (b.) 1.77 pF (c.) 44.25 pF
(d.) 8.85 pF

5. If the parallel plates in question #4 have 100 volts of potential difference between them, how much charge is stored on each plate?

(a.) 177 μC (b.) 2.2 nC (c.) 885 pC (d.) 44.25 nC
(e.) 4.425 nC

6. If the plates in question #5 were immersed in a liquid whose dielectric constant was K = 5, what would be the voltage between the plates?

(a.) 100 v (b.) 20 v (c.) 500 v (d.) 4 v (e.) 2500 v

GENERAL PROBLEMS

Problem 17-1. If you have an isolated conducting sphere, derive the expression for its capacitance. If the sphere was as large as the earth, what capacitance would it have?

Q1: What is the charge on the sphere?
Q2: Then how do I proceed to determine voltage?
Q3: Where is the second conductor?
Q4: What is the potential at the surface of a sphere with charge Q?
Q5: What is the formula for capacitance?
Q6: What is the radius of the earth?

Problem 17-2. You have two spherical conductors, with radii R_1 = 10 cm, R_2 = 5 cm. They are given initial charges Q_1 = 2 nC and Q_2 = -4 nC. They are far enough apart so that they are essentially isolated from each other. What is the voltage between their surfaces? Next they are connected by a long wire, which is then removed. Find the new charges on the spheres, and the new value of the potential at their surfaces.

Q1: What is the potential on their individual surfaces?
Q2: What does connecting them do?
Q3: What determines the new charges on them?
Q4: What determines the new potential at their surfaces?

Problem 17-3. You have a parallel plate capacitor with plate area A = 100 cm^2 and a plate separation of d = 1 mm. It is filled with a slab of pyrex glass whose dielectric constant is 6.0. What is the capacitance of this arrangement? If you connect the plates to a 24 volt battery, how much charge gets stored? If you next disconnect the battery and then remove the slab, find the new values of Q, C, and V.

Q1: What is the expression for the capacitance with the glass?
Q2: What determines the charge stored when the battery is connected?
Q3: When you remove the battery and then the slab, what changes?
Q4: How does C change?

ANSWERS TO QUESTIONS

FUNDAMENTALS:

1. (d.)
2. (a.)
3. (c.)
4. (c.)
5. (e.)
6. (b.)

GENERAL PROBLEMS:

Problem 17-1.

A1: Remember, capacitance doesn't depend on Q.

A2: Postulate *any* Q, find the V this would produce at the surface, and use the definition of capacitance.

A3: An *isolated* sphere means the other conductor is at infinity with $V = 0$.

A4: $V = k_e(Q/R)$, relative to $V = 0$ at infinity.

A5: $C = Q/V$.

Problem 17-2.

A1: $V_1 = k_e(2 \times 10^{-9}\ C)/(10^{-1}\ m)$, $V_2 = k_e(-4 \times 10^{-9}\ C)/(5 \times 10^{-2}\ m)$.

A2: It equalizes their potentials, since it makes them part of *one* conductor, with an equipotential surface.

A3: The conservation of charge requires that $Q_1 + Q_2 = Q_1' + Q_2'$. The equalization of potential requires that $k_eQ_1'/R_1 = k_eQ_2'/R_2$.

A4: *Either* k_eQ_1'/R_1 or k_eQ_2'/R_2.

Problem 17-3.

A1: $C = \epsilon A/d$, where $\epsilon = K\epsilon_o$, with $K = 6.0$.

A2: $Q = CV$.

A3: C changes when you remove the dielectric. With the battery disconnected, Q is trapped with no conducting path through which it can move. So $V = Q/C$ must change.

A4: It simply changes to $C = \epsilon_oA/d$. The voltage is no longer determined by the battery, but adjusts to $V = Q/C$.

CHAPTER 18. DIRECT CURRENT ELECTRICITY

BRIEF SUMMARY AND RESUME

The phenomenon of *electric current* is introduced and mathematically defined. Sources of current, or *EMF's*, such as batteries, are discussed. Currents in *electrolytes* and *metallic conductors* are compared. The concept of *resistance* of a conductor and *Ohm's Law* are introduced and applied to simple *circuits*. The *resistivity* of materials and its *temperature dependence* are examined. *Kirchoff's Rules* are introduced and used to analyze d.c. circuits. The effects of resistors and capacitors in *series* and *parallel* combinations are examined. The power produced by EMF's and consumed by *Joule heating* in resistors is described, and the conservation of energy in circuits is demonstrated.

Material from previous chapters relevant to Chapter 18 includes:

1. Concepts of charge and voltage.
2. Conservation of charge and energy.

NEW DATA AND PHYSICAL CONSTANTS

1. The unit of electric current (I).

   $I = Q/t = \text{coul/sec} = \text{amperes}$.

2. The *faraday* of charge.

   $1 \text{ faraday} = N_A e = 96{,}500 \text{ C}$.

3. The unit of electrical resistance (R).

   $R = V/I = \text{volts/ampere} = \text{ohms } (\Omega)$.

NEW CONCEPTS AND PRINCIPLES

Direct Current in Single Loop Circuits

Understanding this section should enable you to do the following:

1. *Calculate the relation between the current flowing in single loop circuits and the EMF and resistance in the circuit.*
2. *Identify the relation between resistance and resistivity.*
3. *Calculate the resistivity of various materials at different temperatures.*

Electric Current (I)--The rate of flow of charge. Mathematically, $I = Q/t$, where Q is the amount of charge passing a point during the time t. When I is constant in time, we have a *direct current*, or *d.c.* situation.

Important Observations

1. The unit of current is derived to be coul/sec. This is defined as the *ampere*:

   1 coul/sec = 1 ampere (A)

2. Movement of + charges in one direction is equivalent to movement of - charges in the *opposite* direction.
3. In electrolytes, both electrons and positive ions physically move through the solution, thus contributing to the current. In metals, only the electrons are free to move. Nevertheless, *we define I to be in the direction of the movement of + charges.* Thus in metals, I is opposite to the actual physical movement of the electrons.

Seat of EMF ($\mathcal{E}$)--A device which can convert non-electric energy into electric potential energy on a continuing basis. Examples are batteries and generators. When connected to a circuit through which charge can move, this potential energy is converted into the kinetic energy of the charges. The effect of this device is thus to create a continuous *source of voltage* for the charges in the circuit. The unit of EMF is the *volt*.

Electric Circuit--Any arrangement of closed conducting paths in which charges can move in response to EMFs connected to the circuit. A *single loop* circuit has only one closed path, or loop.

Ohm's Law--In a d.c. electric circuit, the applied voltage or EMF and the resulting current are directly proportional to each other. Mathematically, $V = RI$, where the constant of proportionality, R, is called the *resistance* of the circuit.

Important Observations

1. The unit of resistance is derived to be volts/ampere, which defines the *ohm* (Ω). 1 volt/ampere = 1 ohm.
2. Resistance is physically caused by electrons colliding with the atoms in the structure of the conductor. This represents a loss of energy given to the electrons by the EMF.
3. In a drawing, or "schematic" of a circuit, an EMF is drawn as , with the longer line representing the + or higher potential terminal. A resistance is drawn as  . Thus a single loop circuit can be drawn as follows:

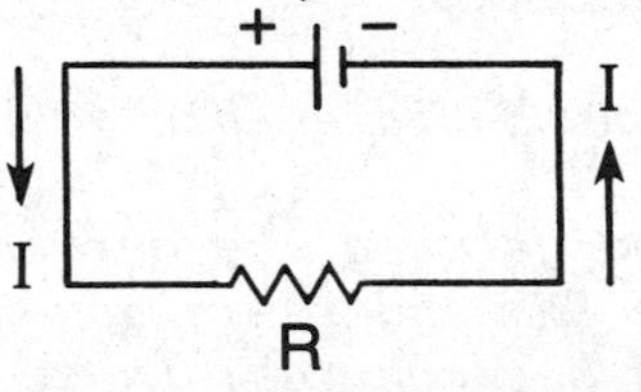

The straight lines are connectors whose resistance is negligible. Notice that the current flow is *out* of the + terminal.

4. The end of the resistor that the current *enters* is at *higher* potential than the other end, since + charges lose energy in passing through R.

Example 18-1. A battery with $\mathcal{E}$ = 12 v is connected in a circuit with a 24 Ω resistor. How much current flows, and how much power is the battery delivering?

S: Ohm's Law would give us $I = V/R$. What do I use for V?

T: Since $\mathcal{E}$ = 12 v is placed across the ends of the *one* resistor, the potential drop from end to end must be V = $\mathcal{E}$.

S: Then I = 12 v/24 Ω = 0.5 v/Ω = 0.5 amperes. How do we find the power from the battery?

T: Remember what an EMF does. It increases the potential energy of a charge q passing through it by ΔPE = q$\mathcal{E}$. The *rate* of increase of PE is the power delivered by the EMF. Thus P = q$\mathcal{E}$/t = I .

S: Then the battery is delivering P = (.5 A)(12 v) = 6 amp-volts of power.

T: It should come as no surprise that *1 amp-volt = 1 watt*, since we are always in SI units. Notice that (amp)(volt) = (C/s)(J/C) = J/s = watts.

Resistivity (ρ)--An inherent property of specific materials, indicating the relative ease or difficulty of electron movement through them. The *resistance* (R) of an actual *piece* of material also depends on its length and cross-sectional area, by the following relation: R = ρ L/A. Values of ρ for various materials are given in Table 18.1 (text).

Temperature Coefficient of Resistivity (α)--The resistivity of metals increases linearly with temperature according to the expression

$$\rho_T = \rho_o(1 + \alpha \Delta T).$$

where $\Delta T = T(^oC) - 0\ ^oC$ and ρ_o = resistivity at T = 0 ^{o}C. Values of α are also shown in Table 18.1 (text).

Important Observations

1. As in the similar expression for thermal *expansion*, α has units of $(C^o)^{-1}$.
2. The unit of resistivity is Ω-m.
3. All pure metals have $\alpha > 0$, so their resistivity *increases* with temperature. Some materials, notably carbon and semiconductors, show a *decrease* in resistivity upon being heated. Their values of α must be *negative*.

Example 18-2. 14 gauge wire, often used in house circuits, has a diameter of 1.63 mm. What length of 14 gauge copper wire would have a resistance of 10.0 Ω at T = 30 ^{o}C? Answer the same for aluminum wire. Which of these lengths would have greater resistance at 0 ^{o}C?

S: From the relation between ρ and R, we have L = RA/ρ . From the table, $\rho_{Cu} = 1.7 \times 10^{-8}$ Ω-m and $\rho_{Al} = 2.6 \times 10^{-8}$ Ω-m.

T: These are values at T = 0°C.

S: Oh, that's right. We have to find ρ_{30} for each. Again from the table, $\alpha_{Cu} = 4.1 \times 10^{-3}\ (C^o)^{-1}$ and $\alpha_{Al} = 3.9 \times 10^{-3}\ (C^o)^{-1}$.

T: So

$$(\rho_{Cu})_{30} = \rho_{Cu}[1 + (4.1 \times 10^{-3}/C^o)(30\ C^o)] = 1.91 \times 10^{-8}\ \Omega\text{-m}$$

and

$$(\rho_{Al})_{30} = \rho_{Al}[1 + (3.9 \times 10^{-3}/C^o)(30\ C^o)] = 2.90 \times 10^{-8}\ \Omega\text{-m}.$$

S: Now we need area A. $A = \pi r^2 = \pi(.815 \times 10^{-3} m)^2 = 2.09 \times 10^{-6}\ m^2$. Then

$$L_{Cu} = (10\ \Omega)(2.09 \times 10^{-6}\ m^2)/(1.91 \times 10^{-8}\ \Omega\text{-m})$$

$$= 1.09 \times 10^3\ m = 1.09\ km!$$

$$L_{Au} = (10\ \Omega)(2.09 \times 10^{-6}\ m^2)/2.90 \times 10^{-8}\ \Omega\text{-m} = 7.21 \times 10^2\ m.$$

Next we have to calculate R for 0^oC.

T: We can, but by *inspection* we can tell which will have greater R. At T = 300 ^oC they're the same, 10 Ω. When lower T, the material with *larger* α will have its R reduced *more*. Looking at the values of α, the aluminum should have more resistance at 0^oC than the copper. Want to check?

S: O.K. At T = 0 ^oC,

$$R_{Cu} = (\rho_{Cu})_0 L_{Cu}/A = (1.7 \times 10^{-8}\ \Omega\text{-m})(1.09 \times 10^3\ m)/(2.09 \times 10^{-6}\ m)^2$$

$$= 8.87\ \Omega.$$

$$R_{Al} = (2.6 \times 10^{-8}\ \Omega\text{-m})(7.21 \times 10^2\ m)/(2.09 \times 10^{-6}\ m)^2 = 8.97\ \Omega.$$

T: Notice the comment in the text about being able to neglect thermal changes of L and A, since the expansion coefficients are generally less than 1% as large as the resistivity coefficients.

Kirchoff's Rules: Multiple Loop Circuits

Understanding this section should enable you to do the following:

1. *Calculate the currents and potential drops in any branch of multiple loop circuits.*
2. *Calculate the equivalent total resistance and capacitance of resistors and capacitors in series and parallel connections.*

Kirchoff's Junction Rule--The sum of currents entering a junction point is equal to the sum of currents leaving that junction point.

Kirchoff's Loop Rule--The algebraic sum of the changes in electric potential encountered in a complete traversal of any closed loop of a circuit is zero.

Important Observations

1. These two rules are merely restatements of the principles of conservation of charge and energy, respectively.

2. To apply the rules, you must label a current in each separate branch, *arbitrarily* showing a direction. If you choose a direction wrong, you will get a minus sign for that current. Obviously you shouldn't pick directions which would require *all* currents to enter or leave a junction.
3. Pick *any* complete loop in the circuit. Traverse it in either direction. The rules for counting potential changes are:

   (a.) EMFs: If you traverse an EMF from the - terminal to the + terminal, the ΔV is $+\mathcal{E}$ and vice versa.
   (b.) Resistances: If you traverse a resistance R *in the direction* of the current you chose, the ΔV is $-IR$. If you traverse it *against* the current, the ΔV is $+IR$.

4. Choose as many loops as is necessary to sample all the circuit elements *once*. If you pick a loop that doesn't involve *at least one new* EMF or resistance, that equation will be *redundant* with your previous equations.
5. Add as many junction equations to the above as it takes to give a total number of equations equal to the number of separate currents in the circuit.

Example 18-3. Find all the currents in the circuit shown in the following figure.

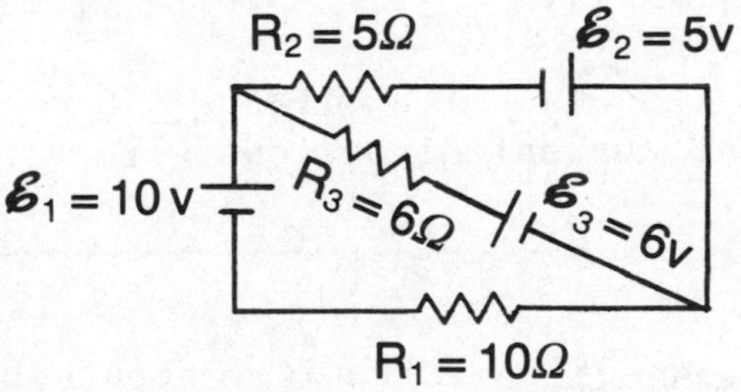

S: I really can pick the currents any way I want?

T: Yes.

S: Then I'll choose as follows:

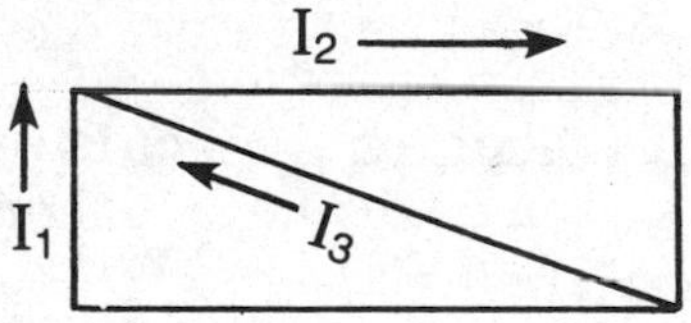

T: Fine. Then the junction rule gives you $I_2 = I_1 + I_3$.

S: I can see 3 different loops. Which shall I take?

T: Any 2 of them. Notice that the third one would *not* sample any *new* information. Also, 2 loop equations plus the junction equation give you the three equations necessary to solve for the 3 currents.

S: Going around the perimeter, I get

$$\mathcal{E}_1 - I_2R_2 + \mathcal{E}_2 - I_1R_1 = 0.$$

In the loop with $\mathcal{E}_2$ and $\mathcal{E}_3$, I get

$$\mathcal{E}_3 - I_3R_3 - I_2R_2 + \mathcal{E}_2 = 0.$$

T: The other loop would give you $\mathcal{E}_1 + I_3R_3 - \mathcal{E}_3 - I_1R_1 = 0$. This is just the difference of your two equations, with nothing new. Putting numbers in your equations, and omitting the symbols,

$$10 - 5I_2 + 5 - 10I_1 = 0.$$

and

$$6 - 6I_3 - 5I_2 + 5 = 0.$$

These have to be solved simultaneously, along with the junction equation.

S: The first step would seem to be to substitute for I_3 in the last equation:

$$6 - 6(I_2 - I_1) - 5I_2 + 5 = 0,$$

or

$$11 - 11I_2 + 6I_1 = 0.$$

Now what?

T: You've reduced the unknowns to two. Now if you multiply this equation by 10/6 and add it to the other loop equation, I_1 will drop out:

$$15 - 5I_2 - 10I_1 + 110/6 - (110/6)I_2 + 10I_1 = 0.$$

or

$$200/6 - (140/6)I_2 = 0, \quad \text{giving } I_2 = 1.43 \text{ A}.$$

Now the other currents fall out easily.

S: Yeah. $6I_1 = 11I_2 - 11$, giving $I_1 = .787$ A.

T: Finally, $I_3 = I_2 - I_1 = 1.43 - .787 = .643$ A.
Your choice of current directions were all correct, since the answers all turned out positive.

<u>Series Connection</u>--Circuit elements that are connected so that the same current must flow through them one after the other *without branching* are connected in *series*.

<u>Parallel Connection</u>--Circuit elements that are connected with one end of each at a common junction and their other ends at another common junction are connected in *parallel*.

<u>Combinations of Resistors</u>--As derived in your text, we have the following results:

a. Resistors in series: $R_{tot} = R_1 + R_2 + R_3 + \ldots$

b. Resistors in parallel: $1/R_{tot} = 1/R_1 + 1/R_2 + 1/R_3 + \ldots$

<u>Combinations of Capacitors</u>--As derived in your text, we have the following results:

a. Capacitors in series: $1/C_{tot} = 1/C_1 + 1/C_2 + 1/C_3 + \ldots$

b. Capacitors in parallel: $C_{tot} = C_1 + C_2 + C_3 + \ldots$

Important Observations

1. All resistors in series carry the same current, and differ in the potential drop across them (unless they are equal R). All resistors in parallel have the same potential drop, but carry different currents.
2. All capacitors in series have the same charge on them, but vary in potential (unless the capacitors are equal). All capacitors in parallel have the same potential drop, but carry various charges.
3. The most common mistake in adding reciprocals is forgetting to take the final reciprocal of that addition, leaving you with $1/R_{tot}$ or $1/C_{tot}$ instead of R_{tot} or C_{tot}.
4. Direct addition, of course, gives a final value *larger* than any of the individual values. Reciprocal addition *reduces* the final value to less than any of the individual values of R or C.

Example 18-4. How much current is drawn from the battery in the circuit shown in the following figure? Take $R_1 = 4\ \Omega$, $R_2 = 8\ \Omega$, $R_3 = 6\ \Omega$, $R_4 = 12\ \Omega$, $R_5 = 6\ \Omega$, and $R_6 = 16\ \Omega$.

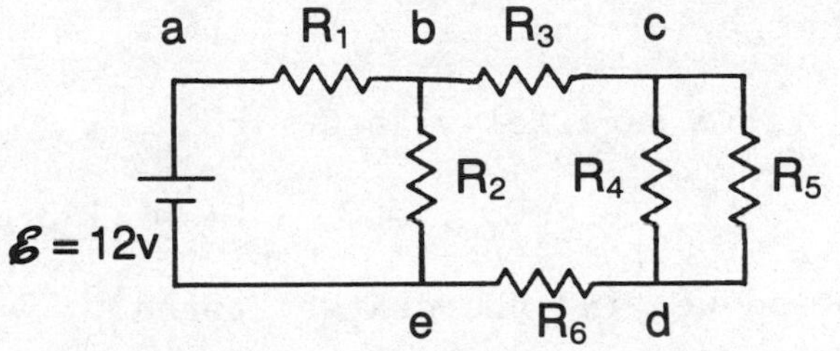

S: Good grief! This would require a lot of equations if we used Kirchoff's Rules.

T: About 6, I reckon. But this arrangement is reducible to a number of series and parallel combinations. For example, R_4 and R_5 are in parallel, so

$$1/R_{cd} = 1/R_4 + 1/R_5 = 1/12 + 1/6 = (3/12)\ \Omega^{-1}$$

Thus

$$R_{cd} = 4\ \Omega.$$

S: And I see this is in series with R_3 and R_6. So

$$R_{be} = R_3 + R_6 + R_{cd} = 6 + 16 + 4 = 26\ \Omega.$$

T: You're *partially* right. These 3 resistors *are* in series, but this is *not* R_{be}. The 26 Ω is in parallel with R_2 to form R_{be}:

$$1/R_{be} = 1/R_2 + 1/26 = 1/8 + 1/26 = (17/104)\ \Omega^{-1}$$

Thus

$$R_{be} = 104/17 = 6.12\ \Omega.$$

S: Then finally, this is in series with R_1:

$$R_{tot} = R_{ae} = R_1 + 6.12 = 10.12\ \Omega.$$

T: Then $I = \mathcal{E}/R_{tot} = 12\ \text{v}/10.12\ \Omega = 1.19\ \text{A}$.

Example 18-5. In the following figure, C_1 = .2 μF, C_2 = .4 μF, and C_3 = .3 μF. How much total charge is stored in this arrangement? (Notice capacitors are drawn with equal lines, unlike EMFs.)

S: There's no current flowing here, is there?

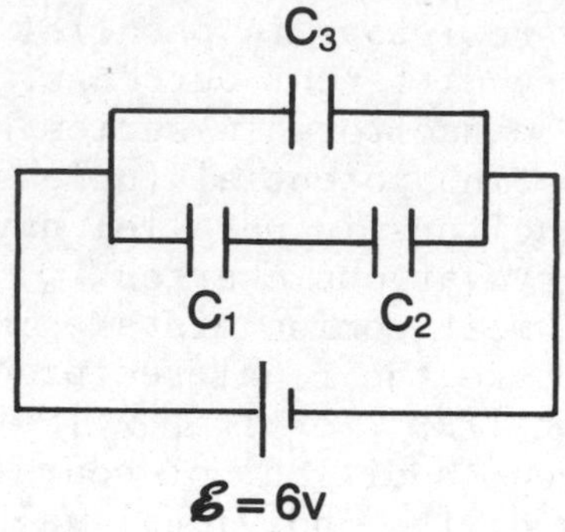

T: When the capacitors are first connected to the EMF, a temporary current flows as they become charged, but this quickly becomes a condition of essentially zero current.

S: C_1 and C_2 are in series. Let's see, that means they add reciprocally:

$$1/C_s = 1/C_1 + 1/C_2 = (1/.2) \times 10^{-6} + (1/14) \times 10^{-6}$$

$$= 3/14 \times 10^{-6}\ F^{-1}, \text{ so}$$

$$C_s = .133 \times 10^{-6}\ F.$$

This *is* smaller than either C_1 or C_2.

T: C_s is in parallel with C_3, so

$$C_{tot} = C_s + C_3 = .133\ \mu F + .3\ \mu F = .433\ \mu F.$$

S: How do we find the charge stored?

T: Remember, C = Q/V, or Q = CV. Here

$$Q = (.433 \times 10\text{-}6\ F)(6\ v) = 2.60 \times 10\text{-}6\ C.$$

S: Is this the charge on *each* capacitor?

T: No. Look back at the Observations. The charge is shared by C_3 and C_s, which both have a potential of 6 v.

S: Do you mean $Q_3 = C_3V$ and $Qs = C_sV$?

T: Yes, along with $Q_3 + Q_s$ = 2.60 μC. So $Q_3/Q_s = C_3/C_s$ = 0.3/0.133 = 2.25, which gives us Qs = 0.8 μC and Q_3 = 1.8 μC. Finally, Q_s is the charge on both C_1 and C_2.

Energy and Power in d.c. Circuits

Understanding this section should enable you to do the following:

1. *Calculate the power delivered to a d.c. circuit by an EMF.*
2. *Calculate the power consumed by resistors in a d.c. circuit.*

Power in d.c. Circuits--The product of current times voltage, IV, is the power associated with any element of a d.c. circuit. Examples:

1. EMFs: The voltage is $\mathcal{E}$, so $P = I\mathcal{E}$. This is the power actually delivered by the energy conversion process in the EMF source.
2. Battery: Here the effective (or terminal) voltage is $V_t = \mathcal{E} - IR_{int}$, as discussed in your text. Thus $P = IV_t$ is the net power delivered to the circuit external to the battery after subtracting the internal power loss.
3. Resistor: The voltage across a resistor carrying current I is V = IR, so the power converted into heat is $P = I^2R$. This is called *Joule heating*.

Important Observations

1. Energy conservation requires that the sum of all power delivered to the circuit by EMFs is balanced by the power consumed as heat by all the resistors.
2. The *energy* delivered by an EMF or consumed by a resistor in a time t is Energy = Pt = $I\mathcal{E}t$ or I^2Rt, respectively.
3. Since electric devices generally consume electric energy at a constant rate when they are operating, the energy unit used in the electric industry is the *kilowatt-hour* (kwh). 1 kwh = 3.6×10^6 J = 860 kcal.
4. If the current is being forced *backward* through an EMF, i.e., from + to - terminal, the power $\mathcal{E}I$ is being *consumed* in working against the potential.

Example 18-6. In Example 18-3, show that the power consumed by the resistors is equal to the power delivered by the EMFs.

S: The power delivered by $\mathcal{E}_1$ is $I_1\mathcal{E}_1$ = (.787 A)(10 v) = 7.87 W. That consumed by R_1 is $I_1^2R_1$ = $(.787\text{ A})^2(10\ \Omega)$ = 6.2 W. These *don't* balance.

T: No, there's no requirement that they balance in each *branch*. Just for the whole circuit. Calculate the other two power sources.

S: From $\mathcal{E}_2$ we have I_2E_2 = (1.43 A)(5 v) = 7.15 W, and from $\mathcal{E}_3$, $I_3\mathcal{E}_3$ = (.643 A)(6 v) =3.86 W.

T: So the total power *delivered* to the circuit is

$$P_{del} = 7.87 + 7.15 + 3.86 = 18.9\text{ W}$$

Now calculate the other two power losses.

S: In R_2 the power loss is $I_2^2R_2 = (1.43\text{ A})^2(5\ \Omega) = 10.2$ W,

and in R_3 it is $I_3^2R_3 = (.643\text{ A})^2(6\ \Omega) = 2.5$ W.

T: Thus P_{loss} = 6.2 + 10.2 + 2.5 = 18.9 W. This power balance requirement serves as a final independent check on whether your current values were correct from Kirchoff's Rules.

QUESTIONS ON FUNDAMENTALS

1. How much current will flow in a circuit whose resistance is 30 Ω if the circuit contains an emf of 10 v?

(a.) 3 amps (b.) .33 amps (c.) 300 amps (d.) 3.33 amps (e.) 30 amps

2. If the resistivity of a certain metal is 4×10^{-8} Ω-m at a temperature of 20°C, and the cross sectional area of a certain wire is .01 mm^2, how long a piece of this wire would be required to make a resistance of 20 Ω if the temperature is 20 °C?

(a.) 500 m (b.) 20 cm (c.) 5 m (d.) 80 m (d.) 50 m (e.) 20 m

3. If the temperature coefficient of resistivity of the material in question #2 was 2.2×10^{-3} /C°, what would be the resistance of the above wire (to 3 significant digits) if the temperature rose to 100 °C?

(a.) 20.2 Ω (b.) 21.8 Ω (c.) 35.2 Ω (d.) 23.5 Ω (e.) 37.6 Ω

4. If you connected four 5 Ω resistors in series, what would be the total resistance of the combination?

(a.) 0.8 Ω (b.) 1.25 Ω (c.) 10 Ω (d.) 20 Ω (e.) 40 Ω

5. If you connected the same four resistors in question #4 together all in parallel, what would the total resistance of the combination be?

(a.) 0.8 Ω (b.) 1.25 Ω (c.) 10 Ω (d.) 20 Ω (e.) 40 Ω

6. If you connected three 5 Ω resistors in parallel and connected that arrangement in series with a fourth 5 Ω resistor, what would the total resistance of the combination be?

(a.) 5.6 Ω (b.) 3.75 Ω (c.) 1.87 Ω (d.) 1.25 Ω
(e.) 6.67 Ω

7. If you connected three 4 μF capacitors in series, what would the total capacitance be?

(a.) 12 μF (b.) 0.75 μF (c.) 1.33 μF (d.) 2.67 μF
(e.) 8.33 μF

8. If you connected three 4 μF capacitors together in parallel, what would the total capacitance be?

(a.) 12 μF (b.) 0.75 μF (c.) 1.33 μF (d.) 8.33 μF
(e.) 0.25 μF

9. How much power does the EMF in question #1 deliver to the circuit?

(a.) 300 W (b.) *3.33 W (c.) 90 W (d.) 0.33 W
(e.) 1.11 W

10. If a 24 v EMF was connected to the combination of resistors in question #6, how much power would be dissipated in the fourth 5 Ω resistor?

(a.) 64.8 W (b.) 86.4 W (c.) 7.2 W (d.) 9.6 W (e.) 90 W

GENERAL PROBLEMS

Problem 18-1. Reduce the circuit in the figure below to a single equivalent resistance and find the current being supplied by the battery.

Q1: What is the equivalent resistance between points BD?
Q2: What is the resistance along the path CDB?
Q3: What is the resistance between CB?
Q4: What is the total resistance of the circuit?

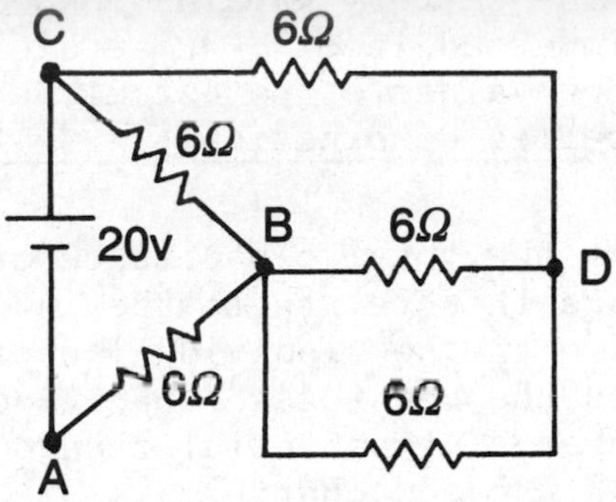

Problem 18-2. A platinum wire resistor is connected in parallel with a copper wire resistor, each having a resistance of 1 at 0 °C. What current will be drawn from a 6 volt battery if the combination is at 220 °C?

Q1: What are the temperature coefficients of resistivity for the two materials?
Q2: What are the individual resistances at 220 °C?
Q3: What is the total resistance at 220 °C?

Problem 18-3. Use Kirchoff's Laws to find the power delivered to the circuit in the figure below by $\mathcal{E}_1$ and the unknown $\mathcal{E}_2$. Show that power delivered = power consumed for the circuit.

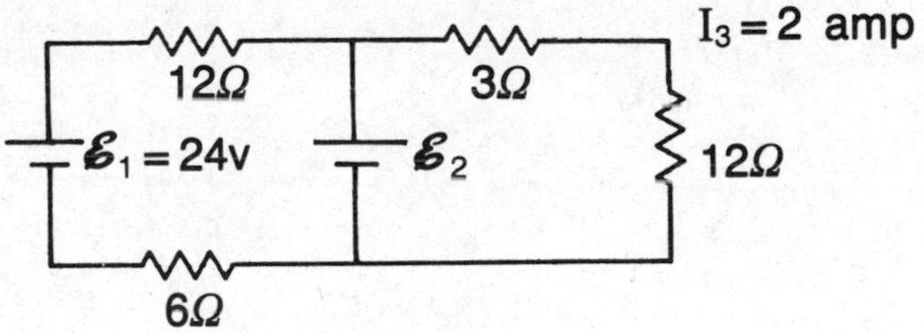

Q1: What is the junction equation at point A?
Q2: What are two loop equations? How do I treat $\mathcal{E}_2$?
Q3: How do I interpret the result for I_1?
Q4: Does this make any difference in the calculation of Joule heating in the resistors carrying I_1?

Problem 18-4. Consider two identical parallel plate capacitors filled with air, connected in *series*. When the combination is charged by a 100 volt battery, a charge of 1 μC becomes stored.

(a.) Find the individual capacitance of each capacitor.
(b.) The battery remains connected, and *one* of the capacitors is immersed in oil, whose dielectric constant is 2.5. Find the new values of potential, charge, and capacitance for the series combination.

Q1: What is the total capacitance of the combination?
Q2: How does this relate to the individual capacitances?
Q3: What are the new individual capacitances, with one in oil?
Q4: What is the new total capacitance, C_{tot}'?
Q5: Have charge or potential changed?
Q6: What is the new charge?

ANSWERS TO QUESTIONS

FUNDAMENTALS:

1. (b.)	2. (c.)	3. (a.)	4. (d.)
5. (b.	6. (c.)	7. (c.)	8. (a.)
9. (b.)	10. (a.)		

GENERAL PROBLEMS:

Problem 18-1.

A1: $1/R_{BD} = 1/6 + 1/6$.

A2: $R_{CDB} = 6 + R_{BD}$.

A3: $1/R_{CB} = 1/6 + 1/R_{CDB}$.

A4: $R_{tot} = 6 + R_{CB}$.

Problem 18-2.

A1: From Table 18.1 (text).

A2: Given by $R_{220} = (1\ \Omega)[1 + \alpha(220\ ^{o}C)]$ for each material.

A3: $(1/R_{tot})_{220} = (1/R_{220})_{Pt} + (1/R_{220})_{Cu}$.

Problem 18-3.

A1: If you pick currents leaving the + terminal of each EMF, it is $I_1 + I_2 = I_3$.

A2: Going around the whole perimeter, $24 - 12I_1 - 6 - 24 - 6I_1 = 0$. Going around the right hand loop, $\mathcal{E}_2 - 6 - 24 = 0$. You just treat $\mathcal{E}_2$ as an algebraic unknown. You *do* know its sign.

A3: Since I_1 turns out negative, it is going the opposite direction, through $\mathcal{E}_1$ from + to - terminal. This represents a power *loss* in the circuit, as mentioned in the Observations.

A4: No. Resistors consume energy no matter which way the current is flowing.

Problem 18-4.

A1: $C_{tot} = Q/V = (10^{-6}\ C)/(10^{2}\ v)$.

A2: $1/C_{tot} = 1/C + 1/C = 2/C$.

A3: In oil, the one capacitor has a new capacitance C' = 2.50. The other stays the same, C.

A4: What is the new total capacitance, C_{tot}'?

A5: Since the battery has remained connected, it *forces* the potential to remain at 100 volts.

A6: $Q' = C_{tot}'V$. The battery can supply or absorb a change in charge.

CHAPTER 19. MAGNETISM

BRIEF SUMMARY AND RESUME

Characteristics of the interaction between *magnetic poles* are summarized, and the concept of *magnetic field* is introduced. The force on an electric current in a magnetic field is described by the *right-hand rule*, and extended to forces experienced by moving charged particles. Applications to the *Hall effect, mass spectrometers*, and the *cyclotron* are analyzed. The manner in which electric currents produce magnetic fields is investigated and the *ampere* is fundamentally defined. *Ampere's Circuital Law* is discussed, as are the magnetic fields of circular coils and *solenoids*. From this is derived the concept of the *magnetic dipole* and *dipole moment*.

Magnetic properties of materials are summarized in terms of *magnetic permeability*. The earth's magnetic field is described.

The magnetic torque on a current-carrying coil is applied to the design of *galvanometers* for the measurement of currents. The use of these devices in circuits in order to operate as voltmeters and ammeters is analyzed.

Material from previous chapters relevant to chapter 19 includes:

1. Newton's 2nd Law and Centripetal Force.
2. Electric current and Charge.
3. Torque.

NEW DATA AND PHYSICAL CONSTANTS

1. The unit of magnetic field (B).

   <u>SI</u>: 1 tesla (T) = 1 N-m/amp

   <u>Common</u>: 1 gauss = 10^{-4} T.

2. Permeability of free space (μ_0).

   $\mu_0 = 4 x 10^{-7}$ T-m/amp.

NEW CONCEPTS AND PRINCIPLES

<u>Summary of Primitive Magnetic Phenomena</u>

1. There are two types of magnetic poles, north (N) and south (S), associated with a magnet.
2. Like poles repel and unlike poles attract.
3. Poles always occur in opposite pairs. A single pole (monopole) does not exist.
4. Certain unmagnetized materials will be attracted by a magnet. Such materials are called ferromagnetic, and include iron, cobalt, and nickel.

<u>Magnetic Field Strength (B)</u>--A vector whose direction at any point is that along which an N-pole would align, and whose magnitude is proportional to the aligning torque on the N-pole.

Magnetic Force on a Current-Carrying Conductor

Understanding this section should enable you to do the following:

1. *Calculate the force per length, both magnitude and direction, on a straight current-carrying conductor placed in a uniform magnetic field.*

Magnitude of Magnetic Force (F_B)--The magnetic force on a current I of length L in a magnetic field B is

$$F_B = BIL \sin \theta = (B_\perp)IL,$$

where θ = the angle between the direction of **B** and **I**.

Direction of Magnetic Force--The magnetic force is *perpendicular* to both **B** and **I**. The *right-hand rule* gives the correct direction: with the fingers of the right hand in the direction of **B**, and the thumb in the direction of **I**, the *palm* of the right hand faces in the direction of $\mathbf{F_B}$.

Important Observations

1. The unit of magnetic field B is derived from the expression for F_B to be N/A-m. This defined to be the *tesla* (T): 1 T = 1 N/A-m. A common unit still in use is the *gauss*: 1 gauss = 10^{-4} T.
2. The magnetic force can be zero in *three* ways:

   (a.) zero current,
   (b.) zero magnetic field, and
   (c.) **B** parallel or anti-parallel to **I**, making $\theta = 0^\circ$ or 180°.

3. $\mathbf{F_B}$ is *not* in the direction of **B**, which is different than the case for the electric force, which is parallel to **E**.

Example 19-1. Use the right-hand rule to determine the direction of **F** in the following situations. x designates a vector *into* the page, and ⊙ designates a vector *out* of the page.

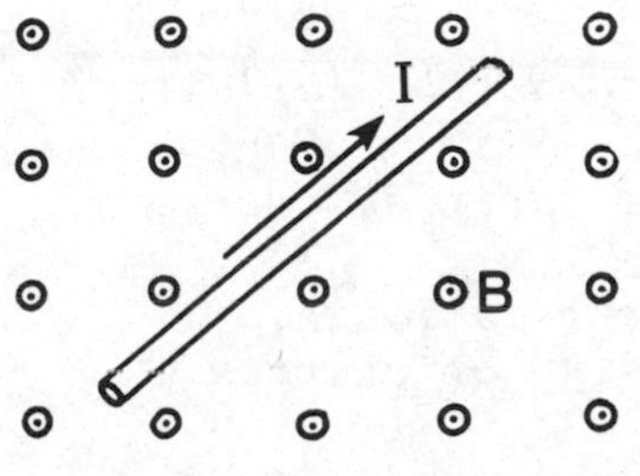

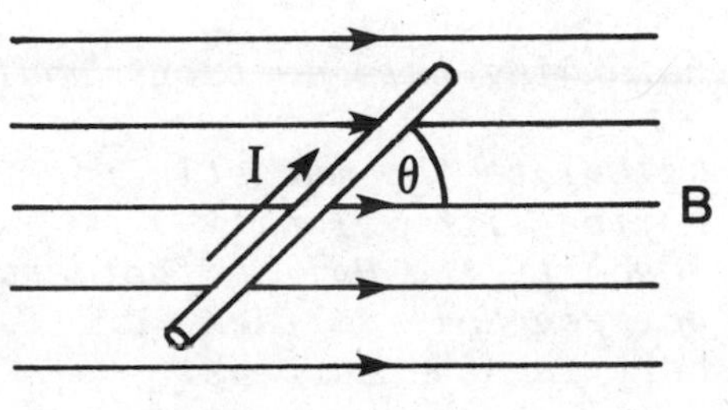

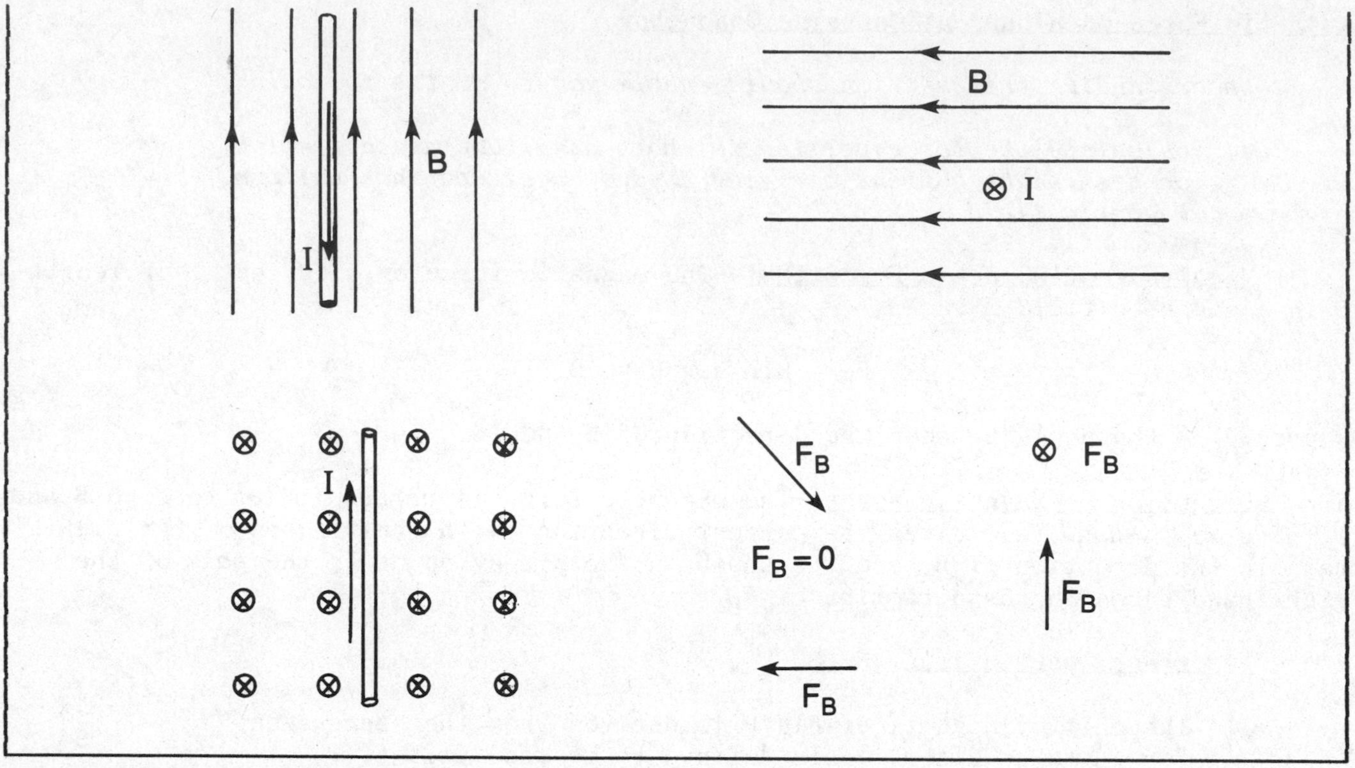

Example 19-2. In the following figure, a current of 5 amps is placed in a uniform magnetic field of 0.2 T, as shown. What is the magnetic force per meter on the current?

S: This seems to be just a plug-in problem. We have $F_B = BIL \sin \theta$, so $F_B/L = (.2\ T)(5\ A)\sin 50^o = .766$ N/m.

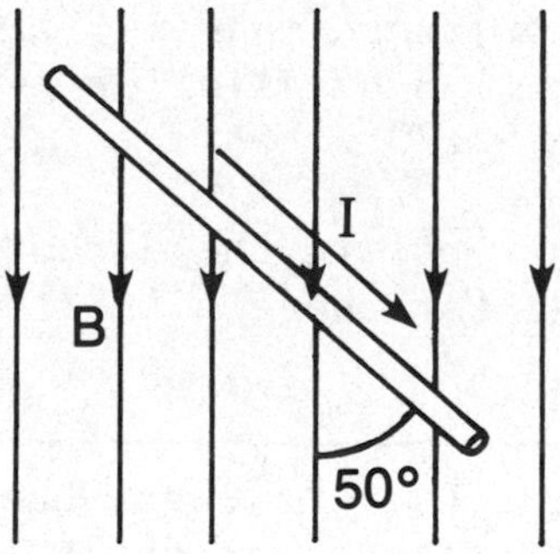

T: Yes, it is quite easy once you identify the angle θ. The direction is always perpendicular to the plane containing I and B. In this case the rule gives the direction *into* the page, distributed uniformly along the conductor.

Magnetic Force on Moving Charged Particles

Understanding this section should enable you to do the following:

1. *Calculate the magnetic force on a particle carrying charge q, moving with velocity* **v**.
2. *Identify the Hall effect and calculate the relationship between Hall voltage, drift velocity and sign of charges carrying the current.*

3. *Calculate the dependence of the radius of circular path of a moving free charge on its momentum and the magnetic field in which it is moving.*

Magnetic Force on a Moving Charge--As derived in Section 19.3 (text), the *magnitude* of the magnetic force on a charge q moving with velocity v is

$$f_B = (B_\perp)qv = Bqv \sin \theta,$$

where

$B_\perp$ = component of **B** perpendicular to **v**, or θ = angle between **v** and **B**.

The *direction* of the magnetic force *on a + charge* is obtained by the right-hand rule: point the thumb of the right hand in the direction of v and the fingers of the right hand along **B**; the *palm* will face in the direction of $\mathbf{F}_B$. For a *negative* charge, the *back* of the right hand will face in the direction of $\mathbf{F}_B$.

Important Observations

1. Since $\mathbf{F}_B$ is always perpendicular to **v**, $\mathbf{F}_B$ *can do no work* on a charged particle, but can *only* turn it in a curved path. If B is constant over the motion, this path is a *circle*, with a radius given by Newton's 2nd Law:

$$R = mv/qB.$$

Example 19-3. Identify whether the charges shown will travel in a clockwise or counter-clockwise direction in the following cases:

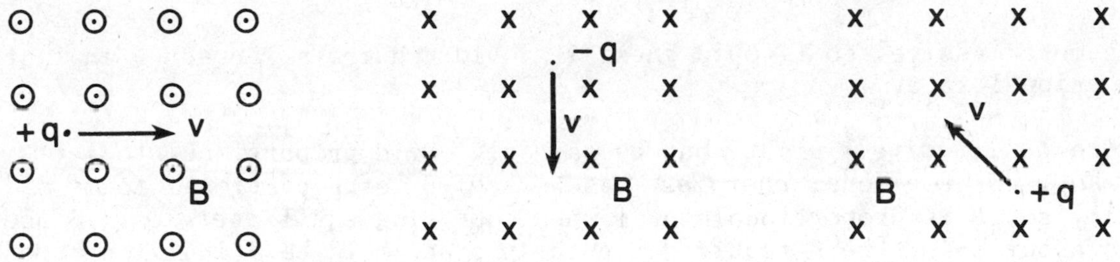

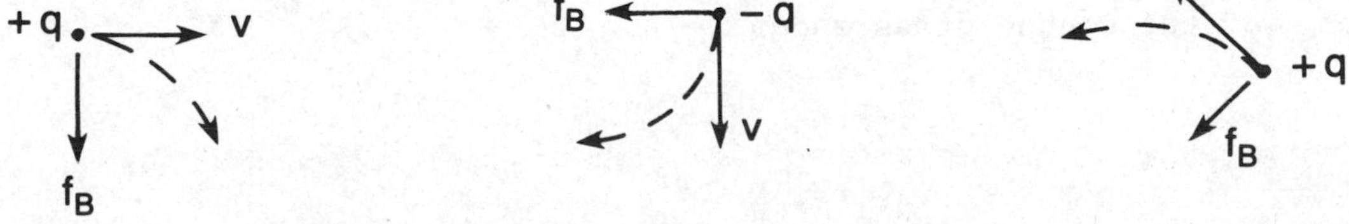

Example 19-4. How strong a magnetic field would be required to turn an electron with 1 eV of kinetic energy in a circle of radius R = 1 cm?

S: The radius doesn't involve energy.

T: Not directly, but of course v implies kinetic energy. In fact

$$(mv)^2 = 2m(\tfrac{1}{2}mv^2) = 2m(KE), \text{ so } mv = [2m(KE)]^{\frac{1}{2}}.$$

S: I can't directly use eV's, can I?

T: No, you have to convert to SI units.

S: We have 1 eV = 1.6×10^{-19} J, and I've looked up the electron mass, $m_e = 9.1 \times 10^{-31}$ kg. Thus

$$mv = 2(9.1 \times 10^{-31} \text{ kg})(1.6 \times 10^{-19} \text{ J}) = 5.4 \times 10^{-25} \text{ kg-m/s}.$$

T: You'll also need $q = -e = -1.6 \times 10^{-19}$ C. Since you're only asked the radius, and not the direction, you don't need the minus sign.

S: The expression for R is R = mv/qB, so

$$B = mv/qR = (5.4 \times 10\text{-}25 \text{ kg-m/s})/(1.6 \times 10^{-19} \text{ C})(10^{-2} \text{ m}) = 3.4 \times 10^{-4}$$

Are these *teslas*?

T: Yes. 1 T = 1 N/A-m = 1 $(\text{kg-m/s}^2)/[(\text{C/s})\text{m}]$ = 1 kg/C-s, as above. These derived units get pretty wild now. The field you found is also 3.4 *gauss*, a few times stronger than the earth's magnetic field. Would a 1 eV proton require a larger or smaller field?

S: It's more massive, so I would guess it would require a larger B. In fact, B is proportional to m.

T: You're *qualitatively* right, but be careful. B is proportional to m *only if* v is the same, not for equal *energies*. As we saw, B is proportional to mv = $[2m(KE)]^{\frac{1}{2}}$, so it is proportional to $\sqrt{m}$ when comparing equal *energies*. Since m_p/m_e is about 1840, the B required for the proton would be $\sqrt{(1840)}$, or about 43 times larger than for the electron.

The Hall Effect--If a current-carrying conductor having rectangular cross-section is placed in a uniform magnetic field **B** perpendicular to the current direction (see the figure below), the charges which form the current are deflected perpendicular to the field and collect on the faces of the conductor as shown.

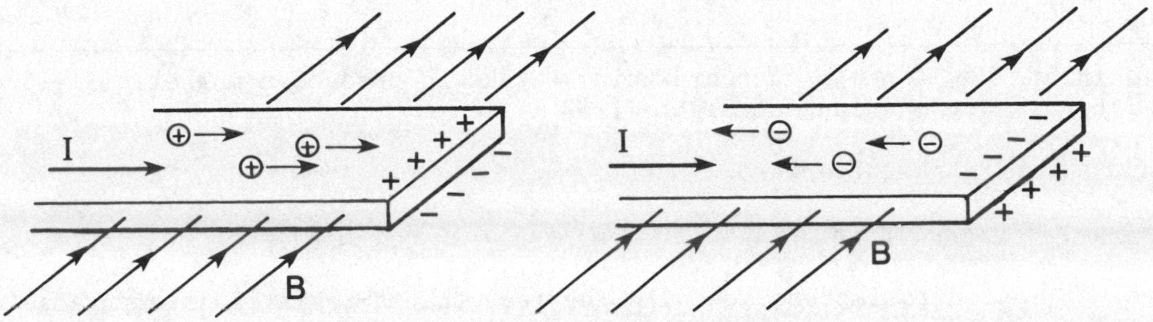

This proceeds until the *electric* field produced by these charges cancels the magnetic force on the remaining current. A voltmeter placed across the sides of the conductor measures the resulting potential difference, called the *Hall voltage*, V_H, given by

$$V_H = vBd$$

where v = the average ("drift") velocity of the charges carrying the current.
d = thickness of the conductor transverse to **B**.

Important Observation

1. As shown in the accompanying figure, the *polarity* of the Hall voltage tells us the sign of the charges responsible for the current. The Hall effect on normal conductors show that negative charges (electrons) move through the conductor. Some semiconductors (see Chapter 30) show movement of positive charges ("holes", not the positive lattice charges).

Magnetic Field Production by Currents

Understanding this section should enable you to do the following:

1. *Calculate the magnetic field produced by a long straight current.*
2. *Calculate the force of interaction between two parallel straight currents.*
3. *Identify the definition of the ampere of current.*
4. *Apply Ampere 's Law to simple current geometries.*
5. *Calculate the magnetic field at the center of a circular coil and inside a solenoid.*
6. *Identify the magnetic dipole moment of a current loop and calculate the torque on it when placed in a uniform magnetic field.*

Magnetic Field Produced by a Long Straight Current--A long straight current produces a magnetic field proportional to the current and inversely proportional to the distance from the current. Mathematically, $B = k_m(I/r) = (\mu_0/2\pi)(I/r)$, where $\mu_0 = 4\pi \times 10^{-7}$ T-m/A. The direction of the field lines are circles concentric with the current in a *clockwise* sense as you look *along* the direction of the current. The right-hand rule given in the text is equivalent: point the right thumb in the direction of the current, and the fingers will curl in the direction of the field around the current.

Example 19-5. How large a current would it take to produce a magnetic field of 1 tesla 1 cm away from a long straight wire?

S: That's very straightforward.

$$I = (2\pi/\mu_o)Br = 2\pi(1\ T)(10^{-2}\ m)/(4\pi \times 10^{-7}\ T\text{-}m/A) = 50{,}000\ A!$$

T: Yes. I wanted to impress you with the fact that a tesla is a very large field value and takes enormous currents to produce in free space.

Force Between Parallel Currents--In this figure, we're looking at two parallel straight currents, I_1 and I_2, coming at us out of the page. They are separated by a distance d. I_1 exerts a force per meter of length on I_2 equal to

$$(F_B/L)_{on\ 2} = B_1 I_2 = (\mu_o/2\pi)(I_1 I_2/d).$$

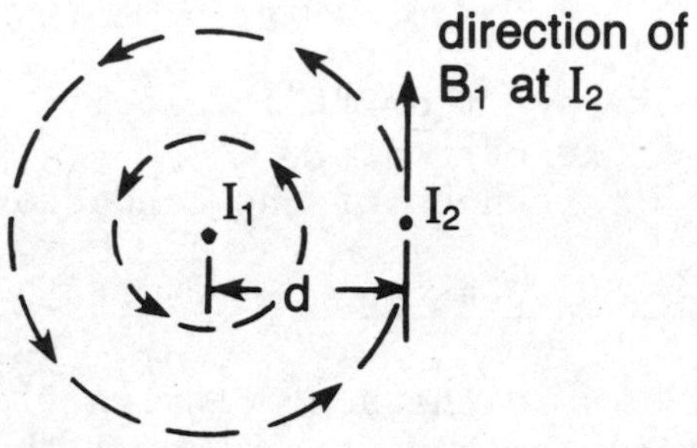

The force, from the right-hand rule, is *attractive*, directed toward I_1.

Important Observations

1. By drawing the field B_2 produced by I_2, a similar calculation shows the *same* force on I_1, directed toward I_2. This is another example of Newton's 3rd Law.
2. If we reverse the direction of one or the other of the currents, so that they are *anti-parallel*, the force directions will reverse, representing *repulsive* forces.
3. This interaction, both in theory and in practice, yields the *fundamental definition of the ampere*: if two equal parallel currents one meter apart experience 2×10^{-7} N per meter of length, the currents are each *1 ampere*. The coulomb of charge is then *derived* from this, as 1 coulomb = 1 ampere -second.

Example 19-6. In the following figure, there is a very long wire carrying a current $I_1 = 50$ A upwards. A square coil 2 cm on a side is placed so that sides AB and CD are parallel to the wire, with side AB 1 cm away. This coil is carrying a current $I_2 = 30$ A clockwise as shown. What is the net magnetic force on the coil?

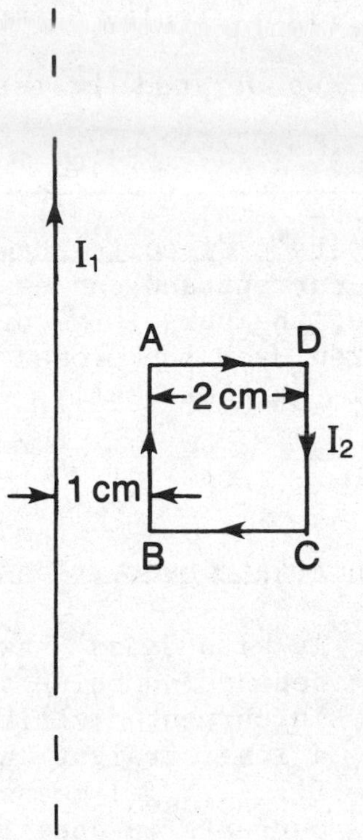

S: As drawn, the magnetic field from I_1 at the location of the coil will be *into* the page. In side AB, I_2 is *parallel* to I_1, and in side CD it is *anti-parallel*.

T: Good observations. That means AB will be attracted to I_1 and CD will be repelled. Check this with the right-hand rule.

S: It looks as if AD will be pulled *upward* and BC *downward*, according to the rule. How about the interactions between the currents I_2 themselves in direct segments of the coil?

T: That's a fair question. It would be difficult to calculate, and we don't have to. Simply remember, an object cannot exert a net force on itself from internal interactions. So any net force must arise from I_1.

S: All segments of AB and CD are a constant distance from I_1, so the field B will be constant over their lengths. But that's not true of sides AD and BC.

T: However, if you break up AD and BC into small segments at equal distances from I_1, you'll see that the forces on each pair of segments cancel, right across the entire sides. So $\mathbf{F}_{AD} = -\mathbf{F}_{BC}$.

S: Then we're left with $\mathbf{F}_{AB} = B_{1(\text{at AB})}I_2L$ *toward* I_1, and $\mathbf{F}_{CD} = B_{1(\text{at CD})}I_2L$ *away* from I_1. What are these values of B_1?

T: For a long straight current, $B = (\mu_o/2\pi)(I/r)$. At AB, r = 1 cm and at CD, r = 3 cm.

S: I see that now.

$$B_1(\text{at AB}) = (2 \times 10^{-7}\ \text{T-m/A})(50\ \text{A})/(10^{-2}\ \text{m}) = 1.0 \times 10^{-3}\ \text{T}$$

and

$$B_1(\text{at CD}) = (1/3)B_1(\text{at AB}) = .33 \times 10^{-3}\ \text{T}.$$

So

$$F_{AB} = (1.0 \times 10^{-3}\ \text{T})(30\ \text{A})(2 \times 10^{-2}\ \text{m}) = 6.0 \times 10^{-4}\ \text{N},$$

and

$$F_{CD} = (.33 \times 10^{-3}\ T)(30\ A)(2 \times 10^{-2}\ m) = 2.0 \times 10^{-4}\ N.$$

Is that right, teslas x amps x meters = newtons?

T: Yes, we *defined* the tesla this way. So

$$F_{net} = F_{AB} - F_{CD} = 4 \times 10^{-4}\ N, \text{ toward } I_1.$$

Ampere's Circuital Law--If we choose some arbitrary path or circuit enclosing electric current and can evaluate the component of B parallel to each segment Δl of this line, the sum of the products $B_{\parallel} \Delta l$ taken over the closed path will equal μ_0 times the total current enclosed by the path. Mathematically,

$$\sum_{\text{closed path}} B_{\parallel} \Delta l = \mu_o \sum I_{\text{enclosed}}$$

Important Observations

1. As with Gauss' Law, the practical ability to use Ampere's Law depends on being able to utilize some simplifying symmetries in current distribution. Your text applies it to the case of a long straight current.
2. $\Sigma(I_{\text{enclosed}})$ means the *algebraic* sum, of course, with currents in opposite directions tending to cancel.

Example 19-7. In the figure below, a coaxial conductor is made of a conducting cylinder of radius R_1 = 1 cm which is insulated from a coaxial cylindrical conducting shell of 0.5 cm thickness. *Both inner and outer conductors are carrying the same current density*, I/Area = j = 20 amps/cm^2, but in opposite directions. Use Ampere's Law to find the magnetic field produced by this conductor:

(a.) 2 cm from the axis
(b.) 1.2 cm from the axis.

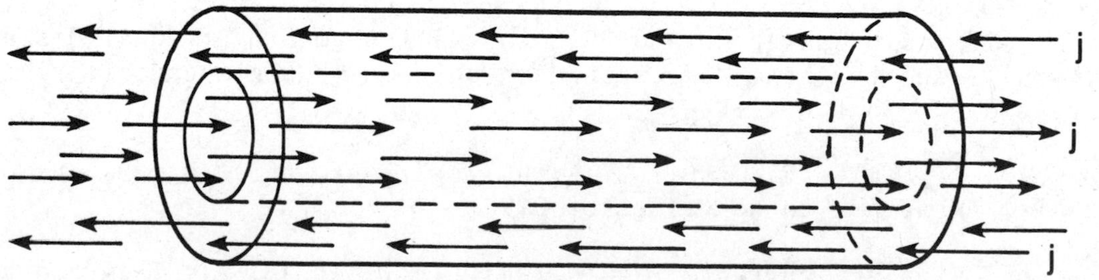

S: Since there is cylindrical symmetry, I'd guess the path to choose would be a *circle* centered on the axis.

T: Right, and of the same radius as the distance at which you want to evaluate B. For (a.) you want a circle of radius 2 cm. By symmetry, B is completely parallel to that circle around the entire circumference, and is constant in magnitude.

S: So $\Sigma B_{\parallel} \Delta l$ is just $B(2\pi r)$?

T: Just as in the text's example for a single wire. This problem differs from the text's example in the calculation of $\Sigma(I_{encl})$.

S: Well, I = j times the cross-sectional area, so the current in the center conductor is

$$I_1 = j(\pi R_1^2) = (20\ \text{amp/cm}^2)(\pi)(1\ \text{cm}^2) = 62.8\ \text{amps}.$$

This is to the *right* in the figure.

T: And an opposite current I_2 is carried to the left in the shell:

$$I_2 = j \times (\text{area of shell}) = j\pi(R_2^2 - R_1^2)$$

$$= (20\ \text{amp/cm}^2)(\pi)(1.5^2 - 1.0^2)\ \text{cm}^2 = 78.5\ \text{amps}.$$

S: Then

$$\Sigma(I_{encl}) = 78.5\ \text{amps} - 62.8\ \text{amps} = 15.7\ \text{amps, to the } \textit{left}.$$

Ampere's Law then says

$$B(2\pi)(2 \times 10^{-2}\ \text{m}) = \mu_o(15.7\ \text{amps}),\ \text{giving } B = 1.57 \times 10^{-4}\ \text{T}.$$

T: The direction is obtained by pointing your right thumb toward the *left* (net current direction) in the figure. The B-field vectors come out of the page at the bottom, curve upward, and enter the page at the top.

S: For part (b.), what is the crucial difference?

T: Just that your path is a circle of r = 1.2 cm, which does *not* enclose *all* of the shell current.

S: So my I_2 enclosed will change to

$$I_2 = (20\ \text{amp/cm}^2)(\pi)(1.2^2\ \text{cm}^2 - 1^2\ \text{cm}^2) = 27.6\ \text{amps}.$$

Since the path still encloses all of the central current, this leaves

$$I_{encl} = 35.2\ \text{amps, to the } \textit{right} \text{ this time.}$$

T: That's correct, which means that the *direction* of the B-field has reversed. We have

$$B = (4 \times 10^{-7}\ \text{T-m/A})(35.2\ \text{A})/(2)(1.2 \times 10^{-2}\ \text{m}) = 5.86 \times 10^{-4}\ \text{T}.$$

Notice that the current *outside* the path *doesn't contribute* to B.

S: At some radius the field must be *zero*, since it switches direction as you go outward from the center.

T: Absolutely! A very good observation. B will be zero if I_{encl} is zero. This will happen at the radius where the enclosed shell current is as large as the central current, or where

$$\pi(r^2 - 1\ \text{cm}^2) = \pi(1\ \text{cm}^2),\ \text{giving}\ \ r = \sqrt{2}\ \text{cm} = 1.414\ \text{cm}.$$

<u>Magnetic Field of Circular Coil</u> --*At the center* of a single circular loop of radius r carrying current I the magnetic field has a magnitude

$$B = (\mu_0 I)/2r.$$

If you wrap the fingers of your right hand in the direction of circulation of the current, your right thumb will point along the axis of the coil in the direction of B.

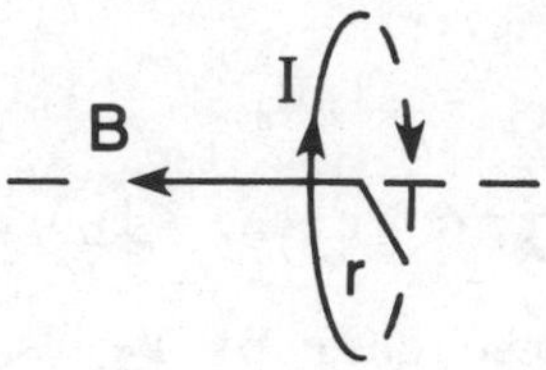

<u>Important Observations</u>

1. If there are N turns of thin wire in essentially the same plane and with essentially the same radius, this simply becomes

$$B = \mu_0 NI/2r.$$

2. The quantity IA or NIA, where A is the area bounded by the loop, is known as the *magnetic dipole moment, a vector in the direction of the B field it produces.*

<u>Torque on a Magnetic Dipole in a Magnetic Field</u>--As derived in Section 19.10 (text), a magnetic dipole placed in an external B-field will experience a *torque tending to align it in the direction of the external field.* The magnitude of this torque is

$$\tau = B(NIA)\sin\theta,$$

where θ = the angle between **B** and the dipole moment.

<u>Important Observations</u>

1. Remember, the direction of the dipole moment is in the direction of the field produced by the coil. Thus the torque tends to line up the coil's field with that of the external field.

<u>Magnetic Field of a Solenoid</u>--A tightly wound cylindrical coil with diameter small compared to its length is called a *solenoid.* Your text finds the field inside a solenoid by the application of Ampere's Law. The result is

$$B = \mu_0 nI,$$

where n = number of turns *per length* along the solenoid axis.

<u>Important Observations</u>

1. The direction of the field is found the same way as for a single loop, since the solenoid is simply a large number of such loops in series, with the same current I going through all of them in the same direction.
2. The expression for B does not contain any position variables, so it represents a *constant* field over the interior volume of the solenoid. It is thus similar to the parallel plate capacitor, in which the *electric* field is constant.

3. If the solenoid is filled with material, the magnetic field is multiplied by a factor K_m, called the *relative magnetic permeability*. See Section 19.7 and Table 19.1 in your text.

Example 19-8. A circular coil of 100 turns is made from 10 meters of wire having a total resistance of 100 Ω. When 120 volts are applied to the coil,

(a.) What is the magnetic field produced at its center?
(b.) What is the magnetic dipole moment of the coil?

S: For a coil, $B = \mu_o NI/2r$, with N = 100 in this example. We can get I from Ohm's Law, but what is r?

T: 10 meters is going into 100 turns, so each turn will have a circumference of 10 cm. Thus r = 10 cm/2π = 1.6 cm.

S: And I = V/R = 120 Volts/100 = 1.2 amps. Then

$$B = (4\pi \times 10^{-7}\ \text{T-m/A})(100\ \text{turns})(1.2\ \text{A})/2(1.6 \times 10^{-2}\ \text{m})$$

$$= 4.8 \times 10^{-3}\ \text{T}$$

The dipole moment is NIA, so we have to find A.

T: This is the area enclosed by the coil, so

$$A = \pi r^2 = \pi(1.6 \times 10^{-2}\ \text{m})^2.$$

Thus the dipole moment is

$$NIA = (100\ \text{turns})(1.2\ \text{amp})(8.0 \times 10^{-4}\ \text{m}^2) = 9.6 \times 10^{-2}\ \text{ampere-m}^2.$$

Example 19-9. Suppose the above coil is placed inside a solenoid so that the plane of the coil is 50° away from the axis of the solenoid. The solenoid carries a current of 15 amps and is wound with 2000 turns of wire along its 75 cm length. What is the torque on the small coil?

S: How do we find the angle between the coil's magnetic moment and $\mathbf{B}_{sol}$?

T: The field if the solenoid is along its axis. The magnetic moment of the coil is perpendicular to its plane. Since the plane of the coil is 50° off the axis, the magnetic moment must be 40° from $\mathbf{B}_{sol}$.

S: Then

$$\tau = B_{sol}(NIA)\sin 40^o, \quad \text{with } NIA = 9.6 \times 10^{-2}\ \text{A-m}^2$$

from before. We still need B_{sol}.

T: That's easy. $B_{sol} = \mu_o nI$, with n = 2000 turns/(.75 m) = 2667 turns *per meter*. So

$$B_{sol} = (4\pi \times 10^{-7}\ \text{T-m/A})(2667\ \text{turns/m})(15\ \text{amps}) = 5.03 \times 10^{-2}\ \text{T}.$$

S: So

$$\tau = (5.03 \times 10^{-2}\ \text{T})(9.6 \times 10^{-2}\ \text{A-m}^2)\sin 40^o = 3.1 \times 10^{-3}\ \text{T-A-m}^2.$$

T: Can you convince yourself that (teslas) x (amps) x $(\text{meters})^2$ are actually *newton-meters*?

QUESTIONS ON FUNDAMENTALS

1. If a current of 12 amps is flowing through a conductor in the -x direction and there is a uniform magnetic field of 0.6 T in the +z direction, in what direction is the magnetic force on the current?

(a.) +y (b.) +z (c.) -y (d.) -z
(e.) the force is zero

2. What is the magnitude of the magnetic force per length on the conductor in question #1?

(a.) zero (b.) 7.2 N/m (c.) 20 N/m (d.) .05 N/m
(e.) 12.6 N/m

3. How large a magnetic field would be required to turn an electron moving at 1% the speed of light in a circle of radius 1 cm?

(a.) 585 mT (b.) 1.71 T (c.) 3.15 T (d.) 1.71 mT

4. What value of magnetic field does a long straight current of 20 amps produce at a distance of 30 cm?

(a.) 21 μT (b.) 70 μT (c.) 42 μT (d.) 6.67 μT
(e.) 84 μT

5. What is the value of the magnetic field produced at the center of a 6 turn coil of radius 30 cm when the coil is carrying a 20 amp current?

(a.) 250 μT (b.) 500 μT (c.) 125 μT (d.) 40 μT
(e.) 833 μT

6. An air-core solenoid is wound with 3000 turns per meter of length. How much current must it carry to produce a magnetic field of 0.2 T in its interior?

(a.) 19 mA (b.) 333 A (c.) 5.3 mA (d.) 3.33 A
(e.) 53 A

GENERAL PROBLEMS

Problem 19-1. Assume there is a uniform magnetic field **B**, directed into the page. Perpendicular to **B** is a wire on which an electrically charged bead is free to slide vertically without friction as shown in the figure. The bead has a charge Q and a mass M. The wire (and with it, the bead) is pulled to the right with a constant velocity **v**. Determine what speed the wire must have in order that the bead does not fall. Assume B = .4 tesla, Q = 5 x 10^{-6} coul, and M = 10 gm.

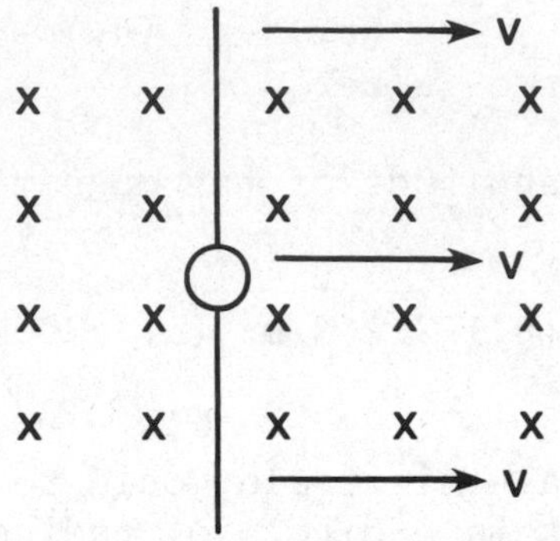

Q1: Is the magnetic force upward?
Q2: What is the expression for the magnetic force?
Q3: What is the gravitational force?
Q4: At what speed will the bead not fall?

Problem 19-2. In the accompanying figure, a long straight current I is at a fixed position b on a table, coming out of the page. Another straight wire is attached to a frame which freely pivots around an axis at position a. Both wires carry the same current I and are of the same length l perpendicular to the page. The currents are in opposite directions, and so *repel* each other. Both wires are a distance d from axis a. The currents are such that the magnetic force holds them apart at an angle θ = 30° as shown in the figure. Using the following data, find I: d = 30 cm, the center of mass of the frame is 20 cm from a, M = 400 grams, and l = 50 cm.

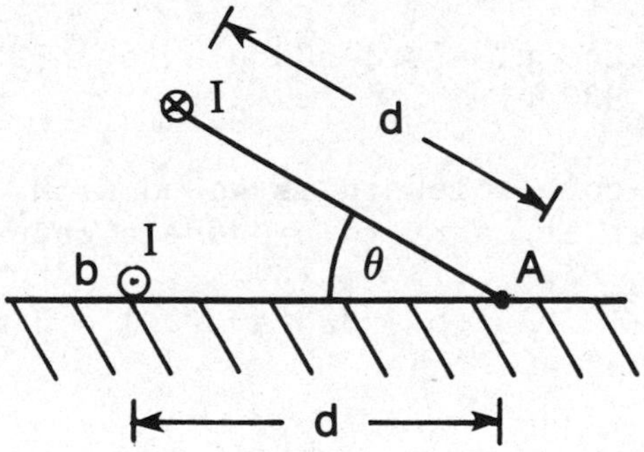

Q1: What is the condition for equilibrium?
Q2: What is the magnetic torque about a?
Q3: What is F_B?
Q4: What is the gravitational torque about a?
Q5: What relation yields the value of I?

Problem 19-3. The figure below shows part of an infinitely wide sheet of current coming out of the page. The current has uniform area density, $j = 10\ A/cm^2$. The sheet has thickness $t = .5$ cm. Using symmetry arguments, apply Ampere's Law to determine the value of **B** (including direction) produced by this current outside of the sheet.

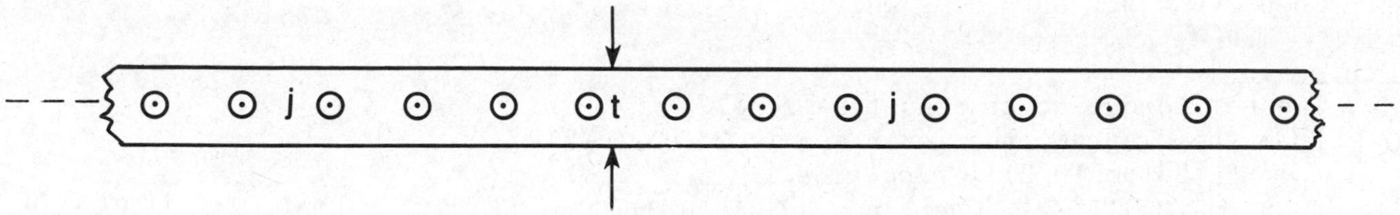

Q1: How do I choose an appropriate Amperian closed path?
Q2: What length and width of path should I choose?
Q3: What are the terms in Ampere's Law applied to this case?
Q4: What expression for B results?

Problem 19-4. In the mass spectrometer shown in the figure below, 1 keV positive ions of some element enter a region of uniform magnetic field directed into the page. The magnetic field strength is 2T. The ions are seen to impact on a screen 2 cm from the entrance slit, as shown. What is the mass of these ions and what element do they represent?

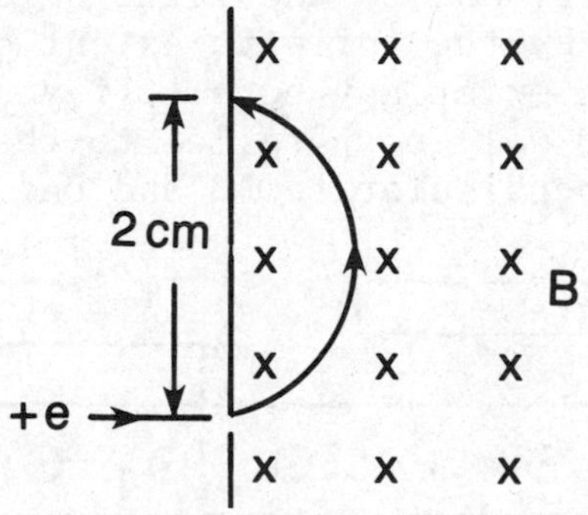

Q1: What is the significance of the 2 cm?
Q2: What determines the radius of a charged particle in a magnetic field?
Q3: What is q?
Q4: What is 1 keV in SI units?
Q5: Once I determine the mass from the above equation, how can I identify the element?

ANSWERS TO QUESTIONS

FUNDAMENTALS:

1. (a.) 2. (b.) 3. (d.) 4. (d.)
5. (a.) 6. (e.)

GENERAL PROBLEMS:

Problem 19-1.

A1: Q is +, so the right-hand rule does give an upward magnetic force.

A2: $F_B = QvB$, upwards.

A3: $F_g = Mg$, downwards.

A4: When $QvB = Mg$, or $v = Mg/QB$.

Problem 19-2.

A1: The net torque about a must be zero.

A2: From the diagram, the lever arm is $2d\cos(\theta/2)$.
$\tau_B = F_B[2d\cos(\theta/2)]$, *clockwise*.

A3: $F_B = (\mu_0/2\pi)(I^2/r)$, where $r = 2d\sin(\theta/2)$.

A4: The lever arm is seen to be $(2/3)d\cos\theta$.
$\tau_g = Mg(2/3)(d\cos\theta)$, *counterclockwise*.

A5: The equilibrium condition:
$Mg(2/3)(d\cos\theta) = (\mu_0/2\pi)[I^2/2d\sin(\theta/2)][2d\cos(\theta/2)]$.
Everything here is known except I.

Problem 19-3.

A1: If you view the current sheet as a large number of parallel currents, each would make a contribution to B to the *left* above the sheet, and to the *right* below the sheet. Due to the symmetry imposed by the infinite extent of the sheet, no other component of B can exist anywhere. If you choose a path such as ABCD shown below, B will be parallel to the path and constant on AB and CD, and perpendicular to BC and DA.

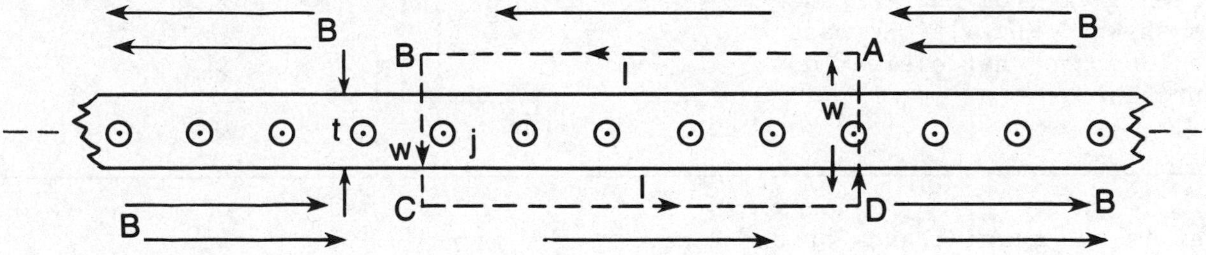

A2: Take any arbitrary values of l and w and see what happens.

A3: $B_\parallel \Delta l = Bl_{AB} + (0)l_{BC} + Bl_{CD} + (0)l_{DA} = 2Bl$; $I_{encl} = jlt$.
(*not* jlw !)

A4: $2Bl = \mu_0 jlt$, or $B = \mu_0 jt/2$.
The direction is evident from A1 above. Notice the result is independent of l or the distance from the sheet.

Problem 19-4.

A1: It is twice the radius of the particle's path in the field.

A2: $R = mv/qB$, and you can use $mv = [2m(KE)]^{\frac{1}{2}}$.

A3: We assume these are *singly* ionized particles, so $q = +e$.

A4: $1\text{ eV} = 1.6 \times 10^{-19}$ J, so $1\text{ keV} = 1.6 \times 10^{-16}$ J.

A5: The mass ought to be close to an integral multiple of $1\text{ u} = 1.66 \times 10^{-27}$ kg. From a table of elements, pick the atomic mass closest to this integer.

CHAPTER 20. ELECTROMAGNETIC INDUCTION

BRIEF SUMMARY AND RESUME

The *induction of an EMF* by the *motion* of a conductor across a magnetic field is described. This is generalized into *Faraday's Law of Induction*, using the concept of *magnetic flux*. *Lenz' Law* is introduced to indicate the *direction* of the induced EMF. *Mutual* and *self-induction* of current-carrying coils is discussed, and the definition of *inductance* is introduced. The energy stored in inductors and capacitors is derived. The effects of inductors or capacitors in series with a resistance when the current in the circuit is made to change is described by the *inductive and capacitive time constants*. *Faraday's Law of Induction* is applied to the operation of electrical *generators* and *motors*.

Material from previous chapters relevant to Chapter 20 includes:

1. Currents, magnetic fields, and the relation between them.
2. The concept of capacitance.
3. Kirchoff' Laws for simple circuits.

NEW DATA AND PHYSICAL CONSTANTS

1. The unit of magnetic flux (Φ).

   Φ = B A = tesla-m^2, defined as the weber (Wb).

   1 tesla-m^2 =1 weber.

2. The unit of mutual or self inductance (M or L).

   M (or L) = $\mathcal{E}/(\Delta I/\Delta t)$ = Volt-sec/amp, defined as the henry (H).

   1 volt-sec/amp = 1 henry.

3. The inductive time constant (τ_L).

   In an R-L series circuit, τ_L = L/R seconds.

4. Capacitive time constant (τ_C).

   In an R-C series circuit, τ_C = RC seconds.

NEW CONCEPTS AND PRINCIPLES

Motional EMFs and Faraday's Induction Law

Understanding this section should enable you to do the following:

1. *Calculate the EMF induced in a conductor moving with a given velocity through a magnetic field.*
2. *Calculate the EMF induced in a circuit by changes in the magnetic flux linking through the circuit.*

Motional EMF--If a metal rod of length L is moved with a velocity **v** perpendicular to its length in a region of magnetic field **B**, an EMF is set up between the ends of the rod. Its *magnitude* is

$$\mathcal{E} = BvL \cos \theta,$$

where θ = the angle between **B** and the *normal* to the plane in which the rod moves.

Important Observations

1. If the moving rod is part of a closed circuit, as in Figure 20.1 (text), this EMF will drive a current through the circuit in accordance with Ohm's Law.
2. If **B** is *perpendicular* to the plane in which the rod is moving, then $\theta = 0^\circ$, and the EMF is *maximum*, $\mathcal{E} = BvL$.
3. This *induced* EMF is a direct consequence of the magnetic force on moving charges. This force causes some of the electrons in the conductor to move toward one end of the rod, making it *negative* and leaving the other end with a surplus *positive* charge. Thus the positive end is at higher potential than the negative end, and a *voltage* results between them.

Example 20-1. In the figure below, a conducting rod of length L = 25 cm is pulled at right angles to a magnetic field B = .5 T.

(a.) How fast would it have to move to produce $\mathcal{E}$ = 12 volts between its ends?

(b.) Which end of the rod would be at higher potential?

(c.) If the rod was hooked up to an external circuit, which way would the current flow?

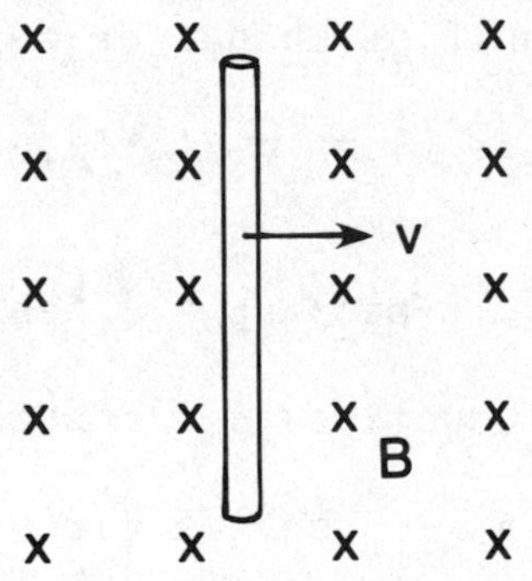

S: It looks like $\theta = 90^\circ$ here, so $\mathcal{E} = BvL$. This gives

$$v = \mathcal{E}/BL = 12\ \text{v}/(.5\ \text{T})(25 \times 10^{-2}\ \text{m}) = 96\ \text{m/s!}$$

How can I tell which end is at higher potential?

T: The end the electrons move toward will be the *low* potential end. The electrons are being forced to the right in the diagram. In what direction is the resulting magnetic force on them?

S: The force diagram of an individual electron would be as shown in the second figure. If it was a + particle, the right-hand rule would give an upward force, so on the electron it would be *downward.*

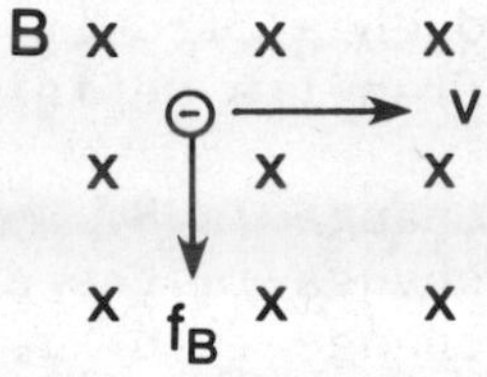

T: Right! It pays to practice with that rule. So an excess of electrons migrate toward the *bottom* end of the rod, leaving the *top* end *positive*, at *higher potential.*

S: Then if the rod was hooked to a circuit, the electrons would be able to continue out of the bottom end, through the circuit and into the top end.

T: That's right. You could also say that *conventional* current would flow from the + end (top) through the circuit, just as we treat the + terminal of any other EMF. As your text shows in Example 20.1, the power consumed by the circuit must be supplied by whatever force is pulling the conductor to produce the EMF.

Magnetic Flux (Φ)--If we have a small area ΔA over which the magnetic field is constant, the *magnetic flux through* ΔA is

$$\Phi = B\,\Delta A \cos\theta = (B_\perp)\,\Delta A,$$

where θ = the angle between B and the *normal* to the area ΔA.

Important Observations

1. The unit of magnetic flux is tesla-meter^2. This is defined to be the *weber* (Wb).

$$1\text{ Wb} = 1\text{ T-m}^2.$$

2. Turning this relation around, we see that $B_\perp = \Delta\Phi/\Delta A$, so the magnetic field strength measures the *area density of magnetic flux*, Wb/m^2. A way to visualize this is the number of lines of magnetic field there are per area perpendicular to the area.
3. If B lies *in the plane* of A, then $\theta = 90^\circ$, and $\Phi = 0$.

Faraday's Law of Induction--If the magnetic flux passing through the area bounded by a circuit *changes in time, an EMF is induced in the circuit which is equal in magnitude to the rate of change of flux.* Mathematically,

$$\xi = -\Delta\Phi/\Delta t.$$

Lenz' Law-- The negative sign above indicates that the induced EMF *is in such a direction that its effect will tend to oppose the change in flux that is taking place.*

Important Observations

1. You should satisfy yourself that Wb/s are indeed equal to volts.

2. If the circuit contains N turns of a coil, the EMFs just add, giving $\mathcal{E} = -N\Delta\Phi/\Delta t$.
3. Flux through a circuit can change in *three* ways, which you can see from its definition.

   a. A change in **B**.
   b. A change in the area of the circuit.
   c. A change in θ, i.e., the *orientation* of B and A.

   Faraday's Law applies equally for *any* reason the flux changes.
4. The meaning of Lenz' Law is that the induced EMF will drive a current in the circuit in such a direction that the magnetic field it produces will *tend to maintain the status quo of the flux condition.* Table 20.1 shows a few examples of this principle.

<u>Table 20.1. Examples of Lenz' Law</u>

(a.) <u>North pole approaching coil</u>.
Flux condition: flux to the *right* is *increasing.* Lenz' Law: coil must create flux to the left to oppose the change. EMF induced drives current as shown.

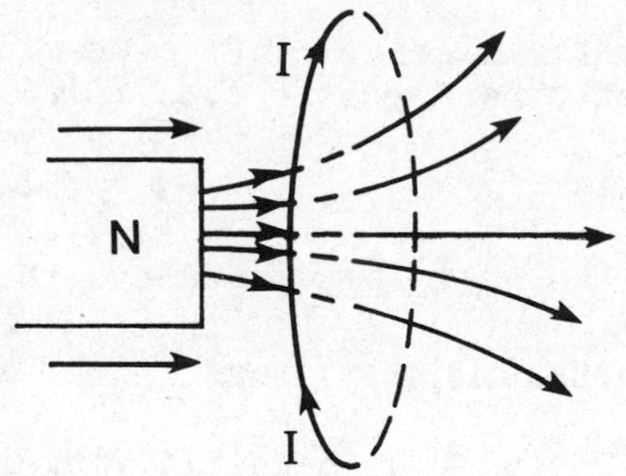

(b.) <u>North pole receding from coil</u>.
Flux condition: flux to the *right* is *decreasing.* Lenz' Law: coil must create additional flux to the right to oppose the change. EMF induced drives current as shown.

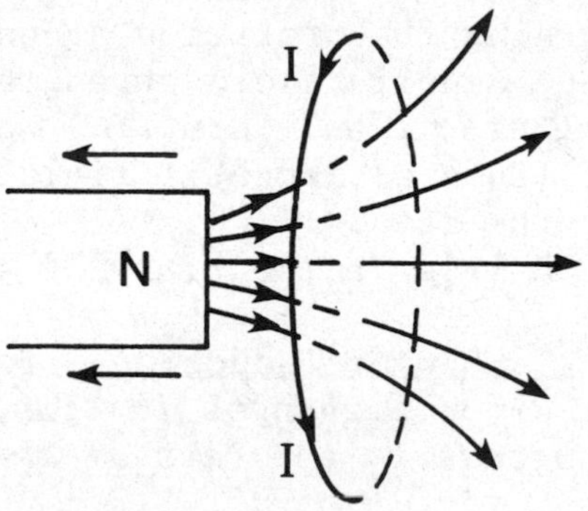

(c.) South pole approaching coil.

Flux condition: flux to the *left* is *increasing*. Lenz' Law: coil must create flux to the *right* to oppose the change. Induced EMF drives current as shown.

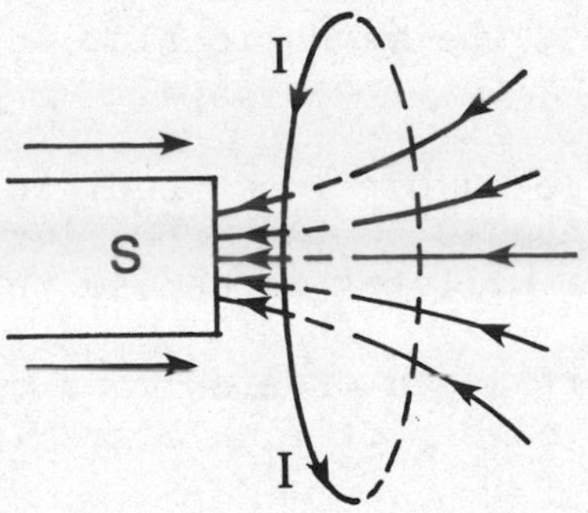

(d.) South pole receding from coil.

Flux condition: flux to the *left* is *decreasing*. Lenz' Law: coil must create additional flux to the left to oppose the change. Induced EMF drives current as shown.

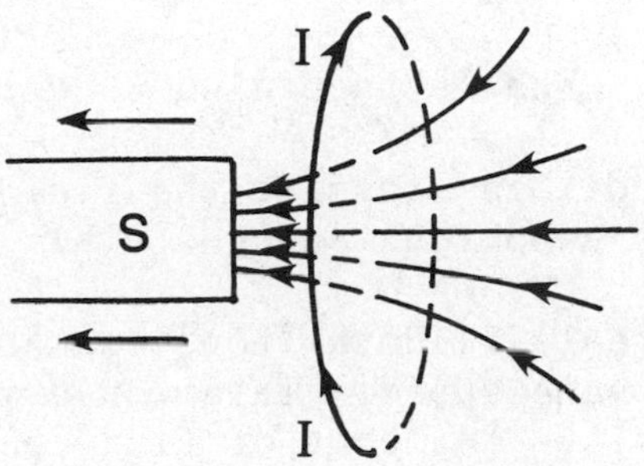

Important Observations

1. The point is not that the external change in flux is totally cancelled out. That only happens in a perfect superconductor. Rather, Lenz' Law tells us the *tendency* to cancel the change and hence the direction of the induced EMF.
2. The coil in these examples merely provides a supply of charges which can react to the EMF. The induced EMF exists in the region *whether or not* a coil or charges happen to be there.

Example 20-2. A *non*-uniform magnetic field is shown in the figure below, directed into the page and increasing in strength to the right (the area density of lines indicates relative field strength). A coil lies in the plane of the page. For each of the following motions of the coil, state the direction of induced current or whether no current is induced.

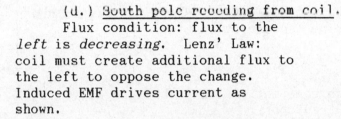

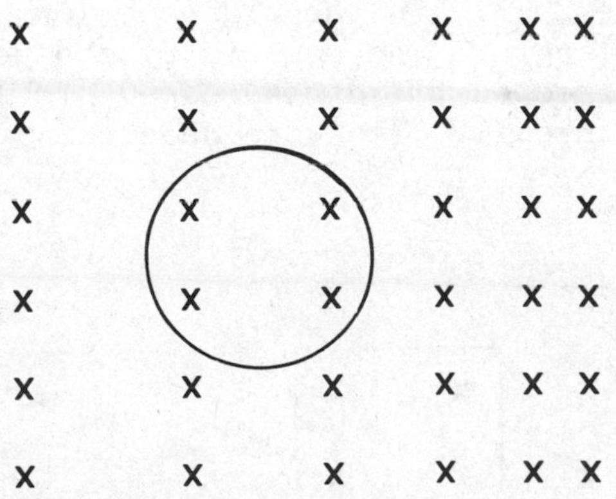

(a.) the coil moves directly to the right along the page.

(b.) the coil moves straight up along the page.

(c.) the coil moves directly *out* ofthe page toward you.

(d.) the coil moves diagonally downward and to the left along the page.

S: In (a.), the magnetic flux through the coil is *increasing*, so the EMF will cause a flux to oppose this?

T: Let's get a bit more specific. The *direction* of the flux is important. Your statement should be that the flux *into the page* through the coil is increasing, so the coil will try to oppose this change. How?

S: Well, it could create lines of flux *out* of the page, to oppose some of the increasing number of flux lines into the page.

T: Exactly. And what current direction would produce that effect?

S: With my right thumb outwards to indicate the directionof B_{coil}, my right-hand fingers curl *counter-clockwise*, which would be the direction of I.

T: Good. What's happening in (b.)?

S: The coil is crossing the field lines, so some motional EMF should result. But there's no flux change. This seems contradictory.

T: Not really. Both "top" and "bottom" parts of the coil are crossing the field with the same speed. This would induce motional EMFs which cancel each other in the circuit. In terms of flux, each time a new line links through the coil, one has been lost at the bottom of the coil, so flux stays constant, and *no current* is induced.

S: In (c.), nothing is happening to the flux within the coil.

T: Quite true, so again no current flows.

S: In (d.), even though the movement is diagonal, it seems the flux into the page is *decreasing*.

T: Right. *Any* component of motion to the left will do this. So the coil reacts oppositely to the way it did in (a.). Current is driven *clockwise* by the induced EMF, to try to *maintain* the flux into the page.

Example 20-3. In the figure below, there is a 10 turn square coil with sides of 10 cm length. The coil is placed perpendicular to a uniform magnetic field, B = 2 T, directed *out* of the page. In 10^{-3} seconds the square coil is changed into a circular coil. The wire does not stretch, but merely changes shape.

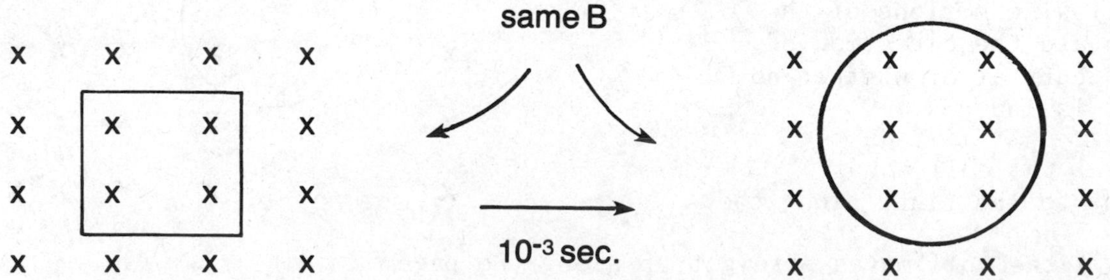

(a.) What is the induced EMF in the coil?
(b.) In what direction will a current flow during this change?

S: Why is an EMF induced here?

T: Check the areas of the square and circle, when the wire length stays the same.

S: You mean the *perimeter* of the coil stays constant. The square has a perimeter of 4 x 10 cm = 40 cm, and an area of $(10\ \text{cm})^2 = 100\ \text{cm}^2$. If the circle's *circumference* is then to be 40 cm, its *radius* must be $r = c/2\pi = 40\ \text{cm}/2\pi = 6.37$ cm. Thus the circle's *area* is $A_{cir} = \pi r^2 = 127\ \text{cm}^2$. There has been an area increase!

T: Right, and thus an increase of magnetic flux directed out of the page. What does Faraday say about this?

S: That EMF = $-N\,\Delta\Phi/\Delta t$. We're given $\Delta t = 10^{-3}$ s. What is the resulting $\Delta\Phi$?

T: Well, $\Phi = BA$ in this case, since **B** is perpendicular to the plane of A, and is uniform. So $\Delta\Phi = B\,\Delta A = B(A_{cir} - A_{sq})$. Areas now have to be in m^2, of course.

S: O.K., then

$$\Delta\Phi = (2\ \text{T})[(127 - 100) \times 10^{-4}]\ \text{m}^2 = 5.4 \times 10^{-3}\ \text{T-m}^2, \text{ or Wb.}$$

Then

$$\text{EMF} = N\,\Delta\Phi/\Delta t = (10)(5.4 \times 10^{-3}\ \text{Wb})/(10^{-3}\ \text{s}) = 54\ \text{Wb/s, or } \textit{volts}.$$

T: Now what about direction?

S: The *change* taking place is an increase in flux out of the page, so the current should create a field opposing the change, i.e., a field *into* the page.

T: So the current will be driven *clockwise* in the diagram. Remember, this is occurring only during the 10^{-3} seconds in which the *change* is taking place.

Mutual and Self Inductance; Inductors

Understanding this section should enable you to do the following:

1. *Identify the definitions of mutual and self inductance and calculate the induced EMF caused by current changes in inductors.*
2. *Calculate the energy stored in an inductor carrying current I.*
3. *Calculate the energy stored in a magnetic field B.*
4. *Calculate the inductive time constant for an R-L series circuit, and the time dependence of the current.*

Mutual Inductance (M)--Two coils placed near each other can interact with each other by induction if the current in one coil is made to vary. The current variation in the first coil causes its magnetic field to vary, part of which may be creating a flux through the second coil. This coil then experiences an *induced EMF* which will be *directly proportional to the rate of change of current in coil #1.* Mathematically,

$$\mathcal{E}_2 = -M(\Delta I_1/\Delta t), \text{ which is equivalent to } M = N_2(\Delta\Phi_2/\Delta I_1),$$

where the proportionality constant M is the *mutual inductance* between the two coils.

Self Inductance (L)--When the current in a single isolated coil is made to vary, it causes a change in its *own* magnetic flux, and therefore induces an EMF *in itself* which tends to oppose the change taking place. The rate of change of flux through the coil, and hence the induced EMF, is directly proportional to the rate of change of the coil's current. Mathematically,

$$\mathcal{E} = -L(\Delta I/\Delta t), \text{ which is equivalent to } L = N(\Delta\Phi/\Delta I).$$

The constant of proportionality L is called the coil's *self inductance.*

Important Observations

1. The unit of inductance in both cases is derived to be volt -second/amp, which is defined to be the *henry* (H).

   1 volt-sec/amp = 1 henry.

2. The mutual inductance will depend on the design of the two coils and their relative orientation.
3. The self inductance of a solenoid is derived in your text, with the following result:

$$L_{sol} = \mu_o N^2 A/l,$$

where N = *total* number of turns
l = solenoid length
A = solenoid cross-sectional area

Example 20-4. In Example 19-6, when the current I_1 is 50 amps, it creates 2.2×10^{-7} Wb of flux through the square coil ABCD.

(a.) What is the mutual inductance between the coil and the long wire?
(b.) If I_1 were to change from 50 amp upward to 50 amp downward in 0.1 sec, what EMF would be induced in the square coil, and in what direction?

S: None of the formulas seem to describe this directly.

T: Since $M = \Delta\Phi_2/\Delta I_1$ (for a single turn coil), and M is a constant, an alternate expression for M is $M = \Phi_2/I_1$, so

$$M = (2.2 \times 10^{-7} \text{ Wb})/(50 \text{ amps}) = 4.4 \times 10^{-9} \text{ H} = 4.4 \text{ nH}.$$

S: In part (b.), $\Delta I_1 = (I_1)_f - (I_1)_i$ = -50 amps - (+50 amps) = -100 amps. The minus sign designates downward in the diagram. So

$$\Delta I_1/\Delta t = -100 \text{ amps}/0.1 \text{ sec} = -1000 \text{ amps/sec}.$$

T: Alright, then

$$\mathcal{E}_2 = -M(\Delta I_1/\Delta t) = -(4.4 \times 10^{-9} \text{ H})(-1000 \text{ amps/s}) = 4.4 \times 10^{-6} \text{ volts}.$$

Can you tell in what direction?

S: First there is a decrease in flux *into* the page through the coil, then an increase of flux *out* of the page.

T: These are both part of the same direction of change which creates a tendency to produce additional flux into the page. Hence the clockwise current in the coil will *increase* during this 0.1 sec.

Inductor--An element of a circuit having self inductance. The circuit diagram symbol for an inductor is  .

Energy Stored in an Inductor--As derived in your text, the energy stored in an inductor carrying current I is

$$W = \tfrac{1}{2}LI^2.$$

Energy Stored in a Capacitor--As derived in your text, the energy stored in a capacitor is

$$W = \tfrac{1}{2}CV^2 = \tfrac{1}{2}QV = \tfrac{1}{2}Q^2/C.$$

Energy Stored in Magnetic and Electric Fields--In regions where B and E exist, the energy per volume, or *energy density* associated with the fields is

$$W/\text{Vol} = w = w_{mag} + w_{elec} = B^2/2\mu_o + \epsilon_o E^2/2.$$

Important Observations

1. You should satisfy yourself that the expressions $\tfrac{1}{2}LI^2$ and $\tfrac{1}{2}CV^2$ have the units of *joules*, and that $B^2/2\mu_o$ and $\epsilon_o E^2/2$ have units of *joules/m³*.
2. The fact that energy in empty space can be associated with magnetic and electric fields gives these fields a *physical reality* rather than just being a mathematical abstraction.
3. The energy needed to charge a capacitor or to establish a current in an inductor is *not* converted to heat as in a resistor, but instead is stored in the fields thus established. This energy is then still available to do work in the circuit.
4. If the B-field region is filled with material whose relative permeability is K_m, then $w_{mag} = B^2/2K_m\mu_o = B^2/2\mu$. If the E -field region is filled with material having dielectric constant K, then $w_{elec} = K\epsilon_o E^2/2 = \epsilon E^2/2$.

Example 20-5. Going back to Example 18-5, how much total energy is stored in that arrangement of capacitors, and how much is stored in each individual capacitor?

S: We found that $C_{tot} = .433 \times 10^{-6}$ F and V = 6 volts.

So

$$W_{tot} = \tfrac{1}{2}CV^2 = \tfrac{1}{2}(.433 \times 10^{-6} \text{ F})(6 \text{ v})^2 = 7.8 \times 10^{-6} \text{ J}.$$

T: Or you could have used $W = \tfrac{1}{2}QV = Q^2/2C$, since you know $Q_{tot} = 2.6 \times 10^{-6}$ coul.

S: C_3 has 6 volts across it directly, so

$$W_3 = \tfrac{1}{2}C_3V^2 = \tfrac{1}{2}(.3 \times 10^{-6} \text{ F})(6 \text{ v})^2 = 5.4 \times 10^{-6} \text{ J}.$$

With C_1 and C_2, V is *not* 6 v.

T: No, but you know that capacitors in series carry equal charges, in this case $Q_1 = Q_2 = Q_s = 0.8$ μC. You don't have to know V to calculate W.

S: So

$$W_1 = \tfrac{1}{2}Q_1{}^2/C_1 = \tfrac{1}{2}(.8 \times 10^{-6} \text{ coul})^2/(.2 \times 10^{-6} \text{ F}) = 1.6 \times 10^{-6} \text{ J}.$$

Also

$$W_2 = \tfrac{1}{2}Q_2{}^2/C_2 = \tfrac{1}{2}(.8 \times 10^{-6} \text{ coul})^2/(.4 \times 10^{-6} \text{ F}) = 0.8 \times 10^{-6} \text{ J}.$$

T: You can see from this that $W_{tot} = W_1 + W_2 + W_3$.

Example 20-6. Suppose you connected the charged capacitor arrangement from the previous example across an inductor. The inductor has 120 turns, and when carrying a 5 amp current has a total magnetic flux of 5×10^{-9} Wb linking through itself. What maximum current will the inductor carry after making the connection?

S: I don't quite see what determines that there will be a maximum current here.

T: Think in terms of available energy.

S: From the previous example, the capacitors had a total stored energy of 7.8×10^{-6} J. I'm beginning to see--this provides the work to establish the current in the inductor.

T: Right. Energy conservation would say that, with no heat loss in resistors, a *maximum* current could be established such that

$$\tfrac{1}{2}LI_{max}{}^2 = 7.8 \times 10^{-6} \text{ J}.$$

Of course this won't be a d.c. current. It will occur at the instant that $Q_{tot} = 0$, since all the capacitor's energy has then been depleted.

S: But we don't know L. Let's see, looking back to Example 20.4, I see something similar. We should be able to write

$$L = N\Phi/I = (120 \text{ turns})(5 \times 10^{-4} \text{ Wb})/(5 \text{ amps}) = 1.2 \times 10^{-2} \text{ H}.$$

T: Very good. Now there's just one step left:

$$I_{max}^2 = 2(7.8 \times 10^{-6} \text{ J})/L = 1.3 \times 10^{-3} \text{ amps}^2 ,$$

giving

$$I_{max} = 3.6 \times 10^{-2} \text{ amps}.$$

You should begin to get the idea that resistors aren't the only circuit elements that determine current, when voltage *changes* are applied.

<u>Series R-L Circuit</u>--A resistance R and an inductance L connected in series is shown schematically in the following figure. Here S is a switch which either connects the battery into the circuit or provides a closed circuit without the battery. From the instant the switch is closed to position ab, the current will not reach its Ohm's Law value of /R immediately, but will have the following behavior:

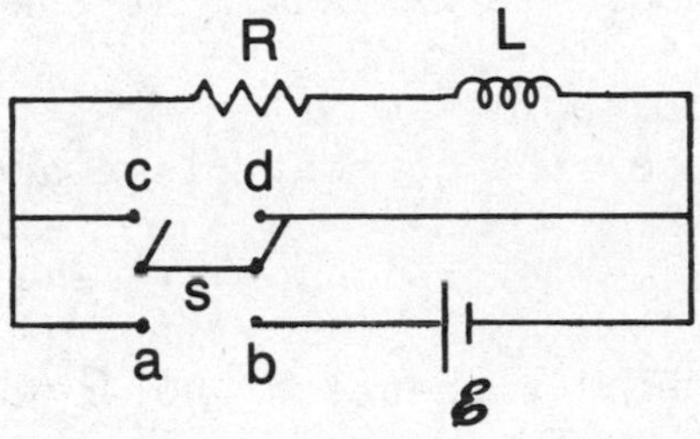

$$I(t) = I_{max}(1 - e^{-t/\tau_L}),$$

where

$$I_{max} = \mathcal{E}/R, \text{ and } \tau_L = L/R.$$

This function is shown graphically in the next figure. If, when some current I_0 is flowing, the switch is thrown to position cd, the current will not become zero immediately, but will decrease exponentially with the following behavior:

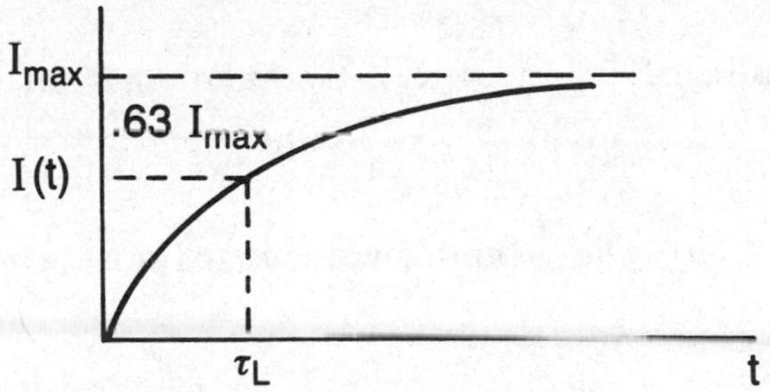

$$I(t) = I_0 e^{-t/\tau_L} .$$

This function is shown graphically in the following figure.

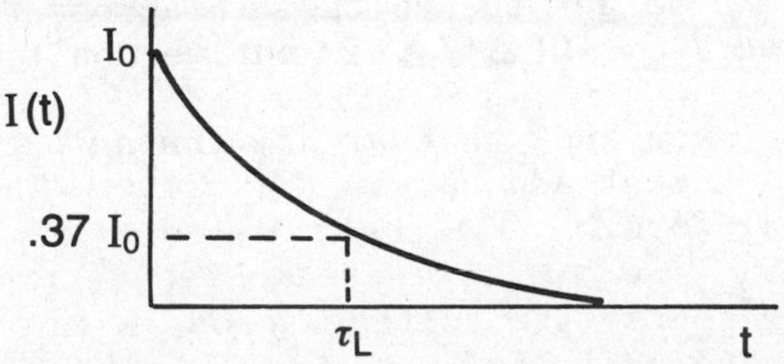

Important Observations

1. On pocket calculators which do not have an e^x key, use inv ln x.
2. τ_L is called the *inductive time constant*, and has units of henries/ohm = seconds.
3. When $t = \tau_L$, $e^{-t/\tau_L} = e^{-1} = .37$. At $t = 2\tau_L$, this becomes $e^{-2} = (.37)^2 = .137$. At $t = n\tau_L$, we have $e^{-n} = (.37)^n$.
4. The larger the time constant, the *slower* the circuit is in responding to imposed changes. This is the case for *large* L, which causes a large back EMF opposing the change, and hence a slow circuit response.

Example 20-7. A 200 resistor is connected in series with a 5 mH inductor. If a 24 volt EMF is connected to the circuit at t = 0, when will the current have reached 80% of its maximum? At that time, what will be the voltage across the inductor?

S: Let's see, the current rises according to the following:

$$I(t) = (\mathcal{E}/R)(1 - e^{-t/\tau_L}).$$

We want to find t when $I = .80(\mathcal{E}/R)$?

T: That's right. Just put that condition into the equation and solve for t.

S: We'll have

$$0.80(\mathcal{E}/R) = (\mathcal{E}/R)(1 - e^{-t/\tau_L}),$$

so

$$e^{-t/\tau_L} = 1 - .80 = .20.$$

How do I get t from this?

T: Remember $\ln e^x = x$. So take the natural logarithm of the equation.

$$\ln e^{-t/\tau_L} = -t/\tau_L = \ln(.20) = -1.61.$$

Notice that the signs turn out right, giving positive values of t.

S: So then

$$t = 1.61\tau_L = 1.61(L/R) = 1.61(5 \times 10^{-3}\ H)/(20\ \Omega) = 4.03 \times 10^{-4}\ s.$$

In order to get the voltage V_L across the inductor, we have $V_L = -L(\Delta I/\Delta t)$, but we don't know $\Delta I/\Delta t$.

T: But you know that *at any instant*, the voltage across R plus the voltage across L must add up to $\mathcal{E} = 24$ v. This is just Kirchoff's Loop Rule in a more general form.

S: Of course the voltage across R is just $V_R = IR$. You're saying that across the inductor it is $V_L = \mathcal{E} - V_R = 24\ v - IR$.

T: Yes, and this is of course an *instantaneous* value, changing continually. At $t = 1.61\tau_L$, $I = .80(\mathcal{E}/R) = .80(24\ v)/(20\ \Omega) = .96$ A. So

$$V_L = 24\ v - (.96\ A)(20\ \Omega) = 4.8 \text{ volts at that instant.}$$

Series R-C Circuit--A resistance R and a capacitance C connected in series is shown schematically in the accompanying figure. From the instant the switch is closed to position ab, the *charge* on the capacitor will increase according to the following equation:

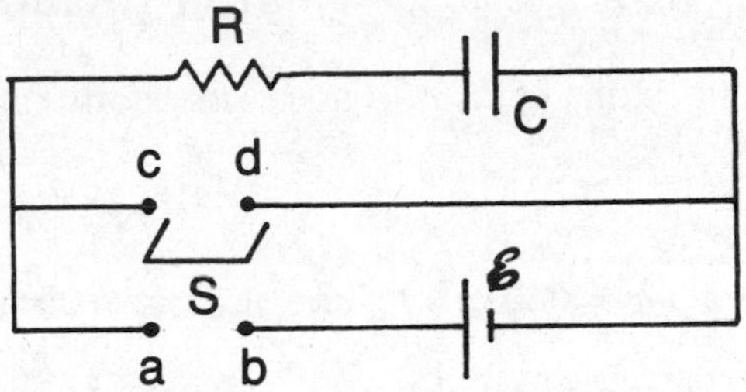

$$Q(t) = Q_{max}(1 - e^{-t/\tau_c}),$$

where $Q_{max} = C\mathcal{E}$, and $\tau_C = RC$. This function is shown graphically in the next figure.

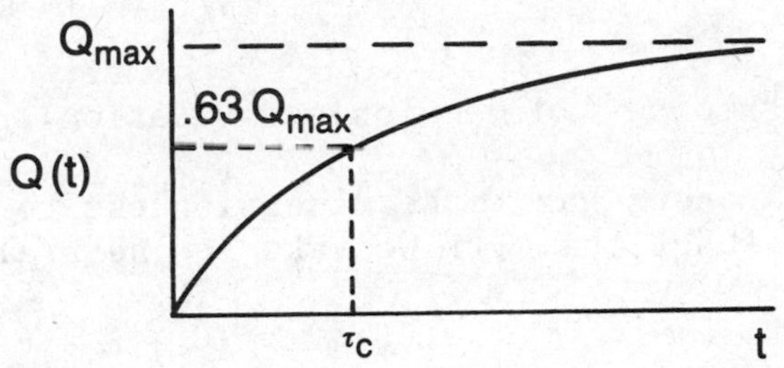

If the switch is thrown to position cd when the capacitor contains some charge Q_0, the charge will decrease exponentially according to the following equation:

$$Q(t) = Q_0 e^{-t/\tau_c}.$$

This function is shown graphically in the following figure.

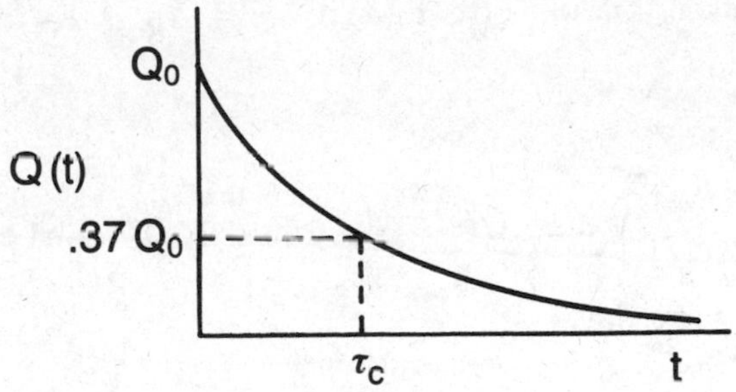

Important Observations

1. τ_C is called the *capacitive time constant*, and has units of ohm-farads = seconds.
2. The behavior of *charge* in the R-C circuit is directly analogous to the behavior of *current* in the R-L circuit.
3. A large capacitance C stores a large charge per volt, and takes more time to "fill up" with charge than does a smaller value of C. Thus τ_C is larger for large C, and the circuit responds more slowly to voltage changes imposed on it.

Example 20-8. An initially uncharged 2 pF capacitor is connected in series with a resistor R. In .05 seconds after a 100 v battery has been switched into the circuit, the charge has risen to 1.25×10^{-4} coul.

(a.) What is the value of R?
(b.) What is the voltage across R at this time?
(c.) If at this instant the switch is thrown to discharge C, how long will it take for its charge to drop to 10^{-6} coul?

S: Now we don't know the time constant. We have the expression

$$Q(t) = C\mathcal{E}(1 - e^{-t/\tau_c})$$

with $Q = 1.25 \times 10^{-4}$ coul at t = .05 sec. Thus

$$1.25 \times 10^{-4} \text{ coul} = (2 \times 10^{-6} \text{ F})(100 \text{ v})[1 - e^{-.05/(2 \times 10\text{-}6)R}].$$

This gives us

$$e^{-.05/(2 \times 10\text{-}6)R} = .375$$

This seems to be a dimensionless equation.

T: Any exponent has to be dimensionless to make any sense. It's the same with log or ln functions. If we take the natural logarithm of both sides, we have

$$-.05/(2 \times 10^{-6})R = \ln(.375) = -.98.$$

Thus

$$R = 2.55 \times 10^4 \ \Omega.$$

S: We know that V = IR, but what is I at this instant?

T: I don't know, but again $\mathcal{E} = V_R + V_C$ at every instant, and *you* do know how to find V_C.

S: Yes,

$$V_C = Q/C = (1.25 \times 10^{-4} \text{ coul})/(2 \times 10^{-6} \text{ F}) = 62.5 \text{ volts}.$$

Then we must have

$$V_R = 100 \text{ v} - 62.5 \text{ v} = 37.5 \text{ volts}.$$

T: Right. Now in part (c.), we have $Q_0 = 1.25 \times 10^{-4}$ coul, and we know that $\tau_C = RC = (2.55 \times 10^4 \ \Omega)(2 \times 10^{-6} \text{ F}) = 5.1 \times 10^{-2}$ sec.

S: So the time for Q(t) to become 10^{-6} coul is obtained from $Q(t) = Q_0 e^{-t/\tau_c}$, which gives

$$10^{-6} \text{ coul} = (1.25 \times 10^{-4} \text{ coul})e^{-t/.051}$$

or

$$e^{-t/.051} = 8.0 \times 10^{-3}$$

T: Then

$$-t/.051 = \ln(8.0 \times 10^{-3}) = -4.83 \ ,$$

giving

$$t = .246 \text{ seconds}.$$

QUESTIONS ON FUNDAMENTALS

1. You want to induce 10 volts per meter of length along a straight conductor by moving it perpendicular to a magnetic field. If you can move the conductor at a maximum speed of 25 m/s, how strong must the magnetic field be?

(a.) 0.4 T (b.) 2.5 T (c.) .04 T (d.) 250 T
(e.) impossible to tell unless length is given

2. A single turn circular coil of radius 50 cm is oriented so that the normal to its plane is at 60° to the direction of a uniform magnetic field. The field strength is 0.3 T. What is the magnetic flux through the coil?

(a.) .382 Wb (b.) .191 Wb (c.) .204 Wb (d.) .236 Wb
(e.) .118 Wb

3. The coil in question #2 changes orientation so that its normal lies along the magnetic field direction. If the change in orientation took place in 2 milliseconds, what was the average value of the induced emf?

(a.) 118 volts (b.) 102 volts (c.) 59 volts
(d.) 191 volts (e.) 95 volts

4. A solenoid 50 cm long is wound with 5000 turns/meter, with a circular diameter of 10 cm. What is its inductance?

(a.) 123 mH (b.) 246 mH (c.) 492 mH (d.) 19.6 mH
(e.) 39.2 mH

5. If the solenoid described in #4 is carrying a steady current of 3 amperes, what is the energy density in its interior? (Refer to the previous chapter).

(a.) 282 J/m^3 (b.) .55 J (c.) 45 J/m^3 (d.) 141 J/m^3
(e.) 1.1 J

6. If the solenoid referred to in #4 has an inductive time constant of 3 milliseconds, what must its resistance be?

(a.) .024 Ω (b.) 41.7 Ω (c.) 2.7×10^3 Ω
(d.) $.37 \times 10^{-3}$ Ω (e.) need more information

7. At the instant 10^{-2} after applying a d.c. voltage, what percentage of the final steady state current would be passing through the solenoid in #6?

(a.) 71.9% (b.) 28.3% (c.) 99.7% (d.) 96.4%
(e.) 3.6%

GENERAL PROBLEMS

Problem 20-1. Consider the R-C circuit shown, in which R_1 = 10 kΩ, R_2 = 20 kΩ, R_3 = 40 kΩ, C_1 = 6 μF, and C_2 = 12 μF. 50 msec after closing the switch, what charge will be on the capacitors? What maximum charge will the capacitors eventually have?

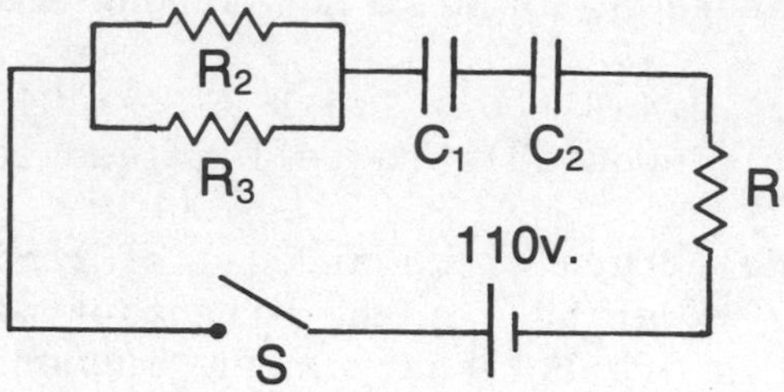

Q1: What determines how fast charge will accumulate?
Q2: What is the equivalent resistance in the circuit?
Q3: What is the equivalent capacitance?
Q4: What is the time constant for the circuit?
Q5: What is Q_{max}?

Problem 20-2. You are asked to design a solenoid with length l = 30 cm, radius r = 1 cm, and resistance R = 1000 Ω. When 120 volts are applied to it, the current is supposed to rise to 1.00 amps in 1.00×10^{-6} sec. How many windings must you use?

Q1: What determines the rate at which the current rises?
Q2: What is the inductance L for a solenoid?
Q3: What is the equation which gives me N?

Problem 20-3. In the figure below, a conducting bar AB falls without friction along two wires, always maintaining contact with them so that there is always a complete electrical circuit ABCD. As the bar falls, it cuts across a uniform magnetic field of 0.2 T, directed into the page. The bar has mass M = 500 grams, length l = 20 cm, and resistance R = .02 Ω. The other wires have negligible resistance compared to this. When the bar reaches terminal speed, v_T, find the current in the bar and v_T.

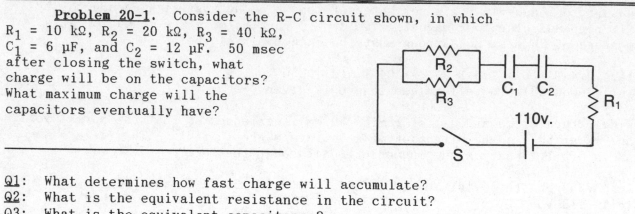

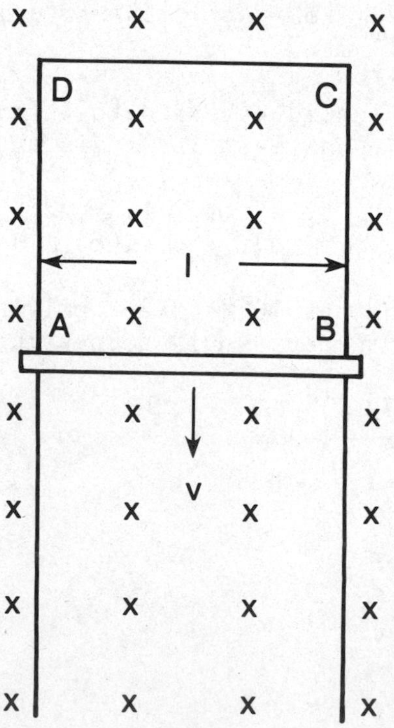

Q1: As the bar falls, in what direction will a current be induced?
Q2: Does this agree with the induction law?
Q3: Why will the bar reach a terminal speed?
Q4: What is the magnetic force on the bar?
Q5: What determines I?
Q6: What is $\xi_{induced}$?
Q7: What is the expression for the rate of flux change?
Q8: What are the final expressions which yield I and v_T?

Problem 20-4. In the figure below, a small 10 turn coil of radius 5 cm is inserted into a solenoid so that its plane is perpendicular to the axis of the solenoid. The solenoid has 1000 turns per meter. The small coil is connected to a sensitive voltmeter which can detect voltages as small as 10^{-6} volts.

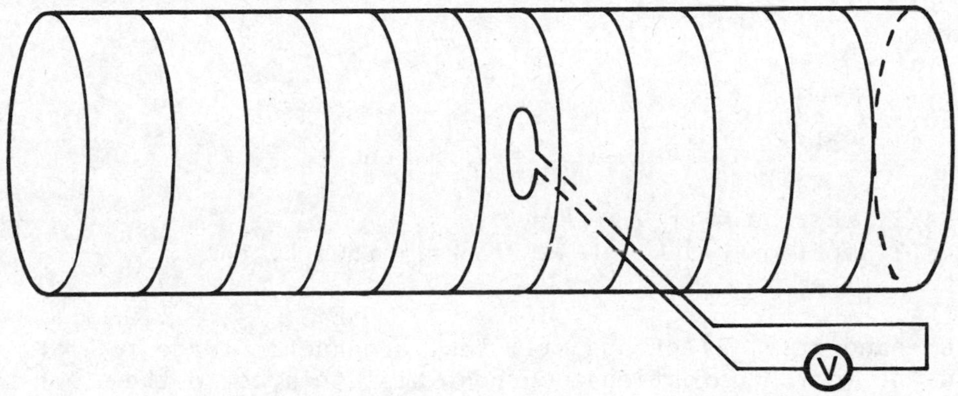

(a.) Find the mutual inductance between the coil and the solenoid.
(b.) What is the *minimum* rate of change of the solenoid current that the coil will be able to detect?

Q1: What is the definition of mutual inductance?
Q2: What determines ξ_{coil}?
Q3: What is $\Delta\Phi_{coil}$?
Q4: What determines ΔB_{sol}?
Q5: What does the expression for M look like when we combine all this?
Q6: What does the condition "minimum rate of change of current" mean?

ANSWERS TO QUESTIONS

FUNDAMENTALS:

1. (a.)	2. (e.)	3. (c.)	4. (a.)
5. (d.)	6. (b.)	7. (d.)	

GENERAL PROBLEMS:

Problem 20-1.

A1: $Q(t) = Q_{max}(1 - e^{-t/\tau_c})$

A2: $R_{tot} = R_1 + [(1/R_2) + (1/R_3)]^{-1}$

A3: $C_{tot} = [(1/C_1) + (1/C_2)]^{-1}$

A4: $\tau_c = R_{tot}C_{tot}$

A5: $Q_{max} = C_{tot}\,\xi$.

Problem 20-2.

A1: $I(t) = I_{max}(1 - e^{-t/\tau_L})$, where $I_{max} = \xi/R$, and $\tau_L = L/R$.

A2: $L_{sol} = \mu_0 N^2 A/l$, where $A = \pi r^2$

A3: $.75 \text{ amps} = 1.2 \text{ amps}(1 - e^{-10^{-6}/\tau_L})$. Find τ_L, then L, then N.

Problem 20-3.

A1: By the right-hand rule, electrons will feel a magnetic force to the left in the rod. Conventional current will then be to the right through the rod and counterclockwise through the circuit.

A2: Sure. More flux links through the circuit into the page as the bar falls. The circuit opposes that change with the counter -clockwise current, producing some flux out of the page.

A3: The magnetic force upward on it will balance the force of gravity.

A4: $F_B = BIl$.

A5: Ohm's Law: $I = \xi_{induced}/R$.

A6: $\xi_{induced} = \Delta\Phi/\Delta t$.

A7: $\Delta\Phi/\Delta t = B(\Delta A/\Delta t) = Bl(\Delta y/\Delta t) = Blv$.

A8: $Mg = BI_T l$ (Force balance), and $\xi_{ind.} = Blv_T = I_T R$ (Induction law)

Problem 20-4.

A1: $\xi_{coil} = M(\Delta I_{sol}/\Delta t)$

A2: The Law of Induction: $\xi_{coil} = N_{coil}(\Delta\Phi_{coil}/\Delta t)$

A3: $\Phi_{coil} = B_{coil}A_{coil}$. Since A_{coil} is constant, $\Phi_{coil} = (\Delta B_{sol})A_{coil}$.

A4: $B_{sol} = \mu_0 n I_{sol}$, so $\Delta B_{sol} = \mu_0 n \Delta I_{sol}$.

A5:

$$M = \frac{\xi_{coil}}{(\Delta I_{sol}/\Delta t)} = \frac{N_{coil}(\mu_0 n A_{coil})(\Delta I_{sol}/\Delta t)}{(\Delta I_{sol}/\Delta t)} = \mu_0 n N_{coil} A_{coil}$$

A6: $(\xi_{coil})_{min} = M(\Delta I_{sol}/\Delta t)_{min}$, and $(\xi_{coil})_{min}$ is determined by the given voltmeter sensitivity.

CHAPTER 21. ALTERNATING CURRENT CIRCUITS

BRIEF SUMMARY AND RESUME

A.C. circuits containing only *resistance* are analyzed and the definitions of *peak-to-peak* and *RMS* values of voltage and current are introduced. Purely *inductive* and purely *capacitive* a.c. circuits are investigated and the *phase relationships* between current and voltage are derived. The concepts of *inductive* and *capacitive reactance* are introduced and their mathematical forms are derived. The oscillatory nature of the current in an *L-C series circuit* is analyzed and compared with the equations describing mechanical SHM. The *series RLC circuit* is introduced, along with the concept of *impedance*, generalizing Ohm's Law to a.c. Circuits. The impedance of this type of circuit is derived mathematically, and the phenomenon of *series resonance* is discussed. The amount of *power consumption* in a.c. circuits is discussed, and the *power factor* of the series RLC circuit is defined.

The operation of *transformers* is analyzed as an application of the law of induction. Their use in *impedance matching* to optimize power transfer between circuits is discussed.

Material from previous chapters relevant to Chapter 21 includes:

1. Ohm's Law.
2. Sinusoidally varying quantities, and phase differences.
3. Power consumption in resistors.
4. Law of magnetic induction.
5. Inductors and Capacitors.

NEW DATA AND PHYSICAL CONSTANTS

1. RMS value of a sinusoidally varying quantity.

   $Q_{RMS} = Q_{max}/\sqrt{2} = .707\ Q_{max}$

2. Inductive impedance (X_L)

   $X_L = 2\pi fL$; units of henries/sec = ohms

3. Capacitive impedance (X_C).

   $X_C = 1/(2\pi fC)$; units of sec/farad = ohms

4. Impedance (Z).

   $Z = V/I$; units of volts/amp = ohms

5. Resonant frequency, series RLC circuit (f_{res}).

   $f_{res} = 1/[2\pi(LC)^{\frac{1}{2}}]$; units of Hz

6. Power factor, series RLC circuit ($\cos\Phi$).

   $\cos\Phi = R/Z$, where Φ is the phase angle between the applied a.c. voltage and the a.c. current

NEW CONCEPTS AND PRINCIPLES

A.C. Circuits with Single Circuit Elements

Understanding this section should enable you to do the following:

1. *Calculate RMS values of current and voltage from the peak-to -peak values and vice versa.*
2. *Identify the phase differences between voltage and current for purely inductive and purely capacitive circuits.*
3. *Calculate the dependence of RMS current in a purely inductive or purely capacitive circuit on frequency and L or C.*

In discussing a.c. circuits, we assume a voltage of the sinusoidal form $V(t) = V_{max} \sin(2\pi ft)$ is applied to the circuit, and want to find the behavior of the current that results. We already have had hints that capacitors and inductors play a role, along with resistors, in determining that current when the voltage changes in time.

A.C. Circuit with Resistance Only--In a resistor, there is no delay between changes in the applied voltage and the current, so at *each instant* Ohm's Law is obeyed, i.e., $I(t) = V(t)/R$. Thus

$$I(t) = (V_{max}/R) \sin(2\pi ft) = I_{max} \sin(2\pi ft).$$

Important Observations

1. Since both $V(t)$ and $I(t)$ are described by the same time dependence, they are *in phase*, i.e., they are simultaneously in the same part of their cyclic behavior.
2. Since a sine function (or a cosine function) is symmetric above and below the $V = 0$ or $I = 0$ axis, the *average* voltage and current over any number of complete cycles is *zero*.
3. The *power* consumed by a resistor is

$$P(t) = I^2(t)R = I_{max}^2 \sin^2(2\pi ft) R.$$

   The squared sine function is always positive, so its average does not vanish. In fact $\sin^2(2\pi ft) = 1/2$ over any number of complete cycles (see Section 21.1, text). Thus the *average* power consumption is $\langle P\rangle = \frac{1}{2}I_{max}^2R$.
4. The *effective* current is that which would make $\langle P\rangle$ the same as in the d.c. case, where $P = I^2R$. So

$$I_{eff} = I_{max}/\sqrt{2} = (.707)I_{max}.$$

   This is also known as the *RMS* value:

$$I_{RMS} = I_{eff} = (.707)I_{max}.$$

5. We can similarly define the RMS voltage, $V_{RMS} = (.707)V_{max}$.
6. The effective (RMS) current and voltage in a purely resistive a.c. circuit obey Ohm's Law:

$$I_{RMS} = V_{RMS}/R.$$

Example 21-1. In a purely resistive circuit, a maximum current of 20 amps occurs when a 50 Hz, 110 volt effective voltage is applied. Find the following:

(a.) The resistance of the circuit.
(b.) The average power consumed by the circuit.
(c.) The instantaneous current at $t = 1.0223$ sec.

S: We either have $I_{RMS} = V_{RMS}/R$ or $I_{max} = V_{max}/R$.

T: So either find I_{RMS} or V_{max}.

S: $I_{RMS} = (.707)I_{max} = 14.1$ amps.
Then $R = V_{RMS}/I_{RMS} = (110 \text{ v})/(14.1 \text{ A}) = 7.8\ \Omega$. The average power is $\langle P\rangle = \frac{1}{2}I^2R = \frac{1}{2}(20 \text{ A})^2(7.8\ \Omega) = 1560$ watts.

T: This is the same as $\langle P\rangle = I_{RMS}{}^2R = (14.1)^2(7.8) = 1560$ watts.

S: The functional form of the current is

$$I(t) = (20 \text{ A}) \sin[2\pi(50 \text{ Hz})t].$$

Putting in $t = 1.0223$ sec, I get $2\pi ft = 321.17$, and $\sin(321.17) = -.627$.

T: Whoa! *Big* mistake! $2\pi ft$ is a number of *radians*, not degrees! $\sin(321.17^\circ) = -.627$, but $\sin(321.17 \text{ rads}) = +.661$. So

$$I(t = 1.0223) = (20 \text{ A})(.661) = 13.2 \text{ amps}.$$

A.C. Circuit with Inductance Only--The induced EMF causes the current to *lag the applied voltage by 1/4 cycle*, or 90° in phase. If the voltage is expressed as a sine function as above, this means the functional form of the current is

$$I(t) = -I_{max} \cos(2\pi ft).$$

Inductive Reactance (X_L)--The ratio of V_{max}/I_{max} or V_{RMS}/I_{RMS} is called the *inductive reactance*, X_L. This reactance is proportional to both the inductance L and the a.c. frequency, f. Mathematically,

$$V_{max}/I_{max} = V_{RMS}/I_{RMS} = X_L = 2\pi fL.$$

Important Observations

1. By graphing $\sin\theta$ and $-\cos\theta$, you can see the latter is always 1/4 cycle behind the former.
2. The reactance $X_L = 2\pi fL$ has the units of henries/sec = *ohms*.
3. The relation $X_L = V_{RMS}/I_{RMS}$ is the generalization of Ohm's Law to this circuit. You can see *qualitatively* that a large f and L should decrease the current. The more frequently you impose a change in current, the more continually you induce an opposing EMF, and the amount of EMF induced is directly proportional to L.
4. Remember, because of the phase difference between I and V, Ohm's Law does *not* apply to their *instantaneous* values. That is, $V(t)/I(t)$ *is not equal to* X_L.

Example 21-2. If a 500 Hz, 30 volt (RMS) generator produces a maximum current I_{max} = 2.6 amps through an inductor, what is the inductance? What RMS current would flow from the same EMF at 60 Hz?

S: The inductance is involved through the reactance $X_L = 2\pi fL$. How do we find X_L?

T: From Ohm's Law, X_L is equal to either V_{RMS}/I_{RMS} or V_{max}/I_{max}.

S: Well, if I_{max} = 2.6 amps, I_{RMS} = (.707)(2.6 amps) = 1.84 amps, so

$$X_L = (30\ \text{v})/(1.84\ \text{A}) = 16.3\ \Omega = 2\pi fL,$$

giving

$$L = (16.3\ \Omega)/2\pi(500\ \text{Hz}) = 5.2\ \text{mH}.$$

T: This is a *fixed* property of the inductor. Its *reactance* varies with the signal frequency. At f = 60 Hz,

$$X_L = 2\pi(60\ \text{Hz})(5.2 \times 10^{-3}\ \text{H}) = 1.96\ \Omega,$$

so

$$I_{RMS} = V_{RMS}/X_L = (30\ \text{v})/(1.96\ \Omega) = 15.3\ \text{amps}.$$

A.C. Circuit with Capacitance Only--The *current* in the circuit *leads the applied EMF by 1/4 cycle*, or 90° in phase. This means the functional form of the current is

$$I(t) = +I_{max}\cos(2\pi ft).$$

Capacitive Reactance (X_C)--The ratio of V_{max}/I_{max} or V_{RMS}/I_{RMS} is called the *capacitive reactance*, X_C. This is *inversely* proportional to both the a.c. frequency and the capacitance C. Mathematically,

$$V_{max}/I_{max} = V_{RMS}/I_{RMS} = X_C = 1/(2\pi fC)\quad .$$

Important Observations

1. Again, by graphing sin θ and cos θ, you can see that the latter is 1/4 cycle *ahead* of sin θ.
2. It is easy to verify that 1/(2πfC) has the units of sec/farads = *ohms*.
3. Qualitatively, we can see that a large C and high frequency f *should* allow more current, as the formula for X_C states. A large C accumulates more charge per volt, and a rapidly varying voltage, i.e., a large value of f, would not allow the capacitor to saturate with charge, and hence would keep a greater current flowing to and from C. Both these effects represent a small X_C.
4. Again note that X_C is *not* equal to the *instantaneous* ratio V(t)/I(t).

Example 21-3. What value of capacitance would allow the same maximum current for the same 500 Hz voltage as in Example 21-2? What current would be drawn at 60 Hz?

S: As in the last example, we would have a reactance of 16.3 Ω. But now it would be a *capacitive* reactance.

T: Right. So $X_C = 16.3\ \Omega = 1/(2\pi fC)$.

S: So

$$1/C = 2\pi(500\text{ Hz})(16.3\ \Omega) = 5.12 \times 10^4\ F^{-1}, \text{ or}$$

$$C = 1.95 \times 10^{-5}\ F = 19.5\ \mu F.$$

T: For the last part, $I_{RMS} = V_{RMS}/X_C = (2\pi fC)V_{RMS}$. Again, C remains fixed, and X_C changes with frequency. At 60 Hz,

$$I_{RMS} = 2\pi(60\text{ Hz})(19.5 \times 10^{-6}\ F)(30\ v) = .221\text{ amps.}$$

Be careful of reciprocals when you're involved with X_C.

A.C. Circuits with Elements Connected in Series

Understanding this section should enable you to do the following:

1. *Calculate the natural frequency of oscillation in an L-C circuit.*
2. *Calculate the impedance of an R-L-C circuit, and its frequency dependence.*
3. *Calculate the resonant condition of an R-L-C circuit, and find I_{max} for the resonance.*
4. *Calculate the phase difference between I and V for an R-L-C circuit, the power factor, and the power dissipated in the circuit.*

L-C Series Circuit--Without a resistance in the circuit, energy is conserved, and oscillates back and forth between stored charge on the capacitor and current in the inductor. As derived in Section 21.4 in your text, the oscillation in current and charge corresponds precisely to mechanical SHM, with a frequency given by

$$f = 1/[2\pi(LC)^{\frac{1}{2}}]$$

Important Observations

1. In the analogy with mechanical systems, the energy of charge storage, $q^2/2C$, corresponds to the *potential* energy of a spring, $\frac{1}{2}kx^2$, and the magnetic energy of the inductor, $\frac{1}{2}LI^2$, corresponds to kinetic energy $\frac{1}{2}mv^2$. In both systems, energy oscillates between the potential and kinetic forms.
2. Further analogies are that q and i $(= \Delta q/\Delta t)$ are related in the same way as x and $v = (\Delta x/\Delta t)$, L corresponds to the "inertia" of the system m, and 1/C corresponds to the elastic constant, k. Thus oscillations occur at the frequencies

$$f_o = (1/2\pi)(k/m)^{\frac{1}{2}} \quad , \text{ or } \quad = (1/2\pi)[(1/C)/L]^{\frac{1}{2}} = 1/[2\pi(LC)^{\frac{1}{2}}]$$

Example 21-4. What oscillation frequency would result from a series connection of the inductor and capacitor mentioned in the previous two examples?

S: We found L = 5.2 mH and C = 19.5 μF. So

$$f_{osc} = (1/2\pi)(5.2 \times 10^{-3}\ H)^{-\frac{1}{2}}(19.5 \times 10^{-6}\ F)^{-\frac{1}{2}}$$

T: Notice two things:
(1) This is the frequency at which the *reactances are equal.*
(2) The units of (henries-farads)$^{-\frac{1}{2}}$ = Hz! Verify this.

Voltage Relations in a Series R-L-C Circuit--To find the relation between applied EMF and current, the *phase differences* between them which occur in capacitors and inductors must be taken into account. A convenient way is to treat the phase differences as geometric angles. The current, common to all circuit elements, is *in phase* with the voltage across the resistor, IR. The voltage on the capacitor, IX_C, *lags* 90° behind this in phase, while the voltage across the inductor, IX_L, *leads* the current by 90°. The result can be made into a vector (or *phasor*) diagram:

As you can see, both the *RMS* and *max* values of the voltages across the three circuit elements add as *vectors* in the diagram.

$$\mathbf{V}_{tot} = \mathbf{V}_R + \mathbf{V}_C + \mathbf{V}_L$$

Mathematically, the *instantaneous* voltages across the three circuit elements add up *as scalars* to the applied voltage:

$$V_{tot}(t) = V_R(t) + V_L(t) + V_C(t).$$

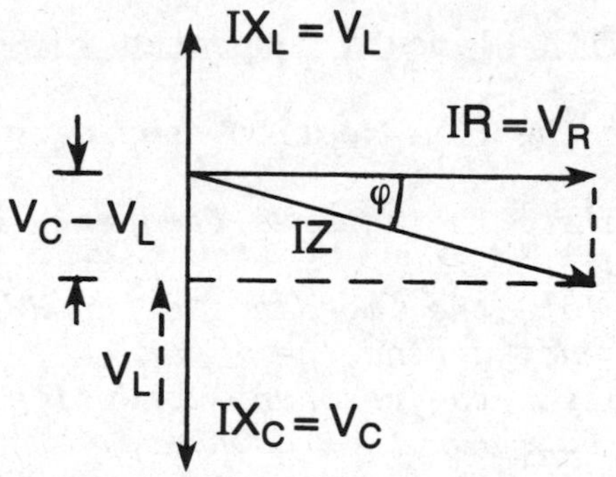

Impedance of an R-L-C Circuit (Z)--The *impedance*, Z, of an a.c. circuit is the ratio of maximum or RMS voltage to the corresponding current. Mathematically,

$$Z = V_{max}/I_{max} = V_{RMS}/I_{RMS}.$$

As derived in your text, for the series R-L-C circuit this is

$$Z = [R^2 + (X_L - X_C)^2]^{\frac{1}{2}}$$

Important Observations

1. As before, the characteristic of any series circuit is that all elements carry the same current, but have different voltages.
2. The unit of impedance is obviously the *ohm*. Notice the differences in the terms resistance, reactance, and impedance, all of which are measured in ohms.
3. The voltages across C and L are 1/2 cycle, or 180° out of phase, and so at any instant they tend to subtract, as seen in the figure above.

4. As before, the *instantaneous* total voltage and current are *not* related by Ohm's Law, but only the RMS or max values.
5. Also, the sum of the RMS or max values of the voltage drops across R, L, and C *do not* equal the applied voltage, but the sum of the instantaneous voltages always does.
6. The angle Φ in the figure above is the phase between current and total voltage in the circuit. Note that $\cos\Phi = R/Z$.
7. If $L = 0$, we have an R-C circuit, where $Z = (R^2 + X_C^2)^{\frac{1}{2}}$. Again $\cos\Phi = R/Z$.
8. If we remove C, $X_C = 0$, because replacing C by a conducting path is the same as having $C = \infty$ in the circuit. Charge could then accumulate with no effect on the circuit, since no voltage would develop. Thus an R-L circuit has $Z = (R^2 + X_L^2)^{\frac{1}{2}}$.

Power in a Series R-L-C Circuit--The only element which converts electrical energy into heat, and thereby represents a *power drain* on the applied EMF, is the *resistance* R. The power dissipated is again given by the formula for joule heating.

$$P = I_{RMS}^2 R = I_{RMS}^2 Z \cos\Phi = I_{RMS} V_{RMS} \cos\Phi.$$

Important Observations

1. This is the same expression as for d.c. circuits, except for the factor of $\cos\Phi$, called the *power factor* of the circuit. The more Z is dominated by the resistance, the closer $\cos\Phi$ is to 1, meaning the voltage and current are close to being in phase.

Example 21-5. A coil having resistance $R = 5\ \Omega$ and inductance $L = 1$ mH is connected in series with a 20 μF capacitor. When the system is connected to a 1000 Hz generator, an RMS current of 2 A is observed. Find the following:

(a.) The RMS voltage of the generator.
(b.) The RMS voltage across the capacitor.
(c.) The RMS voltage across the coil (R and L in series).
(d.) The phase angle between I and V in the circuit.

S: We'll have to know the impedance to find V_{tot}, and

$$Z = [R^2 + (X_L - X_C)^2]^{\frac{1}{2}} .$$

It doesn't seem to make any difference whether we take $(X_L - X_C)$ or $(X_C - X_L)$.

T: Remember, you square *after* you subtract them, so you're right, it doesn't make any difference. The reactances are easy:

$$X_L = 2\pi f L = 2\pi(10^3\ \text{Hz})(10^{-3}\ \text{H}) = 6.28\ \Omega,$$

and

$$X_C = (2\pi f C)^{-1} = [2\pi(1000\ \text{Hz})(20 \times 10^{-6}\ \text{F})]^{-1} = 7.96\ \Omega$$

S: Then the impedance is

$$Z = [(5)^2 + (7.96 - 6.28)^2]^{\frac{1}{2}}\ \Omega = 5.27\ \Omega.$$

So

$$V_{RMS} = I_{RMS}Z = (2\ A)(5.27\ \Omega) = 10.5\ \text{volts}.$$

How can I tell the voltage across the capacitor?

T: Across any element or combination of elements, $V_{RMS} = I_{RMS}Z$. For a pure resistance, $Z = R$; for a pure inductance, $Z = X_L$, for a pure capacitance, $Z = X_C$.

S: Oh, I see. Since $X_C = 7.96\ \Omega$,

$$(V_C)_{RMS} = (2\ A)(7.96\ \Omega) = 15.9\ \text{volts}.$$

For the coil, though, it's not a pure resistance or inductance.

T: No, but it *is* a series R-L element, with

$$Z = [R^2 + X_L^2]^{\frac{1}{2}} = [(5)^2 + (6.28)^2]^{\frac{1}{2}} = 8.03\ \Omega.$$

So

$$(V_{coil})_{RMS} = I_{RMS}Z_{coil} = (2\ A)(8.03\ \Omega) = 16.1\ \text{volts}.$$

S: This looks crazy. Don't we have to have $V_{tot} = V_{coil} + V_C$?

T: For instantaneous values, yes, but we're not dealing with them. Instead, we're dealing with the RMS values, which don't occur at the same instant across the various elements of the circuit. The RMS values are related by a *vector-like* addition in the phasor diagram

$$\mathbf{V}_{tot} = \mathbf{V}_R + \mathbf{V}_L + \mathbf{V}_C,$$

meaning their *magnitudes* are related by Pythagoras' Theorem,

$$V_{tot}^2 = V_R^2 + (V_L - V_C)^2,$$

which defines Z. For the coil alone, we have $V_{coil}^2 = V_R^2 + V_L^2$, which you can easily verify.

S: For part (d.), $\cos \Phi = R/Z = 5\ \Omega/5.27\ \Omega$, so

$$\Phi = \cos^{-1}(.949) = 18.4^\circ.$$

How do I know whether voltage leads or lags the current?

T: Inspect which *reactance* is larger. Here $X_C > X_L$, so the *capacitive* tendency dominates. That is, voltage will *lag behind* the current by 18.4° in this circuit. *Don't forget that this all changes with a different frequency*.

Resonance in the Series R-L-C Circuit--Since X_L is directly proportional to f, and X_C is inversely proportional to f, there *must be some* frequency at which $X_L = X_C$ for any pair of values of C and L. This is shown in the accompanying figure. At this frequency we simply have Z = R, which is the *smallest possible* value of Z at any frequency. This means that the RMS current in the circuit becomes *largest* at this frequency, and the circuit is said to be at *resonance*. Mathematically, this condition is where $2\pi f_r L = 1/(2\pi f_r C)$, or

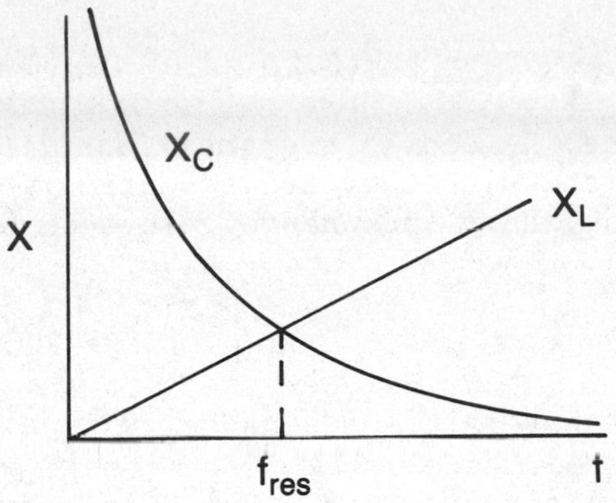

$$f_r = 1/[2\pi(LC)^{\frac{1}{2}}]$$

Important Observations

1. This resonance occurs at the same frequency at which the L-C circuit would oscillate.
2. Figure 21.18 (text) shows the RMS current peak that occurs at f_r. The peak is *not* symmetrical around f_r.
3. The resonant current can be very large if R is small, and since X_L and X_C can be large *individually*, *very large voltages* can develop across the capacitor or inductor separately at resonance.
4. At resonance I and V are *in phase*, i.e., $\Phi = 0^o$.

Example 21-6. Consider a series R-L-C circuit with R = 10 Ω, L = 50 mH, and C = 5 μF.

(a.) Find the frequency of series resonance and the resonant impedance.
(b.) At a frequency 1% higher than f_r, calculate Z and I/I_{res}.
(c.) Find the power consumed as heat in the circuit at f_r and at $(1.01)f_r$, if 20 v RMS is applied to the circuit.

S: At resonance,

$$f_r = [2\pi(LC)^{\frac{1}{2}}]^{-1} = [2\pi(50 \times 10^{-3}\text{ H})^{\frac{1}{2}}(5 \times 10^{-12}\text{ F})^{\frac{1}{2}}]^{-1}$$

$$= 3.18 \times 10^5 \text{ Hz}.$$

What is Z_r?

T: Since X_L cancels X_C at resonance, $Z_r = R = 10\ \Omega$. Now we want to find Z at $f = (1.01)f_r$.

S: O.K., that would be at

$$f = (1.01)(3.18 \times 10^5 \text{ Hz}) = 3.21 \times 10^5 \text{ Hz}.$$

At this frequency,

$$X_L = 2\pi(3.21 \times 10^5 \text{ Hz})(50 \times 10^{-3} \text{ H}) = 1.01 \times 10^5 \ \Omega,$$

and

$$X_C = [2\pi(3.21 \times 10^5 \text{ Hz})(5 \times 10^{-12} \text{ F})]^{-1} = 9.9 \times 10^4 \ \Omega.$$

T: You see, X_L and X_C are *individually* very large here, even at resonance. The frequency at which they are essentially equal is very narrowly defined, i.e., there is a "sharp" resonance. Instead of being zero, now $(X_L - X_C) = 2000 \ \Omega$.

S: Then

$$Z = [10^2 + 2000^2]^{\frac{1}{2}} \ \Omega = 2000 \ \Omega.$$

I see that barely off resonance, R is already insignificant.

T: In this circuit, yes. But R is *everything* at resonance. Since for equal voltages, $I \alpha Z^{-1}$ we immediately find that

$$I/I_{res} = (100 \ \Omega)/(2000 \ \Omega) = .005$$

This extremely abrupt decrease in current only 1% above f_r is a good demonstration of the sharpness of the resonance.

S: The power dissipated in the circuit is given by

$$P = I_{RMS}V_{RMS} \cos \Phi.$$

At *resonance*, $\cos \Phi = R/R = 1$, so $P_r = I_rV \times (1)$, with $I_r = (2.0 \text{ V})/(10 \ \Omega)$. Thus

$$P_r = (2 \text{ amps})(20 \text{ v})(1) = 40 \text{ watts.}$$

Off resonance, P will be much less because of the decreased current.

T: Not only that, but I and V will be considerable out of phase, so cos Φ will be much less than one, further reducing P. At $f = (1.01)f_r$, $Z = 2000 \ \Omega$, so

$$\cos \Phi = R/Z = (10 \ \Omega)/(2000 \ \Omega) = .005, \text{ and}$$

$$I = V/Z = (20 \text{ v})/(2000 \ \Omega) = .01 \text{ amps.}$$

Thus

$$P = IV \cos \Phi = (.01 \text{ A})(20 \text{ v})(.005) = \text{only } .001 \text{ watts!}$$

This is an important point to remember: *this type of circuit absorbs appreciable power from the applied EMF only at resonance.*

Transformers

Understanding this section should enable you to do the following:

1. *Calculate the change in a.c. voltage which occurs in a transformer of given design.*
2. *Identify the principle of impedance matching and its practical uses.*

Transformer--A device which uses mutual induction between two coils to multiply voltages. A typical example is shown in Figure 21.20 (text). The coils are generally linked by a soft iron core whose purpose it is to concentrate and "channel" the magnetic flux so that the same amount links through both coils.

Transformer Equation--If the primary coil has N_1 turns and the secondary has N_2 turns, the ratio of voltages in the coils is given by

$$V_2/V_1 = N_2/N_1.$$

This is derived in Section 21.7 in your text.

Important Observations

1. The coil receiving the *input* voltage is designated as the *primary*; the *secondary* produces the *output* voltage. From the above equation, if $N_2 > N_1$, we have a "step-up" transformer with $V_2 > V_1$; if $N_2 < N_1$ we have a "step-down" transformer with $V_2 < V_1$.
2. Since energy (power) must be conserved, and power = IV, the ratio of voltages in an *ideal* transformer must be accompanied by the *inverse* ratio of *currents*:

   $$I_1/I_2 = V_2/V_1 .$$

   This does not hold true if there are power losses in the transformer itself.
3. A useful analogy is to view the transformer as a "voltage lever." As with a mechanical lever which multiplies force with a proportional decrease in displacement, the transformer multiplies voltage at the expense of proportionally decreased current.
4. Since there is no induction effect with d.c., transformers are strictly a.c. devices.

Example 21-7. The power input to the primary of a transformer is 1 kW from a 110 v. source. The secondary contains a 16 Ω resistance across which there is observed to be a voltage of 125 v. What is the secondary current, the efficiency of the transformer, and the ratio of turns, N_2/N_1?

S: The secondary current should be

$$I_2 = V_2/R_2 = 125 \text{ v.}/16\ \Omega = 7.81 \text{ A}.$$

What is meant by the efficiency?

T: The efficiency is the ratio of power consumed in the secondary to the power delivered by the primary.

S: Then efficiency (e) is

$$e = (V_2{}^2/R_2)/(10^3\ \text{W}) = [(125\ \text{v})^2/(16\ \Omega)]/(10^3\ \text{W}) = .977,\ \text{or}\ 97.7\%$$

To find the ratio of turns, we have $N_2/N_1 = V_2/V_1$ or I_1/I_2. Now, V_2/V_1 = 125 v./110 v. = 1.14. Is this the same as I_1/I_2?

T: We can find I from $P_1 = I_1V_1 = 10^3$ W, giving $I_1 = (10^3\ \text{W})/(110\ \text{v}) = 9.09$ A. Then $I_1/I_2 = 9.09/7.81 = 1.16$.

S: This is *not* the same. What is the *real* turns ratio?

T: In a *non*-ideal transformer like this, the conservation of power has to include those losses in the transformer, so

$$I_1V_1 = I_2V_2 + P_{loss}$$

which is obviously *greater* than I_2V_2. Thus the current ratio is not exactly the reciprocal of the voltage ratio. *The latter correctly gives the turns ratio.*

QUESTIONS ON FUNDAMENTALS

1. An a.c. voltage having amplitude 50 volts is applied to a 20 Ω resistor. What is the rms current which flows in the circuit?

(a.) 2.5 A (b.) 1.77 A (c.) 708 A (d.) 62.7 A
(e.) .565 A

2. What is the rms power dissipated in the resistor in question #1?

(a.) 31.4 W (b.) 708 W (c.) .313 W (d.) 62.7 W
(e.) 125 W

3. What value of capacitance will have a capacitive reactance of 200 Ω when a 60 Hz signal is applied?

(a.) 13.3 μF (b.) 12,000 μF (c.) 83.3 μF (d.) 177 μF
(e.) 26.6 μF

4. What value of inductance will have an inductive reactance of 200 Ω when a 60 Hz signal is applied?

(a.) 13.3 μH (b.) .833 H (c.) 1.41 mH (d.) 3.33 H
(e.) .53 H

5. What is the resonance frequency for a series circuit containing a 20 μF capacitor and a 1 mH inductor?

(a.) 7.96 MHz (b.) 1.13 kHz (c.) 35.6 kHz
(d.) 7.07 kHz (e.) 870 Hz

6. If a 2 Ω resistor was connected in series with the inductor and capacitor of question #5, and a 1200 Hz voltage signal was applied to the R-L-C circuit, what would the impedance of the circuit be?

(a.) 4.83 Ω (b.) 4.11 Ω (c.) 2.2 Ω (d.) 1.7 Ω
(e.) 3.1 Ω

7. What would be the phase angle Φ between the current and voltage in the circuit of question #6?

(a.) 47.7° (b.) 60.9° (c.) 49.8° (d.) 24.6°
(e.) 65.5°

GENERAL PROBLEMS

Problem 21-1. A 100 volt (RMS), 60 Hz generator is connected to a 200 ohm resistor.

(a.) Find the power consumed in the circuit.
(b.) If a 50 μF capacitor was hooked up in series with R, find the power consumed from the same generator.

Q1: What is the effective current in the purely resistive circuit?
Q2: How is this related to the power consumed?
Q3: How does the capacitor alter the problem?
Q4: What determines the new I_{RMS}?
Q5: How is this related to power consumption?

Problem 21-2. An inductor with resistance R = 30 ohms and an unknown inductance L is connected to a 140 v (RMS) generator at a frequency of 1000 Hz. It is observed that the current lags the voltage by 64.5°.

(a.) Calculate the effective current in the circuit.
(b.) What capacitance would be necessary in the series circuit to produce resonance? What would be the resonance current?

Q1: How do I find L?
Q2: What is the relation for I_{RMS}?
Q3: What is the condition for resonance?
Q4: At resonance, what determines I_{RMS}?

Problem 21-3. A 100 v. a.c. generator is applied across a resistance in series with another circuit element. At 500 Hz, 0.735 amps flow. At 600 Hz, the current increases to 0.855 amps. Identify the unknown element and find its value and the value of the resistance.

Q1: What type of circuit element would cause Z to decrease when f increases?
Q2: What is Ohm's Law at 500 Hz?
Q3: What is Ohm's Law for 600 Hz?
Q4: Is there an easy way to solve these?

Problem 21-4. An L-R series circuit has a time constant $\tau_L = 1.5 \times 10^{-2}$ s. When a d.c. voltage of 12 v. is applied, the current is 0.3 amps. After a capacitor has been added, an effective 12 v., 60 Hz a.c. generator delivers 1.8 watts of power to the circuit. What is the capacitance that was added?

Q1: What does τ_L tell me?
Q2: What is the significance of the d.c. data?
Q3: When the capacitor is added, what happens?
Q4: What good is there in knowing the power?
Q5: I don't know Z or I or Φ.

ANSWERS TO QUESTIONS

FUNDAMENTALS:

1. (b.)
2. (d.)
3. (a.)
4. (e.)
5. (b.)
6. (c.)
7. (d.)

GENERAL PROBLEMS:

Problem 21-1.

A1: $I_{RMS} = V_{RMS}/R = 100\ v/200\ \Omega$.
A2: $P_{RMS} = I_{RMS}V_{RMS}$. The power factor is 1 for pure resistance.
A3: It introduces a reactance, causing a reduced current and a phase difference between current and applied voltage.

A4: $I_{RMS} = V_{RMS}/Z$, where $Z = [R^2 + X_C{}^2]^{\frac{1}{2}}$ and $X_C = (2\pi fC)^{-1}$.
A5: $P = I_{RMS}V_{RMS} \cos \Phi$, where $\cos \Phi = R/Z$.

Problem 21-2.

A1: $\cos \Phi = R/Z$, and $Z = [R^2 + X_L{}^2]^{\frac{1}{2}}$.
With $X_L = 2\pi fL$, you can find L.
A2: $I_{RMS} = V_{RMS}/Z$.
A3: $X_L = X_C$ at resonance. You know f = 1000 Hz, so you can find C from this.
A4: $(I_{RMS})_{res} = V_{RMS}/Z_{res}$, where $Z_{res} = R$.

Problem 21-3.

A1: Since R is frequency *independent*, Z must contain the reactance which shows this behavior. This is the *capacitive* reactance, $X_C = (2\pi fC)^{-1}$.
A2: $I = .735\ \text{amp} - V/Z = (100\ v)[R^2 + X_C{}^2]^{-\frac{1}{2}}$, where $X_C = [2\pi(500)C]^{-1}$
A3: $.855\ \text{amp} = (100\ v)[R^2 + X_C{}^2]^{-\frac{1}{2}}$, with $X_C = [2\pi(600)C]^{-1}$
A4: R and C are the two unknowns. If you square the equations, you can arrange them thus:

$$R^2 + X_{C(500)}{}^2 = (100/.735)^2\ ; \qquad R^2 + X_{C(600)}{}^2 = (100/.855)^2.$$

Subtracting the two will eliminate R and allow you to solve for C. Then substitute C in either one to find R.

Problem 21-4.

A1: The ratio L/R.
A2: The inductance isn't reactive for d.c., i.e., $X_L = 0$ at $f = 0$. So $R = (V_{d.c.})(I_{d.c.})$. Combined with τ_L, you know R and L.
A3: The impedance becomes $Z^2 = R^2 + (X_L - X_C)^2$, where X_C is unknown.
A4: There are alternate expressions for power:
$P = IV \cos \Phi = (V^2/Z)\cos \Phi$.
A5: But $\cos \Phi = R/Z$, so $P = (V^2/Z)(R/Z)$, which involves *only* the unknown Z when you're given P. Find Z, then X_C, then C.

CHAPTER 22. ELECTROMAGNETIC WAVES

BRIEF SUMMARY AND RESUME

Maxwell's Equations are introduced and discussed as a summary and extension of the laws of Gauss, Ampere, and Faraday, and as the theoretical foundation of all electromagnetism, including *electromagnetic waves.* The unique contribution of Maxwell in introducing the concept of *electric flux* and *displacement current* is discussed. Electromagnetic waves are described in terms of the interaction of *time-dependent* electric and magnetic fields. Various types of E-M waves are summarized in the E-M wave *spectrum.* The theoretical prediction of the *speed of light* is derived from Maxwell's equations, and various ways of measuring this speed are described. The energy carried by E-M waves is analyzed, and radio wave generation and detection is described.

Material from previous chapters relevant to Chapter 22 includes:

1. Gauss', Ampere's, and Faraday's Laws.
2. The concept of magnetic flux.
3. Dielectric Constant.

NEW DATA AND PHYSICAL CONSTANTS

1. Speed of Light (c).

In vacuum: $c = (\epsilon_0 \mu_0)^{-\frac{1}{2}} = 2.998 \times 10^8$ m/s.

In dielectrics: $v = c/\sqrt{K}$, where K = dielectric constant.

NEW CONCEPTS AND PRINCIPLES

Maxwell's Equations

Understanding this section should enable you to do the following:

1. *Identify the relation of Maxwell's equations to the work of Coulomb, Gauss, Ampere, and Faraday which preceded them.*
2. *Calculate the displacement current from a time-varying electric field.*

Maxwell's First Equation--This describes how the electric field is related to the charges which produce it, and is essentially the same as Gauss' Law:

$$\sum_{\text{surface}} E_\perp \Delta A = 4\pi k_e Q_{encl}.$$

Maxwell's Second Equation--This is a statement about magnetic fields analogous to Gauss' Law for electric fields, recognizing that there are no point sources or isolated "poles" of magnetism to provide the source of B in the way that charges are point sources of E. Mathematically,

$$\sum_{\text{surface}} B_\perp \Delta A = 0.$$

Maxwell's Third Equation--As derived in your text, Faraday's Induction Law ultimately is a statement saying that when a magnetic field changes in time, it produces an electric field. The electric field encircling an area A through which the magnetic flux is changing is given by

$$E = -(A/2\pi r)(\Delta B_{\perp}/\Delta t).$$

Maxwell's Fourth Equation--This is based on Ampere's Circuital Law, with the addition of Maxwell's *displacement current* to any actual charge flow linking the closed path. Mathematically,

$$\sum_{\text{closed path}} B_{\parallel}\,\Delta l = \mu_o(I + I_D)_{encl}.$$

Maxwell's Displacement Current (I_D)--If a time-varying electric field links through a closed path, it has an effect in producing a magnetic field equivalent to a current

$$I_D = \epsilon_o A(\Delta E_{\perp}/\Delta t).$$

This is called a "displacement current." Here A is the area bounded by the closed path in Ampere's Law, and $E_{\perp}$ is the component of electric field perpendicular to the plane of A.

Important Observations

1. Along with the two force equations, $F_{el} = qE$ and $F_m = qvB$, Maxwell's four equations describe *all* of electromagnetic phenomena.
2. The addition of the displacement current in the fourth equation was Maxwell's unique contribution, and *symmetrized* the dependence of electric and magnetic fields on each other. It predicted that a magnetic field could be induced by a *changing electric flux*, in the same way that an electric field can be induced by changing magnetic flux.

Example 22-1. In a cylindrical region of uniform electric field, shown in the accompanying figure, you notice that the voltage between points 2 cm apart is increasing by 2 kV per second. At a distance 5 cm from the axis of the cylinder, what magnetic field is being induced?

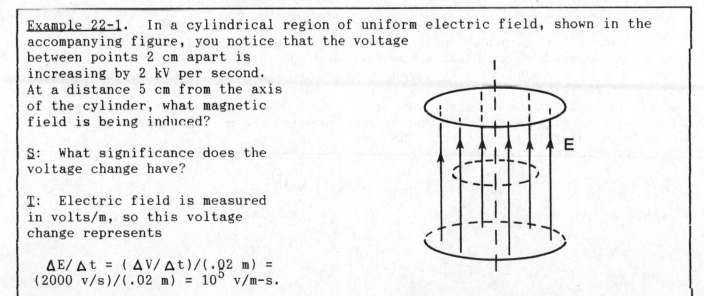

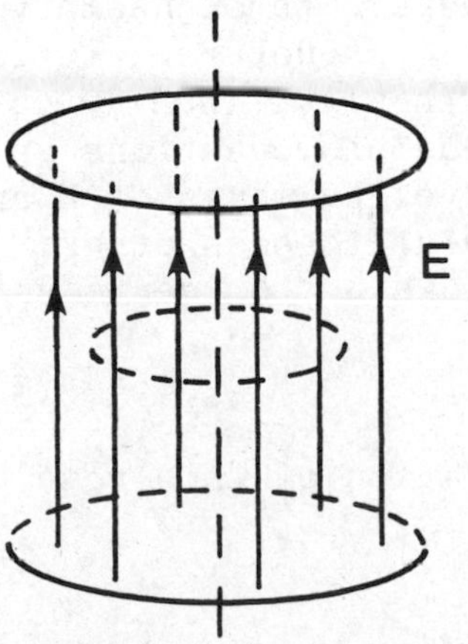

S: What significance does the voltage change have?

T: Electric field is measured in volts/m, so this voltage change represents

$$\Delta E/\Delta t = (\Delta V/\Delta t)/(.02\text{ m}) = (2000\text{ v/s})/(.02\text{ m}) = 10^5\text{ v/m-s}.$$

S: To use Maxwell's fourth equation, we need to know I and A.

T: There is no actual charge flow in this region, so I = 0. A is defined by our choice of an Amperian path around the region, which will be a circle of radius 5 cm, centered on the axis of symmetry. This is an example of a magnetic field being produced without any real current present. Of course, something is causing E to change, but is external to the region.

S: Then the source of the magnetic field has to be the displacement current,

$$I_D = \epsilon_o A(\Delta E_\perp/\Delta t) = (8.85 \times 10^{-12}\ C^2/N\text{-}m^2)(\pi)(.05\ m)^2(10^5\ V/m\text{-}s)$$

$$= 6.9 \times 10^{-9}\ C^2\text{-}v/N\text{-}m\text{-}s$$

Is this *really* 6.9×10^{-9} *amperes*?

T: It had better be! It looks messy, but the key is that a volt is a joule/coul. Then the cancellation of units is easy. Now the fourth equation of Maxwell says simply that in this case

$$B(2\pi r) = \mu_o I_D, \text{ or}$$

$$B = (4\pi \times 10^{-7}\ N/A^2)(6.9 \times 10^{-9}\ A)/\{2\pi(.05\ m)^2\} = 5.52 \times 10^{-14}\ T.$$

Again, you'd better check out these units.

The Speed of Light and the Nature of Electromagnetic Waves

Understanding this section should enable you to do the following:

1. *Calculate the speed of light in vacuum and in dielectric media from the permittivity and permeability values.*
2. *Calculate the frequency of an electromagnetic wave from its wavelength, and vice versa.*
3. *Identify which part of the E-M spectrum a wave belongs to, based on its wavelength or frequency.*

Speed of Light--The fact that time varying electric and magnetic fields can produce each other by induction leads to the ability of sinusoidal waves of the fields to propagate through empty space without the need of a material medium to support the wave structure. It is enough that the two fields continue to reinforce each other as they *continually* vary in time. In fact, Maxwell's last two equations can be combined into equations known as *wave equations*, whose solutions describe sinusoidal traveling waves of E and B, *transverse* to the direction of wave motion, and perpendicular to each other. The constants ϵ_o and μ_o appear in the equation in a way that allows us to identify the *speed* of these *electromagnetic* waves as

$$c = (\epsilon_o \mu_o)^{-\frac{1}{2}} = 2.998 \times 10^8\ m/s.$$

A simple derivation of this result is given in your text.

Important Observations

1. The value of c predicted by Maxwell's Equations agreed very well with experimental measurements of the speed of *light* which had been made up to that time. Prior to this, light was the subject of *optics*, and it was not suspected that there was any connection with electric and magnetic fields. Thus Maxwell's work predicted the existence of an extremely wide range, or *spectrum*, of E-M waves of various wavelengths, of which visible light was only a small part.
2. Light and all other E-M waves travel through empty space at the same speed, differing in frequency and wavelength according to the expression $c = \lambda f$.
3. When passing through material, the speed is reduced to

$$v = (\mu\epsilon)^{-\frac{1}{2}} ,$$

where $\epsilon = K\epsilon_o$, and $\mu = K_m\mu_o$. For most materials, $K_m = 1$, so

$$v = c/\sqrt{K} ,$$

4. Be careful! The dielectric constant K is frequency *dependent*, so these values of K are *not* the same as the d.c. values we used in treating capacitors, except for low frequency radio waves. At visible (or *optical*) frequencies, K is quite different.
5. The frequency dependence of K means that various wavelengths travel at *different speeds through materials*, a phenomenon which is called *dispersion*. We'll encounter this in succeeding chapters as the reason a prism separates colors of light, for example.

Electromagnetic Spectrum--The range of wavelengths (or frequencies) of E-M waves that have been observed is extremely large. It is known as the *electromagnetic spectrum*, and has been arbitrarily divided into various wavelength categories. The categories do not have really sharply defined wavelength limits. Roughly, this E-M spectrum is as follows: (also see Figure 22.9, text)

Type of Wave	Approximate Wavelength Range (m)
Radio	$>3 \times 10^{-1}$
Microwave	10^{-3} to 3×10^{-1}
Infrared	7×10^{-7} to 10^{-3}
Visible	4×10^{-7} to 7×10^{-7}
Ultraviolet	10^{-9} to 4×10^{-7}
X-Rays	10^{-11} to 10^{-9}
Gamma Rays	$<10^{-11}$

Important Observations

1. Notice what a very narrow band of wavelengths accounts for all of visible light. The *short* wavelength limit is the color violet, and the *long* wavelength limit is red.

2. Various wavelength regions of the spectrum are produced by quite different phenomena. For example, radio waves are produced by electronic circuits; infrared, visible, and ultraviolet are produced by heated objects and electric arcs; X-rays and Gamma rays are caused by energetic electrons colliding with matter and radioactive emissions from nuclei. All the wavelengths can also be produced by the spiraling of charged particles in magnetic fields.

<u>Example 22-2</u>. The oscillations of current in a series L-C circuit are sinusoidal in form and can be used to feed an antenna and produce E-M waves. If you have L = .70 μH and C = 3.5×10^{-12} F, what wavelength will be produced? What type of wave is this?

<u>S</u>: Is the E-M wave frequency the same as the L-C oscillation frequency?

<u>T</u>: That's the basic idea. The frequency of any kind of wave is the same as the frequency of its source.

<u>S</u>: Well, the frequency of an L-C circuit is

$$f = 1/\{2\pi(LC)^{\frac{1}{2}}\} = (2\pi)^{-1}(.70 \times 10^{-6}\ H)^{-\frac{1}{2}}(3.5 \times 10^{-12}\ F)^{-\frac{1}{2}}$$

$$= 1.02 \times 10^{8}\ Hz.$$

The corresponding wavelength is

$$\lambda = c/f = (3 \times 10^{8}\ m/s)/(1.02 \times 10^{8}\ Hz) = 2.94\ m.$$

<u>T</u>: You can see from the preceding table that this is a *radio* wave. If you check with Table 22.3 (text), you'll see it lies specifically in the FM region of the radio spectrum.

<u>Example 22-3</u>. Ethyl alcohol has a low frequency dielectric constant K = 26, while its value at optical frequencies is $K_{opt} = 1.85$.

(a.) What are the speeds of radio and light waves in alcohol?
(b.) What would be the wavelength of a 100 MHz radio wave in alcohol?

<u>S</u>: We have that $v = c/\sqrt{K}$ in any material. So light waves (optical frequencies) would have a speed

$$v_{opt} = c/(K_{opt})^{\frac{1}{2}} = (3 \times 10^{8}\ m/s)/(1.85)^{\frac{1}{2}} = 2.2 \times 10^{8}\ m/s.$$

<u>T</u>: Right. Radio waves are slowed down more,

$$v_{radio} = c/(K_{d.c.})^{\frac{1}{2}} = (3 \times 10^{8}\ m/s)/(26)^{\frac{1}{2}} = 5.88 \times 10^{7}\ m/s.$$

<u>S</u>: Is the 100 Mhz the frequency *in* the alcohol?

<u>T</u>: The wave *doesn't change frequency* from one medium to another. It changes *speed*, and hence *wavelength*.

S: Then

$$\lambda(\text{in alcohol}) = v/f = (5.88 \times 10^7 \text{ m/s})/(100 \times 10^6 \text{ Hz})$$
$$= 0.588 \text{ meters.}$$

T: In air or vacuum, this wavelength would be

$$\lambda(\text{vacuum}) = c/f = (3 \times 10^8 \text{ m/s})/(100 \times 10^6 \text{ Hz}) = 3 \text{ meters.}$$

Energy in E-M Waves: Intensity

Understanding this section should enable you to do the following:

1. *Calculate the power per area, or intensity, carried by an E-M wave from a knowledge of the amplitudes of B and E.*

E-M Wave Intensity (S)--The intensity of a wave is the average rate of energy passing through an area A perpendicular to the direction of wave travel, divided by A. For E-M waves, as derived in your text, this turns out to be

$$\langle S\rangle = \text{Energy}/(A\,\Delta t) = \tfrac{1}{2}\epsilon_o E_{max}^2 c = E_{max}B_{max}/2\mu_o = B_{max}^2 c/2\mu_o$$

Important Observations

1. The units of intensity are watts/m^2.
2. These results assume the wave is traveling in vacuum, or air.
3. As derived in the text, the relation which must hold for an E-M wave is that $E_{max} = cB_{max}$.

Example 22-4. A typical laboratory laser might emit a beam 1 mm. in diameter with 1 mW of power. What are the approximate values of the electric and magnetic field amplitudes in this beam?

S: The area of the beam would be $A = \pi d^2/4 = 7.85 \times 10^{-7}\ m^2$.
What is the intensity?

T: We have by definition

$$\langle S\rangle = \text{Power/Area} = (10^{-3} \text{ W})/(7.85 \times 10^{-7} \text{ m}^2) = 1.27 \times 10^3 \text{ W/m}^2.$$

By the way, sunlight at the top of the earth's atmosphere has an intensity of about 1.4 kW/m^2, very comparable to this little laser beam!

S: We also have

$$S = \tfrac{1}{2}\epsilon_o c E_{max}^2 \text{ , so}$$

$$E_{max}^2 = 2(1.27 \times 10^3 \text{ W/m}^2)/(3 \times 10^8 \text{ m/s})(8.85 \times 10^{-12} \text{ C}^2/\text{N-m}^2)$$

This gives

$$E_{max} = 9.78 \times 10^2 \text{ v/m.}$$

T: To find the amplitude of B, the easiest way is to use $B_{max} = E_{max}/c$. So

$$B_{max} = (9.78 \times 10^2 \text{ v/m})/(3 \times 10^8 \text{ m/s}) = 3.26 \times 10^{-6} \text{ teslas.}$$

QUESTIONS ON FUNDAMENTALS

1. Suppose an electric field of 50,000 v/m is set up in a cylindrical region of radius 10 cm. The field is made to decrease to zero at a constant rate in a time of 13.9 ns. What is the value of the effective displacement current in this region?

(a.) .32 A (b.) 1 mA (c.) 1 A (d.) 3 A (e.) 3 mA

2. What is the value of the magnetic field induced around the perimeter of the cylindrical region by the displacement current in question #1?

(a.) 2 μT (b.) 1 μT (c.) 20 μT (d.) 0.67 μT
(e.) 0.5 μT

3. The intensity of sunlight at the top of the earth's atmosphere (called the solar constant) is 1353 W/m^2. What is the value of the amplitude of the electric field in this sunlight?

(a.) 1.02×10^6 v/m (b.) 714 v/m (c.) 570 v/m
(d.) 1010 v/m (e.) 0.51×10^6 v/m

GENERAL PROBLEMS

Problem 22-1. A capacitor is made of parallel plates having an area of 1.5 m^2, placed 1 mm apart. The plates are connected in series with a 50 kΩ resistor. A 120 v. battery is suddenly switched into the circuit. What magnetic field is produced 0.2 milliseconds after the switch is closed, at a distance of 9 cm from the line joining the center of the plates?

Q1: Is there real current between the plates?
Q2: Is there a displacement current?
Q3: What is $\Delta E/\Delta t$?
Q4: What is C?
Q5: How do I find I(t)?
Q6: What is Q(t)?
Q7: What is B?

Problem 22-2. A major radio station broadcasts a signal of 50 kW of power. If we assume this is radiated isotropically (equally in all directions), what electric and magnetic field amplitudes would strike your radio set 60 miles from the station?

Q1: How does the intensity of the radio waves depend on distance?
Q2: What would the intensity be at 60 miles?
Q3: What electric field amplitude does this represent?
Q4: What magnetic field amplitude is there?

ANSWERS TO QUESTIONS

FUNDAMENTALS:

1. (c.) 2. (a.) 3. (d.)

GENERAL PROBLEMS:

Problem 22-1.

A1: No.

A2: Yes. $I_D = \epsilon_o A(\Delta E/\Delta t)$.

A3: Between the plates, $E(t) = V(t)/d = Q(t)/Cd$, so $\Delta E/\Delta t = (\Delta Q/\Delta t)/Cd = I(t)/Cd$.

A4: $C = \epsilon_o A/d$. Notice this makes $I_D = I$, the circuit current.

A5: At any instant t, 120 v. = $I(t)R + Q(t)/C$.

A6: $Q(t) = \epsilon C(1 - e^{-t/\tau_c})$, where $\tau_C = RC$.

A7: Ampere's Law gives $B(2\pi r) = \mu_o I(t)$, where this problem asks for B at r = 9 cm.

Problem 22-2.

A1: Isotropic radiation falls off in intensity as the inverse square of distance.

A2: 60 miles = 96 km. $\langle S\rangle = (50{,}000\ W)/(96 \times 10^3\ m)^2$

A3: Given by $\langle S\rangle = \frac{1}{2}\epsilon_o c E_{max}^2$.

A4: Given either by $B_{max} = E_{max}/c$ or $\langle S\rangle = cB_{max}^2/2\mu_o$.

CHAPTER 23. GEOMETRICAL OPTICS: MIRRORS AND LENSES

BRIEF SUMMARY AND RESUME

The principle of *total internal reflection* is investigated. The formula for *critical incident angle* is derived and applications are discussed. Refraction and dispersion of light by a *prism* are discussed.

Image formation by reflection from *spherical mirrors* is analyzed, and *principal ray diagrams* for locating and identifying images are described. The nature of *real* and *virtual* objects and images is discussed. The *mirror equation* which describes the relation between image and object positions and *focal length* is derived, and the *magnification* is defined.

The same concepts are applied to *thin lenses*, with the formation of images by refraction instead of reflection. The *lens* equation is derived and found to be the same as for spherical mirrors, with differences in the *sign convention* noted. Image formation by *lens combinations* are discussed. The "power" of a lens is defined in terms of its *diopter number*. Aberrations in the quality of the image are briefly discussed.

Material from previous chapters relevant to Chapter 23 includes:

1. The laws of reflection and refraction.

NEW DATA AND PHYSICAL CONSTANTS

1. Critical angle for total internal reflection (θ_c).

   Total internal reflection occurs for all incident angles greater than θ_c, given by

   $\sin \theta_c = n_2/n_1$, where n_2 must be less than n_1.

2. Diopter number.

   diopter number = 1/f , where f is in *meters*

NEW CONCEPTS AND PRINCIPLES

Total Internal Reflection

Understanding this section should enable you to do the following:

1. *Identify the circumstances under which total internal reflection can take place.*
2. *Calculate the critical incident angle for this phenomenon from the indices of refraction for the two media.*

Critical Angle for Total Internal Reflection (θ_c)--When a ray of light refracts into a medium which is optically *less* dense than the incident medium, it is refracted *away* from the *normal*, i.e., *toward the boundary* between the materials. The incident angle which produces a refracted angle of *90°* is called the *critical angle*, θ_c, since this is obviously the largest possible refraction angle. From Snell's Law,

$$\sin \theta_c = n_2/n_1 \ , \quad (n_2 < n_1).$$

For incident angles larger than θ_c, refraction is *impossible*, and *all* the light energy must be in the *reflected* ray.

Important Observations

1. Notice the similarity of the expressions for θ_c and for Brewster's angle, θ_B. The differences are:

   (a.) θ_B exists for *any* combination of n_1 and n_2, but θ_c exists *only* for $n_2 < n_1$.
   (b.) For $n_2 < n_1$, θ_c is always *greater* than θ_B. Pick some values of n_1 and n_2 and try it.

2. This internal reflection is truly 100%, better than the most highly polished metal surfaces.

Example 23-1. Will a right angle glass prism work as a 90° total reflection element under water? See the following figure.

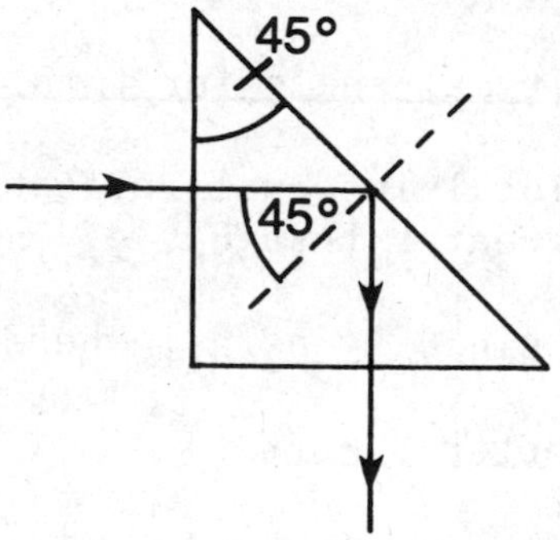

S: What glass is the prism made of?

T: Well, a common crown glass prism has n = 1.52. In air, its critical angle would be $\theta_c = \sin^{-1}(1/1.52) = 41.1°$. Since the incident angle on the glass-air boundary is 45°, it gives a *total* reflection of 90°.

S: Underwater, $\sin \theta_c = 1.33/1.52$, giving $\theta_c = 61.0°$. So this wouldn't work.

T: *Some* of the incident light would be reflected at 90°, but much would be refracted into the water at an angle given by

$$\sin \theta_r = (n_1/n_2) \sin 45° = (1.52/1.33)(.707), \text{ or}$$

$$\theta_r = 53.9°.$$

Is there a type of glass that could be used?

S: To get θ_c smaller we need a *higher* index of refraction. The largest one listed in the text for glass is 1.89 for dense flint glass. Under water, this glass would have

$$\theta_c = \sin^{-1}(1.33/1.89) = 44.7^\circ.$$

Is this close enough?

T: Yes. The angles are taken to be very precise here, and the effect of total reflection is abrupt and complete once θ_c is exceeded by the *slightest* amount.

Refraction of Light by a Prism

Understanding this section should enable you to do the following:

1. *Calculate the index of refraction of a prism from tbe prism angle and angle of minimum deviation.*
2. *Calculate the angle of minimum deviation for a given prism surrounded by various media.*

Angle of Minimum Deviation (δ_m)--Following Figure 23.9 (text), the smallest angle of deviation of light through a prism, δ_m, is given by the expression

$$\sin\tfrac{1}{2}(A + \delta_m) = n \sin(A/2),$$

where A = vertex angle of the prism and n = refractive index of the prism *relative* to the surrounding medium.

Important Observations

1. The ray direction that undergoes minimum deviation is the one which passes through the prism *symmetrically*, perpendicular to the bisector of the vertex angle A.
2. Here $n = n_2/n_1$. Of course, if the prism is in air, then $n_1 = 1.00$ and $n = n_2 = n_{prism}$.
3. This gives a practical way of finding n, since A and δ_m can be easily measured and the sines calculated.

Example 23-2. A prism with a 30° vertex is made out of flint glass (n = 1.65).

(a.) In air, what will be the minimum angle of deviation?
(b.) When immersed in a certain fluid, the minimum angle of deviation becomes 4.0°. What is the index of refraction of the fluid?

S: We have to solve for δ_m from $\sin \frac{1}{2}(A + \delta_m) = n \sin(A/2)$. How do we start?

T: You can calculate the right hand side of the equation:

$$n\sin(A/2) = (1.65)\sin(30^\circ/2) = .427$$

Thus you have

$$\sin\tfrac{1}{2}(A + \delta_m) = .427$$

S: I see. Then we can find $\frac{1}{2}(A + \delta_m) = \sin^{-1}(.427) = 25.3^\circ$. So

$$\delta_m = 50.6^\circ - 30^\circ = 20.6^\circ.$$

Why does δ_m change whe the prism is placed in the fluid?

T: We must now treat n as n_2/n_1, and the deviation at each boundary will be less since this is smaller than n in the previous case.

S: With $\delta_m = 4.0^\circ$, we have

$$\sin\tfrac{1}{2}(A + \delta_m) = \sin 17^\circ = .292,$$

and

$$\sin(A/2) = \sin 15^\circ = .259.$$

So $n = .292/.259 = 1.13$.

T: And $n_1 = n_2/n = 1.65/1.13 = 1.46$, for the fluid.

Spherical Mirrors: Image Formation by Reflection

Understanding this section should enable you to do the following:

1. *Calculate, for an object of given size and location, the image size and location formed by a spherical mirror of given focal length. Do this by principal ray diagrams and by using the mirror equation.*
2. *Identify the difference between real and virtual images.*
3. *Use the sign convention properly for the mirror equation, and interpret the meaning of algebraic sign in the results.*

The following terms which describe spherical mirrors are illustrated in Figures 23-19a, 23.19b, and 23-20.

Radius of Curvature (R)--The distance from the mirror surface to the center of the sphere of which the mirror is part.

Convex Mirror--A mirror in which the reflecting surface is facing *away* from its center of curvature.

Concave Mirror--A mirror in which the reflecting surface is facing *toward* its center of curvature.

Principal Axis--A line normal to the mirror surface at its center. This line extends through the center of curvature, C.

Focal Point--Incident rays *parallel* to and *close* to the principal axis, upon reflection either *converge toward* (concave mirror) or appear to *diverge from* (convex mirror) a certain point on the principal axis. This is the *focal point* of the mirror.

Focal Length (f)--The distance along the principal axis from the mirror to the focal point. For spherical mirrors, this is equal to one-half the radius of curvature.

$$f = R/2.$$

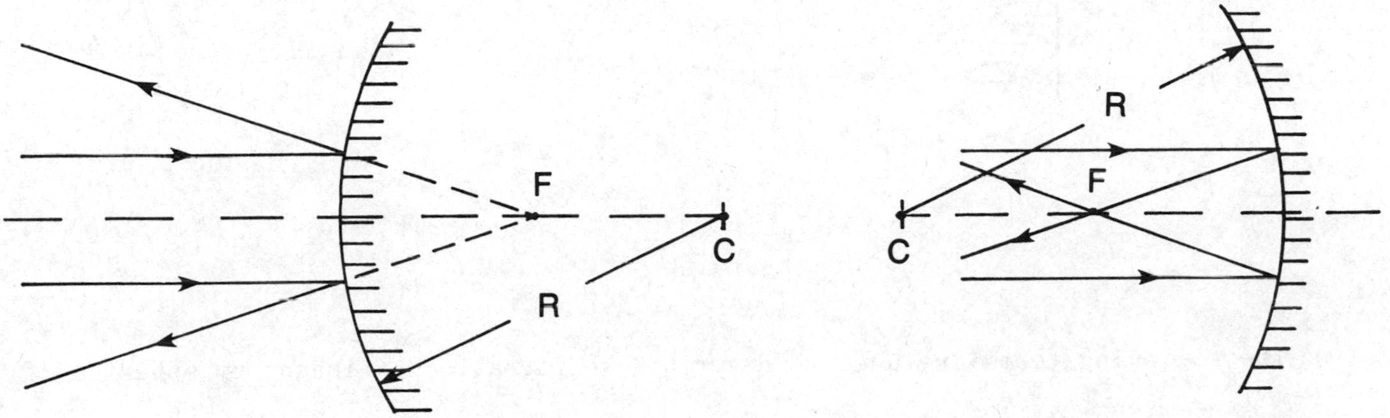

Principal Rays--We can draw the reflection of *any* ray striking the mirror by using the law of reflection. There are some rays, however, that are especially easy to do this with, and are called *principal rays*, illustrated in the figure below. They are the following:

(a.) An incident ray parallel to the principal axis. It is reflected either *through* the focal point (concave) or in a direction *away from* the focal point (convex).

(b.) An incident ray *through* (concave) or *toward* (convex) the focal point. It is reflected back parallel to the principal axis.

(c.) An incident ray *through* (concave) or *toward* (convex) the center of curvature. Since this is normal to the mirror, it is reflected back on itself.

(d.) A ray incident at the *vertex* of the mirror. It is reflected on the opposite side of the principal axis at an angle equal to its incident angle.

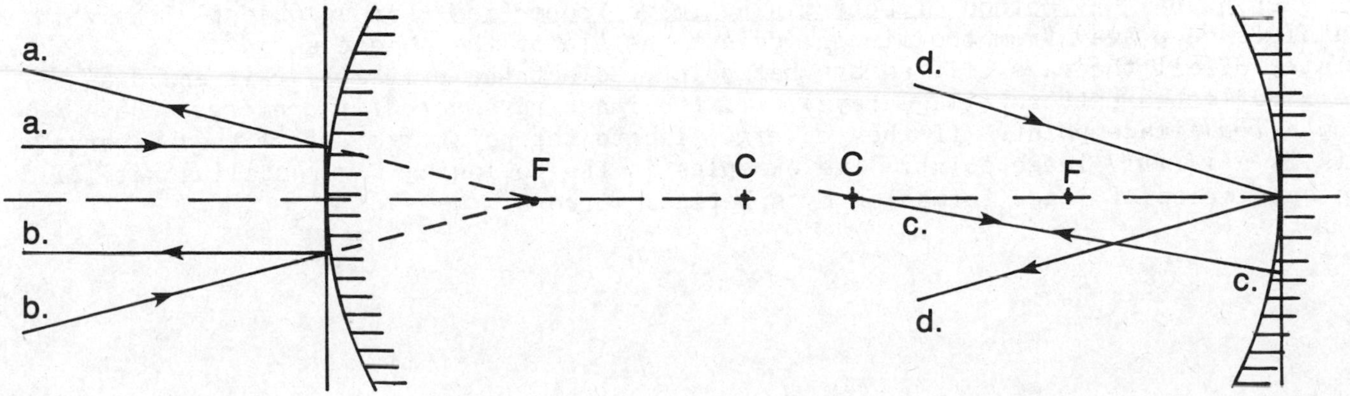

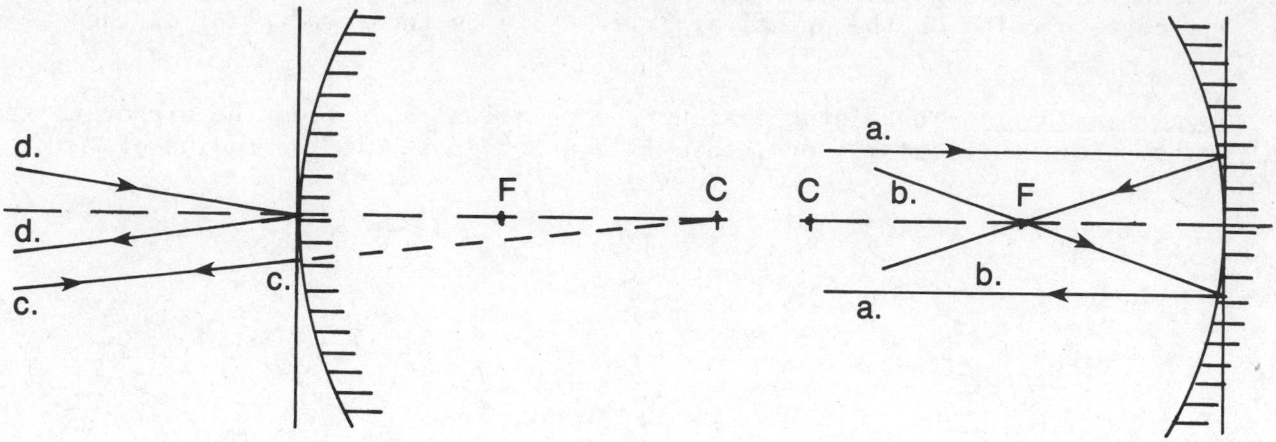

The following terms are used to describe the formation of images by spherical mirrors:

Real Object Point--A source of light rays (object rays) *diverging* in a radial pattern, and striking the reflecting surface of the mirror.

Object Distance (p)--If a line is drawn from the object point perpendicular to the principal axis, the object distance is the distance from that line to the mirror, measured along the principal axis.

Real Image Point--The point to which the reflections of the object rays *converge*. This point is in front of the mirror.

Virtual Image Point--The point from which the reflections (image rays) *appear to diverge*. This point is in back of the mirror.

Image Distance (q)--Real or virtual, this is the distance from the mirror to the image, measured along the principal axis in the same way as described for the object distance.

Principal Ray Method of Determining Image Properties--For an object of height h a distance p away from the mirror, select the tip of the object as an object point. Of all the rays emitted by that point, select the *principal rays* and draw their reflected pattern (image rays). If the image rays *actually converge*, they form a *real* image point. If they *diverge*, locate the point *from which* they diverge. This is a *virtual* image point. The examples in the following figures illustrate all the categories of image formation by spherical mirrors.

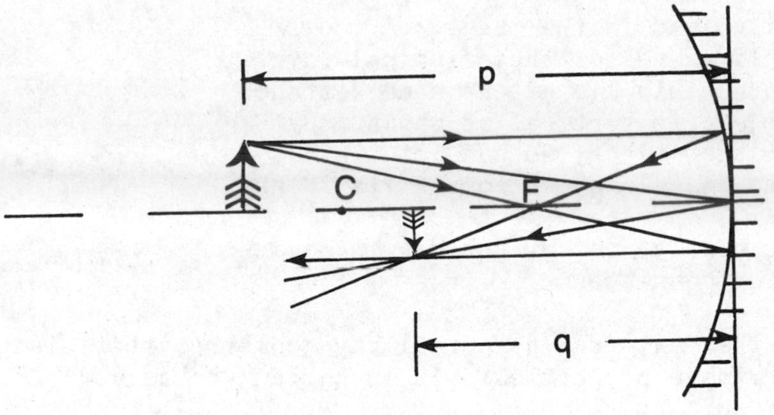

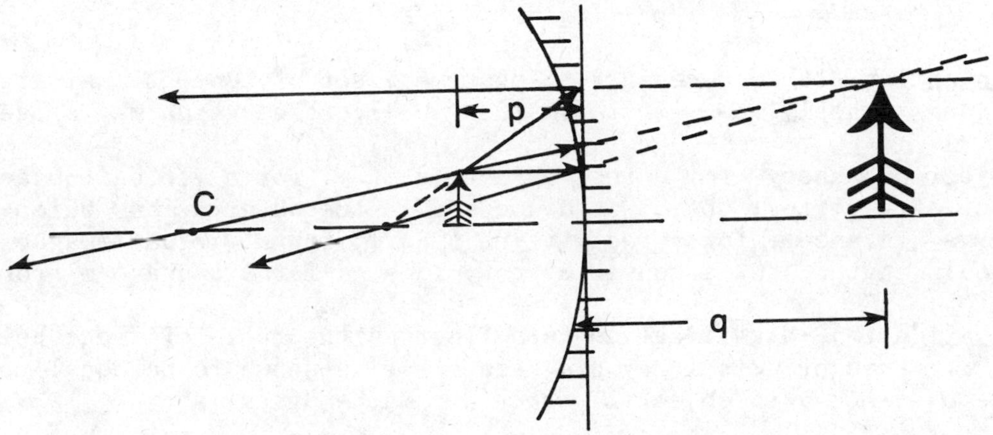

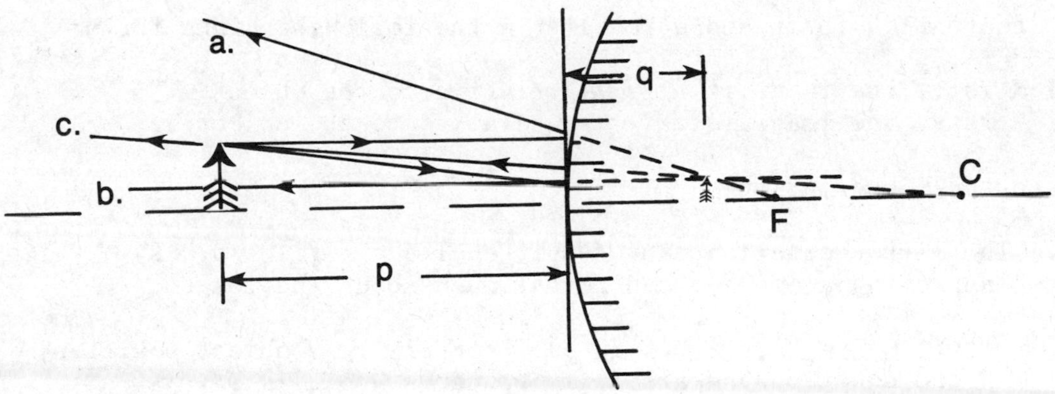

Important Observations

1. You'll notice that we have used the mirror *plane* tangent to the vertex for our principal ray constructions. The reason is that for this ideal treatment of images, we have to restrict ourselves to rays almost parallel to the axis and striking close to the vertex of the mirror. The curvature of the mirror is thus exaggerated in our drawings relative to

the positions of the rays involved in the image.

2. *Any* point on the object could serve for the principal ray diagram. All the image points would lie at the same distance from the mirror when the object is vertical as shown. By taking the *extreme* point on the object, we can immediately draw the whole extent of the image once we locate its image point.
3. Only *two* principal rays are really needed, but drawing a third one is a good check for consistency.

Mirror Equation--As derived in the text from principal ray geometry, there is an *analytical* way of finding the same image properties. It is called the *mirror equation*, and states that

$$1/p + 1/q = 1/f = 2/R.$$

Sign Convention--It is necessary to observe a set of conventional signs for p, q, and f in order to apply the mirror equation to all cases which may arise:

(a.) Object distance: for a real object, p is +; for a virtual object, p is - (The concept of a virtual object is discussed in the observations which follow.)
(b.) Image distance: for a real image, q is +; for a virtual image, q is -.
(c.) Focal length: for a convex mirror, f is -; for a concave mirror, f is +.

Magnification (M)--Magnification is defined as the ratio of image height h' to object height h. From principal ray diagrams, this is shown to be equal to the ratio of image *distance* q to object *distance* p. Mathematically,

$$M = h'/h = -q/p .$$

The sign convention for p and q results in the following signs for M:

If M is +, the image is *upright* relative to the object.
If M is -, the image is *inverted* relative to the object.

Important Observations

1. The mirror equation can be written $1/q = 1/f - 1/p$. By inspection alone, we can reveal much about the image:

   (a.) Convex mirror (f is -). For *any real* object position, the equation has all *negative* terms, so q is *always* negative, and the image *virtual*. A convex mirror receiving diverging rays can only increase the rate of divergence of them.

   (b.) Concave mirror (f is +). *For p > f*, the right hand side, and hence q, is positive. Thus for objects *outside* the focal point, a *real* image is formed. *For p < f*, the right hand side, and hence q, is *negative*. A *virtual* image is thus formed for objects *nearer* than the focal point.

2. It is possible that rays reflected from one mirror might *converge* on a second mirror. Such a ray pattern cannot be produced by a *real* object, so we characterize the object as *virtual*, with its location at the convergent point. The rays never *reach* that point, but its location describes the rate

of convergence. If the object distance in this case is treated as *negative*, the mirror equation still gives the correct image properties.

Example 23-3. An object 3 cm high is placed 40 cm in front of a convex mirror whose radius of curvature is 80 cm. Describe the position, size, and nature of the image. What if the mirror was concave?

S: What is the mirror's focal length?

T: We're given that $f = R/2 = 40$ cm. For a *convex* mirror we must use $f = -40$ cm.

S: Then

$$1/q = 1/f - 1/p = 1/(-40) - 1/40 = -2/40.$$

So

$$q = -20 \text{ cm}.$$

This means the image is *virtual*, located behind the mirror. What else?

T: The magnification is $M = -q/p = -(-20)/40 = +1/2$.

S: So the image is *1.5 cm high* and *upright*.

T: Right. The situation is like the one in Example 23.7 in your text.

S: For a concave mirror, $f = +40$ cm. Then

$$1/q = 1/40 - 1/40 = 0.$$

What's going on here?

T: This means that $q = \infty$. The reflected rays are *parallel*, since the object is at the focal plane.

Example 23-4. Take the concave mirror in the previous example and find the image properties when $p = 30$ cm and 50 cm.

S: Well, at $p = 30$ cm, $1/q = 1/40 - 1/30 = -1/120$. This gives $q = -120$ cm. The image is *virtual*. Also, $M = -(-120)/40 = +3$, so the image is *upright* and *9 cm high*.

T: Very good. This is the type of image shown in Example 23.6 in your text.

S: For $p = 50$ cm, we get $1/q = 1/40 - 1/50 = 1/200$. This gives $q = +200$ cm. This time the image is *real*, located *in front* of the mirror.

T: And $M = -(200)/40 = -5$. This means the image is again *15 cm high*, but *inverted*. This type of ray diagram is shown in Example 23.5 of your text.

Thin Lenses: Image Formation by Refraction

Understanding this section should enable you to do the following:

1. *Calculate, for an object of given size and location, the image size and location formed by a lens of given focal length.*
2. *Identify from its shape whether a lens is converging or diverging.*
3. *Use the sign convention properly for the lens equation and interpret the algebraic sign of the result.*

Much of the vocabulary for image formation by lenses is the same as for mirrors. Only slight changes are needed in some interpretations.

Thin Lens--A lens whose thickness is negligible compared to its focal length.

Principal Axis--The line through and perpendicular to the center of the lens faces.

Focal Points and Focal Length--These have the same meaning as for mirrors. The one difference is that a lens has focal points on *both sides*, since a ray can pass through the lens in either direction. For a thin lens they are symmetrically located.

Converging Lens--A lens which converges incident rays which are parallel to the principal axis to a focus at the focal point on the far side of the lens. See Figure 23.34(a.) (text).

Diverging Lens--A lens which causes incident rays parallel to the axis to diverge and *appear* to radiate from the focal point in front of the lens. See Figure 23.34(b.) (text).

Principal Rays--Again, of all the rays being refracted by the lens, some are easiest to deal with, called *principal rays*. We can treat the thin lens as a single refracting plane, with the following principal rays incident upon that plane:

(a.) A ray parallel to the principal axis refracted *through* the far focal point (convergent lens) or radiating *from* the near focal point (diverging lens).
(b.) A ray which passes through the front focal point and is refracted parallel to the axis (convergent lens).
(c.) A ray which is directed toward the far focal point and is refracted parallel to the axis (divergent lens).
(d.) A ray incident at the center of the lens, and passes through without refracting (both types). The following figures illustrate these.

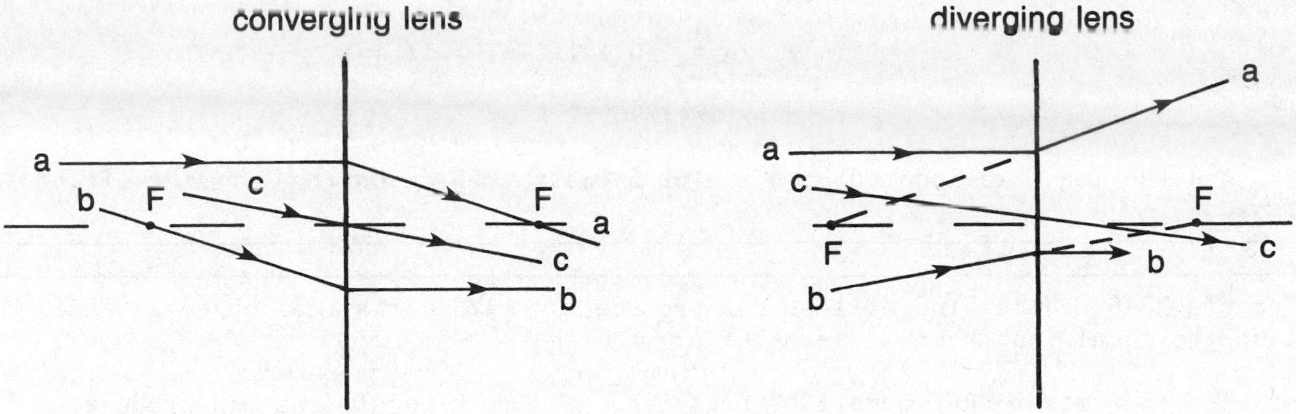

Thin Lens Equation--From the principal ray diagrams the following relation between p, q, and f can be derived: $1/p + 1/q = 1/f$.

Sign Convention--The sign convention is essentially the same as for mirrors:

1. Object distance (p). For real objects, p is (+); for virtual objects, p is (-).
2. Image distance (q). For real images, q is (+); for virtual images, q is (-).
3. Focal length (f). For converging lenses, f is (+); for diverging lens, f is (-).

Magnification (M)--The same definition and sign interpretation holds as for the case of mirrors.

Important Observations

1. The following summary might prove useful:

 Real object: diverging incident rays; located in front of lens.
 Virtual object: converging incident rays; located behind lens.
 Real image: converging refracted rays; located behind lens.
 Virtual image: diverging refracted rays; located in front of lens.

Example 23-5. How far from a converging lens of focal length 50 cm must an object be placed if the image is to be real and 4 times as large as the object?

S: We don't know either p or q, so we can't use the lens equation alone.

T: That's right. Exactly what do we know about the problem?

S: We are given f = +50 cm and M = 4. It doesn't say whether this is + or - 4.

T: Ah, but we know the image is *real*. Since p and q are both +, it means M = -4, an inverted image. From this, we find that q = 4p.

S: This gives us the second relation between p and q that we need. Substituting into the lens equation:

$$1/f = 1/p + 1/q = 1/p + 1/(4p) = 5/(4p),$$

so

$$p = (5/4)f = 62.5 \text{ cm}$$

T: And the image is located at q = 4(62.5 cm) = 1250 cm on the far side of the lens.

Example 23-6. What kind of lens can produce M = +1/3 for a real object? What must the focal length be in terms of p?

S: M = +1/3 means that q must be negative. *Either* kind of lens can produce a virtual image.

T: But how about a *reduced* virtual image? Use *all* the information you're given.

S: I don't see it yet. Let me use the lens equation to find f, with q = -p/3:

$$1/f = 1/p + 1/(-p/3) = (1 - 3)/p \ .$$

This gives f = -p/2. So this is a *diverging* lens. Is the result general?

T: Yes. The virtual image of a real object produced by a diverging lens is *always reduced.* One formed by a converging lens (which requires $p < f$, of course), is *always enlarged.*

Combinations of Lenses

Understanding this section should enable you to do the following:

1. *Locate object and image positions for a succession of lenses and determine the total magnification for given lens combinations.*
2. *Calculate the equivalent focal length for thin lenses in contact.*
3. *Identify the diopter number of a lens of given focal length.*

Two lens combinations--To find the final image produced by *two* lenses, we use the lens equation for *each lens* separately and in sequence. The image from the first lens becomes the *object* for the second lens. The *magnification* of the combination is the *product* of the magnifications produced by the individual lenses. Mathematically,

$$M_{tot} = M_1 M_2 .$$

Important Observations

1. This method can be used step by step for *any number* of lenses in combination.
2. Combinations of lenses and mirrors mixed can be treated the same way.

<u>Example 23-7</u>. A lens L_1 with $f_1 = +30$ cm is 60 cm in front of a second lens L_2 with $f_2 = -20$ cm. An object is placed 70 cm in front of L_1. Determine the location and characteristics of the final image, and its magnification.

<u>S</u>: The first lens will produce an image at q_1 given by

$$1/q_1 = 1/f_1 - 1/p_1 = 1/30 \text{ cm} - 1/70 \text{ cm} = +.019 \text{ cm}^{-1}.$$

Thus

$$q_1 = +52.5 \text{ cm, a } \textit{real} \text{ image.}$$

<u>T</u>: And $M_1 = -q_1/p_1 = -(52.5/70) = -0.75$.

So the first image is inverted and reduced to 3/4 the size of the object.

<u>S</u>: How do we find the object distance, p_2, for the second lens?

<u>T</u>: The lenses are separated by 60 cm, so the first image lies 60 - 52.5 cm = 7.5 cm *in front* of L_2. This means the rays have converged to their real image point *before* they reach L_2. They pass through that point, emerging in a *divergent* pattern, striking L_2 as rays from a *real* object point would. So $p_2 = +7.5$ cm.

<u>S</u>: Then the second image position is given by

$$1/q_2 = 1/f_2 - 1/p_2 = 1/(-20 \text{ cm}) - 1/(7.5 \text{ cm}) = -.183 \text{ cm}^{-1}.$$

Thus $q_2 = -5.45$ cm. This is a *virtual* image.

<u>T</u>: Yes, located that far *in front of* L_2. The magnification of this lens is $M_2 = M_1M_2 = (-5.45)/(+7.5) = +.727$. Thus the total magnification is

$$M_{tot} = M_1M_2 = (-.75)(.727) = -.545.$$

This shows the final image is *inverted* and *reduced* relative to the original object. A diagram of the situation is shown in the following figure:

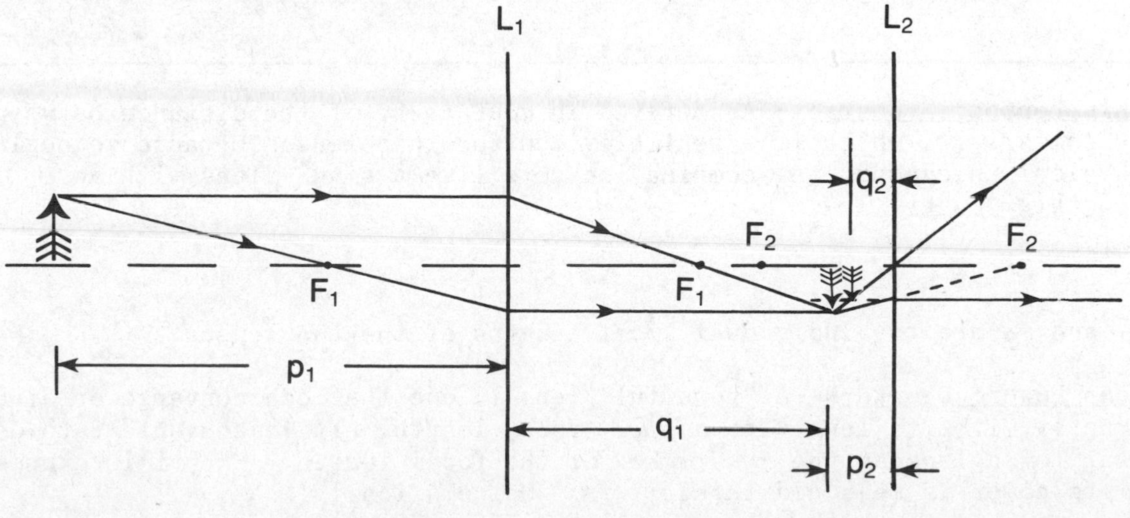

Example 23-8. Assume the same situation as in the previous example, except that L_2 is moved up to a position only 40 cm in back of L_1. Again find the final image location, characteristics, and magnification.

S: q_1 will have the same value, 52.5 cm in back of L_1, and M_1 will be the same. What has changed?

T: This position now places the first image 12.5 cm *behind* L_2. This means that the rays from L_1 are *still converging* when they strike L_2. As observed before, this pattern of rays *cannot* represent a *real* object, because real objects are *always* a source of divergent rays. We can still use the lens equation, however, if we treat this as a *virtual* object point, located at $p_2 = -12.5$ cm.

S: Then

$$1/q_2 = 1/f_2 - 1/p_2 = 1/(-20 \text{ cm}) - 1/(-12.5 \text{ cm}) = +0.3 \text{ cm}^{-1}.$$

So $q_2 = +33.3$ cm. This image is real, located in back of L_2.

T: You see that under these conditions this diverging lens still allows the rays to converge to a real image, but at a more distant point.
Also, $M_2 = -(33.3)/(-12.5) = +2.66$. So $M_{tot} = (.727)(2.66) = +1.94$. This final image is *real*, *upright*, and *enlarged*. The following figure shows this.

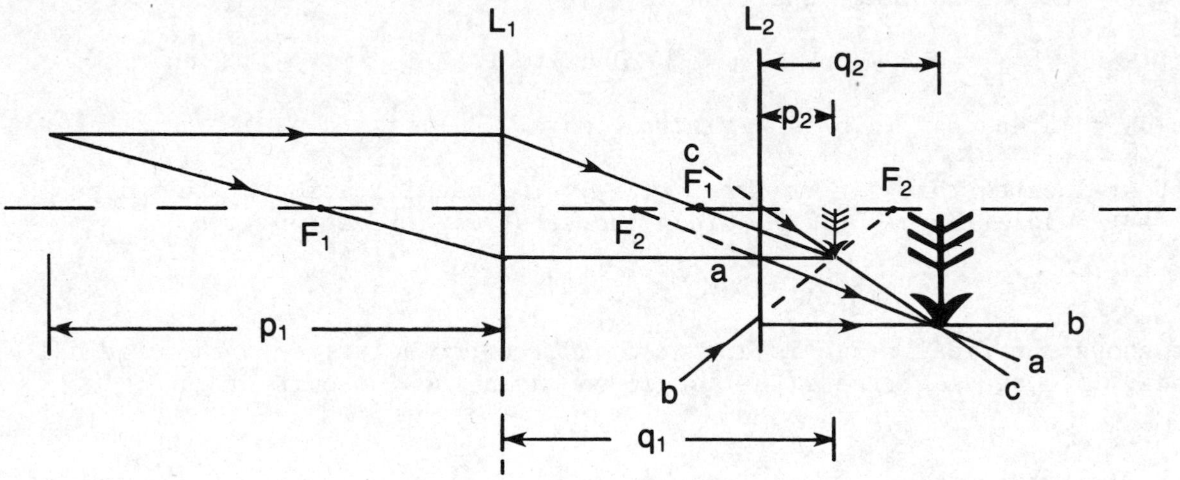

Thin Lenses in Contact--As derived in your text, if the distance between lenses becomes *zero*, which it essentially can for thin lenses because we neglect their physical thickness, the combination acts like a single lens with an *equivalent* focal length given by

$$1/f_{eq} = 1/f_1 + 1/f_2$$

where f_1 and f_2 are the individual focal lengths of the two lenses.

Lens Power: Diopters--A "powerful" lens is one that can converge or diverge rays abruptly, i.e., a lens with a *short* focal length. It is convenient to define the "power" of a lens as the *reciprocal* of the focal length. When f is expressed in meters, its power is measured in *diopters*. Mathematically,

$$\text{Diopters (D)} = 1/f \quad \text{(f in meters)}.$$

Important Observations

1. If there are more than two thin lenses in contact, the above equation can be generalized to

$$1/f_{eq} = 1/f_1 + 1/f_2 + 1/f_3 + \ldots \text{ etc.}$$

From a practical standpoint, this requires that the physical thickness of the combination is negligible compared to any of the focal lengths involved.

2. Obviously, the diopters of lenses in contact simply add algebraically to give the diopters of the equivalent lens. Remember that diverging lenses have *negative* diopters.

Example 23-9. At a certain distance from an object, a lens of +3.5 diopters forms a real image 125 cm away from it. When cemented to a second lens, the combination forms an image of the same object at a distance of 175 cm. Find the power of the second lens and the position of the object.

S: We can find the object position from the data of the first lens alone. We have

$$1/p = 1/f_1 - 1/q_1 = 1/f_1 - 1/125 \text{ cm}.$$

Now I have to find f_1 from the diopter number.

T: You could, but to get used to thinking in reciprocals (diopters), let's do it a little differently. Note that

$$1/q_1 = (1/1.25)\ m^{-1} = .8\ m^{-1}.$$

Then

$$1/p = +3.5\ m^{-1} - 0.8\ m^{-1} = 2.70\ m^{-1}, \text{ so } p = .370\ m.$$

Now use this in the equation for the combination lens.

S: O.K. We have the lens equation,

$$1/f_{eq} = 1/p + 1/q_2, \quad \text{where } 1/q_2 = 1/(1.75)\ m^{-1} = .571\ m^{-1}.$$

So

$$1/f_{eq} = 2.70\ m^{-1} + .571\ m^{-1} = +3.27\ m^{-1}.$$

This is the diopter number of the combination, right?

T: Yes, and remember, diopters add very simply:

$$D_{eq} = +3.27 = D_1 + D_2 = +3.5 + D_2.$$

So $D_2 = -.23$, showing that L_2 is a *diverging* lens, with $f_2 = 1/(-.23) = -4.35$ m.

QUESTIONS ON FUNDAMENTALS

1. What is the smallest angle of incidence for which a ray of light in crown glass will be totally reflected by striking a glass-water boundary?

(a.) at no angle of incidence will this happen
(b.) 41.1° (c.) 48.7° (d.) 53.1° (e.) 61.0°

2. What is the minimum angle of deviation for light passing through a flint glass prism whose vertex angle is 45°? Assume air surrounds the prism.

(a.) 33.8° (b.) 78.8° (c.) 39.4° (d.) 5.6°
(e.) 56.3°

3. Suppose you have a spherical mirror whose radius of curvature is 50 cm. Where will the image of an object placed 150 cm in front of the mirror be located?

(a.) 33.3 cm behind the mirror (b.) 30 cm in front
(c.) ∞ (d.) 21.4 cm in front (e.) 100 cm in front

4. Repeat question #3 if the object is placed 20 cm in front of the mirror?

(a.) 11.1 cm in front (b.) 33.3 cm behind
(c.) 100 cm behind (d.) 33.3 cm in front
(e.) 100 cm in front

5. Suppose you have a lens whose focal length is 60 cm. How far in front of the lens must an object be placed in order that a real image be formed at a distance of 100 cm from the lens?

(a.) 37.5 cm (b.) 42.9 cm (c.) 15 cm (d.) 150 cm
(e.) 160 cm

6. If you placed the lens in question #5 in contact with a lens whose focal length is -120 cm, where would an image be formed if an object was placed 60 cm. from the lens combination?

(a.) 120 cm in front of the lens (b.) 60 cm in front
(c.) 120 cm behind (d.) 40 cm behind (e.) 40 cm in front

7. What is the focal length of a lens whose diopter number is 3?

(a.) 3 meters (b.) 3 cm (c.) 33.3 cm (d.) 3.33 mm

GENERAL PROBLEMS

Problem 23-1. A diver is at the bottom of a pond. What is the *total* angle through the vertical that he must turn his head to see objects on opposite shores of the pond? If he looks at a lower angle than this, what will he see?

Q1: For an object on shore, what direction will a light ray take to reach the diver's eye?
Q2: What is the critical angle for the water-air boundary?
Q3: If he looks beyond θ_c from the vertical, what path does a ray take to reach his eye?

Problem 23-2. You have an object 5 cm high and want to use a mirror to project an enlarged 30 cm image on a screen five meters from the object. What characteristics should the mirror have and where should it be placed?

Q1: What kind of mirror does it take to project a *real* image?
Q2: What focal length must the mirror have?
Q3: What information about distances does the magnification give?
Q4: The screen is 5 meters from the object. What does that say mathematically?
Q5: What are p and q?

Problem 23-3. In order to measure the radius of curvature of a convex spherical mirror, we place a +2 diopter lens 40 cm in front of the mirror. When parallel rays of light are incident on the lens, a final image is produced 50 cm in front of the mirror. Notice that the light passes through the lens *twice*. Find the radius of curvature of the mirror.

Q1: What sort of image is formed in front of a mirror?
Q2: Let's make a sketch of this arrangement.
Q3: Where would the initial rays focus, neglecting the mirror?
Q4: What is the object distance for the mirror that corresponds to this image position?
Q5: Where does the mirror form *its* image?
Q6: What is q_3, the final image position formed by the lens?
Q7: What object position does this q_3 require?
Q8: Now what is q_2, the image distance formed by the mirror?
Q9: What is the mirror's radius of curvature?

Problem 23-4. You have a fixed lens, and can move an object along its principal axis. You find that the image is 3 times the object size when the object is at position A, and that $|M| = 5$ when the object is at point B, 25 cm *closer* to the lens. What are the characteristics of the lens?

Q1: What does the question mean by "lens characteristics"?
Q2: How many unknowns do we have?
Q3: We have two pieces of information from magnifications. What do they tell us?
Q4: What is the direct relation given between p_A and p_B?
Q5: We need two more relations, involving f in at least one.
Q6: When we combine all these, what do we find for p_A and p_B?

ANSWERS TO QUESTIONS

FUNDAMENTALS:

1. (e.)	2. (a.)	3. (b.)	4. (c.)
5. (d.)	6. (a.)	7. (c.)	

GENERAL PROBLEMS:

Problem 23-1.

A1: As shown in this figure, it is the reverse of the critical ray for total internal reflection:

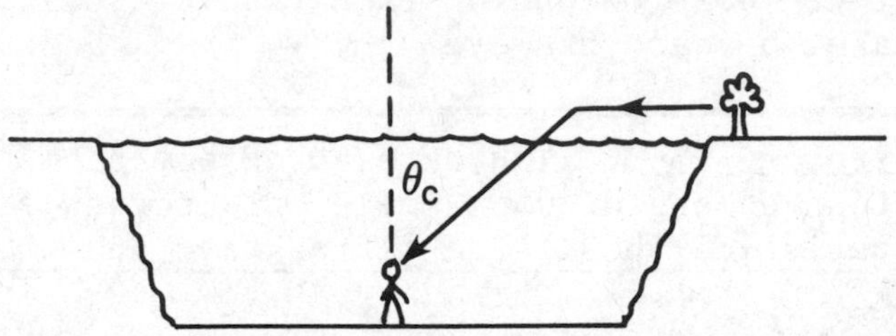

A2: $\theta_c \sin^{-1}(1/1.33)$.

A3: It has to be the reflection of some feature on the bottom of the pond.

Problem 23-2.

A1: Concave.

A2: Given by $1/f = 1/p + 1/q$, all of which are positive, and none of which are known.

A3: $|M| = 5 = q/p$, so $q = 5p$.

A4: $p + q = 5$ meters.

A5: $p = 5/6$ meters, $q = 25/6$ meters.

Problem 23-3.

A1: A *real* image.

A2: Here is the arrangement:

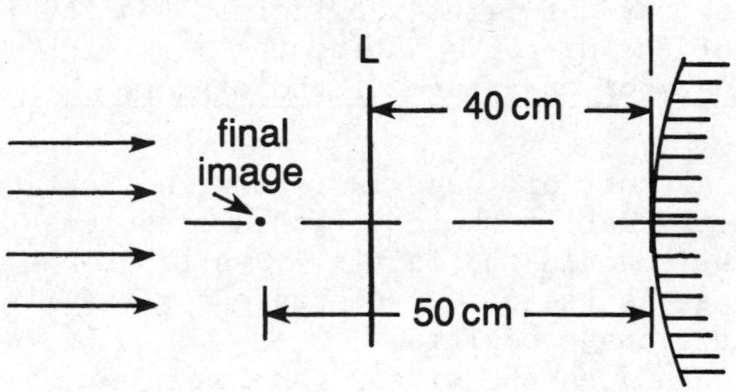

A3: At the focus of the lens. $q_1 = f_L = +50$ cm.

A4: q_1 is 10 cm *past* the mirror, so $p_2 = -10$ cm, a *virtual* object. A5: We don't know its focal length, f_M. So we have to turn to the known final image position q_3 and work back to the mirror. A6: $q_3 = +10$ cm, a *real* image formed by the second passage of the light through the lens.

A7: $1/p_3 = 1/f_L - 1/q_3 = 1/50 - 1/10$. This gives $p_3 = -12.5$ cm. The reflected rays *converge* on the lens, creating a *virtual* object *beyond* the lens.

A8: $q_2 = 12.5 \text{ cm} + 40 \text{ cm} = +52.5$ cm from the mirror.

A9: $R = 2f_M$, where $1/f_M = 1/p_2 + 1/q_2 = 1/(-10) + 1/52.5$.

Problem 23-4.

A1: Find its focal length and identify the type of lens.

A2: *Five.* p_A, p_B, q_A, q_B, and f.

A3: $q_A/p_A = 3$ and $q_B/p_B = 5$.

A4: $p_A - p_b = 25$ cm.

A5: The lens equation applies to both positions:
$1/f = 1/p_A + 1/q_A = 1/p_B + 1/q_B$.

A6: $p_A = +250$ cm and $p_B = +225$ cm.

CHAPTER 24. THE WAVE NATURE OF LIGHT: INTERFERENCE, DIFFRACTION, AND POLARIZATION

BRIEF SUMMARY AND RESUME

Huygens' Principle is introduced as a method of wavefront construction. The laws of *reflection* and *refraction* are defined, and *Snell's Law* of refraction is derived. The *diffraction* of light through apertures is described. *Interference* of light from double and multiple slits is analyzed, including the operation of the *diffraction grating*. Diffraction and interference are compared. Interference in *thin films* and *wedges* is described and applications of these phenomena are discussed. *Polarization* of light waves is investigated, and the equation for *Brewster's angle* is derived.

Material from previous chapters relevant to Chapter 24 includes:

1. The principles of superposition and interference of waves.

NEW DATA AND PHYSICAL CONSTANTS

1. Index of Refraction (n). The ratio of light speed in vacuum to that in a given material:

$$n = c/v.$$

Values for some materials are given in Table 23.1 (text).

2. Brewster's Angle (θ_B). The angle of incidence at which the reflected light is 100% polarized.

$$\theta_B = \tan^{-1}(n_2/n_1)$$

NEW CONCEPTS AND PRINCIPLES

Reflection and Refraction of Light

Understanding this section should enable you to do the following:

1. *Calculate the angles of reflection and refraction for given incident angles and for various materials.*

Huygens' Principle--Every point on a wavefront acts as the source of spherical wavelets which spread out at the wave speed in the medium. The shape of the wavefront a short time later is the surface tangent to these wavelets.

Important Observations

1. Huygens' Principle gives us a method (however tedious!) to construct the propagation of any particular wave shape. Simple examples are shown in Figures 24.2 and 24.3 of the text.
2. Notice in these and other figures that the "rays" of light represent the direction of wave travel, and are always normal to the wavefronts.

<u>Laws of Reflection</u>--At the boundary between different materials, part or all of the incident wave is reflected back into the incident medium according to the *laws of reflection*:

1. The incident ray, reflected ray, and normal to the boundary at the point of incidence are all in the same plane.
2. The angle of reflection, r, is equal to the angle of incidence, i.

<u>Refraction</u>--The changing in direction of a light wave as it travels from one medium into another, caused by a difference in *wave speed* in the two materials.

<u>Index of Refraction (n)</u>--The ratio of light speed in vacuum (c) to that in a given material (v):

$$n = c/v.$$

<u>Law of Refraction: Snell's Law</u>--As light passes across a boundary between two materials, the product of *index of refraction* times the *sine* of the *angle* the ray makes with the normal to the boundary *remains constant*. This is known as *Snell's Law*, and has the mathematical form

$$n_1 \sin \theta_1 = n_2 \sin \theta_2$$

This is illustrated in Figure 23.9 of the text.

<u>Important Observations</u>

1. The "incident ray" is *any* ray which typifies the incoming wave. What happens to that particular ray tells us what is happening to that part of the wave.
2. The angles of incidence and reflection or refraction are always measured *relative to the normal to the boundary*. Obviously, this normal is the one drawn to the *point* of incidence, where the ray strikes the boundary.
3. The use of Huygens' Principle to derive these laws is demonstrated in the text in Figures 24.3 and 24.4.
4. The index of refraction is a measure of the *optical density* of a material. The *larger* the n, the *more slowly* light moves, and hence the *more dense* the material is in its optical properties.
5. Any of these ray diagrams are *reversible* in direction. The ray follows the same path either way.
6. Snell's Law indicates that a ray will be bent *toward* the normal when entering an *optically denser* material, and *away* from the normal when entering an *optically less dense* material.
7. Generally, reflection and refraction occur *simultaneously*. That is, *part* of the energy of the light is reflected back into the incident medium at an angle equal to the incident angle, and the rest of the energy is refracted into the second medium at an angle given by Snell's Law.

<u>Example 24-1</u>. In the following figure, a light ray is incident on an air-water boundary at 30° relative to the normal. The refracted ray is then incident upon a water-crown glass boundary.

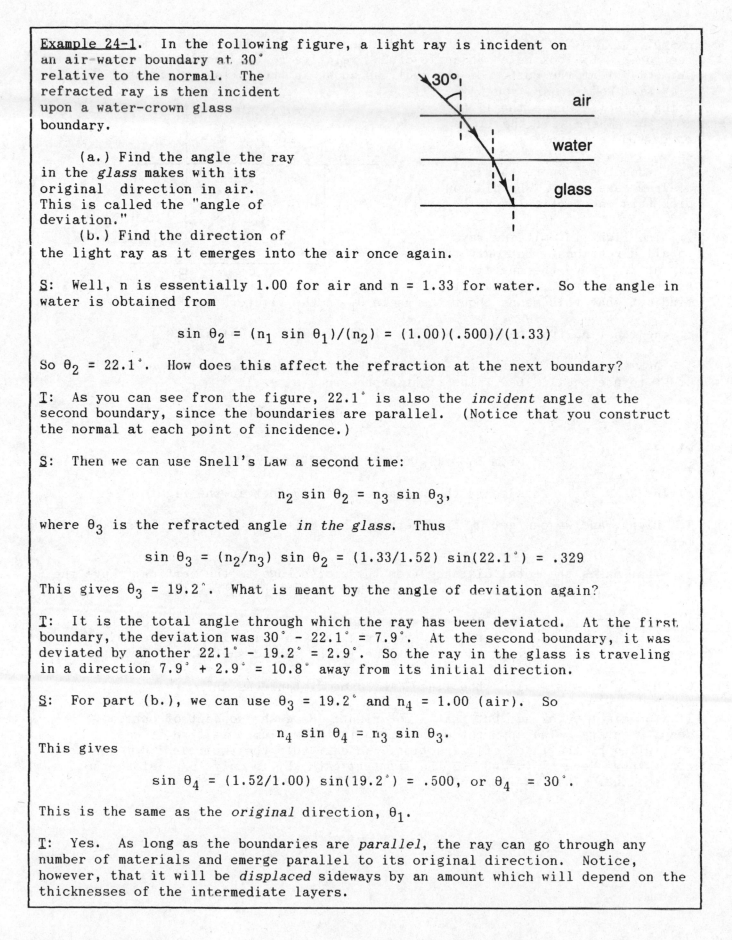

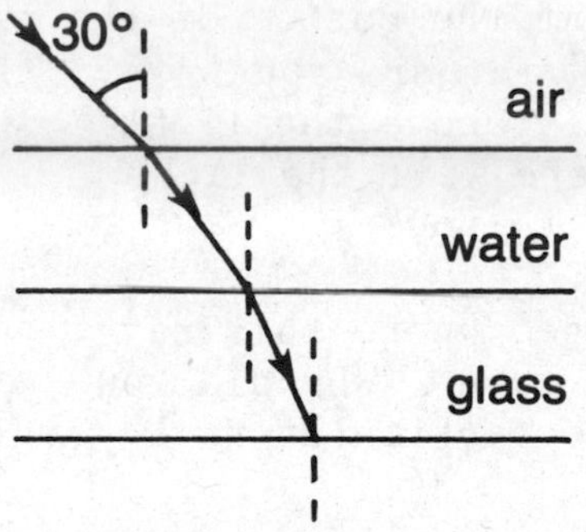

(a.) Find the angle the ray in the *glass* makes with its original direction in air. This is called the "angle of deviation."

(b.) Find the direction of the light ray as it emerges into the air once again.

<u>S</u>: Well, n is essentially 1.00 for air and n = 1.33 for water. So the angle in water is obtained from

$$\sin \theta_2 = (n_1 \sin \theta_1)/(n_2) = (1.00)(.500)/(1.33)$$

So $\theta_2 = 22.1°$. How does this affect the refraction at the next boundary?

<u>T</u>: As you can see fron the figure, 22.1° is also the *incident* angle at the second boundary, since the boundaries are parallel. (Notice that you construct the normal at each point of incidence.)

<u>S</u>: Then we can use Snell's Law a second time:

$$n_2 \sin \theta_2 = n_3 \sin \theta_3,$$

where θ_3 is the refracted angle *in the glass*. Thus

$$\sin \theta_3 = (n_2/n_3) \sin \theta_2 = (1.33/1.52) \sin(22.1°) = .329$$

This gives $\theta_3 = 19.2°$. What is meant by the angle of deviation again?

<u>T</u>: It is the total angle through which the ray has been deviated. At the first boundary, the deviation was 30° - 22.1° = 7.9°. At the second boundary, it was deviated by another 22.1° - 19.2° = 2.9°. So the ray in the glass is traveling in a direction 7.9° + 2.9° = 10.8° away from its initial direction.

<u>S</u>: For part (b.), we can use $\theta_3 = 19.2°$ and $n_4 = 1.00$ (air). So

$$n_4 \sin \theta_4 = n_3 \sin \theta_3.$$

This gives

$$\sin \theta_4 = (1.52/1.00) \sin(19.2°) = .500, \text{ or } \theta_4 = 30°.$$

This is the same as the *original* direction, θ_1.

<u>T</u>: Yes. As long as the boundaries are *parallel*, the ray can go through any number of materials and emerge parallel to its original direction. Notice, however, that it will be *displaced* sideways by an amount which will depend on the thicknesses of the intermediate layers.

Example 24-2. You are 6 feet tall and standing on the edge of a swimming pool 15 feet deep. By looking at an angle of 20° relative to the pool's surface, you can see a light at the bottom of the pool, as shown in the following figure.

(a.) How far are you from being directly above the light?

(b.) Where will the light *appear* to be to you?

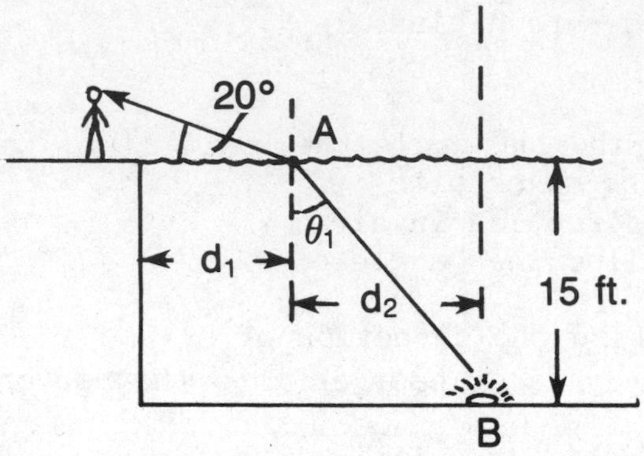

S: I don't see where this ray diagram came from. Why did you pick this one particular ray?

T: The light is emitting rays in all directions. But only the ray at 20° above the surface reaches *your* eyes. You have to find out what this means about the angle θ_1 in the figure.

S: We can find that easily. We have $(1.33)\sin\theta_1 = (1.00)\sin(20^{\circ})$.

T: *No*, that's *not* right. You have to take angles relative to the *normal* in order to use Snell's Law. The right expression is

$$(1.33)\sin\theta_1 = (1.00)\sin(70^{\circ}).$$

S: So

$$\sin\theta_1 = (1.00/1.33)\sin(70^{\circ}) = .707,$$

giving $\theta_1 = 45^{\circ}$. This means the distance d_2 is 15 feet in the figure.

T: Right, and we can get d_1 by observing that $6\text{ ft}/d_1 = \tan 20^{\circ}$. Thus $d_1 = 16.5$ feet.

S: That makes the total distance from the pool's edge to the vertical above the light $d_1 + d_2 = 31.5$ feet. I don't know how to interpret part (b.)

T: Well first, your eye interprets the position of objects to be along the straight line in the direction of the light rays entering it. So the object would *apparently* be along the dotted line in the second sketch.

S: O.K., but *where* along the line? The same distance as AB?

T: You might think so, but that's not right. Here the concept of optical density enters. The apparent path length in the medium *acts* like it was multiplied by the index of refraction. We call this the *equivalent optical path*, $L_{opt} = nL$. Here $L_{opt} = AB$, so the *apparent* path, L, is only AB/n, along the dotted line. Thus $L = AB/1.33$.

S: Then let's see, AB = $d_1/\sin 45^\circ$, so

$$L = (d_1/\sin 45^\circ)/1.00 = d_1/(1.00)\sin 70^\circ .$$

Looking at the figure, this puts the apparent position B' directly above the actual position!

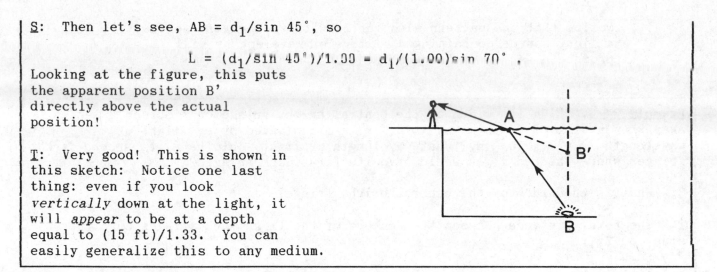

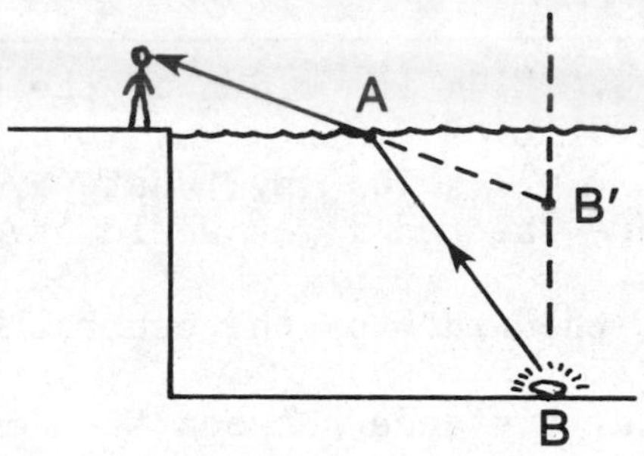

T: Very good! This is shown in this sketch: Notice one last thing: even if you look *vertically* down at the light, it will *appear* to be at a depth equal to (15 ft)/1.33. You can easily generalize this to any medium.

Diffraction of Light

Understanding this section should enable you to do the following:

1. *Calculate the separation between dark fringes in the diffraction pattern of light through slits of various widths, and the dependence of this separation on wavelength.*

Diffraction--Diffraction is the apparent bending of waves into the shadow region behind an obstacle or aperture. The effect is greatly enhanced when the dimension of the aperture is comparable to or smaller than the wavelength, and when the waves are *coherent*.

Coherence--Waves which are in phase, and essentially of a single frequency (monochromatic).

Single-slit Diffraction Pattern--A very long ($l \approx \infty$) slit of width w is illuminated by coherent, monochromatic light. Due to interference effects from wavelets in various parts of the slit, as described in the text and shown in Figures 24.23 and 24.24, a pattern of alternating bright and dark areas (called "fringes") are observed on a screen a large distance L (L >> w) from the slit. The centers of the dark fringes are located at distances y_m from the central bright fringe given by the expression

$$y_m = m(\lambda L/w), \quad \text{where } m = 1, 2, 3, \ldots \text{ etc.}$$

Important Observations

1. Diffraction is strictly a result of *wave interference*, and contradicts a *particle* model of light, which thinks of light as a stream of particles which would cast a totally *sharp-edged* shadow, with no fringes.
2. The narrower the slit, the wider is the spread of the diffraction pattern.
3. Bright fringes are due to *constructive* interference, where wavelets reach the screen *in phase*. Dark fringes are due to *destructive* interference, where wavelets reach the screen *180° out of phase*, thereby cancelling.

4. Notice that the pattern also varies with *wavelength*, so that a light source containing a mixture of wavelengths will produce a series of fringes for each color.

Example 24-3. Find the width of the central bright fringe on a screen 2 meters away when a slit of width $w = 2 \times 10^{-5}$ m is illuminated by red light of wavelength $\lambda = 7 \times 10^{-7}$ m. What wavelength of light would have its *second* dark fringes where the red light would have its first dark fringes?

S: What is the width of the central bright fringe?

T: The total distance between the centers of the first dark fringes on either side, or $2y_1$.

S: Then

$$2y_1 = 2(\lambda L/w) = 2(7 \times 10^{-7}\ \text{m})(2\ \text{m})/(2 \times 10^{-5}\ \text{m}) = .14\ \text{meters}$$

For the second part, the *second* dark fringes of the unknown wavelength would be separated by a total distance equal to the above, .14 meters.

T: Good, so

$$.14\ \text{m} = 2y_2 = 2(2\lambda L/w).$$

Notice the second factor of 2 represents m = 2. This gives

$$\lambda = (.14\ \text{m})(2 \times 10^{-5}\ \text{m})/4(2\ \text{m}) = 3.5 \times 10^{-7}\ \text{m}.$$

This is in the *ultraviolet* region of the spectrum.

Interference Phenomena

Understanding this section should enable you to do the following:

1. *Calculate the positions of bright and dark fringes in the interference pattern from a double slit, and the angles for brightness maxima for a diffraction grating.*
2. *Calculate the conditions for constructive and destructive interference for light reflecting off thin films.*

Double-slit Interference--In the following figure, coherent wavefronts are drawn emerging from 2 narrow slits a distance d apart. You can see that there are directions in which wave crests coincide, producing bright fringes on the screen a distance L away. There are other directions in which a wave trough from one slit and a crest from the other coincide, producing cancellation and hence dark fringes. This results in a *double-slit interference* pattern on

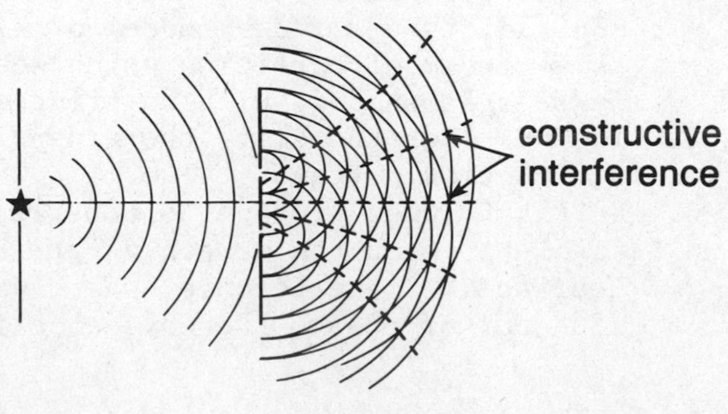

the screen. The geometry is shown in Figure 24.7 (text). The mathematical conditions for the centers of the fringes are:

<u>Bright Fringes</u>: $\sin\theta_m = m\lambda/d$, where ($m = 0, 1, 2, 3, \ldots$), or

$y_m = m\lambda L/d$ if θ_m is small.

<u>Dark Fringes</u>: $\sin\theta_m = (m + \frac{1}{2})\lambda/d$ or

$y_m = (m + \frac{1}{2})\lambda L/d$ if θ_m is small.

<u>Important Observations</u>

1. As before, the meaning of small θ_m is where $\sin\theta = \tan\theta$ to *your* desired accuracy. Tables of sine and tangent can answer this for you. As an example, sin 10° and tan 10° are equal to within 1.5%.
2. The conditions for *constructive* interference are simply that in those directions the difference in travel distance from the two slits is an *integral multiple* of λ. For *destructive* interference, this distance difference must be an *odd* multiple of $\lambda/2$.
3. As the distance between slits is *reduced*, the fringe pattern *spreads out*, and vice versa.
4. For a given slit separation, the fringe pattern is *wider* for *longer* wavelengths, and vice versa.

<u>Example 24-4</u>. Suppose you have a source of light containing wavelengths of 500 nm (green) and 640 nm (red), which you shine on a double-slit arrangement.

(a.) What must the slit separation be so that only the 1st and 2nd order bright fringes appear on the screen?
(b.) With this separation, what will be the *angular separation* between the centers of the 1st order red and green bright fringes? Are these fringes truly separated?

<u>S</u>: Why should there be a limit on the order of fringes on the screen?

<u>T</u>: Successive orders of interference, with higher values of m, occur at larger angles. If the angle requirement equals or exceeds 90°, that fringe can't appear on the screen. This is like saying the distance y_m would have to be infinite. But notice, you can't use the small angle formulas when talking about large angles.

<u>S</u>: Then we want θ_3 to be ≥90°. For which wavelength?

<u>T</u>: The longer wavelength will produce fringes at wider angles, so if you require that $\theta_3 = 90°$ for the *green* line, you can be sure that $\theta_3 > 90°$ for the *red*.

<u>S</u>: So we want

$$\sin\theta_3 = 1 = 3\lambda/d = 3(500 \text{ nm})/d.$$

This gives us d = 1500 nm. In part (b.), what is the "angular separation" being asked for?

T: It's just the difference in θ_1 for the two wavelengths.

S: Well, for the red light,

$$\sin\theta_1 = \lambda/d = (640\text{ nm})/(1500\text{ nm}) = .427$$

This gives

$$\theta_1 = 25.3^\circ.$$

T: And for the green fringe, $\sin\theta_1 = (500\text{ nm})/(1500\text{ nm}) = .333$. This gives $\theta_1 = 19.5^\circ$. Thus the *centers* of the fringes are separated in angle by $25.3^\circ - 19.5^\circ = 5.8^\circ$. One question remains: does this separation really separate the fringes? What are the *widths* of the fringes?

S: We could calculate the angles at which the 1st and 2nd *dark* fringes occur. They bracket the first bright fringe.

T: Exactly. Let's *do* it.

S: The first dark fringe for red occurs for m = 0, at the angle given by

$$\sin\theta_1 = (0 + \tfrac{1}{2})\lambda/d = \tfrac{1}{2}(640\text{ nm})(1500\text{ nm}) = .214.$$

So $\theta_1 = \underline{12.3^\circ}$. The second dark fringe for red is where

$$\sin\theta_2 = (1 + \tfrac{1}{2})\lambda/d = .664, \text{ giving } \theta_2 = \underline{41.6^\circ}.$$

T: You could do the same for the green wavelength, but it's clear that the fringes are *very wide* compared to the separation of their centers. There is a lot of overlap.

Diffraction Grating--When the number of parallel slits is increased, the intensity becomes more concentrated in increasingly narrow bright fringes. Utimately, when there are thousands of slits per cm, the arrangement is known as a *diffraction grating*, and the directions in which there is constructive interference are very sharply defined by the *grating equation*:

$$\sin\theta_m = m(\lambda/d), \text{ with } m = 0, 1, 2, 3, \ldots$$

Important Observations

1. The grating equation is the same as for the centers of bright fringes from a double slit. The difference is that with the grating the widths of these fringes are *very* narrow and you can achieve angular separation of wavelengths that are nearly equal in value. This is called high *angular resolution*, which is what was missing in Example 24-4.
2. d is the separation between *adjacent* slits or grooves in the grating.

<u>Example 24-5</u>. The fringe width is so narrow in a diffraction grating with 5000 grooves per cm that the 589.0 and 589.6 nm yellow wavelengths from a sodium vapor lamp can be resolved in both 1st and 2nd order. What are the angular separations in both cases?

<u>S</u>: The 1st order angles would be

$$\sin \theta_1 = (589.0 \text{ nm})/d \quad \text{and } (589.6 \text{ nm})/d$$

What is d?

<u>T</u>: With 5000 lines/cm, or 5.000×10^5/m, the separation is

$$d = 1/(5.000 \times 10^5 \text{ m}^{-1}) = 2.000 \times 10^{-6} \text{ m.}$$

<u>S</u>: Then

$$\sin \theta_1 = (589.0 \text{ nm})/(2000 \text{ nm}) \quad \text{and } (589.6 \text{ nm})/(2000 \text{ nm}) ,$$

giving $\theta_1 = 17.13^\circ$ and 17.15°.

This is an angular separation of .02°.

<u>T</u>: For second order, $\sin \theta_2 = 2 \sin \theta_1$, so $\theta_2 = 36.09^\circ$ and 36.13°, a separation of .04°.

<u>Example 24-6</u>. Using the grating in Example 24-5, you observe a blue line at θ = 58.43°. What is the wavelength and the order of this line?

<u>S</u>: Since $\sin \theta_m = m\lambda/d$, and m and λ are both unknown, we don't have enough information.

<u>T</u>: In a way, you're right. But "blue" restricts the wavelength possibilities to somewhere in the range of 440-500 nm. Let's see what we can do with this.

$$\sin 58.43^\circ = .8520 = m\lambda/(2 \times 10^{-6} \text{ m}),$$

so

$$m\lambda = 1.874 \times 10^{-5} \text{ m} = 18,740 \text{ nm.}$$

<u>S</u>: Obviously, m = 1 or 2 give wavelengths that are infrared. m = 3 gives a red line. m = 4 gives λ = 18,740/4 = 468.5 nm, which is in the blue.

<u>T</u>: And m = 5 would give λ = 3748 nm. This is ultraviolet, so m = 4 is the only correct answer.

<u>Thin Film Interference</u>--When light shines at nearly normal incidence on a thin film of a substance surrounded by another material, the rays reflected by the top and the bottom boundaries interfere constructively or destructively, depending on:

1. How much phase difference occurs because of the extra distance traveled by the ray that goes through the film, reflecting from the bottom surface. This phase change is the ratio of the *round trip optical path* in the film to the vacuum wavelength. If t is the film's thickness, this is 2 (nt)/λ .
2. The 180° phase change of the ray which reflects from the more optically dense region. If n_{film} is *less* than n of the

surrounding material, the *bottom* ray will undergo this 1/2 cycle phase change. If n_{film} is *greater* than the surroundings, the *top* reflected ray has this change. Either way, there is a difference of 180° in phase between the two rays because of this.

Mathematically, this is all summarized by the following:

Constructive Interference: $2nt = (m + \frac{1}{2})\lambda$.

Destructive Interference: $2nt = m\lambda$.

Important Observations

1. These equations assume that the film is surrounded on *both* sides by material either more optically dense or less. If the film has denser material on one side and less dense material on the other, *both* rays undergo either a 1/2 cycle phase change or no change. This would result in the above equations being interchanged.
2. Out of "white" light, some wavelengths will meet the requirements for constructive interference and will be most intense in the reflected light, while others will be cancelled and not be reflected.

Example 24-7. A thin film of alcohol (n = 1.35) is sandwiched between two flat glass plates (n = 1.65). When illuminated by "white" light at normal incidence, it is found that the visible wavelengths of 412 nm and 577 nm were most intense in the reflected beam. What is the minimum thickness of the film?

S: Do I have to worry about the glass plates?

T: No. Both rays from the top and bottom of the film pass through the glass identically, so no net phase difference is introduced by the glass.

S: For constructive interference, we have $2\ nt = (m + \frac{1}{2})\lambda$. What's m?

T: That's to be found out. Since t is fixed, m must be different from each of the wavelengths.

S: Yeah, I see that. We have the two equations

$$2(1.35)t = (m + \tfrac{1}{2})(412\ \text{nm}),\ \text{and}$$

$$2(1.35)t = (m' + \tfrac{1}{2})(577\ \text{nm}).$$

T: The easiest thing to do is to take the ratio of these:

$$1 = (m + \tfrac{1}{2})(412\ \text{nm})/(m' + \tfrac{1}{2})(577\ \text{nm})\ ,$$

This gives us

$$(m' + \tfrac{1}{2})577/412 = m + \tfrac{1}{2}, \quad \text{or } 1.4m' + .2 = m.$$

S: How do you find both m and m' from *one* equation?

T: You must use the fact that they are *integers*. Let's multiply the last equation by 5 to get integral coefficients:

$$7m' + 1 = 5m.$$

Now make a table:

m'	1	2	3	4	5	6	7	etc.
5m	8	15	22	29	36	43	50	etc.

Only two of the values of m' shown have an *integral* solution for m, namely m' = 2 and 7, although you could find an infinite number of values if you went through all the whole numbers.

S: I see why the problem asks for the *minimum* thickness. That's obtained from the *smallest* solutions for m and m'.

T: That's right. The snallest numbers that fit are m' = 2 and m = 3. We can get t from either one:

$$t = (7/2)(412 \text{ nm})/2(1.35) = (5/2)(577 \text{ nm})/2(1.35) = 534 \text{ nm}.$$

You should verify that m = 1 or 4 give wavelengths outside the visible spectrum.

Example 24-8. If 100 nm of magnesium fluoride (n = 1.38) is coated on a piece of glass (n = 1.52), find the phase difference between the light reflected from the top and bottom of the film for normally incident green light when λ = 540 nm.

S: The bottom reflection travels an effective distance of 2(1.38)(100 nm) further than the top reflection. What phase difference does this amount to?

T: One whole cycle of phase is equal to a distance λ . What part of λ is the distance you have there?

S: The fraction is 2(1.38)(100 nm)/(540 nm) = .511. This is just over 1/2 of a cycle.

T: We can express this in degrees, since 360° = 1 cycle. You have .511(360°) = 184° of phase difference from the path length. Is this the *total* phase shift?

S: The top reflection undergoes 180° shift in phase when it encounters the denser MgF_2. The bottom reflection seems to do likewise, since the glass has higher n than MgF_2.

T: That's correct, so the phase changes due to reflection off of optically denser materials cancel out. *The total phase change in this case is 184°.* The bottom ray lags behind the top one by this much, so they almost completely cancel.

Polarization of Light

Understanding this section should enable you to do the following:

1. *Know how light can become polarized and how to test for it.*
2. *Calculate the intensity of polarized light passing through a polarizer whose polarizing axis is at an angle θ relative to the light's plane of polarization.*
3. *Calculate the angle at which light becomes totally polarized upon reflection from various materials.*

Unpolarized Light--A light wave in which the electric field oscillations are distributed randomly in directions normal to the direction of wave travel.

Linearly (or Plane) Polarized Light--A light wave in which the electric field oscillations are all oriented in *one plane* normal to the direction of wave travel.

Polarizer--Any device which produces polarized light from unpolarized light. This can be done by selectively absorbing all oscillations of the electric field along one direction, as in Polaroid material, or by reflecting field oscillations only in one direction, as in Nicol prisms.

Analyzer--A polarizer used to determine the plane of polarization of light incident upon it. The intensity of polarized light passing through an analyzer depends on the angle θ between the polarizing axis of the analyzer and the plane of polarization of the light. Mathematically,

$$I_{transmitted} = I_{incident}\cos^2\theta.$$

Important Observations

1. Even a *perfect* polarizer will transmit only 50% of the intensity of *unpolarized* light incident upon it, since one of the two possible components of transverse E-field oscillations is being eliminated. Actually, the transmission of polaroid plates is about 40% in the visible spectrum.
2. The polarizability of light is proof of its *transverse* wave nature (E and B oscillating transverse to the direction of propagation), since *longitudinal* waves cannot be polarized.

Example 24-9. Three "perfect" polarizing plates are stacked one on top of another. The first and the third are crossed, i.e., they have their polarizing axes perpendicular. The middle plate has its axis 70° relative to the first. What fraction of incident unpolarized light will the stack transmit?

S: You should get *no* transmission through crossed polarizers. What effect can the middle polarizer have on that?

T: Plate #2 essentially causes the plane of polarization of the light that it transmits to be rotated 70°, so that it no longer is perpendicular to #3.

S: Well, the 1st plate will cut 50% of the unpolarized incident light intensity, so the amount it lets through will be $I_1 = \frac{1}{2}I_o$. Then this is incident on plate #2, which will transmit an intensity $I_2 = I_1 \cos^2(70°) = (.117)I_1 = (.0585)I_o$.

T: And finally, this intensity is incident on #3, which is rotated 20° relative to #2. So the final transmitted intensity is:

$$I_3 = I_2 \cos^2(20°) = I_2(.883) = (.0517)I_o.$$

About 5% of the original intensity gets through.

Polarization by Reflection--As shown in Figure 24.34 (text), light becomes *partially plane polarized* upon reflection from a dielectric. This happens except at normal incidence. The E-field oscillations are of larger amplitude parallel to the plane of the reflecting surface.

Total Polarization: Brewster's Angle (θ_B)--At the incident angle for which the reflected and refracted rays are *perpendicular*, this polarization becomes total. This incident angle is known as Brewster's angle. The derivation in your text shows it to be given by

$$\tan \theta_B = n_2/n_1,$$

where

n_1 = refractive index for the incident medium

n_2 = refractive index for the refracting medium.

Example 24-10. Unpolarized light passing through an unknown liquid is incident upon a flint glass plate (n = 1.65). When the angle of the *refracted* ray in the glass os 42.3°, you notice the *reflected* light is completely polarized. What is the index of refraction of the liquid?

S: Without n_2, I can't calculate the incident angle, and without the incident angle I can't determine n_2.

T: *Think* a little before you give up! For complete polarization there is a definite relation between the directions of the reflected and refracted rays.

S: Let's see. The rays must be perpendicular, so $\theta_{reflect} + \theta_{refract} = 90°$. This gives me $\theta_{reflect} = 47.7°$. Is this equal to Brewster's angle?

T: Sure, because the angle of reflection and angle of incidence are equal (Law of Reflection). Thus $\theta_B = 47.7°$, and $\tan \theta_B = 1.099 = n_2/n_1$. So $n_1 = n_2/1.099 = 1.65/1.099 = 1.50$. The liquid could possibly be benzene.

QUESTIONS ON FUNDAMENTALS

1. If a light ray in water is incident at an angle of 20° relative to the normal to a boundary with flint glass, what is the angle of the refracted ray?

   (a.) 25.3° (b.) 15.9° (c.) 64.7° (d.) 23.1°
   (e.) 40°

2. How fast does light travel in flint glass?

   (a.) .60c (b.) 1.66c (c.) c (d.) .66c

3. If light of 600 nm wavelength in vacuum passes into water, what is its wavelength in the water?

   (a.) 800 nm (b.) 600 nm (c.) 400 nm (d.) 900 nm
   (e.) 450 nm

4. If a double slit has separation between slits of 2 μmeters, what is the angle at which you would observe the second dark fringe in the interference pattern produced by light of wavelength 600 nm?

   (a.) 36.9° (b.) 64.2° (c.) 26.7° (d.) 8.6°
   (e.) 17.5°

5. Light whose vacuum wavelength = 500 nm strikes a thin film of water 4500 nm thick. What is the effective optical thickness of this film, in terms of the number of wavelengths in water which fit between the two boundaries of the film?

   (a.) 9 wavelengths (b.) 12 wavelengths
   (c.) 6.75 wavelengths (d.) 18 wavelengths
   (e.) 13.5 wavelengths

6. At what angle of incidence for light in air striking the surface of a diamond would the reflected ray be 100% polarized?

   (a.) 67.5° (b.) there is no angle for diamond at which this would happen
   (c.) 22.5° (d.) 45° (e.) 24.4°

GENERAL PROBLEMS

Problem 24-1. In the following figure, a rectangular glass block ($n = 1.52$) is surrounded by water ($n = 1.33$). A light ray strikes the top face of the block at an angle θ_1. The refracted ray strikes the vertical face at Brewster's angle, so that the ray reflected back into the glass is 100% polarized.

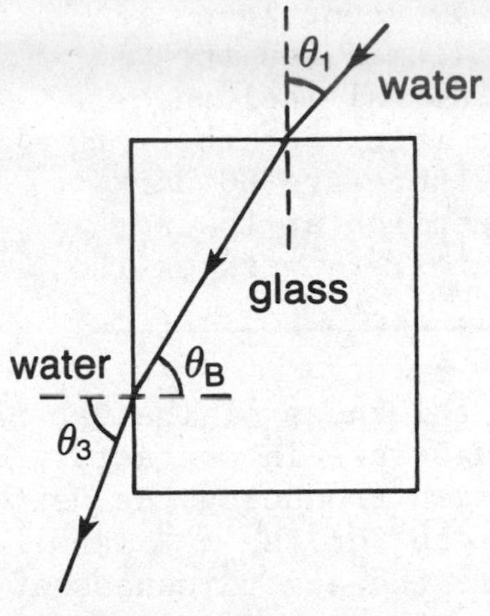

(a.) What must the incident angle, θ_1, be?

(b.) At what angle θ_3 does the refracted ray enter the water at the vertical face?

Q1: What is Brewster's angle for the glass-water boundary?
Q2: How is this related to the refraction angle, θ_2, at the top face?
Q3: What expression gives θ_1?
Q4: How do I find θ_3?

Problem 24-2. White light is incident on a diffraction grating with 9000 grooves per cm. The 1st order diffracted light falls on a glass plate ($n = 1.60$) parallel to the grating, as shown in the figure. How wide is the angular dispersion of the visible spectrum (400 nm to 700 nm) in the glass?

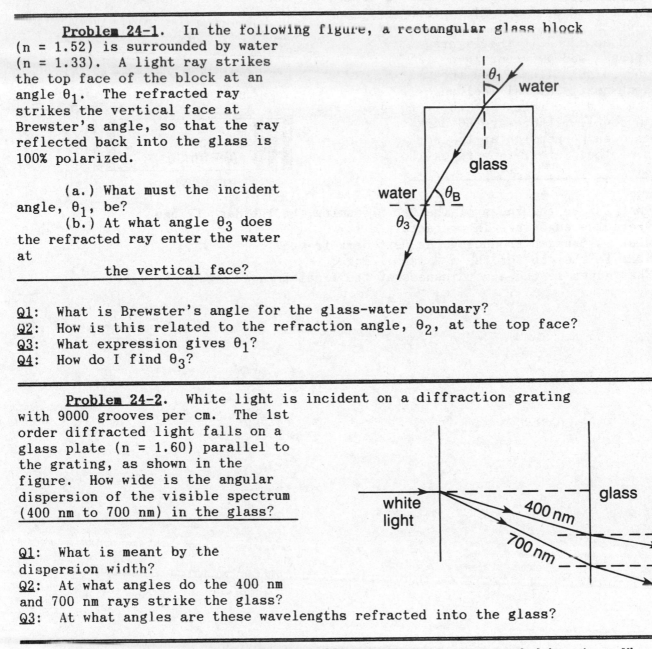

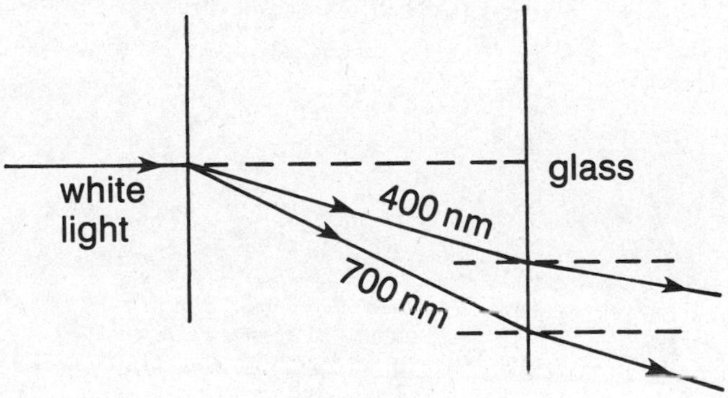

Q1: What is meant by the dispersion width?
Q2: At what angles do the 400 nm and 700 nm rays strike the glass?
Q3: At what angles are these wavelengths refracted into the glass?

Problem 24-3. A glass block, 1,000 mm thick, is surrounded by air. When red light, wavelength 600.0 nm, is incident normal to its surfaces, constructive interference between the top and bottom reflected rays is observed. How much temperature change would it take to change this to destructive interference? Take the coefficient of thermal expansion of the glass to be $\alpha = 9 \times 10^{-6}$ /C°, and $n = 1.5$.

Q1: What is the condition for constructive interference at the original temperature?
Q2: What must happen to change from a condition of constructive to destructive interference?
Q3: What does this mean mathematically?
Q4: What determines ΔT?

Problem 24-4. A wedge of air is formed by placing a thin hair between the edges of two flat glass plates which are in contact at the other end as shown in the figure. Light from a sodium vapor lamp (λ = 589.3 nm) illuminates the plates at nearly normal incidence. Starting with the fringe at the ends in contact, there are 66 dark fringes seen perpendicular to the length of the plates. How thick is the hair?

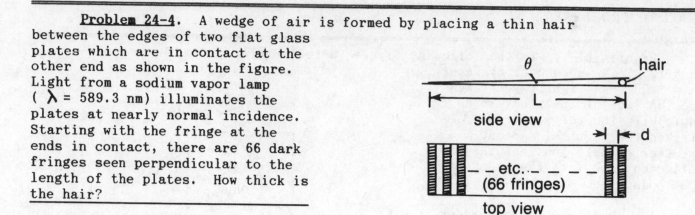

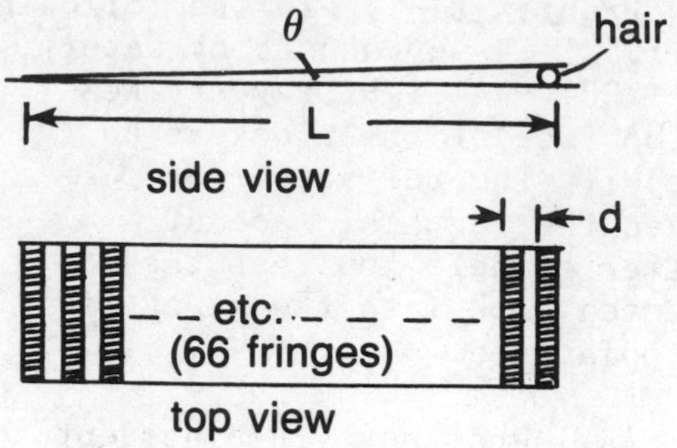

Q1: What is the thickness of the gap producing the 1st dark fringe where the edges are in contact?

Q2: What is the gap producing the next dark fringe?

Q3: What is the gap for the mth dark fringe?

Q4: What gives me the air thickness at the final fringe?

ANSWERS TO QUESTIONS

FUNDAMENTALS:

1. (b.) 2. (a.) 3. (e.) 4. (c.)
5. (b.) 6. (a.)

GENERAL PROBLEMS:

Problem 24-1.
A1: $\tan \theta_B = n_2/n_1 = 1.33/1.52$
A2: $\theta_B + \theta_2 = 90^\circ$
A3: Snell's law: $n_1 \sin \theta_1 = n_2 \sin \theta_2$ with $n_1 = 1.33$ and $n_2 = 1.52$.
A4: Either Snell's Law, $n_2 \sin \theta_B = n_1 \sin \theta_3$, or the fact that $\theta_B + \theta_3 = 90^\circ$.
Check them both.

Problem 24-2.
A1: The angular separation between the extreme wavelengths in the visible spectrum.
A2: The same as their angles of emergence from the grating, given by $\sin \theta = \lambda/d$.
A3: At angles given by Snell's Law for their incident angles: $n_1 \sin \theta_1 = n_2 \sin \theta_2$, where $n_1 = 1.00$, $n_2 = 1.60$. The difference between these refracted angles is the angular dispersion asked for.

Problem 24-3.
A1: $(m + \frac{1}{2}) \lambda = 2nt$
A2: The block must either increase or decrease in thickness so that the *round trip* optical path increases or decreases by $\lambda/2$.
A3: That $2n(\Delta t) = \lambda/2$.
A4: $\Delta t/t = \alpha \Delta T$.

Problem 24-4.
A1: Zero. The destructive interference is from the reflective phase change at the bottom plate.
A2: Since $n = 1.00$ for the air gap, $2t = \lambda$.
A3: $2t = m\lambda$.
A4: The above expression, with $m = 66$.

CHAPTER 25. OPTICAL INSTRUMENTS

BRIEF SUMMARY AND RESUME

Cameras are discussed as a straightforward application of the lens equation. The dependence of *image brightness, depth of field,* and *exposure* on *f-number* are also discussed. The *human eye* is compared to the camera, and its physical characteristics and common defects are investigated. The performances of *simple magnifiers*, *compound microscopes*, and *telescopes* are analyzed in terms of their *magnifying* and *resolving powers.* Specialized microscopes are described briefly. The principles of operation of *Michelson* and *Fabry-Perot interferometers* are analyzed.

Material from previous chapters relevant to Chapter 25 includes:

1. Focal lengths, types of objects and images, magnification, and the lens and mirror equations.
2. Lenses in combination.
3. Optical path length and the interference of light.

NEW DATA AND PHYSICAL CONSTANTS

1. <u>f - Number</u>. A dimensionless quantity equal to the ratio of the focal length of a lens to its diameter, $f^{\#} = f/d$.

2. <u>Rayleigh's criterion for resolution</u>. The smallest angle resolvable by an imaging system is given by

$$\theta_R(\text{radians}) = 1.22\ \lambda/d.$$

Here λ = wavelength of light being imaged, and d = aperture diameter.

NEW CONCEPTS AND PRINCIPLES

<u>Cameras and the Human Eye</u>

Understanding this section should enable you to do the following:

1. *Calculate the focussing requirements of a simple camera and the human eye for objects at various distances, and calculate the magnification.*
2. *Calculate the dependence of image brightness and exposure on the f-number and exposure time for a simple camera.*
3. *Identify common defects of the eye and calculate the power of corrective lenses needed to remedy near- and far-sightedness.*

<u>Cameras with Lenses</u>--As shown in Figure 25.2 (text), the simple camera consists of a movable lens of fixed focal length, which creates a *real* image at the film plane. We assume the thin lens formula is applicable to this system. There is also an adjustable aperture, or *iris*, which controls the effective lens diameter.

<u>Image Brightness and f-Number</u>--The *intensity* of light at the image depends on how much light enters the camera and to what extent that light is spread out, i.e., the size of the image. From the text's derivation,

$$\text{Brightness} \propto (d/f)^2 = (\text{f-number})^{-2}$$

Here f = focal length of the lens
d = diameter of the lens aperture
and f/d = the *f-number* of the lens.

Exposure--The *total* amount of light incident on the film determines the amount of *exposure*. The amount of light is the product of the *image brightness* (intensity) times the *time* of exposure. Mathematically,

$$\text{Exposure} = (1/f^{\#})^2 t.$$

Thus, for example, if you *double* the exposure *time*, you must also *increase* the $f^{\#}$ by a factor of $\sqrt{2}$ to maintain the same amount of film exposure.

Magnification (M)--Since the lens travel in a camera is very slight, the image distance remains *almost* equal to f for all object distances. So to a good approximation, the magnification produced by a camera is

$$M = q/p = f/p\ .$$

Important Observations

1. Notice that increasing the $f^{\#}$ *narrows* the lens aperture, and vice versa.
2. The conventional "f-stop" settings on cameras are a sequence of f-numbers, each of which is $\sqrt{2}$ larger than the previous number. Thus each "f-stop" is equivalent to doubling or halving the exposure.
3. There are a number of practical trade-offs in choosing exposure settings. Rapid object motion requires short exposure time and consequently lower $f^{\#}$. This means a *wide* aperture, which makes the *depth of field* shallow. This is the range of object distances which have image focus that is acceptable.
4. A *slide projector* is essentially a camera in reverse. The film (slide) is the object, and a movable lens forms an *enlarged real* image at various screen distances.

Example 25-1. The housing on a certain camera lens reads f/2.8, 9 cm. The focussing ring is calibrated to focus objects from 4 feet distant to infinity.

(a.) What is the wide-open aperture diameter of the lens? What would it be at f/11?

(b.) How much lens travel is necessary to accommodate the entire range of object distances?

S: Assuming the f/2.8 description refers to the wide open lens, its diameter would be

$$d = f/f^{\#} = (9\text{ cm})/2.8 = 3.2\text{ cm} = 1.26\text{ inches}.$$

At f/11, the aperture would be narrower by a factor of 2.8/11, or .255 times the above result.

T: This is about 1/4 the aperture, or 0.8 cm. This amounts to 4 "f-stops", from f/2.8 to f/4, f/5.6, f/8, and f/11. This will decrease the exposure by $1/(2)^4 = 1/16$.

S: To focus on an object at infinity, the lens should be a focal length in front of the film, or q = 9 cm. For objects 4 feet (122 cm) away, we need

$$1/q = 1/f - 1/p = 1/9 - 1/122 = .103\ \text{cm}^{-1}$$

or

$$q = 9.72\ \text{cm}.$$

T: So the lens must move a distance of .72 cm to focus this range of object distances. Notice that as the object moves nearer, the lens moves *out toward* it to maintain focus.

The Human Eye--The eye is essentially a camera, with the *retina* taking the place of the film. There is one main difference: the lens to retina distance (q) is fixed at about 1.7 cm, and the lens is flexible enough to *change its focal length* to accommodate different object distances. These range from the *far point* at infinity to a *near point* which is normally at around 25 cm. See Figure 25.4 (text).

Common Defects of the Eye

(a.) Near-sightedness (Myopia). The lens is too *strong* to focus distant objects, and images them *in front* of the retina. *Diverging* corrective lenses are needed. See Figure 25-6 (text).

(b.) Far-sightedness (Hyperopia). The lens is not strong enough, focussing near objects *in back* of the retina. *Converging* corrective lenses are needed. See Figure 25-7 (text).

Example 25-2. A student wears corrective contact lenses of power -8.5 diopters. The student can give an extra +3.5 diopters to eye lens when accommodating to his near point. His lens is 1.7 cm from his retina.

(a.) What defect does the student's eye have?
(b.) What is the power of his eye lens when relaxed?
(c.) Where is the student's near point with and without the correction?

S: Requiring a diverging corrective lens indicates *myopia*. When is the eye "relaxed"?

T: When focussing on the far point, $p = \infty$. Let D_{far} be the power of the relaxed lens. Then the formula for lenses in contact gives us

$$D_{far} + D_{corr} = D_{far} - 8.5 = \quad = 58.8\ \text{m}^{-1}.$$

S: Solving for D_{far} gives $D_{far} = 58.8 + 8.5 = +67.3$ diopters. Now we have to find D_{near}. That's essentially given to us:

$$D_{near} = D_{far} + 3.5 = +70.8\ \text{diopters}.$$

T: So his near point *without* correction is found from

$$1/(p_{near}) = D_{near} - 1/(.017\ \text{cm}) = 70.8 - 58.8 = 12.0\ \text{m}^{-1}.$$

Thus

$$p_{near} = 8.3 \text{ cm.}$$

And corrected,

$$1/(p_{near}) = (D_{near} + D_{corr}) - 1/q = (70.8 - 8.5) - 58.8 = 3.5 \text{ m}^{-1},$$

giving

$$p_{near} = 28.5 \text{ cm.}$$

The Simple Magnifier

Understanding this section should enable you to do the following:

1. *Calculate the angular magnification of a simple magnifier.*
2. *Identify the essential principle of operation of the simple magnifier and the type of image it produces.*

Angular Size of an Object--As shown in following figure, the angle subtended at P by an object of height h at a distance d is

$$\tan \Phi = h/d.$$

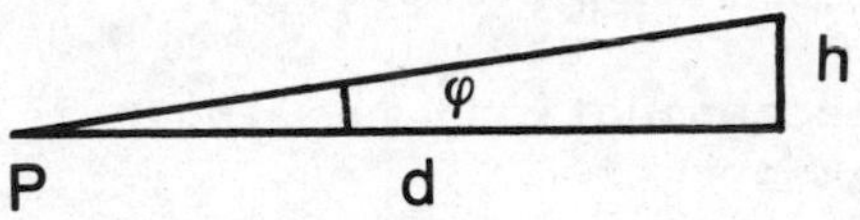

If h << d, the small angle approximation is

$$\Phi \text{ (radians)} = h/d.$$

Angular Magnification (M_Φ)--Also known as *magnifying power*, this is the ratio of the angular size of the image produced by a magnifier (Φ') to that of the object in the *absence* of the magnifier (Φ). Mathematically,

$$M_\Phi = \Phi'/\Phi.$$

Magnifying Power of a Simple Magnifier--As shown in Example 23-7, a converging lens produces an upright *enlarged* virtual image of an object placed inside its focal point. By placing such a lens close to the eye, your eye can focus on this enlarged image of a small object in front of the lens, since the image is at or beyond the eye's near point (see Figure 25.13, text). Comparing the angular size of this image with that of the object placed at the near point *without* the lens, we have for the relaxed eye,

$$M = 25/f(\text{cm}) = .25D, \text{ with the image at } \infty,$$

or for the eye at full accommodation,

$$M = 1 + 25/f(\text{cm}) = 1 + .25D, \text{ image at the near point.}$$

Important Observations

1. The entire purpose of the magnifier is to give your eye an equivalently stronger lens, allowing it to focus on and examine objects placed only a few cm away instead of being limited to 25 cm.

2. The two magnifications listed above represent the limits of M_Φ for a given lens. Any M_Φ between these is possible, depending on where the virtual image is. The general expression is

$$M = 25(1/f + 1/|q|),$$

where $|q|$ can be anything from 25 cm to ∞.

Example 25-3. In the following, assume an eye with normal ability to accommodate.

(a.) What is the magnifying power of a lens of +50 diopters when adjusted for minimum eyestrain? For maximum accommodation?

(b.) If the lens and the eye are close enough to be considered in contact, how close to the lens can an object be placed and still be focussed on the retina 1.8 cm away? What is the magnification in this case?

S: For minimum eyestrain, the virtual image is at infinity, and

$$M_\Phi = 25/f(\text{cm}) = .25\ D = .25(50) = 12.5\ X.$$

For maximum accommodation, the image is at 25 cm, the normal near point. The formula gives us

$$M_\Phi = 1 + .25\ D = 13.5\ X.$$

I don't quite understand what this means.

T: It's quite easy. You're comparing the angle subtended by an object of height h with that subtended by an image of height $h' = h(q/p)$, both 25 cm from the eye. We know that, since $q = -25$ cm,

$$1/p = 1/f - 1/q = 1/f + 1/25 = (25 + f)/25f\ .$$

So

$$M_\Phi = -q/p = (25)(25 + f)/25f = 1 + 25/f.$$

Another perspective on this is given in part (b.). Treat the magnifier and the eye lens as a single strong converging lens.

S: Well, the equivalent power would be $D_{eq} = D_{eye} + D_{lens}$. What is D_{eye}?

T: For maximum accommodation, where p = 25 cm, we have

$$D_{eye} = 1/.25 + 1/.018 = +59.6 \text{ diopters.}$$

S: Then

$$D_{eq} = 59.6 + 50 = 109.6 \text{ diopters.}$$

So

$$1/p_{min} = D_{eq} - 1/.018 = 109.6 - 55.6 = 54.0\ m^{-1}.$$

Thus

$$p_{min} = 1/54.0 = .0185\ m = 1.85\ cm.$$

T: The angle an object would subtend at this distance is 25/1.85 times larger than it would be at 25 cm, so

$$M_\Phi = 25/1.85 = 13.5.$$

Notice that this is the same as the previous result. The additional lens has allowed the eye to bring the object 13.5 times closer and still form a retinal image. Notice that p = 1.85 cm is *inside* the 2 cm focal length of the magnifier. So this is consistent with the approach involving an intermediate virtual image. The object must be placed between 1.85 and 2 cm in front of the magnifier, which is held next to the eye.

The Microscope and Telescope

Understanding this section should enable you to do the following:

1. *Calculate the magnifying power and resolving power of compound microscopes and telescopes.*
2. *Identify the Rayleigh criterion for resolution of images.*

The Compound Microscope--As shown in Figure 25.14 (text), the compound microscope consists of two converging lenses a distance L apart. The object to be magnified is placed just outside the focal point of the first lens, called the *objective*. This produces a *real*, inverted, and *enlarged* image which lies just *inside* the focal point of the second lens, called the *eyepiece* or *ocular*. This in turn produces an upright, *virtual*, and *enlarged* image which serves as a magnified object for the eye. The magnification is given to a good approximation by

$$M = M_1M_2 = (1/f_1)(L - f_2)(25/f_2) ,$$

where all lengths are in cm.

Important Observations

1. Notice that the ocular serves as a simple magnifier with which we can view the image produced by the objective. This is generally adjusted for viewing with the relaxed eye.
2. Two important assumptions are made in this expression for M_1: that the object distance is essentially f_1 and that the real image lies barely inside f_2. In a crude microscope these may not be true, and you would have to use the basic lens equation to find $M = (q_1/p_1)(25/f_2 + 25/|q_2|)$.
3. In many commercial microscopes, L = 18 cm and $f_2 << L$, so we can make a further approximation:

$$M = (18/f_1)(25/f_2) = 450/f_1f_2.$$

Example 25-4. A homemade microscope has two lenses, each with f = 5 cm, separated by 30 cm. An object is placed 6.2 cm in front of the objective lens. Find where the final image is located, and calculate the total magnification.

S: Can we use the standard expression for M?

T: At the outset, we can't be sure. We can do that calculation and compare it with a more basic approach:

$$M = (1/f_1)(L - f_2)(25/f_2) = 25\ X.$$

S: To find the position of the 1st image,

$$1/q_1 = 1/f_1 - 1/p_1 = 1/5 - 1/6.2,$$

giving $q_1 = +25.8$ cm. Thus

$$M_1 = q_1/p_1 = 25.8/6.2 = 4.16\ X.$$

What's next?

T: Next we need the object distance for the ocular, p_2. Obviously, $p_2 = 30$ cm - 25.8 cm = 4.2 cm. Notice this *is* inside the focal point of L_2.

S: Then to find the final image,

$$1/q_2 = 1/f_2 - 1/p_2 = 1/5 - 1/4.2,$$

giving $q_2 = -26.25$ cm. Now can't we just use $M_2 = q_2/p_2$?

T: It's correct to use the *linear* magnification, q_1/p_1, for the first stage since it tells how large an object L_2 is imaging, but we're interested in the *angular* magnification for the last stage. q_2 is almost at the 25 cm near point, so we could use $M_2 = 1 + 25/f_2$ as a good approximation. This would give $M_2 = 6$ X. The more accurate expression is

$$M_2 = 25(1/f_2 + 1/|q_2|) = 5.95\ X.$$

S: So M = (4.2)(5.95) = 24.99 X. This *is* equal to the standard result!

T: So it is. But the approximations built in to that formula weren't really satisfied, so it was a bit of luck. If in doubt, use the approach we went through here.

Telescopes--As shown in Figure 25.15 (text), the telescope is composed of a long focal length objective (either lens or mirror) and an ocular, separated by a distance equal to the sum of their focal lengths, $L = f_1 + f_2$. The magnification is given simply as the ratio of the focal lengths, $M = f_1/f_2$. Here f_1 = objective, and f_2 = ocular.

Resolution--Because of diffraction effects, images of point objects have a certain width. If two images of object points situated very close to each other overlap, it can become impossible to distinguish, or *resolve*, the separate object points. The *minimum angular separation* between object points which can be definitely resolved is called the *resolution* or *resolving power* of the lens or mirror system. This resolving power is proportional to the wavelength forming the image and inversely proportional to the aperture diameter of the lens or mirror. Mathematically, this is given by *Rayleigh's Criterion*, derived in your text:

$$\sin\theta_R \approx \theta_R\ (\text{radians}) = 1.22(\lambda/d).$$

Important Observations

1. The concept of resolution applies equally to distinguishing actual separation of point sources (such as close stars) and to distinguishing adjacent detail on an object (such as crater diameters on the moon or the dimensions of minute structures on a microscope slide).
2. The Rayleigh criterion is the *absolute limit* of detail resolution. Imperfect lenses or turbulent atmosphere (in astronomy) can of course make the resolution worse.
3. Since visible light has such a narrow wavelength range, using any average visible wavelength will suffice to calculate θ_R. Your text uses λ = 500 nm.

Example 25-5. The periods on this page are approximately 1 mm in diameter.

(a.) At the near point of the eye, what angle does this subtend?
(b.) With a human eye able to resolve 0.7 minutes of arc, how far away would the periods be resolvable?

S: It seems that the small angle approximation is valid here. We have a height of 1 mm and a distance of 25 cm, so

$$\theta = (10^{-1}\text{ cm})/(25\text{ cm}) = .004\text{ radians.}$$

T: Yes, which is equal to $.004(180/\pi) = .229^\circ = 13.8$ minutes.

S: For part (b.), we have to convert 0.7 minutes to radians first.

$$(0.7\text{ min})(1^\circ/60\text{ min})(180/\pi\text{ rad}/^\circ) = 2.04 \times 10^{-4}\text{ rad.}$$

T: If we want this to equal the resolving limit, we have

$$2.04 \times 10^{-4}\text{ rads} = h/d_{max},$$

giving

$$d_{max} = (-10^{-1}\text{ cm})/(2.04 \times 10^{-4}\text{ rad}) = 4.9 \times 10^2\text{ cm} = 4.9\text{ meters.}$$

Example 25-6. Many rumors circulate about how much "spy" satellites can see. If it was claimed that a satellite camera could read license plate numbers from an orbit 150 miles above the earth, what minimum lens diameter must it have?

S: We would have to know the angles it is required to resolve.

T: Exactly. We should be able to make a reasonable estimate of what they would be.

S: Well, we have the distance, and a license plate is about a foot long and maybe half a foot wide.

T: Good, but to *read* the plate numbers, which seem to be about 2 inches wide, you would have to resolve down to *at least* an inch, and probably even finer. But let's use one inch as a crude estimate of what's required of the camera.

S: O.K. What angle does 1 inch subtend at the camera 150 miles away?

$$\theta = (1 \text{ inch})/(150 \text{ mi} \times 5280 \text{ ft/mi} \times 12 \text{ in/ft}) = 1.05 \times 10^{-7} \text{ radians.}$$

How does this tell us the required lens diameter?

T: The Rayleigh criterion requires that in our case

$$1.22(\lambda/d) = 10^{-7} \text{ rad, with } \lambda = 500 \text{ nm.}$$

So

$$d = (1.22)(5.00 \times 10^{-7} \text{ m})/(10^{-7} \text{ rad}) = 6.1 \text{ meters} = 240 \text{ inches!}$$

This would be as large as the largest telescope aperture in the world, which is obviously not in orbit!

Resolving Power of Microscopes--Since the object for a microscope is placed essentially at the focal point of the objective, the angle it subtends at the lens is $\theta = s/f$, where s is the object's diameter. Rayleigh's criterion then gives us the *minimum size of detail* resolvable with the microscope:

$$s_{min} = 1.22\lambda(f/d) = 1.22\lambda(f^{\#}).$$

As shown in the text (Figure 25.22), the diameter of the objective's aperture is related to the sine of half the angle subtended by the aperture at the object, yielding an *alternate* expression of s_{min}:

$$s_{min} = 1.22\lambda/(2 \sin\alpha) ,$$

where α = the half angle described above.

Numerical Aperture (NA)--If the region surrounding the objective is filled with fluid having an index of refraction n, the wavelength of light in the fluid is reduced to λ/n. Then the resolution is *improved* to

$$s_{min} = 1.22\lambda/(2n \sin\alpha) = 0.61\lambda/NA$$

where $NA = n \sin\alpha$ = is the *numerical aperture* of the objective.

Important Observations

1. The following factors contribute to greater resolving power:

   a. Shorter wavelength of illumination.
   b. Wider aperture (smaller $f^{\#}$).
   c. Larger index of refraction of material surrounding the objective outside the microscope.

2. The resolution s_{min} cannot be much smaller than the illuminating wavelength.
3. The instrument may resolve detail too small for the eye to see. Various oculars may bring out additional detail to view. For example, fine photographic emulsions can record detail far too small to "see" until enlarged.

Example 25-7. A certain commercial microscope ($L = 18$ cm) has an objective with focal length of 3.2 mm and a diameter of 4 mm.

(a.) When illuminated with yellow light ($\lambda = 580$ nm), what is the smallest object that can be resolved?

(b.) What must be the focal length of the ocular to allow this object to be seen by the observer?

S: The smallest object would have a dimension

$$s_{min} = .61(\lambda/NA), \text{ with } \lambda = 580 \times 10^{-9} \text{ m}, \text{ and } n = 1.00.$$

How do we calculate the numerical aperture (NA)?

T: It is $NA = n \sin \alpha$. We have $n = 1.00$, and so

$$NA = \sin \alpha = (d/2)/f = (2 \text{ mm})/(3.2 \text{ mm}) = .625 \quad ,$$

which is a dimensionless number.

S: So then

$$s_{min} = (.61)(580 \times 10^{-9} \text{ m})/(.625) = 5.66 \times 10^{-7} \text{ m}.$$

I don't understand what part (b.) calls for.

T: Well, assume the observer has good eyes, that can resolve, say, 0.7 minutes of arc. Can she see such a small object at her near point unaided?

S: Obviously not. It would subtend only

$$(5.66 \times 10^{-5} \text{ cm})/(25 \text{ cm}) = 2.26 \times 10^{-6} \text{ rad} = 7.82 \times 10^{-3} \text{ minutes.}$$

T: How much does the microscope magnify this angle?

S: For a standard length of $L = 18$ cm,

$$M = (1/f_1)(L - f_2)(25/f_2).$$

We don't know M or f_2.

T: Don't we? We know what M *must be* in order to barely see the object:

$$M = \Phi'/\Phi = (0.7 \text{ min})/(7.82 \times 10^{-3} \text{ min}) = 89.5 \text{ X}.$$

S: Aha! And we can get the required f_2 from

$$89.5 = (1/.32)(18 - f_2)(25/f_2).$$

This gives $f_2 = 8.40$ cm.

The Michelson Interferometer

Understanding this section should enable you to do the following:

1. *Calculate the fringe shift occurring in a Michelson interferometer when the optical path in one arm changes by a given amount relative to the other arm.*

The Michelson Interferometer--Figure 25.27 (text) shows the basic arrangement of the interferometer, which is discussed in detail in your text. In essence, a single monochromatic beam is split into two paths, and the phase shift of one beam relative to the other upon recombining depends on the relative optical lengths of the two arms, D_1 and D_2. If the arms have no refractive material in them, the conditions for interference are:

$$\text{Constructive:} \quad 2|D_1 - D_2| = m\lambda \quad , \text{(m an integer)}$$

$$\text{Destructive:} \quad 2|D_1 - D_2| = (m + \tfrac{1}{2})\lambda \quad , \text{(m an integer)}$$

Important Observations

1. Across the width of the beam, a number of the above conditions can alternately occur which produces a series of light and dark fringes. If M_1 and M_2 are precisely perpendicular, these will have the shape of concentric circles. As M_2 moves, the circular fringes will appear to move toward or away from the center, with the central spot becoming alternately bright and dark. A full fringe cycle corresponds to a movement of M_2 equal to $\lambda/2$.
2. Most often, one mirror is *slightly* tilted, producing *straight* fringes across the recombined beam diameter. Then a reference crosshair can be used to count the number of fringe cycles (fringe "shifts") which cross it, each corresponding to a travel of $\lambda/2$ for mirror M_2.
3. There will be a fringe shift even with fixed arms if material of refractive index n is inserted into one arm. Such material produces a shortened wave length, λ/n, and hence a *greater* phase shift than the empty arm produces. The effective length *change* produced by the insertion is $nt - t = (n - 1)t$, where t is the physical thickness of the inserted matter.

Example 25-8. The movable mirror in a Michelson interferometer is mounted on a micrometer screw. The interferometer beam consists of red light from cadmium with $\lambda = 643.85$ nm. An observer counts 4162.5 fringes crossing the reference mark when the screw is extended an *indicated* 1.331 mm. What is the % error of the micrometer screw?

S: The mirror travels $\lambda/2$ for each fringe, so if there are 4162.5 fringes, the mirror has traveled a distance

$$d = 4162.5(\lambda/2) = \tfrac{1}{2}(4162.5)(643.85 \times 10^{-9} \text{ m}) = 1.340 \text{ mm}.$$

T: So the error in the screw is 100 x (1.340 - 1.331)/1.340 = 0.67%.

QUESTIONS ON FUNDAMENTALS

1. If you obtained a correct camera exposure of a subject with settings of f/8 at 1/60 sec, which of the following exposure times would give the best exposure of the same subject at an f-number of f/2.8?

(a.) 1/175 sec (b.) 1/8 sec (c.) 1/500 sec
(d.) 1/20 sec (e.) 1/250 sec

2. What is the focal length of a simple lens that would produce a magnification of 5 with its image at the near point of the eye?

(a.) 16 cm (b.) 5 cm (c.) 6.25 cm (d.) 20 cm
(e.) 12.5 cm

3. What is the smallest diameter lens which could resolve detail to .01° in an object illuminated by light of 500 nm?

(a.) 3.5 mm (b.) 3.5 inches (c.) 2.86 cm
(d.) 2.86 mm (e.) 1.1 cm

4. In a commercial microscope with the lenses 18 cm apart, and an objective focal length of 5 mm, what focal length should the ocular have in order that the total magnification of the microscope be 40 X?

(a.) 8 mm (b.) 22.5 cm (c.) 6.25 cm (d.) 18 cm
(e.) 8 cm

5. If your telescope has an objective with a focal length of 1200 mm, what is its magnifying power if you use an eyepiece having 25 mm focal length?

(a.)300 (b.) 1225 (c.) 24.5 (d.) 125 (e.) 48

GENERAL PROBLEMS

Problem 25-1. Given that the angle subtended at the earth by the sun's diameter is 32 minutes of arc (32'), what focal length must a lens have to produce a real image of the sun 2 cm in diameter?

Q1: Where will the image position be?
Q2: What angle will the image subtend at the lens' center?
Q3: What relation does this angle have to the 32' subtended by the object?
Q4: Then what must f be?

Problem 25-2. A photographer wishes to photograph a building that is 15 meters tall from a distance of 20 meters. In order for the image on the film to be 20 mm high, what must be the focal length of the camera lens?

Q1: What is the desired magnification?
Q2: What is q?
Q3: What determines f?

Problem 25-3. What is the angular magnification of a reading glass whose focal length is 12.5 cm if it is held 10 cm from a book, and if the observer's eyes are 30 cm from the lens?

Q1: What defines angular magnification?
Q2: What is the object height?
Q3: What is the image height relative to object height?
Q4: What is the value of M?
Q5: What is Φ?
Q6: How far is the image from the eye?
Q7: What is Φ'?

Problem 25-4. Suppose you build a microscope with an objective focal length of +0.3 cm and an ocular focal length of +2.0 cm. The separation between the lenses is to be 20 cm.

(a.) When adjusted for a relaxed eye, what is the magnifying power to 3 significant figures?

(b.) How much greater magnification can be obtained by adjusting to the fully accommodated normal eye?

Q1: In (a.) where is the final image?
Q2: Where is the intermediate image formed by the objective?
Q3: How far from the objective must the object be placed?
Q4: What is the linear magnification of the objective?
Q5: What is the angular magnification of the ocular?
Q6: For full accommodation, how does this change?
Q7: Does M_1 change also?
Q8: What effect does that have on p_1 and q_1?

Problem 25-5. Go back to the homemade microscope in Example 25-1. If the objective lens was 1.5 cm in diameter,

(a.) What resolving power would the microscope have?
(b.) Could the eye, capable of 0.7 minutes resolution, detect the smallest resolvable objects while looking through the microscope?
(c.) If the objective was immersed in oil with refractive index of 1.6, answer part (b.).

Q1: What is the applicable expression for s_{min}?
Q2: How big is the intermediate image of this small detail, as produced by the objective?
Q3: How big an angle would this subtend if 25 cm away from the naked eye?
Q4: How much does the ocular enlarge this angle?
Q5: With oil immersion of the objective, how does this change?
Q6: Could the eye see the full detail?
Q7: Can this be remedied?

ANSWERS TO QUESTIONS

FUNDAMENTALS:

1. (c.) 2. (c.) 3. (a.) 4. (b.)
5. (e.)

GENERAL PROBLEMS:

Problem 24-1.
A1: The sun, 93 million miles away, can be considered to be at $p = \infty$, so q will be equal to f.

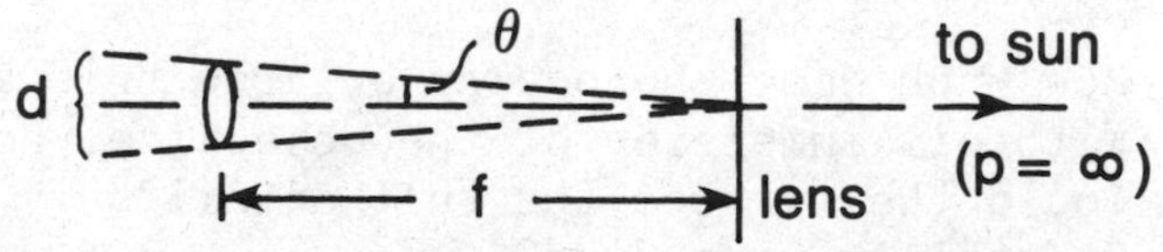

A2: An angle 2θ, where $\sin\theta = (d/2)/f$, as shown in this sketch:
A3: The angles subtended by the object and the image at the center of the lens are equal. The principal rays through the lens center show that, as in the second sketch. So $2\theta = 32'$.
A4: Using the small angle formula, $2\theta = d/f$, in radians. You have to convert 32' to radians, then find f.

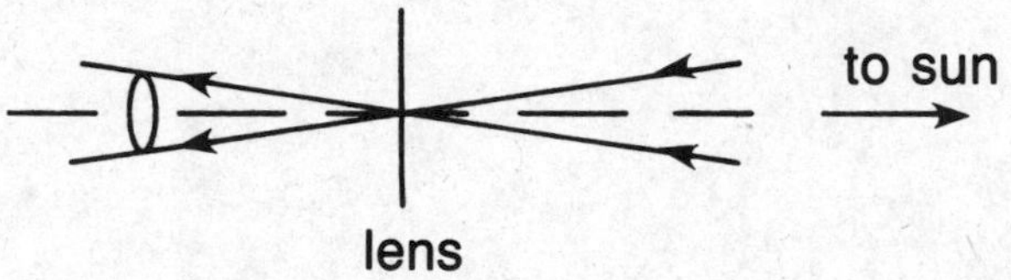

Problem 24-2.
A1: $|M| = h'/h = 20 \times 10^{-3}/15 = 1.33 \times 10^{-3}$.
A2: $|M| = h'/h = 1.33 \times 10^{-3} = q/p$, and p = 20 meters. So q = 2.67 cm.
A3: $1/f = 1/p + 1/q$.

Problem 24-3.
A1: $M\Phi = \Phi'/\Phi$ = angle subtended at eye by image divided by the angle subtended at eye by object at 25 cm.
A2: Just call it h. It might be the height of a letter on the page, for example.
A3: $h' = Mh$, where $M = -q/p$.
A4: $1/q = 1/f - 1/p = 1/12.5 - 1/10$, so q = -50 cm, and M = 5.
A5: $\Phi = h/25$ radians, using the small angle approximation.
A6: $p + |q| = 80$ cm.
A7: $\Phi' = h'/80 = 5h/80$.

Problem 24-4.
A1: $q_2 = -\infty$.
A2: A distance f_2 from the ocular, or 18 cm behind the objective.
A3: $1/p_1 = 1/f_1 - 1/q_1 = 1/0.3 - 1/18$. This gives $p_1 = .305$ cm.
A4: $M_1 = q_1/p_1 = 59.0$ X.
A5: For relaxed vision, $M_2 = 25/f_2 = 12.5$ X.
A6: $M_2 = (25/f_2) + 1 = 12.5$ X.
A7: Slightly. With $q_2 = 25$ cm, the lens equation shows that p_2 becomes 2.17 cm instead of 2.00 cm.
A8: q_1 is shortened to 20 - 2.17 = 17.83 cm, but p_1 is still .305 cm to 3 significant figures. So $M_1 = 17.83/.305 = 58.5$ X.

Problem 24-5.

A1: s_{min} = (1.22 λ)/(2n sin a), where sin a = (d/2)/p_1, since p_1 is *not* equal to f_1. For n = 1, this gives

$$s_{min} = 1.22(5 \times 10^{-7} \text{ m})(6.2 \text{ cm})/(1.5 \text{ cm}) = 2.5 \times 10^{-6} \text{ meters.}$$

A2: M_1 = 4.16, so the image of this is 4.16(2.5 x 10^{-6} m) = 10^{-5} m.

A3: An angle Φ' = s'/d = 10^{-5} m/.25 m = 4.16 radians, or 0.14 minutes.

A4: M_2 = 5.95, so Φ = 5.95(.14 min) = 0.85 minutes. Your eye can just make out this fine detail.

A5: s_{min} is reduced by n, and so are all of the subtended angles. s_{min} would be (2.5 x 10^{-6} m)/1.6 = 1.56 x 10^{-6} m.

A6: No. The final image of s_{min} would subtend only .53 min at your eye.

A7: Yes, with a shorter focal length ocular, which would increase M_2, revealing to your eye the finer detail that the objective has resolved.

CHAPTER 26. THE THEORY OF RELATIVITY

BRIEF SUMMARY AND RESUME

Galilean, or *classical* relativity is discussed, *inertial reference frames* are defined, and the equations for *coordinate transformations* for this situation are examined. The concept of the ether and the attempts of Michelson and Morley to measure motion relative to it are analyzed. The postulates of Einstein's *Special Theory of Relativity* are introduced. The transformation equations which describe these postulates are compared with those for Galilean relativity. The *contraction* of *moving lengths* and *time dilation* of moving clocks which result from these equations are investigated. The relativity of *simultaneous events* is analyzed. The dependence of mass on motion and the connection between *mass* and *energy* are examined. A number of *experimental confirmations* of the predictions of Special Relativity are discussed. Finally, there is a brief discussion of the concepts of *General Relativity.*

Material from previous chapters relevant to Chapter 26 includes:

1. All the concepts of mechanics, including mass, velocity, energy, momentum, and time.
2. The Michelson interferometer.

NEW DATA AND PHYSICAL CONSTANTS

There are none for Chapter 26.

NEW CONCEPTS AND PRINCIPLES

Galilean Relativity

Understanding this section should enable you to do the following:

1. *State the definition of an inertial frame of reference.*
2. *Calculate the Galilean transformation of positions, times, velocities, and accelerations from one inertial frame to another.*

Inertial Frame of Reference--A coordinate system in which the Law of Inertia (Newton's 1st Law) is valid. Such a system *cannot* be accelerating.

Galilean Transformation Equations--If one inertial system S_2 is moving with constant velocity v along the positive x-direction relative to system S_1, the coordinates and times of events in the systems are related by

$$x_2 = x_1 - vt;\ y_2 = y_1;\ z_2 = z_1;\ t_2 = t_1.$$

As a consequence of $t_1 = t_2$, the transformation of velocities, obtained by dividing the above by t_1 or t_2, is:

$$v_{2x} = v_{1x} - v;\ v_{2y} = v_{1y};\ v_{2z} = v_{1z}.$$

Since v is *constant*, the transformation of accelerations is obviously

$$a_2 = a_1.$$

Important Observations

1. Even though positions, speeds, and therefore momenta and energies have different values in different inertial frames, *changes* in them are *not* different. Lengths, time intervals, and *accelerations* are the same. The latter are involved in the *Laws of Mechanics*, so these Laws remain *invariant* in any transformation between inertial frames.
2. This principle implies that mass, length, and time, which are the basic dimensions of mechanics, are independent of any *inertial* motion of an observer.

Example 26-1. In reference frame S_1, an object with mass m is dropped from rest from a height y_{1i}. As it falls, mechanical energy is conserved, so that $\frac{1}{2}mv^2 = mgy_{1i}$ when it reaches $y_1 = 0$. Show that in a second frame S_2, moving upward with speed v relative to S_1, mechanical energy is also conserved. Assume the origins of S_1 and S_2 coincide at t = 0, when the object is released.

S: What equations connect positions and speeds in the two reference frames?

T: It's the same as in the text, except that the velocity of S_2 is along the y-axis:

$$x_2 = x_1,\ z_2 = z_1,\ y_2 = y_1 - vt \text{ and } v_2 = v_1 - v.$$

The latter becomes $v_2 = v_1 + v$ if we choose downward to be positive. We haven't labelled t, since $t_2 = t_1 = t$. Find the initial KE in S_2.

S: At t = 0, $y_{2i} = y_{1i}$, and $v_{2i} = 0 + v$. So $KE_{2i} = \frac{1}{2}mv_2$. Now we have to find the initial PE.

T: Since the origins coincided at t = 0, it is the same as for S_1:

$$PE_{2i} = PE_{1i} = mgy_{1i}.$$

What is the KE_2 when the object reaches $y_1 = 0$?

S: In general, we can write $KE_2 = \frac{1}{2}mv^2 = \frac{1}{2}m(v_1 + v)^2$. So when $y_1 = 0$,

$$KE_{2f} = \tfrac{1}{2}m\{2gy_{1i} + 2(2gy_{1i})^{\frac{1}{2}}v + v^2\}.$$

T: Now the final PE_2, when $y_1 = 0$ is

$$PE_{2f} = mgy_{2f} = mg(0 - vt),$$

where t = time of fall $= (1/g)(2gy_{1i})^{\frac{1}{2}}$. So

$$PE_{2f} = -mv(2gy_{1i})^{\frac{1}{2}}.$$

Adding PE to KE, we have

$$PE_{2i} + KE_{2i} = mgy_{1i} + \tfrac{1}{2}mv^2, \text{ and}$$
$$PE_{2f} + KE_{2f} = mgy_{1i} + \tfrac{1}{2}mv^2.$$

The total mechanical energy in S_2 is thus conserved.

The Ether and the Postulates of Special Relativity

Understanding this section should enable you to do the following:

1. *State the difficulties in applying Galilean relativity to electromagnetism.*
2. *State and interpret the postulates of the ether theory, why they were made, and the experimental evidence for or against the theory.*
3. *State and interpret Einstein's postulates in the Special Theory of Relativity.*
4. *Calculate the transformations in position and time between two inertial frames according to Einstein's postulates.*

The Ether and Electromagnetic Theory--The following summarizes the developing crisis in understanding electromagnetic theory and experiment which had arisen by Einstein's time:

1. Galilean relativity predicts that observers in different inertial frames would measure different speeds of light, according to their motion relative to the medium carrying the light wave.
2. From Maxwell's theory, the speed of light should be a *universal* constant, since $c = 1/\mu_0\epsilon_0$.
3. The *ether* was postulated as *the* stationary reference frame in which the above was true. That meant that an observer could measure his or her *absolute* speed in space by measuring a *different* speed of light than that given above.
4. A different light speed would mean that in your frame of reference, ϵ_0 and μ_0 would have different values, and hence would *not* be universal constants. Furthermore, since they are involved in the basic expressions for the electric and magnetic forces, the force laws would be quantitatively different, and "correct physics" could only be discovered in the ether's frame, instead of *any* inertial frame.
5. Since astronomical bodies maintain their motion without perceptible drag, the ether had to be extremely *tenuous*. At the same time, it had to have extreme *rigidity* in order to propagate a transverse wave as fast as light.
6. Michelson and Morley, as well as many after them, have never been able to measure *any* change of light speed for various motions of the apparatus, even though the experiment was clearly sufficiently accurate and precise to record such an effect if present.

The Michelson-Morley Experiment--Fully described in Section 26.3 of your text, certain aspects of the Michelson-Morley experiment deserve special emphasis.

1. The interferometer they used had equal arms. The ether theory predicted that the times of travel of the two beams would nevertheless be unequal, creating a *delay* of the upstream-downstream beam relative to the other, causing a phase difference upon recombining.
2. The above phase difference wasn't directly observable without a zero velocity frame as a reference. But *interchanging* the roles of the two arms by *rotating the apparatus through 90°* was the crucial strategy. The ether theory would predict a definite and measurable phase shift during this rotation.
3. Furthermore, the experiment was repeated many times as the earth was traveling in different directions through space in its orbit around the sun. The apparatus could not have been at rest in *all* those directions of motion.
4. The *null* result of this experiment and the many versions of it since the original attempts demonstrates that it is not possible to detect *absolute* motion or rest, *even in principle*!

<u>Einstein's Postulates</u>--The bases of the Special Theory of Relativity are two postulates which remove the crises of theory and experiment.

Postulate 1: The speed of light is the *same* in *all* inertial reference frames.

Postulate 2: *All* the laws of physics have the same form in *all* inertial reference frames.

Taken together, these state that all inertial reference frames are *equivalent*, that there is *no* physical way of establishing that a given frame is at *absolute rest*, and that one can only observe *relative* motion between them.

<u>The Lorentz Transformation</u>--The coordinate transformation equations which satisfy Einstein's two postulates are known as the *Lorentz equations*. Again we assume an inertial frame S_2 moving with constant velocity v toward the positive x-axis in an inertial frame S_1. The positions and times are related by

$$x_2 = \gamma(x_1 - vt_1);\ y_2 = y_1;\ z_2 = z_1;\ \text{and } t_2 = \gamma(t_1 - vx_1/c^2),$$

where

$$\gamma = (1 - v^2/c^2)^{-\frac{1}{2}}.$$

<u>Important Observations</u>

1. There are two differences of enormous consequence between these Lorentz equations and the Galilean transformation:

   (a.) There is a difference in *time* and *time scales* for the two systems! Time had always been assumed to have a universal, absolute value for all observers.

   (b.) There is an additional velocity dependent factor, γ, connecting x_2 and x_1, as well as t_2 and t_1. This factor becomes very large as the speed between the two reference frames approaches the speed of light. See Figure 26.14 (text).

<u>Example 26-2</u>. In an inertial system S_1 an event is recorded at x_1 = 500 km when $t_1 = 10^{-3}$ sec. The same event is recorded in a second inertial reference frame at $t_2 = .90 \times 10^{-3}$ sec. S_2 is moving toward the + x-axis in S_1, and S_1 and S_2 coincided at $t_1 = t_2 = 0$.

(a.) How fast is S_2 moving relative to S_1?
(b.) Where was the event recorded in S_2?

<u>S</u>: We have $t_2 = .90 \times 10^{-3}$ s, and have to find v from the time transformation equation.

$$.90 \times 10^{-3}\ \text{s} = \gamma[10^{-3}\ \text{s} - (5 \times 10^5\ \text{m})v/c^2]$$

This is going to be complicated, since γ is also dependent on v.

T: Well, its going to be a quadratic equation for v, so we have to be careful with the algebra and the exponents. Start by putting in the expression for γ, squaring the equation, and cross-multiplying. Then collect terms in v, v^2, and constants. You should get

$$2.19 \times 10^{-17} v^2 + 1.11 \times 10^{-8} v - .19 = 0.$$

We've used $c = 3 \times 10^8$ m/s, and omitted mention of the units for now.

S: The solution has the form

$$v = \frac{-11.8 \times 10^{-8} \pm \sqrt{(1.11 \times 10^{-8})^2 - 4(2.19 \times 10^{-17})(-.19)}}{2(2.19 \times 10^{-17})}$$

$$v = (-1.11 \times 10^{-8} \pm 1.18 \times 10^{-8})/(4.38 \times 10^{-17}).$$

Which of these solutions makes sense?

T: We have assumed v to be positive when we wrote down the equations, so we must use the positive solution here.

S: Then

$$v = (+.07 \times 10^{-8})/(4.38 \times 10^{-17}) = 1.60 \times 10^7 \text{ m/s}.$$

This is about 35 *million* miles per hour!

T: Yes. The effect of different times only occurs to a significant degree when speeds are *very* high. This is why our "common" sense is surprised. Such speeds are not part of our intuitive experience. This result means that $v/c = (1.6 \times 10^7)/(3 \times 10^8) = .053$, or a little over 5% the speed of light. What does that make γ?

S: $\gamma = 1/(1 - .053^2)^{\frac{1}{2}} = 1.0014$. This is still very close to 1.

T: Finally, the event takes place at the position x_2 given by

$$x_2 = \gamma(x_1 - vt_1) = (1.0014)(500 \text{ km} - 1.6 \times 10^4 \text{ km/s} \times 10^{-3} \text{ s})$$

$$= 485 \text{ km}.$$

Length Contraction and Time Dilation; Simultaneity

Understanding this section should enable you to do the following:

1. *Calculate the difference between proper and moving measurements of length and time intervals.*
2. *Be able to show that two events cannot appear simultaneous in two systems moving relative to each other.*

Proper Measurements of Length and Time--If the distance between two points or the time interval between two events are measured in a system where the points or events are *stationary*, those measurements are said to be *proper* measurements.

Length Contraction--If the observer measures the distance L between points that are moving with speed v relative to the observer, this distance will be *shorter* than the proper distance L_o. The Lorentz equations show that

$$L = (1 - v^2/c^2)^{\frac{1}{2}} L_o = L_o/\gamma \ .$$

Time Dilation--If an observer measures the time t between two events which ara moving at a speed v relative to the observer, this time interval is expanded, or *dilated*, compared to the proper time interval, t_o. The Lorentz equations show that

$$\Delta t = (1 - v^2/c^2)^{-\frac{1}{2}} \Delta t_o = \gamma \Delta t_o .$$

Important Observations

1. The difference between the length and time measurements do *not* imply *incorrect* measurements. The results of your measurements depend on how fast you are moving relative to the objects or events you're observing.
2. This concept does not appear in classical (Newtonian) physics. Differences in L or Δt become substantial only as v approaches c.
3. These relations are *symmetric* for observers in S_1 and S_2. *Both* would claim that meter sticks they measured in the other's system would be shorter than 1 meter, and that clocks in the other's system appeared to run more slowly than their own.

Example 26-3. A space station is built so its length is 3 km. A fast space ship passes by at constant speed, and an astronaut notices explosions on opposite ends of the station occurring .167 sec apart. He also measures the length of the station to be 1750 meters. What would a space station observer record regarding the explosions?

S: We don't know the speed of the ship relative to the station.

T: But we know the 1750 meter length is related to the proper length of 3 km by

$$L/L_o = (1 - v^2/c^2)^{\frac{1}{2}} = 1750/3000 = .5833 \ .$$

S: This gives $1 - v^2/c^2 = (.583)^2 = .340$, so we get

$$v^2 = .660\ c^2, \text{ or } v = .812\ c.$$

The astronaut also made an "improper" measurement of the time interval, right?

T: Sure--the events were moving in his frame of reference. So

$$(\Delta t)_{ship} = (\Delta t)_{station}(1 - v^2/c^2)^{-\frac{1}{2}}.$$

Thus on the station the explosions occurred only

$$(\Delta t)_{station} = (.583)(.167 \text{ sec}) = .097 \text{ seconds apart.}$$

Simultaneity--Because of the difference in time for observers in relative motion, events which happen at the *same time* in system S_1 *cannot* appear to be simultaneous when observed from a system moving relative to S_1. If in S_1 we observe $t_1 = t_1'$, an observer in S_2 will observe

$$t_2 - t_2' = \gamma(v/c^2)(x_1' - x_1).$$

Important Observations

1. Events can appear simultaneous whether or not they are stationary relative to you. The point is, if they *are* simultaneous in your frame of reference, they *cannot* appear so for *any* observer moving relative to you.
2. It also follows that the *order* of two events, A and B, depends on the motion of the observer relative to them. One observer may see them as simultaneous, another may see A before B, and yet another may see B before A. However, if A *caused* B, it turns out that this *causality is preserved regardless of the motion of the observer.* No observer in that case could move in such a way that she could observe B before A.

Example 26-4. A stationary spaceship, stranded halfway between Earth and Mars sees flashes of light simultaneously from the two planets and concludes they happened at the same time. A second spaceship flashes by the same point at that instant, going toward Mars at one-fourth the speed of light. Assume Mars and Earth to be 66 million miles apart when this happens. What does the moving astronaut observe?

S: Which ship should we label S_1?

T: If we picture Earth on our left and Mars on our right, then the second ship is moving at $v = .25c$ toward the +x-axis. This agrees with our previous equations if we designate the second ship S_2 and the Earth-Mars-First ship system as S_1.

S: Then the times of the flashes in S_1 are equal, $t_1 = t_1'$. In S_2, the time interval between the flashes is

$$t_2 - t_2' = \gamma(v^2/c^2)(x_1' - x_1).$$

T: Here let's say the primed quantities denote Mars, and unprimed ones Earth. So $x_1' - x_1 = 66 \times 10^6$ mi $\times 1.6 \times 10^3$ m/mi.

S: And $v = .25c$, so $\gamma = (1 - .25^2)^{-\frac{1}{2}} = 1.03$.

T: Thus

$$t_2 - t_2' = (.25/3 \times 10^8)(1.03)(66 \times 10^6 \times 1.6 \times 10^3) = 90.6 \text{ sec.}$$

Notice that $t_2 > t_2'$, so the signal from Earth arrives 90.6 seconds *after* the one from Mars.

Mass and Energy

Understanding this section should enable you to do the following:

1. *Calculate the mass of a moving object and its momentum.*
2. *Calculate the rest mass energy of an object.*
3. *Calculate the kinetic energy of a mass moving at close to the speed of light, and calculate the speed of an object from its kinetic energy.*

Speed Limit of the Universe--In the expressions for length and time intervals, if v was greater than c, the factor would involve the square root of a *negative* number. Such a number is called *imaginary*, and cannot represent a physically measurable quantity. We are therefore led to the strange conclusion that *no frames of reference can move faster than the speed of light relative to each other.* The speed of light is thus the *fastest speed possible* in the universe in any frame of reference.

Using the Lorentz equations to find the transformation of velocities from one inertial frame to another, we find that they are related by

$$v_{2x} = (v_{1x} - v)/(1 - v_{1x}v/c^2) \quad , \text{ or } \quad v_{1x} = (v_{2x} + v)/(1 + v_{2x}v/c^2)$$

Example 26-5. A radioactive sample simultaneously emits two electrons in opposite directions, each with a speed of 0.9c relative to the emitter. What would be the speed of one electron relative to the other?

S: It *seems* as though it should be 1.8c, but that would make the Lorentz equations imaginary. How do we view this?

T: Assume you are moving to the left with one electron, so you are in the reference frame S_1 of this electron. The reference frame of the *emitter*, S_2, is then moving at the speed of 0.9c away from you. The other electron has a speed of 0.9 *measured in* S_2. That is, v_{2x} = 0.9c. You have to calculate what this speed is in S_1, that is, calculate v_{1x}.

S: From the results of the Lorentz equations above, we have

$$v_{1x} = (v_{2x} + v)/(1 + v_{2x}v/c^2) = (.9c + .9c)/\{1 + (.9)(.9)\}$$

$$= (1.80)c/(1.81) = 0.994c$$

The formula gives this, but I can't feel that it's right.

T: That's true of most of the consequences of Einstein's relativity. Strange things happen near the speed of light. The Lorentz equations, which are necessary to preserve the equivalence of all inertial systems, unambiguously demand that velocities add this way. However, unless v and v_{2x} are near c, the denominator is essentially 1, and you have the Galilean addition of velocities.

Relativistic Mass--Since $v = at = (F/m)t$ for a constant force, *mass must increase with speed* to be consistent with the requirement that $v < c$, no matter how long the force acts. Mathematically, the Lorentz equations, along with Newton's 2nd Law, give us

$$m = \gamma m_o = m_o(1 - v^2/c^2)^{-\frac{1}{2}} ,$$

where m_o = *rest* mass .

Mass Energy--Another consequence of Einstein's Relativity was the recognition that *mass represents energy.* Even at *rest*, an energy

$$E_o = m_o c^2$$

is associated with the existence of the mass m_o. When moving, the mass increases, and the total mass energy includes the kinetic energy:

$$E = \gamma m_o c^2 = mc^2.$$

Relativistic Kinetic Energy--The kinetic energy is just the difference in the mass energy due to motion:

$$KE = E - E_o = (m - m_o)c^2 = (\gamma - 1)m_o c^2 .$$

This is *not* $\frac{1}{2}mv^2$, unless $v \ll c$.

Important Observations

1. The Newtonian idea that mass is the "quantity of matter" contained in an object is no longer a valid concept. The operational definition $m = F/a$, i.e., the concept of mass being a measure of *inertia* is correct. An object *does not* gain in its number of atoms or amount of material as v approaches c, but its inertia increases, since greater inputs of work produce diminishing increases in velocity. Hence mass is not a fixed property of an object, but depends on the speed of the object relative to the observer.
2. Momentum is also affected by the velocity dependence of m. Relativistic momentum is $p = mv = \gamma m_o v$, and the Principle of Momentum Conservation is obeyed at all speeds.
3. The speed of a relativistic object can no longer be obtained from $\frac{1}{2}mv^2 = KE$. *Even if you use the relativistic mass*, this does not represent the KE. You must find v from $KE = (\gamma - 1)m_o c^2$, if the KE is a few percent or more of the rest energy $m_o c^2$.
4. The rest energy $m_o c^2$ includes all *potential* or *structural* energies arising from forces which hold matter together. This will become clear in the chapter dealing with nuclear structures.

Example 26-6. Calculate the speed of an electron which has been accelerated through a megavolt of electric potential.

(a.) Find the classical result.
(b.) Find the relativistic result.
(c.) Find the mass of this electron.

S: The electron would have 1 MeV of kinetic energy.

T: Right. And 1 eV = 1.6×10^{-19} J. Also $m_e = 9.1 \times 10^{-31}$ kg at rest.

S: Classically, $\frac{1}{2}m_e v^2 = 1.6 \times 10^{-13}$ J.

$$v = \left\{2(1.6 \times 10^{-13}\ \text{J})/(9.1 \times 10^{-31}\ \text{kg})\right\}^{1/2} = 5.93 \times 10^{8}\ \text{m/s}.$$

That's nearly *twice* the speed of light!

T: And it is *not* the speed we actually *observe* for such an electron.

S: I don't understand yet how to find v relativistically.

T: The speed is contained in the factor γ in the expression for KE.

$$KE = 1.6 \times 10^{-13}\ \text{J} = (\gamma - 1)m_0 c^2.$$

S: Oh, I see now. We solve for γ, and then for v.

$$(\gamma - 1) = (1.6 \times 10^{-13}\ \text{J})/(9.1 \times 10^{-13}\ \text{kg})(3 \times 10^{8}\ \text{m/s})^2 = 1.954.$$

Thus

$$\gamma = 2.954 = (1 - v^2/c^2)^{-\frac{1}{2}}.$$

T: So

$$1 - v^2/c^2 = 1/(2.954)^2 = .1146,$$

giving

$$v^2/c^2 = .885, \text{ and } v = .941c.$$

S: That's pretty neat! But is it right?

T: Atomic particles can easily be given energies that classically would have them moving *much* faster than c. Yet the *measurement* of their speeds has always agreed with the prediction of Einstein's Theory. It is a very direct test. Now what's the electron's mass?

S: Oh, yeah. $m = \gamma m_0 = 2.954\ m_0 = 2.69 \times 10^{-30}$ kg.

T: You see, the electron has almost tripled its inertia in *our* frame of reference.

Example 26-7. How much more work would it take to accelerate a 14 pound bowling ball from rest to v = .8c than Newtonian mechanics would predict?

S: Let's see, 14 pounds would be a mass of 14 lbs. x 1 kg/2.20 lbs. = 6.36 kg, here on Earth. The classical KE would be

$$KE_{class} = \tfrac{1}{2}mv^2 = \tfrac{1}{2}(6.36\ \text{kg})(2.4 \times 10^{8}\ \text{m/s})^2 = 1.83 \times 10^{17}\ \text{J}.$$

That's a lot of energy!

<u>T</u>: It is almost the entire energy produced by a major 1000 MW power plant running for 6 years! But we actually need *even more* energy than that, since the mass of the ball will increase with speed.
For v = .8c, γ = 1/(.600) = 1.667 and we need to calculate the work necessary to raise its kinetic energy to

$$KE_{rel} = (\gamma - 1)m_oc^2.$$

<u>S</u>: This comes out to be

$$KE_{rel} = (.667)(6.36 \text{ kg})(3 \times 10^8 \text{ m/s})^2 = 3.82 \times 10^{17} \text{ J}.$$

<u>T</u>: This is more than *twice* the classical prediction.

QUESTIONS ON FUNDAMENTALS

1. If you observe a spaceship whose proper length is 200 m to be passing you at a speed of 0.8c, what length would *you* measure it to be?

(a.) 333 m (b.) 72 m (c.) 160 m (d.) 120 m (e.) 128 m

2. While the spaceship in question #1 is passing you, you cause two lights separated by 1000 meters in your frame of reference to be flashed simultaneously. What time interval between the flashes does a person on the spaceship observe?

(a.) 4.44 μs (b.) 1.33×10^3 s (c.) 1.06×10^3 s (d.) 1.60 μs
(e.) 1.33 μs

3. How much rest mass energy does a 1000 kg car have?

(a.) 3×10^{11} J (b.) 9×10^{16} J (c.) 9×10^{19} J
(d.) 3×10^{5} J (e.) 9×10^{22} J

4. How fast would an object (*any* object) have to be moving in order for its kinetic energy to equal its rest mass energy?

(a.) .75c (b.) .25c (c.) .33c (d.) .625c (e.) .866c

GENERAL PROBLEMS

Problem 26-1. An instrument in Detroit detects an explosion in New York City and 3 msec later detects one in Chicago, 1200 km away from New York. In principle, could an instrument moving from New York to Chicago record that the Chicago explosion *preceded* the one in New York? If so, how fast must the instrument be going?

Q1: How much time does it take for light to travel 1200 km?
Q2: Could the event in New York have *caused* the event in Chicago?
Q3: What is the expression for the time interval observed by the moving instrument?
Q4: What will determine the speed at which the moving instrument will detect the Chicago explosion first?
Q5: Can't that always be done in any case?

Problem 26-2. In our reference frame, a stationary neutron is struck by another neutron having velocity .88c. The neutrons stick together after collision. How fast are they traveling? What would classical theory predict?

Q1: Is momentum conserved?
Q2: What is the system's momentum before collision?
Q3: What is the system's momentum after collision?
Q4: What result would classical theory give?

Problem 26-3. Electrons move through a high energy linear particle accelerator 2 miles long at speeds which make their masses 5×10^4 times their rest masses. What are the travel times in the lab reference frame and as "seen" by the electron?

Q1: How far do the electrons travel in the lab frame?
Q2: What is their speed?
Q3: How much time does the trip take in the lab frame?
Q4: How long is the accelerator in the electron's reference frame?
Q5: What time does the trip take as the electron "sees" it?

Problem 26-4. A hypothetical spaceship travels away from the earth at a velocity of $v = 2 \times 10^8$ m/s. A 1000 kg projectile is shot forward from the spaceship at $v_1 = 2.5 \times 10^8$ m/s relative to the spaceship. Compare its KE relative to earth and to the ship.

Q1: What is the speed of the projectile relative to earth?
Q2: What is the KE of the projectile relative to the ship?
Q3: What is the KE of the projectile relative to the earth?

Problem 26-5. If you could form a "gas" of electrons, at what temperature would the average electron have a velocity of c/3?

Q1: What defines an "average" electron?
Q2: How is the electron's speed related to this energy?
Q3: What specific expression yields the temperature?

ANSWERS TO QUESTIONS

FUNDAMENTALS:

1. (d.) 2. (a.) 3. (c.) 4. (e.)

GENERAL PROBLEMS:

Problem 26-1.

A1: 4 msec.

A2: No. No signal or physical influence could have reached Chicago from New York in 3 msec.

A3: If primed quantities refer to Chicago and unprimed ones to New York, we have $(t_2' - t_2) - \gamma(t_1' - t_1) + \gamma(v/c^2)(x_1 - x_1')$, where $t_1' - t_1 = 3$ msec, and $x_1 - x_1' = -1200$ km.

A4: At low speeds, t_2' will be greater than t_2 (Chicago's event after New York). The crucial speed is that which makes $t_2' = t_2$. Speeds greater than that will make $t_2' - t_2$ negative, i.e., $t_2 > t_2'$.

A5: No. v cannot be greater than c. If $t_1' - t_1$ is larger than the travel time for light, you cannot make $t_2 > t_2'$ unless $v > c$, a physical impossibility.

Problem 26-2.

A1: Yes.

A2: $P_{tot} = \gamma m_o v = (2.11)(1.0087\ u)(.88c)$.

A3: The neutrons will move together with speed v', making $\gamma' = (1 - v'^2/c^2)^{-\frac{1}{2}}$. The momentum will be $P_{tot}' = \gamma'(2m_n)v'$. v' is then found by equating $P_{tot} = P_{tot}'$.

A4: Classical theory would erroneously give $v' = .44c$.

Problem 26-3.

A1: The length is the *proper* length, $L_o = 2$ miles.

A2: You get it by solving $(5 \times 10^4)m_o = m_o$. That is, what speed v makes $\gamma = 5 \times 10^4$?

A3: $\Delta t = L_o/v$.

A4: It is shortened: $L = L_o/\gamma = L_o/(5 \times 10^4)$.

A5: $\Delta t' = L/v = (.2 \times 10^{-4})\Delta t$.

Problem 26-4.

A1: $v_2 = (v + v_1)/(1 + v_1 v/c^2) = 2.89 \times 10^8$ m/s $= .964$ c.

A2: $(KE)_1 = (\gamma_1 - 1)m_o c^2$, where $\gamma_1 = (1 - v_1^2/c^2)^{-\frac{1}{2}} = 1.81$.

A3: $(KE)_2 = (\gamma_2 - 1)m_o c^2$, where $\gamma_2 = (1 - v_2^2/c^2)^{-\frac{1}{2}} = 3.76$.

Problem 26-5.

A1: One that has average kinetic energy, (3/2)kT.

A2: The correct relativistic expression is $KE = (\gamma - 1)m_e c^2 = (3/2)kT$.

A3: $T = (2/3k)(\gamma - 1)m_e c^2$, where $\gamma = (1 - 1/9)^{-\frac{1}{2}} = 1.061$.

CHAPTER 27. THE EXPERIMENTAL BASIS OF QUANTUM MECHANICS

BRIEF SUMMARY AND RESUME

The concept of the *photon*, or *particle* theory of light is introduced, and four separate experiments which agree with its predictions are reviewed. They are:

1. The Blackbody Radiation Spectrum.
2. The Photoelectric Effect.
3. X-Ray Spectra.
4. The Compton Effect.

The *energy* of a photon, called a *quantum* of light energy, is related to the light's wavelength and frequency. The paradoxical *duality* of the wave and particle nature of light is discussed, and extended to the *wave properties of material particles*. The experimental evidence for these properties is demonstrated in the results of *electron diffraction* and *interference*, and in the development of *electron microscopes*. Finally, the *Heisenberg Uncertainty Principle* is introduced.

Material from previous chapters relevant to Chapter 27 includes:

1. Radiation of energy from a heated object.
2. The electron-volt of energy.
3. Momentum and energy conservation in collisions.
4. Diffraction and interference of waves.

NEW DATA AND PHYSICAL CONSTANTS

1. Planck's Constant (h).

$$h = 6.62 \times 10^{-34}\ \text{J-s} = 4.14 \times 10^{-15}\ \text{eV-s}.$$

2. Compton Wavelength (λ_C).

$$\lambda_C = h/m_0c = .00243\ \text{nm (for the electron mass)}.$$

NEW CONCEPTS AND PRINCIPLES

The Particle Nature of Light

Understanding this section should enable you to do the following:

1. *Identify experimental results which were in direct conflict with predictions of the wave theory of light.*
2. *Identify how each of the above results was explained by the particle theory of light.*
3. *Calculate the energy density in a given small wavelength interval of the radiation emitted by a heated object.*
4. *Calculate the energy of a photon of a given wavelength or frequency.*
5. *Calculate the relation between the work function of a metal and the kinetic energies of photoelectrons emitted by light of various wavelengths striking the metal.*
6. *Calculate the highest frequency photon produced by a given voltage in an X-ray tube.*

7. *Calculate the momentum of a photon and the wavelength shift occurring when they scatter off free electrons at various angles.*

Blackbody Radiation--The radiation emitted by a perfect radiator (or absorber) when heated to a temperature T. Chapter 14 dealt with two experimental characteristics of this radiation:

Stefan's Law: $P = \sigma A T^4$, where $\sigma = 5.67 \times 10^{-18}$ W/m^2-K^4

Wien's Law: $\lambda_{max.\ intensity} = (2.898 \times 10^{-3}$ m-K)/T

The theory of the *spectrum* of this radiation was developed by Planck. *Per unit volume* of space, the energy within a small spread of wavelength $\Delta\lambda$ centered around the wavelength λ is given by

$$E(\lambda)\Delta\lambda = \frac{8\pi(h\nu)}{\lambda^4(e^{(h\nu/kT)} - 1)}$$

Here ν = the frequency (= c/λ) of the radiation
k = Boltzmann's Constant = 1.38×10^{-23} J/K
h = Planck's Constant = 6.62×10^{-34} J-s

Photon Energy--In the above, Planck postulated that electromagnetic radiation occurs in "bundles" of *quanta* of energy proportional to the frequency of the radiation. The constant of proportionality, h, is called *Planck's constant*. Mathematically,

$$E_{quantum} = h\nu = hc/\lambda \ .$$

These quanta are known as photons.

Important Observations

1. The experimental blackbody spectral curves like those shown in Figure 27.4 (text) couldn't be theoretically explained by classical electromagnetic theory. The key lay in Planck's hypothesis that E-M energy was emitted in *quantized* amounts proportional to the frequency. These photons act like individual packets or particles of energy. The higher the radiation frequency, the greater the photon's energy. Planck first determined the value of h by seeing what value would produce the best fit of his theory to the experimentally observed spectra.
2. With the methods of calculus it is straightforward to show that this theoretical expression yields both the form of the Wien and Stefan laws and the values of the empirical constants they contain.
3. The volume in which $E(\lambda)$ is the energy spectrum is the region of the blackbody in which the radiation is in thermal equilibrium with the material at temperature T. This could be the cavity of the blackbody shown in Figure 27.2 (text), or the photosphere (the outer few hundred kilometers) of a star, for example. As the energy is radiated away from the blackbody it of course diminishes in intensity, but the relative amounts of energy at various wavelengths remains the same as the $E(\lambda)$ spectrum.

Example 27-1. What is the radiation energy density from a 2000 K blackbody within the wavelength range 395 nm to 405 nm (violet) relative to that within 5 nm on either side of the wavelength at which the *maximum* energy density exists?

S: The Planck formula for $E(\lambda)$ looks pretty tricky. Any suggestions to make it easier?

T: It helps to explicitly calculate $h\nu$ and $h\nu/kT$ right at the outset, using in the middle of the wavelength interval. For $\lambda = 400$ nm, the photon energy is

$$hc/\lambda = (6.62 \times 10^{-34}\ \text{J-s})(3 \times 10^{8}\ \text{m/s})/(4 \times 10^{-7}\ \text{m})$$
$$= 4.97 \times 10^{-19}\ \text{J} = 3.11\ \text{eV}$$

Also

$$kT = (1.38 \times 10^{-23}\ \text{J/K})(2000\ \text{K}) = 2.76 \times 10^{-20}\ \text{J} = .173\ \text{eV}.$$

Then $(hc/\lambda)/kT = 18.0$. Now you can easily plug these in.

S: O.K., here goes: at $\lambda = 400$ nm and with $\Delta\lambda = 10$ nm,

$$E(400\ \text{nm})\Delta\lambda = 8\pi(4.97 \times 10^{-19}\ \text{J})(10^{-8}\ \text{m})/(4 \times 10^{-7}\ \text{m})^4(e^{18.0} - 1)$$

With $e^{18} = 6.57 \times 10^{7}$, this becomes

$$E(400\ \text{nm})\Delta\lambda = 7.44 \times 10^{-8}\ \text{J/m}^3.$$

T: Notice that, since each photon has an energy of 4.97×10^{-19} J, there are $7.44 \times 10^{-8}/4.97 \times 10^{-19}$ = *1.5 x 10^{11} photons per m^3* within this small wavelength range in the volume of the blackbody. The large exponential factor in the denominator severely reduces the number of photons when their energies are much greater than kT.

S: What is the wavelength at which $E(\lambda)$ will be a *maximum*?

T: That's given by Wien's Law as

$$\lambda_{max} = (2.898 \times 10^{-3}\ \text{m-K})/(2000\ \text{K}) = 1.45 \times 10^{-6}\ \text{m}.$$

This is 1450 nm, well into the infrared.

S: These photons have energy given by

$$hc/\lambda = (6.62 \times 10^{-34}\ \text{J-s})(3 \times 10^{8}\ \text{m/s})/(1.45 \times 10^{-6}\ \text{m})$$
$$= 1.37 \times 10^{-19}\ \text{J} = .856\ \text{eV}.$$

Also

$$(hc/\lambda)/kT = .856\ \text{eV}/.173\ \text{eV} = 4.95.$$

So

$$E(\lambda) = 8\pi(1.37 \times 10^{-19}\ \text{J})(10^{-8}\ \text{m})/(1.45 \times 10^{-6}\ \text{m})^4(e^{4.95} - 1)$$
$$= 5.52 \times 10^{-5}\ \text{J/m}^3.$$

> T: This is almost *750 times the energy density at 400 nm.* And it represents $5.52 \times 10^{-5}/1.37 \times 10^{-19}$ = *1×10^{14} photons per meter^3* within this range of wavelengths, about 2700 times as many as in the violet spectrum. Now if we picked longer wavelengths than this, we would find that $E(\lambda)$ *decreases* again, due to the λ^4 factor in the denominator.

The Photoelectric Effect--The process by which electrons are ejected from metallic surfaces when illuminated by light of certain frequencies is called the photoelectric effect. A detailed description is given in Section 27.3 of your text. To summarize,

1. Unless the light is above a certain threshold frequency ν_o, which depends on the specific metal target being used, *there is no effect* regardless of light intensity or duration of illumination.
2. If the light has a frequency above ν_o, there is an effect *instantly*, no matter how weak the intensity. The *amount* of photocurrent, i.e., the number of electrons released, is proportional to the light intensity.
3. The ejected electrons exhibit a range of kinetic energy from zero up to a certain *maximum* which is directly proportional to the amount by which the frequency exceeds the threshold. Mathematically,

$$KE_{max} = h(\nu - \nu_o).$$

This maximum does not depend on the light intensity.

Important Observations

1. As Table 27.1 (text) shows, *every* result of the experiment except the proportionality of the photocurrent to light intensity was in *disagreement* with the classical picture of the electrons gradually absorbing *wave* energy until they could break free from the metal.
2. The crucial point in Einstein's explanation of the effect was that the mechanism of energy transfer to the electrons was by a *collision* of the electrons with photons, in which a fixed amount of energy, $h\nu$, was given to an electron in a single instant. Either this was enough to free the electron, or it wasn't ejected.
3. A different amount of energy is required to release an electron from various metals. This is called the metal's *work function*, Φ. The photon energy must be *at least* equal to Φ in order to release an electron. Thus the threshold frequency is given by $h\nu_o = \Phi$.
4. A more *intense* light beam doesn't contain more *energetic* photons, but instead contains a greater *number* of photons, each with $E = h\nu$.
5. Some electrons lying deeper in the metal expend more energy than Φ in getting out, and hence would have less KE after release than the maximum amount observed.

Example 27-2. In a classroom laboratory, when light of 350 nm strikes the surface of a metal, the measured stopping voltage for the most energetic electrons is .572 volts. For light of 390 nm wavelength, it is .205 volts. What value of Planck's constant would be obtained from this data? What was the metal, according to Table 27.2 (text)?

S: You're saying we can't assume the value of h, and we don't know Φ. Is Φ the same for both wavelengths?

T: Sure. It's determined entirely by the metal. Write the photoelectric equation for both wavelengths.

S: We have

$$e(.572 \text{ v}) = h(3 \times 10^8 \text{ m/s})/(3.50 \times 10^{-7} \text{ m}) - \Phi,$$

and

$$e(.205 \text{ v}) = h(3 \times 10^8 \text{ m/s})/(3.90 \times 10^{-7} \text{ m}) - \Phi .$$

What units are most convenient when I'm dealing with volts and electron charges?

T: This is ready made for the use of eV's of energy, in which you treat e as having a value of 1 (1 electron charge). h will then come out in eV-sec. Subtracting the equations eliminates Φ:

$$(.572 \text{ eV} - .205 \text{ eV}) = h(3 \times 10^8 \text{ m/s})(1/3.5 - 1/3.9) \times 10^7 \text{ m}^{-1}$$

S: Then

$$h = 4.17 \times 10^{-15} \text{ eV-s.}$$

T: That's within 1% of the accepted value. Now you can go back to either equation to find Φ.

S: The first equation gives us

$$\Phi = (4.17 \times 10^{-15} \text{ eV-s})(3 \times 10^8 \text{ m/s})/(3.50 \times 10^{-7} \text{ m}) - .572 \text{ eV}$$
$$= 3.00 \text{ eV.}$$

T: The metal would seem to be *aluminum*, at least from those listed in Table 27.2.

X-Rays--When high energy electrons (typically 10 - 100 keV) collide with a metal target, electromagnetic radiation is produced with wavelengths of less than 1 nm. This radiation is very penetrating and is called *X-rays*. The spectrum has the following characteristics:

1. A broad continuous spectrum with a sharply defined *minimum* wavelength, called the *"cut-off" wavelength*. This is independent of which metal is used as a target, and is equal to $\lambda_o = 1.24 \times 10^{-6}/V$ meters. Here V is the voltage through which the electrons were accelerated, and represents the energy of the electrons in eV.

2. A number of sharp intensity peaks superimposed on the continuous spectrum, which do depend on the target material. These will be discussed in Chapter 28.

Important Observations

1. Classical electromagnetic theory predicts radiation from decelerated electrons. But the sharp wavelength cut-off is more evidence for the photon theory. Notice that $hc = 1.24 \times 10^{-6}$ eV-m, and so the expression for λ_o is just

$$\lambda_o = hc/E_{electron}.$$

The electron cannot create electromagnetic radiation of higher frequency than the photons which represent its total energy.

X-Ray Diffraction--Since X-ray wavelengths are of the same order of magnitude or smaller than typical atomic separations in solid crystals (< .1 nm), such crystals can serve as a *diffraction grating from X-rays* reflected by the crystal. This situation is shown in Figure 27.22 (text). The angles for constructive iterference in the reflected beam are sharply defined and are given by the *Bragg equation*:

$$2d \sin \theta_n = n\lambda \text{ , where } n = 1, 2, 3, \ldots$$

Here d = the distance between adjacent atoms, or the *lattice spacing* of the crystal.
Q = the angle of incidence and reflection *relative to the plane* of the crystal.

Important Observations

1. In this case, the angles are *not* measured from the normal to the plane, but from the plane itself.
2. Notice the similarity of the Bragg equation to the thin film interference equation for normal incidence, $2t = (m + \frac{1}{2})\lambda$. If you put $\theta = 90^\circ$ in the Bragg equation, and forget the 1/2 cycle phase shift at one surface of the thin film, the equations are really describing the same kind of phenomena even though in quite different physical situations.

Example 27-3. It is found that a certain crystal produces constructive interference at 90° and two other angles for X-rays having $\lambda = .140$ nm.

(a.) What is the lattice spacing of the crystal?
(b.) What are the other two angles for constructive interference?

S: How do we know what order the 90° beam is?

T: The lowest order corresponds to the *smallest* angle. Since there are only three observed intensity peaks, the 90° angle must represent the third order, n = 3.

S: So

$$2d \sin 90^\circ = 2d = 3\lambda.$$

Then we get $d = (3/2)\lambda = .210$ nm. The other angles must have n = 1 and n = 2 then?

T: Right. Using $d = (3/2)\lambda$, we have

$$2(3/2)\lambda \sin\theta_2 = 2\lambda, \text{ giving } \theta_2 = 41.8^\circ,$$

and

$$2(3/2)\lambda \sin\theta_1 = \lambda, \text{ giving } \theta_1 = 19.5^\circ.$$

The Compton Effect--When X-rays are scattered off *free* electrons, the scattered radiation has an increased wavelength which depends on the angle of scattering. This phenomenon is called the *Compton Effect.* Compton explained this as the result of an *elastic collision* between a photon and an essentially free electron. The laws of momentum conservation and energy conservation result in the following:

$$\Delta\lambda = (h/m_0c)(1 - \cos\theta) = \lambda_C(1 - \cos\theta),$$

where $\Delta\lambda$ = the change in wavelength of the photon.
$\lambda_C = h/m_0c$ = the Compton wavelength = .00243 nm for the electron's mass.
θ = the angle of the scattered photon relative to its incident direction.

Important Observations

1. A photon has energy $E_{ph} = hc/\lambda$ and momentum $P_{ph} = h/\lambda$, but *always* travels at the speed c. When some energy and momentum are transferred to the electron in a collision, the photon cannot slow down as a material particle would. It can only increase its *wavelength* (or equivalently, decrease its *frequency*).
2. In the photoelectric effect, the photon could be absorbed and cease to exist because the metal atoms were involved in the conservation of momentum. For a collision off of a *free* electron, the photon must retain some momentum (and therefore its very existence) to satisfy momentum conservation.
3. Even though the electrons are in a solid target, their binding energy is negligible compared to the energy of the X-ray photons, so they act as if free when the collision occurs.

Example 27-4. X-rays having wavelength .1500 nm strike free electrons and at a certain angle are observed to have a wavelength of .1512 nm.

(a.) What is the scattering angle being observed?
(b.) What fraction of the original photon energy is transferred to the recoiling electrons?
(c.) What is the longest wavelength observed for the scattered photons at *any* angle?

S: We have $\Delta\lambda = \lambda_C(1 - \cos\theta)$. Solving for cosθ, we get $\cos\theta = 1 - \Delta\lambda/\lambda_C$. We're given that $\Delta\lambda = .1512 - .1500$ nm = .0012 nm Thus

$$\cos\theta = (1 - .0012 \text{ nm})/(.00243 \text{ nm}) = .494 ,$$

giving

$$\theta = 60.4^\circ .$$

What determines the KE of the electrons?

T: The energy loss represented by the increase in λ must have been transferred to the electron:

$$\Delta E_{ph} = hc(1/\lambda_2 - 1/\lambda_1) = -\Delta E_{elect}.$$

Again it's the usual choice to use eV of energy here.

S: Since $hc = 1.24 \times 10^{-6}$ eV-m,

$$E_{elect} = (1.24 \times 10^{-6} \text{ eV-m})(1/.1500 - 1/.1512) \times 10^9 \text{ m}^{-1} ,$$

$$= 64.8 \text{ eV}.$$

T: The original photons had energy

$$E_1 = hc/\lambda_1 = (1.24 \times 10^{-6} \text{ eV-m})/(.1500 \times 10^{-9} \text{ m})$$

$$= 8.27 \times 10^3 \text{ eV}.$$

Thus the *fraction* of energy transferred is small:

$$E_{elect}/E_1 = (64.8 \text{ eV})/(8.27 \times 10^3 \text{ eV}) = 7.8 \times 10^{-3} = 0.78\%.$$

S: Would we observe the longest wavelength for photons scattered at 90°, where $\cos\theta = 0$?

T: No, that's *not* right. Remember, $\cos\theta = -1$ at $\theta = 180^\circ$, which would represent head-on collisions, with the photons scattered back along the incident direction. They would have the largest $\Delta\lambda$, $= 2\lambda_C = .00486$ nm. So their wavelength would be $\lambda_2 = .1500 + .00486 = .1549$ nm, to four significant figures.

Example 27-5. For Compton scattering of the X-rays in the last example at $\theta = 180^\circ$, what fraction of the original photon momentum is taken up by the recoil electron? How fast are these electrons moving?

S: With momentum we have a vector problem.

T: In general, yes, but for scattering at $\theta = 180^\circ$, there is only *one* dimension to the velocities of photon and electron. What is the original momentum?

S:

$$P_1 = h/\lambda_1 = (6.62 \times 10^{-34} \text{ J-s})/(.1500 \times 10^{-9} \text{ m}) = 4.4 \times 10^{-24} \text{ J-s/m}.$$

We need to find P_{ph}, which is equal to P_{elect}.

T: Notice that, since P_{ph} α λ^{-1} we can quickly find P_2:

$$P_2 = P_1(\lambda_1/\lambda_2) = (4.4 \times 10^{-24} \text{ J-s/m})(.1500/.1549)$$

$$= 4.26 \times 10^{-24} \text{ J-s/m}.$$

So

$$P_{ph} = -.14 \times 10^{-24} \text{ J-s/m} = -P_{elec}.$$

S: Then

$$P_{elec}/P_1 = (.14 \times 10^{-24})/(4.4 \times 10^{-24}) = .032 = 3.2\%.$$

How do we find the speed of the electron? Do we have to use relativity?

T: A *very good* question. A quick way to estimate whether classical physics is valid is to see if the KE is a significant fraction of the *rest* energy, m_0c^2. If it is, you have to use the relativistic expressions for KE. For the electron, m_0c^2 = 511 keV. In our problem, its KE is only of the order of 100 eV. So we very definitely have a *classical* problem, where you can use KE = $\frac{1}{2}m_0v^2$ and p = m_0v.

S: Then

$$v = P_{elec}/m_0 = (14.0 \times 10^{-26} \text{ J-s/m})/(9.1 \times 10^{-31} \text{ kg}) = 1.54 \times 10^5 \text{ m/s}.$$

T: You see, this is only about .0005 times the speed of light, so our classical approach was justified.

The Wave Nature of Particles

Understanding this section should enable you to do the following:

1. *Calculate the wavelength associated with a particle's momentum.*
2. *Calculate the diffraction effects produced by particles impinging on crystals having certain inter-atomic separation.*
3. *Calculate the relationship between uncertainties in momentum and position and in energy and time.*

deBroglie Waves--In 1924 L. deBroglie made the postulate that material particles possess *wave* properties, described by a wavelength related to the particle's momentum in the same way as for the photon. Mathematically,

$$\lambda = h/p = h/mv .$$

Electron Diffraction--It is very easy, by accelerating electrons through a hundred volts or so, to produce electron wavelengths comparable to X-rays. Such electrons behave in the same way as X-rays when they strike a crystalline target, exhibiting maximum *numbers* of reflected electrons at angles given by the Bragg equation, using the deBroglie formula for their wavelength.

Important Observations

1. Because of the extreme smallness of h on our scale of experience, particle momenta must be very small to make their wave nature observable. Thus it is only the elementary particles, with extremely small masses, for which this

duality of behavior is significant.

2. A handy reference formula for *electrons* is that when KE is measured in *eV's*, the deBroglie wavelength is

$$\lambda = (1.23)/(KE)^{\frac{1}{2}} \text{ nanometers.}$$

See Example 27.7 (text).

<u>Example 27-6</u>. Suppose you accelerate electrons through 3 kv of potential. Some of them strike a metal plate and create X-rays. The X-rays with the shortest wavelength are reflected off a crystal whose atoms are .300 nm apart. The other 3 kV electrons impinge directly on the same crystal. How do the first order interference effects compare?

<u>S</u>: Wouldn't the wavelengths and therefore the interference effects be the same?

<u>T</u>: Not at all. The cut-off wavelength for X-ray production is proportional to 1/voltage, whereas the wavelength for particles (non-relativistic) goes as $1/(\text{voltage})^{\frac{1}{2}}$.

<u>S</u>: Hmm. Well, for the X-rays, $\lambda_o = (1.24 \times 10^{-6})/(3000)$ meters = .41 nm.

<u>T</u>: So the first order interference maximum would occur at

$$2(.300 \text{ nm})\sin\theta_1 = \lambda_o = .41 \text{ nm},$$

or

$$\theta_1 = \sin^{-1}(.683) = 43.1^\circ.$$

<u>S</u>: According to the formula in the observations above, the electrons will have a wavelength of

$$\lambda = (1.23)/(3000)^{\frac{1}{2}} = .0224 \text{ nm}.$$

That *is* a difference!

<u>T</u>: The difference lies in the fact that photons are by their very nature completely relativistic. If we were dealing with extremely fast (> 10 MeV) electrons, they would show the same dependence of wavelength on energy from the relativistic equations as the photons have.

<u>S</u>: These electrons would show a first order interference maximum at

$$2(.300 \text{ nm})\sin\theta_1 = .0224 \text{ nm},$$

or

$$\theta_1 = \sin^{-1}(.0373) = 2.14^\circ.$$

<u>T</u>: This crystal lattice is too *large* to give good order resolution for these short wavelengths. The second order would be at

$$\theta_2 = \sin^{-1}(.0746) = 4.28^\circ.$$

On the other hand, you can see that the X-rays show only the first order maximum.

The Heisenberg Uncertainty Principle--The wave-particle duality of the properties of light and particles is so fundamental that Heinsenberg described it in terms of an unavoidable level of inherent uncertainty in determining the position and momentum of anything simultaneously. The process of measuring *one* of these quantities very precisely *must* disturb or change the system so that the other quantity is *less precisely determinable*. This "trade-off" in determination *cannot* be totally overcome by more and more precise instruments or measuring techniques. Heisenberg's *Uncertainty Principle* states that there is an ultimate limit to *simultaneous* precision, given by

$$(\Delta x)(\Delta P_x) \geq h/2\pi.$$

There is an equivalent minimum uncertainty of the energy of a system and the time during which the system has that energy:

$$(\Delta E)(\Delta t) \geq h/2\pi.$$

Important Observations

1. Once again, the smallness of h means that this principle is significant only on the atomic scale, where the quantities involved are very small. For macroscopic objects, these ultimate uncertainties are well below any conceivable limit of attainable precision, and so we do not perceive that it is impossible, even in principle, to measure as precisely as we wish.

Example 27-7. Consider a beam of 10 eV electrons along the z direction, impinging on a slit of width x_s perpendicular to the beam.

(a.) What uncertainty in P_x does the slit give to the electrons for $x_s = 10^{-6}$ m, and for $x_s = 10^{-8}$ m?
(b.) Through what angle would you expect the electron beam to spread as it emerges from the slit?

S: How does the slit give an uncertainty to the momentum?

T: By *localizing* the electron in the x-direction within $x = x_s$. The Uncertainty Principle (U. P.) states that when that is done, the x-momentum cannot be precisely known. It was assumed that $P_x = 0$ in the original beam.

S: Well, following the U.P., we would have

$$(\Delta P_x)(\Delta x_s) \geq h/2\pi, \quad \text{so} \quad P_x \geq h/2\pi x_s.$$

For $x_s = 10^{-6}$ m, this gives $p_x \geq 1.05 \times 10^{-28}$ J-s/m

And for $x_s = 10^{-8}$ m, we get $P_x \geq 1.05 \times 10^{-26}$ J-s/m.

So what? Isn't this small enough to be neglected?

T: Not hardly. The electrons are also small, and have small momentum. What is their momentum, P_z?

S: Let's see, we have KE = 10 eV, so the electrons are not relativistic. So

$$P_z^2 = 2m_e(KE) = 2(9.1 \times 10^{-31} \text{ kg})(10 \times 1.6 \times 10^{-19} \text{ J})$$

Thus

$$P_z = (2.9 \times 10^{-48} \text{ kg-J})^{\frac{1}{2}} = 1.71 \times 10^{-24} \text{ J-s/m}.$$

T: You're doing pretty well with those units. Are you guessing, or what? Anyway, you can now see that

$$(\Delta P_x)/P_x \geq .006 \text{ for } x_s = 10^{-8} \text{ m, and}$$

$$\geq .00006 \text{ for } x_s = 10^{-6} \text{ m.}$$

Notice that these are the *angles* (in radians) of the spread of the beam in the x-direction.

S: Converting to degrees, we have

$$\theta = .34^\circ \text{ and } .0034^\circ, \text{ respectively.}$$

These are also very small.

T: Now calculate the electron's wavelength.

S: We have $\lambda = 1.23/(10\text{eV})^{\frac{1}{2}} = .389$ nm.

T: You see, we're looking at *diffraction* effects here, even though the effects are small because the aperture is still quite wide compared to λ. For comparison, at what angle would the first dark fringe occur for *light* of this wavelength passing through the 10^{-8} m slit?

S: From Chapter 24, we have

$$y_1/L = \theta_1 = \lambda/x_s = (.389 \text{ nm})/(10 \text{ nm}) = .039 \text{ radians.}$$

I see what you're getting at. The spread in ΔP_x for the electrons is comparable with the spreading of the diffraction pattern for the same wavelength of light.

T: Right. And furthermore, the spreading in momentum gets larger for narrower apertures. The U.P. gives only a *crude* picture of what happens. A total wave analysis of the electrons predicts an intensity pattern for the electrons passing through the slit that is *identical* to that for light of the same wavelength, and experiment shows that this is indeed true.

QUESTIONS ON FUNDAMENTALS

1. At what temperature would a black body radiate its peak intensity at a wavelength of 1 A° (X-Rays)?

(a.) 29 million K (b.) 29 billion K
(c.) 2.9 million K (d.) 29 thousand K
(e.) 2.9 million K

2. How much radiated power would the black body in question #1 emit per square meter of its area?

(a.) 1.5×10^{32} W (b.) 4×10^{12} W (c.) 1.64×10^{8}
(d.) 4×10^{15} W (e.) 1.64×10^{6} W

3. What is the longest wavelength of 0.200 A° X-rays which could be observed to scatter from collisions with free electrons?

(a.) 0.176 A° (b.) 0.151 A° (c.) 0.200 A°
(d.) 0.224 A° (e.) 0.249 A°

4. How fast would electrons have to travel in order to have a de Broglie wavelength of 500 nm?

(a.) 1460 m/s (b.) 8300 m/s (c.) 230 m/s
(d.) 1.5 m/s (e.) 3×10^{7} m/s

5. In order for an electron to be located to within an accuracy of 500 nm, what would be the *minimum* value of its speed?

(a.) 1440 m/s (b.) 230 m/s (c.) 2.2 m/s
(d.) 110 m/s (e.) 1310 m/s

ANSWERS

1. (a.)
2. (b.)
3. (e.)
4. (a.)
5. (b.)

CHAPTER 28. THE STRUCTURE OF ATOMS

BRIEF SUMMARY AND RESUME

Empirical data on the *spectrum* of light emitted by *hydrogen* atoms is reviewed, and summarized by the *Rydberg equation.* The experiments of *Rutherford*, confirming the *nuclear model* of the atom, are described. *Bohr's postulates* about the electron orbits in hydrogen are introduced, and the resulting *energy levels* are shown to explain the hydrogen radiation spectrum. The absorption of energy by hydrogen in resonance with the energy level structure is discussed, and the *Franck-Hertz* experiment and the operation of the *laser* are described as evidences supporting the theory.

The complete set of *quantum numbers* specifying a definite state of the electron is introduced, and the periodic table of the elements is examined by application of the *Pauli Exclusion Principle* involving these quantum numbers. The *characteristic X-ray spectra* are analyzed and interpreted by the Bohr theory. The *deBroglie wave theory* is applied to the electron of the hydrogen atom and found to agree precisely with Bohr's postulates. Finally, the limitations of the Bohr model are briefly discussed.

Material from previous chapters relevant to Chapter 28 includes:

1. The photon theory of light.
2. The Coulomb force and electric potential energy.
3. Centripetal force and circular orbits.
4. The deBroglie particle wave theory.

NEW DATA AND PHYSICAL CONSTANTS

1. <u>The Rydberg Constant (R)</u>

   $R = 1.0968 \times 10^7\ m^{-1}$

2. <u>The Bohr Ground State Energy (E_1)</u>

   $E_1 = m_e k_e^2 e^4/2h^2 = -13.606$ eV

3. <u>The First Bohr Radius (r_1)</u>

   $r_1 = h^2/m_e k_e e^2 = .0528$ nm.

NEW CONCEPTS AND PRINCIPLES

The Hydrogen Spectrum

Understanding this section should enable you to do the following:

1. *Identify the wavelength regions of the Lyman, Balmer, and Paschen series in the hydrogen spectrum.*
2. *Calculate the wavelengths of these spectral lines from the Rydberg formula.*

<u>The Rydberg Formula</u>--Hydrogen atoms emit a spectrum of light which a spectrometer reveals to be a complex series of discrete wavelengths, or *spectral lines*. The lines are grouped into *series* by rather wide gaps of separation in wavelength. There is one series in the ultraviolet, one in the visible spectrum, and a number of them in the infrared. The empirical values of the wavelengths are summarized by the *Rydberg formulas*:

Lyman series (ultraviolet):

$$1/\lambda_n = R(1/1^2 - 1/n^2), \quad \text{where } n = 2, 3, 4, \ldots$$

Balmer series (visible):

$$1/\lambda_n = R(1/2^2 - 1/n^2), \quad \text{where } n = 3, 4, 5, \ldots$$

Paschen series (infrared):

$$1/\lambda_n = R(1/3^2 - 1/n^2), \quad \text{where } n = 4, 5, 6, \ldots$$

Here R is the same constant for all series, called the *Rydberg constant*, with an experimental value of $R = 1.0968 \times 10^7 \text{ m}^{-1}$.

<u>Important Observations</u>

1. Each of these series has the same pattern. Starting with the *longest* wavelength, the lines converge with decreasing separation to a short wavelength *series limit*. There are an *infinite* number of lines in each series, since n has no limit, but the lines become *infinitesimally* separated as the series limit is reached.
2. The series limit is the wavelength corresponding to $n = \infty$ in each series. Thus $1/\lambda_\infty = R$, $R/4$, and $R/9$ respectively for the three series listed above.
3. The *longest* wavelength in each series is obviously the one with the smallest value of n. These are:

$$\text{Lyman}: \quad 1/\lambda_2 = (3/4)R$$

$$\text{Balmer}: \quad 1/\lambda_3 = (5/36)R$$

$$\text{Paschen}: \quad 1/\lambda_4 = (7/144)R$$

These series *do not overlap*.

<u>Example 28-1</u>. Calculate the *range* of wavelengths occupied by each of the three series discussed above.

<u>S</u>: The three series *limits* are:

<u>Lyman</u>: $\lambda_\infty = 1/(1.0968 \times 10^7 \text{ m}^{-1}) = 91.174 \text{ nm}$

<u>Balmer</u>: $\lambda_\infty = 4/(1.0968 \times 10^7 \text{ m}^{-1}) = 364.7 \text{ nm}$

<u>Paschen</u>: $\lambda_\infty = 9/(1.0968 \times 10^7 \text{ m}^{-1}) = 820.6 \text{ nm}$

T: The Balmer limit is barely into the near ultraviolet. Usually the limit of the visible spectrum is taken to be 400 nm. How about the longest wavelengths?

S: For the Lyman series, λ_2 = 4/(3R) = 121.6 nm, the Balmer series, λ_3 = 36/(5R) = 656.5 nm, and the Paschen series, λ_4 = 144/(7R) = 1875.6 nm.

T: So the series ranges are:

Lyman:	91.17 nm to 121.6 nm
Balmer:	364.7 nm to 656.5 nm
Paschen:	820.6 nm to 1875.6 nm

The Bohr-Rutherford Atom

Understanding this section should enable you to do the following:

1. *State the basic postulates of the Bohr model of electron orbits in the hydrogen atom.*
2. *Calculate the electronic energy levels and orbital radii in the Bohr model.*
3. *Identify the electron transitions which produce the various features of the hydrogen spectrum.*

Bohr's Postulates--With Rutherford's demonstration that atoms consist of a small, dense, positively charged nucleus, Bohr set out to explain how this model could be stable and what mechanism was responsible for the observed line spectrum. The postulates he used were as follows:

1. The electrons can only exist in a set of *discrete* orbits, governed by the rule that their *angular momentum* L = mvr is an integral number times $h/2\pi$. Mathematically,

$$L = n(h/2\pi) = mvr, \text{ with } n = 1, 2, 3, \ldots$$

2. While in any of these *states*, the atom is *stable* and emits no radiation.
3. Energy is radiated from the atom when an electron makes a transition from a higher energy (E_i) to a lower energy (E_f) orbit. When this happens, the energy appears as a *single photon* with energy $h\nu = E_i - E_f$, i.e., with a frequency $\nu = (E_i - E_f)/h$.

Important Observations

1. Notice that the units of h, Planck's "quantum of action," are the units of angular momentum. The first postulate above states that the electron's angular momentum in orbit around the nucleus can only be some multiple of this basic quantum.
2. According to classical electromagnetic theory, the nuclear model of the atom should be extremely *unstable*, since an orbiting (hence accelerating) charge should *continually* radiate energy, spiralling into the nucleus in less than 1 µsec. Furthermore, the radiation should have a broad spectrum, not the discrete wavelengths observed experimentally.

Consequences of the Bohr Postulates--As derived in your text, the Bohr postulates predict the following

1. The *orbital radii* of the electrons are *quantized* also. The allowed radii are given by

$$r_n = n^2 r_1, \quad \text{with } n = 1, 2, 3, \ldots$$

Here r_1 is the 1st Bohr radius, given by

$$r_1 = h^2/m_e k_e e^2 = .0528 \text{ nm}.$$

2. The electron *energies* are *quantized*, and can have only the following values:

$$E_n = -E_1/n^2, \quad \text{with } n = 1, 2, 3, \ldots$$

where $$E_1 = k_e e^2/2r_1 = 13.606 \text{ eV}.$$

Important Observations

1. Combining his postulates with Newton's equations of motion involving the Coulomb force, Bohr was able to derive expressions for r_1 and E_1 in terms of *fundamental constants of nature*, namely e, k_e, m_e, and h.
2. By choosing the potential energy of the electron to be zero at $r = \infty$, all the bound electron orbits have *negative* energy values. The *smallest radius* has the *lowest* (most negative) energy, and is the most *tightly bound* by the attraction of the protons in the nucleus.
3. The lowest energy state, E_1, is called the *ground state* of the hydrogen atom.
4. The value of $|E_n|$ is the amount of energy required to remove the electron completely from the atom. $|E_1|$ in eV is called the *ionization potential* for hydrogen.

Explanation of the Rydberg Formula--The results of the preceding section become more astonishing when the *third* postulate of Bohr is applied. This predicts a photon of frequency $\nu = (E_i - E_f)/h$ or
$1/\lambda = (Ei - Ef)/hc$ would be created when the electron somehow changes states from E_i to E_f. Using the Bohr results of E_n, this means this is the same *form* as the Rydberg formula, and allows a *calculation* of the empirical constant R in terms of fundamental constants. The result is

$$R = E_1/hc = k_e^2 m_e e^4/4\pi h^3 c = 1.0974 \times 10^7 \text{ m}^{-1}.$$

Important Observations

1. A graphic picture of the hydrogen spectrum now results. As Figure 28.7 (text) indicates, the spectral series are explained as follows:

   Lyman: The electron makes a transition from any excited state (n = 2, 3, 4, etc.) to the *ground* state (n = 1).

Balmer: The electron makes a transition to the *1st* excited state (n = 2) from any state of higher energy (n = 3, 4, 5, etc.).

Paschen: The electron makes a transition to the *2nd* excited state from any higher energy state.

2. Bohr's value of R is within .05% of the experimental value! Even this small disparity can be accounted for if one treats the electron and the proton as moving about a common center of mass, rather than the proton being stationary. The electron mass m_e is then replaced by a "reduced" mass, μ:

$$\mu = m_e\{m_p/(m_e + m_p)\}.$$

3. "Hydrogen-like" atoms are ionized elements with only one remaining electron, such as He^+, Li^{++}, etc. These atoms display hydrogen-like spectra. The Bohr model applies directly to them if their correct nuclear charge, Ze, is used in all the Bohr formulas. Thus E_1 and R are *increased* by a factor of Z^2 while r_1 is decreased by 1/Z. For additional refinement, the reduced electron mass appropriate to the particular atom being considered should be used. As the nucleus becomes heavier, however, this becomes less important, since μ is then closer to m_e.

Example 28-2. Classical theory would predict that an orbiting charge would radiate at a frequency equal to its orbital frequency. What orbital frequency does an electron have in the ground state of hydrogen? In the first excited state?

S: The frequency will be the reciprocal of its period, $T = 2\pi r/v$, so $f = v/2\pi r$. What is v?

T: The simplest procedure is to note that $v = p/m_e$, and the angular momentum is $L = pr = nh/2\pi$.

S: So $f = p/2\pi m_e r = nh/4\pi^2 m_e r^2$.

T: Well, almost. You haven't gotten all the dependence on n yet. Remember, the r you have is *quantized*, with values $r_n = n^2 r_1$. So

$$f_n = nh/4\pi^2 m_e n^4 r_1^2 = (h/4\pi^2 m_e r_1^2)n^{-3}.$$

S: Then we get

$$f_1 = h/4\pi^2 m_e r_1^2$$

$$= (6.62 \times 10^{-34}\ \text{J-s})/4\pi^2(9.1 \times 10^{-31}\ \text{kg})(.0528 \times 10^{-9}\ \text{m})^2$$

$$= 6.59 \times 10^{15}\ \text{Hz}.$$

The little devil is really moving!

T: And $f_2 = f_1/2^3 = 6.59 \times 10^{15}/8 = 8.24 \times 10^{14}$ Hz. For comparison, the transition between these two states produces the line λ_2 in the Lyman series, with λ_2 = 121.6 nm. This is a frequency of $\nu = c/\lambda_2$, = 2.47 x 10^{15} Hz. This is somewhere *between* the orbital frequencies we've calculated.

Example 28-3. In mass units, the electron has m_e = .000549 u, the proton has m_p = 1.007276 u, and the nucleus of "heavy hydrogen," the deuteron, has m_D = 2.013553 u. If you had a spectrometer capable of measuring wavelengths to 5 figure precision, would you be able to measure the difference in the series limit of the Lyman and the Balmer series for the two species of hydrogen?

S: The series limit is given by

$$\lambda_\infty = 1/R \text{ (Lyman) and } \lambda_\infty = 4/R \text{ (Balmer).}$$

Why should the two species emit different wavelengths? R is a universal constant, isn't it?

T: R depends on the value of the electron mass, and as observed previously, you can't neglect the fact that the nucleus is not totally stationary. Without proof, it was stated that for ultimate accuracy, the *reduced* mass of the electron should be used instead of m_e. This is

$$\mu = m_e\{m_p/(m_e + m_p)\} \quad \text{(ordinary hydrogen)}$$

and

$$\mu = m_e\{m_D/(m_e + m_D)\} \quad \text{(deuterium)}$$

S: Then R becomes slightly smaller with μ in place of m_e. For hydrogen, $\mu_H = m_e(1.007276/1.007825) = .999455\, m_e$. This would make

$$R_H = (.999455)(1.0974 \times 10^7 \text{ m}^{-1}) = 1.0968 \times 10^7 \text{ m}^{-1}.$$

This is the empirical result.

T: Right, and experiment shows λ_∞ = 91.174 nm (Lyman) and λ_∞ = 364.70 nm (Balmer).

S: For deuterium, $\mu = m_e(2.013553/2.014102) = .99973\, m_e$. This is going to be close!

T: It gives $R_D = (.99973)(1.0974 \times 10^7 \text{ m}^{-1}) = 1.0971 \times 10^7 \text{ m}^{-1}$.

S: So λ_∞ = 91.149 nm (Lyman) and λ_∞ = 364.60 nm (Balmer).

T: You see, the heavier nucleus *shortens* the wavelengths, and the difference shows up in the *fourth* significant figure.

Energy Absorption and Emission by Atoms

Understanding this section should enable you to do the following:

1. *State the basic mechanisms by which electrons absorb energy in atoms.*
2. *Calculate the wavelengths of photons which can be absorbed by atoms with a given energy level structure.*
3. *Interpret the results of the Franck-Hertz experiments.*

Mechanisms for Energy Absorption by Atoms--There are three basic mechanisms by which an atom absorbs energy from its surroundings:

(a.) Photon Absorption. If the photon energy is *less* than the *ionization energy* of the atom, only those photons can be absorbed whose wavelengths correspond to energies equal to allowed transitions by electrons to higher states. Mathematically, if ΔE is such an energy jump, a photon of wavelength $\lambda = 1.24 \times 10^{-6}/(\Delta E)$, with ΔE in eV and λ in meters, could be absorbed.

(b.) Inelastic Collisions with Particles. In this case the particle is not absorbed, so it can excite electrons in the atom as long as it has *at least* as much energy as the electron transition requires. This energy can come from the particles in a *monoenergetic beam* such as electrons or protons accelerated through a given voltage.

(c.) Thermal Collisions Between Atoms. The thermal distribution of energies in a gas at a given temperature will result in a certain *fraction* of collisions being energetic enough to excite the electrons out of the ground state. This fraction increases rapidly with higher temperatures.

Important Observations

1. The average KE of an atom at room temperature is only about .04 eV, so the vast majority of such atoms have their electrons in the *ground* state, since even the *first* excited state is several eV above it.
2. The wavelengths of light *absorbed* by atoms are precisely the same as those *emitted*, since the same energy level structure is involved. When white light passes through a cool gas, an *absorption line* spectrum is created, with the gas transparent to all wavelengths except those absorbed by electron excitations.
3. If the photon's energy is *greater* than the ionization energy, a *continuum* of wavelengths can be absorbed, since the freed electrons do not have discrete, quantized energy levels, but can take on *any* amount of energy.

The Franck-Hertz Experiment--A monoenergetic particle beam of electrons can be used as a probe into atomic structure, as in the experiments of Frank and Hertz, described in Section 28.5 (text). The beam is directed through a gas and the beam current is monitored as the energy of the particles is increased. As the particles collide with the atoms, *none of their energy can be transferred* to the atom until the particle energy reaches the *first excitation energy* of the atom's electrons. Then, and at *each successive* excitation energy, the beam current diminishes abruptly as beam energy is lost to the atoms in *inelastic* collisions. (See Figures 28.10 and 28.11, text).

Coordinated with each beam current drop is the appearance of the *emission* line of the *same* excitation energy, as the excited electrons return to the ground state.

Example 28-4. Franck and Hertz found that electrons had to pass through 16.6 volts before the beam current showed its first drop in passing through neon gas. In what region ot the spectrum would photons be able to excite neon from its ground state?

S: This means the first excited energy level is 16.6 eV above the ground state for neon. This corresponds to a photon whose wavelength is

$$\lambda = (1.24 \times 10^{-6})/(16.6) \text{ meters} = 7.47 \times 10^{-8} \text{ m.}$$

T: This is 74.7 nm, in the *extreme ultraviolet*. At the time Hertz and Franck observed this result, spectroscopy in this spectral range hadn't been developed. This neon line has subsequently been observed.

Example 28-5. Suppose you irradiated mercury vapor with an intense light having = 255 nm, and at the same time performed a Franck-Hertz experiment with an electron beam passing through the vapor. For what three lowest electron energies would you expect to find drops in the beam current? See Figure 28.12 (text) for data on mercury.

S: The energy of the photons is

$$E(\text{eV}) = (1.24 \times 10^{-6} \text{ eV-m})/(2.55 \times 10^{-7} \text{ m}) = 4.86 \text{ eV.}$$

This is the same as the 1st excitation energy for mercury. Won't this cause many Hg atoms to be in the 1st excited state?

T: Exactly so. If the light is intense enough, a significant fraction of the can be "pumped" into that state.

S: What does that have to do with the beam current?

T: Remember, the beam current drops whenever the electrons in the beam can lose energy by kicking the electrons into a higher energy state.

S: Oh yeah, and many electrons are *already* 4.86 eV above ground state. From there, they can reach the *second* excited state with only $\Delta E = 6.67 - 4.86 = 1.81$ eV of additional energy. So *1.81 volt* electrons should show the first current drop.

T: Very good. And ...?

S: To jump from level 2 to level 4 would require 8.84 - 4.86 = 3.98 eV. So *3.98 volts* should produce the second drop in beam current.

T: The third drop should occur where Franck and Hertz originally found it, at *4.86 volts*, because the ground state won't be completely vacant. Many atoms will still be able to absorb energy from 4.86 volt electrons.

Quantum Numbers and the Periodic Table

Understanding this section should enable you to do the following:

1. *Identify the four quantum numbers that specify an electron's state.*
2. *Calculate the number of states which correspond to a given principal quantum number.*
3. *Use the Pauli Exclusion Principle to determine the ground state electron configuration for various elements.*

Quantum Numbers--The four quantum numbers which specify an electron state are summarized in Section 28.7 and Table 28.2 (text). They are:

Principal (n): The number which quantizes the energy levels in the Bohr model. Can take on any positive integral value.

Orbital Angular Momentum (l): Electrons with different l, same n, differ slightly in energy due to the magnetic interaction between the orbiting electron and the nucleus. l can have positive integral values from 0 to (n - 1).

Magnetic (m_l): Specifies quantized directions the electron's angular momentum can assume relative to an *external* magnetic field. Can have integral values between -l and +l.

Spin (m_s): Specifies the *two* orientations in which the electron can *spin* relative to its orbital angular momentum. Can have one of two values, ± 1/2.

Pauli Exclusion Principle--No two electrons in an atom can have identical sets of the four quantum numbers.

Important Observations

1. Spectroscopic notation designates the l quantum numbers by letters. The first five are: s(l = 0), p(l = 1), d(l = 2), f(l = 3), and g(l = 4).
2. Electrons with the same n are said to be in the same *shell*. By the exclusion principle, there can be *2(2l + 1)* electrons in a sub-shell, and n sub-shells in a shell. When either of these numbers is filled by electrons, the subshell or shell is said to be *closed*. This gives the following sub-shell capacities:

sub-shell	s	p	d	f	g
max. # of electrons	2	6	10	14	18

Example 28-6. The noble gases are $_2He$, $_{10}Ne$, $_{18}Ar$, $_{36}Kr$, $_{54}Xe$, $_{86}Rn$. Knowing that they have to have closed shells or sub-shells, determine their ground state electron configurations.

S: Table 28.3 (text) shows us He and Na. Ne would be like Na without the one 3s electron. It's just a counting game from there, right?

T: We'll see.

S: Beyond Ne, we have to fill the (3s) subshell with 2, then (3p) with 6. That gives us argon already:

$$_{18}Ar: \ (1s)^2(2s)^2(sp)^6(3s)^2(3p)^6$$

T: Right. To go to $_{36}Kr$, we have to add 18 more, ending with a filled sub-shell or shell.

S: We can do that with 10 in (3d), 2 in (4s), and 6 in (4p).

T: That's right.

$$_{36}Kr: \ (1s)^2(2s)^2(sp)^6(3s)^2(3p)^6(3d)^{10}(4s)^2(4p)^6$$

S: Now for $_{54}Xe$, we have to add 18 more. The next sub-shells in order are (4d)10 and (4f)14. But that would result in atomic number *60*. We would overshoot Xe without closing a sub-shell at 54.

T: As we get to more complex atoms, the *order* of filling the sub-shells is not so straightfoward as you thought. See what happens when you don't fill the 4f shell right away.

S: Then we would have to start on the (5s) with 2 and the (5p) with 6. This gives 54, all right.

T: So we have

$$_{54}Xe: \ [_{36}Kr](4d)^{10}(5s)^2(5p)^6$$

S: I don't know which way to go now.

T: It takes a little trial and error, maybe. Go back and fill the (4f) sub-shell now.

S: O.K., that's 14 more. Next it would seem I need 10 in (5d). If I then go to (5f), I'll go past 86 without filling a shell.

T: So we'll go to (6s) with 2 and (6p) with 6, and forget the (5f) sub-shell. This adds up to 86.

$$_{86}Rn: \ [_{36}Kr](4d)^{10}(5s)^2(5p)^6(4f)^{14}(5d)^{10}(6s)^2(6p)^6.$$

Characteristic X-ray Spectra

Understanding this section should enable you to do the following:
1. Explain how characteristic X-ray spectra are interpreted as the Bohr model of the atom.

X-Ray Line Spectra--As discussed in your text, the characteristic X-ray line spectra are *peaks* in intensity above the continuous spectrum, and have the following characteristics:

(a.) They do not depend on the chemical state of the element, i.e., whether it is in molecular form, solid, liquid, or gas.

(b.) As you compare element with element, new lines appear as new electron *shells* are added. Otherwise only slight shifts to higher frequencies occur with increasing atomic number.

(c.) The wavelengths correspond to photon energies of several *keV*, as opposed to optical photons of a few eV.

Interpretation by the Bohr Model--X-rays are produced when a tightly bound electron from an *inner shell* is knocked out of the atom. The other electrons can then proceed to make transitions into the inner shell vacancy, jumps which amount to thousands of eV's in heavy elements. The n = 1, 2, 3, and 4 shells of an atom are called the K, L, M, and N shells, respectively. If a K-shell electron is ejected, a *series* of characteristic lines, called the K_α, K_β, K_γ, etc. lines, are produced as electrons from the L, M, N, etc. shells fall into the K shell. Similarly, the L series (L_α, L_β, L_γ, etc.) is produced by M and N electrons falling into a vacancy in the L shell.

Important Observations

1. In each series, the N-line is the *least* energetic, and hence has the *longest* wavelength, since it is produced by a transition from the shell *adjacent* to the vacated one.
2. X-ray line spectra are far less complex than optical spectra, with each series containing only a *few* lines, rather than an infinite number. This is because only a few shells above the K or L shell are occupied by electrons, even in the heaviest elements.

Example 28-7. Two of the K series X-ray lines for the element molybdenum ($_{42}Mo$) are found at .071 nm and .063 nm.

(a.) Is this the complete K series for $_{42}Mo$? (Example 28-6 above should help).

(b.) What is the energy difference between the two shells that contributed electrons to these spectral lines?

S: How do we determine the nunber of lines a series has?

T: By finding the *number of occupied shells* (not sub-shells) in the atom's ground state. The K series involves electrons *ending* in the K shell. How many shells *above* the K shell are occupied (not necessarily filled) in $_{42}Mo$?

S: From Example 28-6, as we added electrons after $_{36}Kr$, the next 10 went into the (4d) sub-shell. So at atomic number 42, the n = 1, 2, 3, and 4 shells (K, L, M, and N) have electrons.

T: Very good. So when a K electron is removed, a K series line can be produced by L->K, M->K, and N->K transitions, giving *3 lines to the series.* The ones listed happen to be the K_α and K_β lines.

S: As for the energy level difference between the L and M shells, it would be the difference in the energies of the K_α and K_β photons.

T: Super! That's easy to figure:

$$E(K_\alpha) = (1.24 \times 10^{-6} \text{ eV-m})/(.071 \times 10^{-9} \text{ m}) = 1.75 \times 10^4 \text{ eV}$$

and

$$E(K_\beta) = (1.24 \times 10^{-6} \text{ eV-m})/(.063 \times 10^{-9} \text{ m}) = 1.97 \times 10^4 \text{ eV}$$

So the energy level separation between the L and M shells in $_{42}Mo$ is about *2.2 keV*. Notice how X-ray spectra can reveal detail about the *inner* structure of complex atoms.

Wave Mechanics and the Bohr Atom

Understanding this section should enable you to do the following:

1. Interpret the quantization of electron orbits in tbe Bohr model by applying the deBroglie wave properties of the electron.

Conditions for Stable Electron States--The deBroglie model views the stability of electron orbits as a *wave interference* problem. If the electron wave can *constructively* interfere with itself around the *circumference* of an orbit, the orbit is *stable*. If not, the electron wave would interfere *destructively* and could not exist. The constructive interference condition requires an *integral number* of electron wavelengths to fit the circumference precisely. Mathematically,

$$n\lambda = 2\pi r, \text{ where } \lambda = h/p, \quad \text{and } n = 1, 2, 3, \ldots$$

Thus

$$nh/2\pi = pr = mvr,$$

which is exactly Bohr's postulate.

Important Observations

1. This approach points out that the wave nature of the electron *in this situation* is its *dominant* property, rather than its point-like particle nature.

Example 28-8. The kinetic energy of the ground state electron in hydrogen is 13.6 eV. From this and the deBroglie wave theory, calculate the radius of this orbit.

S: I thought 13.6 eV was the *binding* energy of hydrogen.

T: It is. The *potential* energy of the ground state electron is actually -27.2 eV. With 13.6 eV of *KE*, the ground state still is bound by -13.6 eV. In general for these circular orbits,
$KE = -\frac{1}{2}PE = -\text{Binding Energy}$.

S: How does the KE determine r? Should I go back to Newton's equations?

T: Good grief, no! Find the momentum of the electron.

S: Since it is not relativistic,

$$p = \{2m_e(KE)\}^{\frac{1}{2}} = \{2(9.1 \times 10^{-31} \text{ kg})(13.6 \text{ eV} \times 1.6 \times 10^{-19} \text{ J/eV})\}^{\frac{1}{2}}$$

$$= 2.0 \times 10^{-24} \text{ kgm/s.}$$

T: Now find the deBroglie wavelength.

S: $\lambda = h/p = (6.62 \times 10^{-34} \text{ J-s})/(2.0 \times 10^{-24} \text{ kg-m/s}) = 3.33 \times 10^{-10}$ m.

T: What do you suppose the *lowest* interference order is?

S: I'd guess n = 1, meaning *one* wavelength in $2\pi r$.

T: That's right. So $r_1 = (3.33 \times 10^{-10} \text{ m})/2\pi = .0529$ nm.
You see, by measuring the *Binding Energy* of an electron, you can actually find the size of its orbit from deBroglie's wave theory.

QUESTIONS ON FUNDAMENTALS

1. What is the *second* longest wavelength in the Balmer series of the hydrogen spectrum?

(a.) 365 nm (b.) 656 nm (c.) 485 nm
(d.) 433 nm (e.) 1020 nm

2. What is the energy of the *fifth* allowed electron orbit in the Bohr model of hydrogen?

(a.) -0.544 eV (b.) -3.4 eV (c.) -1.51 eV
(d.) -0.85 eV (e.) -2.72 eV

3. What is the frequency of the photon that an atom would absorb in an electron energy jump of 3 eV?

(a.) 1240 nm (b.) 716 nm (c.) 138 nm (d.) 413 nm
(e.) 826 nm

4. According to the Pauli Exclusion Principle, how many electrons could occupy the n = 4 energy state?

(a.) 4 (b.) 8 (c.) 32 (d.) 16 (e.) 18

GENERAL PROBLEMS

Problem 28-1. One of the spectral lines in the Balmer series has λ = 410.1 nm. Which electron energy levels are involved, and how much binding energy does the electron have when in the upper state?

Q1: What expression gives the Balmer series?
Q2: What value of n gives λ_n = 410.1 nm?
Q3: What is the energy of the nth state?

Problem 28-2. How many electrons are required to fill the entire N-shell?

Q1: What is the principal quantum number of the N-shell?
Q2: What subshells exist for the N-shell?
Q3: How many electrons does each subshell accommodate?

Problem 28-3. A 1000 kg satellite circling the earth 100 miles above the surface has a period of 88 minutes. According to the Bohr postulate, what would its principal quantum number be? What would be the difference between this orbit and the next larger allowed one?

Q1: What is the Bohr postulate?
Q2: What is the satellites orbital speed?
Q3: What is the angular momentum of the satellite?
Q4: What is n, the quantum number?
Q5: How does the radius depend on n in the Bohr model?
Q6: What is the difference between adjacent allowed radii?

Problem 28-4. It is found that electrons in excess of 20 keV are necessary to produce the K_α line in the X-ray spectrum of molybdenum, yet this line is due to photons of only 17.5 keV energy, as found in Example 28-7. Explain this.

Q1: What transition gives rise to the K_α line?
Q2: When can this happen?
Q3: Why does that take more energy than 17.5 keV?

ANSWERS TO QUESTIONS

FUNDAMENTALS:

1. (c.) 2. (a.) 3. (d.) 4. (c.)

GENERAL PROBLEMS:

Problem 28-1.

A1: $1/\lambda_n = R(1/4 - 1/n^2)$, with $n = 3, 4, 5, \ldots$

A2: $1/n^2 = 1/4 - 1/(R\lambda_n)$.

A3: $E_n = -E_1/n^2 = -13.6 \text{ eV}/n^2$. This is the binding energy of the nth state.

Problem 28-2.

A1: $n = 4$.
A2: s, p, d, and f.
A3: 2, 6, 10, and 14.

Problem 28-3.

A1: The angular momentum is $= nh/2\pi$, where n is some integer.
A2: $v = 2\pi r/T$, where $r = R_E + 100$ mi. = 4100 mi. So v = 7810 m/s.
A3: $mvr = 5.12$ J-s.
A4: $n = mvr/(h/2\pi)$.
A5: $r_n = n^2$, so $r_{(n+1)} = (n+1)^2$.
A6: In ratio form, $(\quad r)/r_n = \{(n+1)^2 - n^2\}/n^2 = 2/n$, since $n >> 1$.
So $(\Delta r) = (2/n)r_n$.

Problem 28-4.

A1: The L-shell to K-shell electron transition.
A2: When there is a vacancy in the K-shell.
A3: Because the L- and M-shells in Mo are filled. The K electron must be given *at least* enough energy to go into the unfilled N-shell, which is 20 keV above the K-shell. Then the 3 transitions mentioned in Example 28-7 can all occur.

CHAPTER 29. NUCLEAR PHYSICS

BRIEF SUMMARY AND RESUME

The discovery of the *neutron-proton model* of the nucleus is discussed, along with the dependence of *nuclear size* on mass number. Nuclear *binding energies* are analyzed by mass energy calculations. *Natural radioactivity* is described and the mathematical equations governing *activity* and *half-life* are introduced. Simple conservation laws governing *nuclear reactions* are discussed. *Nuclear fission* and *fusion* are examined in detail and discussed as practical sources of energy. *Biological effects of radiation* are investigated and quantities of *dose* and *relative biological effectiveness* (RBE) are defined. Methods of *detecting radiation* are summarized.

Material from previous chapters relevant to Chapter 29 includes:

1. Mass-energy equivalence.
2. Atomic mass units.

NEW DATA AND PHYSICAL CONSTANTS

1. Unified Mass Unit (u). A mass exactly equal to 1/12 the mass of the ^{12}C atom.

$$1\ u = 931.49\ MeV/c^2 = 1.66056 \times 10^{-27}\ kg.$$

2. Radioactive Decay Constant (λ).

$$\lambda = .693/T_{\frac{1}{2}}\ ,$$

where $T_{\frac{1}{2}}$ = the half-life of a radioactive species

3. Units of Ionizing Radiation.

(a.) Nuclear Activity.

1 becquerel (Bq) = 1 nuclear disintegration/sec.

1 curie (Ci) = 3.7×10^{10} Bq

(b.) Absorbed Radiation Energy (Dose).

1 gray (Gy) = 1 J/kg

1 rad = 10^{-2} Gy

(c.) Relative Biological Effectiveness (RBE).

	RBE
Gamma rays, X-rays, electrons	1
Protons, alphas, fast neutrons	10
Thermal neutrons	3
Heavy Ions	20

(d.) Actual Biological Damage.

$$1 \text{ sievert (Sv)} = 1 \text{ Gy} \times \text{RBE}$$

$$1 \text{ rem} = 10^{-2} \text{ Sv}$$

NEW CONCEPTS AND PRINCIPLES

Nuclear Structure and Binding Energies

Understanding this section should enable you to do the following:

1. *Calculate the size of the nucleus of an atom of given mass number.*
2. *Calculate the total nuclear binding energy and the binding energy per nucleon for a given nucleus from its mass defect.*

Nuclear Structure--After 1932 and the discovery of the neutron, the nucleus was determined to be constructed of Z protons and N = (A - Z) neutrons, where A = the mass number of the nucleus. The properties of the proton and neutron, along with the electron, are summarized in Table 29.1 (text). The symbolism used to identify a specific nucleus is ${}^{A}_{Z}X$ where X is the alphabetic symbol for a particular chemical element.

Important Observations

1. X is redundant with the proton number, Z. That is, the proton number completely specifies the element.
2. For a given element, there are a number of nuclear species with different N (and hence different A) which may exist. These are called the various *isotopes* of that element.

Nuclear Size (r_o)--The *volume* of the nucleus increases in *direct proportion to mass number A*. If we assume a spherical shape, this means the nuclear radius r_o depends on $A^{(1/3)}$. Mathematically,

$$r_o = (1.2 \times 10^{-15} \text{ m})\, A^{(1/3)}$$

Important Observations

1. This size behavior is *not* like atomic sizes, which vary *periodically* with the filling of electron shells.
2. 10^{-15} meters is called 1 fermi (Fm): 1 Fm = 10^{-15} m

Example 29-1. Determine the ratio of the size of the gold nucleus to that of aluminum. Determine the ratio of their *atomic* sizes and compare the results.

S: The ratio of nuclear sizes would be just

$$(r_o)_{Au}/(r_o)_{Al} = (A_{Au}/A_{Al})^{1/3}$$

What are the mass numbers?

T: Both elements consist almost 100% of a *single* isotope. They are ${}^{27}_{13}Al$ and ${}^{197}_{79}Au$.

S: So

$$(r_o)_{Au}/(r_o)_{Al} = (197/27)^{1/3} = 1.94.$$

How do I find atomic size?

T: In the solid state, we can view the atoms as about one atomic diameter apart with the outer electrons just overlapping. So the volume taken up by n atoms is proportional to the cube of the atomic radius times n, $V \approx r^3 n$. If we assume the same proportionality constant (same type of crystal structure), we can compare the solid volumes of *equal numbers* of the atoms. A convenient number would be N_A, Avogadro's number. Then

$$(r_{Au}/r_{Al}) = (V_{Au}/V_{Al})^{1/3}$$

S: How do we find the volumes?

T: From mass density tables. We find $\rho_{Au} = 19.3$ gm/cm^3 and $\rho_{Al} = 2.70$ gm/cm^3. You also know that N atoms have a mass equal to their atomic mass number.

S: O.K., I see. $\rho = m/V$, so

$$(V_{Au}/V_{Al}) = (m/\rho)_{Au}/(m/\rho)_{Al} = (197/27)(2.70/19.3) = 1.02.$$

Their volumes are about the same!

T: That's true. The electron shells just keep packing in more closely to the nucleus, but each nucleon added to the nucleus requires more space. Taking the cube root of the volume ratio gives

$$r_{Au}/r_{Al} = (1.02)^{1/3} = 1.007.$$

Mass Defect (Δm)--There is a difference between the sum of the individual masses of the protons and neutrons making up a nucleus and the actual mass of the nucleus itself. This is known as the *mass defect* of the nucleus.

Nuclear Binding Energy--The reason a mass defect exists lies in the fact that *bound* nucleons are in a *lower energy state* than are *free* nucleons. This lower energy must correspond to a *lower mass*, according to the equivalence between energy and mass. Mathematically,

$$\text{Binding Energy} = (\Delta m)c^2.$$

Binding Energy per Nucleon--This is just (Binding Energy)/A. The results of this for the most common isotopes of the elements are shown in Figure 29.3 (text). Notice the following features:

(a.) Most nucleons are bound in the nucleus with about *7 to 9 MeV* of energy.
(b.) The ${}^{56}_{26}Fe$ nucleus has the *tightest* binding per nucleon.
(c.) The nuclei lighter than ${}^{23}_{11}Na$ are bound with progressively *less* energy per nucleon except for ${}^{4}_{2}He$, which shows an unusually tightly bound structure among light nuclei.

Important Observations

1. Mass defects occur in principle for *chemical* binding also. But the energies involved are in the range of eV's or less, making m unmeasurably small. In *nuclear* binding, the energies range into MeV, and the mass defects are significant.
2. For mass defect calculations, it is important to utilize *all* the seven or eight significant figure precision with which masses can be measured.

Example 29-2. Calculate the binding energy per nucleon for the following nuclei, all of which have the same number of nucleons:

${}^{20}_{8}O$ (m = 20.00408 u), ${}^{20}_{9}F$ (m = 19.99999 u), ${}^{20}_{10}Ne$ (m = 19.99244 u), ${}^{20}_{11}Na$ (m = 20.00890 u), and ${}^{20}_{12}Mg$ (m = 20.0170 u). Which of these should be the most stable mixture of protons and neutrons for this A?

S: We need the accurate masses of the *free* proton and neutron. From Table 29.1 (text) they are: m_p = 1.00728 u, m_n = 1.00866 u.

For ${}^{20}_{8}O$, Δm = 20.00408 - 8(1.00728) - 12(1.00866) = -.15808 u.

For ${}^{20}_{9}F$, Δm = 19.99999 - 9(1.00728) - 11(1.00866) = -.16079 u.

For ${}^{20}_{10}Ne$, Δm = 19.99244 - 10(1.00728) - 10(1.00866)= -.16696 u.

For ${}^{20}_{11}Na$, Δm = 20.00890 - 11(1.00728) - 9(1.00866) = -.14912 u.

For ${}^{20}_{12}Mg$, Δm = 20.0170 - 12(1.00728) - 8(1.00866) = -.1402 u.

T: To get B.E. per nucleon, we multiply each of these by 931.49 MeV per u, and divide by A = 20. The results:

${}^{20}_{8}O$: -7.3625 MeV/nucleon

${}^{20}_{9}F$: -7.4887 MeV/nucleon

${}^{20}_{10}Ne$: -7.7761 MeV/nucleon

${}^{20}_{11}Na$: -6.9452 MeV/nucleon.

${}^{20}_{12}Mg$: -6.530 MeV/nucleon.

You can see that the *most tightly* bound and therefore the most *stable* arrangement is ${}^{20}_{10}Ne$.

S: The atomic masses of these elements as they occur in the periodic table are: oxygen 15.9994, fluorine 18.9984, sodium 22.9898, and magnesium 24.305. The only one close to mass number 20 is neon with 20.079. Do the other nuclei exist?

T: Not as *stable* structures. They all have a tendency to change into the more stable $^{20}_{10}Ne$ by a process called *natural radioactivity*, discussed in the next section.

Natural Radioactivity

Understanding this section should enable you to do the following:

1. *Identify the various types of radioactive processes and how the nucleus changes in each process.*
2. *State the relation between the decay constant and the half-life.*
3. *Calculate the fraction of original nuclei remaining and the activity of a sample at any time t, given the half-life of the substance.*

Types of Radioactivity--If a nucleus can become *more stable* (more tightly bound) by changing its mixture of protons and neutrons, it will do so *spontaneously* with a certain statistical probability. The original nucleus is called the *parent*, the process of change is called *radioactive decay*, and results in a final nucleus called the *daughter*. The following are examples of types of natural decay processes:

α-decay: A $^{4}_{2}He$ nucleus (alpha particle) is emitted, resulting in a daughter nucleus with Z reduced by 2 and A reduced by 4. If X represents the parent and Y represents the daughter, this process is

$$(\alpha\text{-decay}): \quad {}^{A}_{Z}X \rightarrow {}^{4}_{2}He + {}^{A-4}_{Z-2}Y.$$

ß-decay: Either a regular electron (e^-) or an anti-electron (e^+) is emitted by the parent. The daughter has a value of Z either increased by 1 or decreased by 1, respectively. The nucleon number A does not change:

$$(\beta^-\text{-decay}): \quad {}^{A}_{Z}X \rightarrow {}^{0}_{-1}e + {}^{A}_{Z+1}Y$$

$$(\beta^+\text{-decay}): \quad {}^{A}_{Z}X \rightarrow {}^{0}_{+1}e + {}^{A}_{Z-1}Y$$

γ-decay: The parent nucleus emits a gamma ray, which reduces its energy, but does not change its neutron-proton content. Very often γ radiation occurs simultaneously with ß or α decay.

Rules Governing Natural Radioactivity--

1. Conservation of charge. The sum of the proton numbers of the daughter plus decay particle must equal the proton number of the parent.
2. Conservation of nucleon number, A. The sum of the nucleon numbers of the daughter plus decay particle must equal that of the parent. Notice that e^- and e^+ have $A = 0$.
3. For spontaneous decay to occur, the mass of the daughter plus that of the decay particle must be *less* than the mass of the parent. Thus radioactivity is a natural way to *increase stability*, releasing energy in the process.
4. The energy release is equal to

$$(\Delta E) = (M_{parent} - M_{daughter} - M_{\alpha \text{ or } \beta})c^2.$$

Example 29-3. Suppose you have made a nucleus consisting of 9 protons and 11 neutrons.

(a.) What chemical symbol would this have?
(b.) What daughters would result if it underwent α decay? β^-decay? β^+ decay?
(c.) Is any of these processes possible, according to the rules, or is the original nucleus stable?

S: With 9 protons, you would have the element *fluorine*, ${}^{20}{}_{9}F$. Since flourine has atomic mass *19*, I suspect the A = 20 isotope is unstable.

T: Good observation. We'll see. If it lost an alpha particle, what nucleus would be left?

S: Z would go from 9 to 7, and A from 20 to 16. So an isotope of *nitrogen* would result:

$${}^{20}{}_{9}F \rightarrow {}^{4}{}_{2}He + {}^{16}{}_{7}N.$$

T: Could this happen, according to the rules?

S: I'd have to know the masses of all three particles.

T: Indeed! They are tabulated in a *chart of the nuclides* and in many handbooks.

$$M({}^{4}{}_{2}He) = 4.00260 \text{ u.}$$
$$M({}^{20}{}_{9}F) = 19.99999 \text{ u.}$$
$$M({}^{16}{}_{7}N) = 16.00610 \text{ u.}$$

S: Well $M({}^{4}{}_{2}He) + M({}^{16}{}_{7}N) = 20.0087$ u, which is *more* than $M({}^{20}{}_{9}F)$, so this process would *require* energy and wouldn't happen spontaneously.

T: Quite right. For β^- decay, we would have ${}^{20}{}_{9}F \rightarrow {}^{20}{}_{10}Ne + {}^{0}{}_{-1}e$.
How about β^+ decay?

S: Well, I'd guess the anti-electron would have the symbol ${}^{0}{}_{+1}e$, right? So the rules would give ${}^{20}{}_{9}F \rightarrow {}^{20}{}_{8}O + {}^{0}{}_{+1}e$.

T: That's very good. Here are some additional masses:

$$m_{e^-} = m_{e^+} = .00055 \text{ u.}$$
$$M({}^{20}{}_{8}O) = 20.00408 \text{ u.}$$
$$M({}^{20}{}_{10}Ne) = 19.99244 \text{ u.}$$

S: For the β^+ decay, the mass of ${}^{20}{}_{8}O$ is greater even without adding m_{e^+}. So again this would not happen. For the β^- decay, we have
$m_{e^-} + M({}^{20}{}_{10}Ne) = 19.99299$ u. This *is less* than $M({}^{20}{}_{9}F)$, so it *can* occur.

T: And *does*, with a half-life of 11.6 seconds (see next section). ${}^{20}{}_{10}Ne$ is the *only* A = 20 nucleus that is stable, having less mass than any other mixture of protons and neutrons totalling 20.

Activity and Decay Constant (λ)--The *activity* of a radioactive species is simply the *rate* at which the nuclear disintegrations are occurring. At any instant, the activity is proportional to the number of that species of nucleus in the sample. The proportionality constant, λ , is called the *decay constant* of that particular species. Mathematically,

$$(\Delta N)/(\Delta t) = -\lambda N(t).$$

With this behavior, the expression for the *fraction* of the *original* number N_0 in the sample remaining after time t is given by the expression

$$N(t)/N_0 = e^{-\lambda t}.$$

Half-Life ($T_{\frac{1}{2}}$)--In a case of exponential decay, the original amount is continually reduced by half in equal periods of time, as shown in Figure 29.7 (text). These periods of 50% reduction are known as the species' *half-life*, symbolized by $T_{\frac{1}{2}}$. $T_{\frac{1}{2}}$ is related *inversely* to the decay constant by the expression

$$T_{\frac{1}{2}} = .693/\lambda\ .$$

Important Observations

1. All radioactive processes follow this type of behavior, although the half-lives for various species span a tremendous range of values, from small fractions of a second to billions of years. Some examples are listed in Table 29.2 (text).

Example 29-4. Radium ($^{226}_{88}Ra$) has a half-life of 1590 years, and decays by emitting an alpha particle.

(a.) What fraction of a radium sample disintegrates in 1 sec?
(b.) What fraction disintegrates in 300 years?
(c.) What is the activity of an original sample of 1 mg of Ra after 600 years?
(d.) How long would it take for the activity to get down to 1 microcurie?

S: I don't remember any expression for part (a.).

T: Maybe not directly. Then you have to think. What it asks for is the quantity $\Delta N/N$ during *one second*, which is an *instant* compared to $T_{\frac{1}{2}}$. Since N is practically constant in such a small Δt, we can get this directly from the decay constant:

$$\Delta N/\Delta t = -\lambda N, \quad \text{so} \quad \Delta N/N = -\lambda(\Delta t).$$

S: I see, for $\Delta t = 1$ sec, $\Delta N/N$ is just $-\lambda$.

T: This obviously wouldn't work if Δt was comparable to or larger than $T_{\frac{1}{2}}$.

S: Well, λ isn't so hard: $\lambda = .693/1590 \text{ yr} = 4.36 \times 10^{-4} \text{ yr}^{-1}$.

T: Or, converting to sec^{-1}:

$$\lambda = 4.36 \times 10^{-4} \text{ yr}^{-1} \times (1 \text{ yr}/3.15 \times 10^7 \text{ sec}) = 1.38 \times 10^{-11} \text{ sec}^{-1}.$$

This is the fraction of *any* sample of radium that disintegrates in one second.

S: In part (b.), we no longer can treat Δt as an instant, so what approach do we take?

T: The easiest thing to do is to find the fraction that is *left* after 300 years.

S: All right, $N(300\ \text{yr}) = N_0 e^{-\lambda(300)}$. So

$$N(300)/N_0 = e^{-(.000436/\text{yr})(300\ \text{yr})} = e^{-.131} = .877.$$

T: So even after 300 years, 87.7% is left.

S: The activity, $\Delta N/\Delta t$, is just $-\lambda N(t)$, so we have to find how much is left after 600 years for part (c.).

$$N(600\ \text{yr}) = N_0 e^{-(.000436)(600)} = N_0 e^{-.262} = .770\ N_0$$

What is N_0?

T: We have 1 mg of a material whose atomic mass number is 226. For this calculation, we're only using three significant figures, so we don't need the precision we used for mass defect calculations.

S: So $N_0 = (10^{-3}\ \text{gm}/226\ \text{gm/mole}) \times N_A$ atoms/mole $= 2.66 \times 10^{18}$ atoms. Thus after 600 years,

$$\Delta N/\Delta t = -(1.38 \times 10^{-11}\ \text{sec}^{-1})(.770)(2.66 \times 10^{18}) = -2.83 \times 10^{7}/\text{sec}.$$

Why the minus sign?

T: Because N is *decreasing*. An activity of 1 disintegration/sec is the unit we call the *becquerel*.

$$\Delta N/\Delta t = 2.83 \times 10^{7}\ \text{Bq}.$$

S: How many Bq is a microcurie?

T: $1\ \text{Ci} = 3.7 \times 10^{10}$ Bq, so $1\ \mu\text{Ci} = 3.7 \times 10^{4}$ Bq.
We have to set $\lambda N(t) = \lambda N_0 e^{-\lambda t} = 3.7 \times 10^{4}$ Bq and solve for t.

S: This means that

$$3.7 \times 10^{4}\ \text{Bq} = (1.38 \times 10^{-11}\ \text{s}^{-1})(2.66 \times 10^{18}) \times e^{-(.000436/\text{yr})t}$$

Or,

$$e^{-(.000436/\text{yr})t} = 1.008 \times 10^{-3}.$$

T: Taking the natural log of both sides,

$$(4.36 \times 10^{-4}\ \text{yr}^{-1})t = -6.90$$

which gives

$$t = 15{,}800\ \text{years}.$$

Example 29-5. A radium nucleus (atomic mass = 226) emits an alpha particle of 4.79 MeV.

(a.) How much recoil energy is given to the daughter nucleus?
(b.) What fraction of the original mass was converted into energy?

S: This process changes Z from 88 to 86, and A from 226 to 222. So the daughter nucleus is $^{222}_{86}Rn$. What determines how the energy is split?

T: Momentum conservation, as in the recoil problems of Chapter 7. We assume the Ra nucleus was stationary, so $P_{tot} = 0$. Then $P_\alpha = -P_{Rn}$. Then we need to know how the KE is related to the momentum.

S: I thought KE = $P^2/2m$.

T: *Non-relativistically*, that's true. Is that the case here?

S: I think so. An alpha has a mass of about 4 u, which is 4(931) = 3700 MeV. So a KE of 4.79 MeV is a small fraction of the rest mass energy.

T: You're right. So momentum conservation states that

$$m_\alpha E_\alpha = m_{Rn}E_{Rn}, \text{ or } E_{Rn} = (m_\alpha/m_{Rn})E_\alpha = (4/222)(4.79 \text{ MeV}) = .086 \text{ MeV}.$$

S: We still have 226 mass units left. How do we find the mass converted to energy without more precise mass values?

T: Here we are directly given the energy released by the α-decay: (4.79 + .086) MeV = 4.88 MeV. The rest of the energy is still in the masses m_α and M_{Rn}. 4.88 Mev is equivalent to 4.88/931.5 = .0052 u of mass. Thus .005/226 x 100 = .002% of the radium's mass was converted into the KE of the alpha plus the radon.

Induced Nuclear Reactions

Understanding this section should enable you to do the following:

1. *Identify various types of induced nuclear reactions, and calculate the energies released or consumed by them.*
2. *Define tbe difference between nuclear fusion and fission.*
3. *Define a "breeding reaction" and what is meant by a fissile material.*

Induced Radioactivity--The bombardment of stable nuclei by protons, neutrons, deuterons, or alpha particles can create new nuclear species, which are generally unstable and undergo subsequent radioactive decay. The same rules apply as for natural radioactivity, except that energy can be *supplied* to the reaction, enabling the creation of *more* massive species. Generally, an induced reaction is symbolized by X + a -> Y + b, or X(a,b)Y.

Important Observations

1. *Nuclear* masses are Zm_e *smaller* than the neutral *atomic* masses. When calculating the mass defects for any nuclear reactions, you must be consistent *one way or the other*. For example, in the reaction $^2_1H + ^3_1H \rightarrow ^4_2He + ^1_0n$, you can use the *atomic* masses on both sides, which include the mass of two electrons. But if you used the *deuteron* mass and the alpha particle mass, which are the *nuclei* of ^2_1H and ^4_2He, you would have to subtract m_e from the mass of ^3_1H to get its nuclear mass for consistency.

Nuclear Fusion--Strictly speaking, a "fusion" reaction is any one where lighter nuclei are combined, or *fused*, into a more complex nucleus. The binding energy curve (Figure 29.3, text) reveals that energy will be *released* in such reactions as nuclei up to $^{56}_{26}Fe$ are built up, and beyond that energy would be *consumed* in producing a fusion. More commonly, fusion refers to reactions involving isotopes of hydrogen fusing into isotopes of helium. These release large amounts of energy because of the great binding energy of helium, and are the reactions involved in the sun's core and in the H-bomb. A number of these reactions are listed in Section 29.7 (text).

Example 29-6. The sun's energy is produced primarily by the "proton-proton" cycle of fusion as follows:

$$^1_1p + ^1_1p \rightarrow ^2_1H + ^0_{+1}e + \text{energy}$$

$$^1_1p + ^2_1H \rightarrow ^3_2He + \text{energy}$$

$$^3_2He + ^3_2He \rightarrow ^4_2He + 2(^1_1p) + \text{energy}$$

(a.) Calculate the total energy released in the three steps.

(b.) Calculate the mass of hydrogen nuclei (protons) which must undergo fusion each second to provide the observed power output of the sun, 4×10^{26} watts.

S: We'll need accurate masses or all these isotopes.

T: Since these are nuclear reactions, we'll emphasize that fact by using *nuclear* masses throughout (see comments in the previous observation). They are:

$$m_e = .00055u;\ m_p = 1.00728\ u;\ m_D = 2.01355\ u;\ m_\alpha = 4.00151\ u$$

and

$$M(^3_2He \text{ nucleus}) = 3.01493\ u.$$

S: O.K. The first reaction releases energy given by

$$\Delta E_1 = (2m_p - m_D - m_e)c^2 = (2.01456 - 2.01355 - .00055)c^2$$

$$= (.00046\ u)(931.49\ \text{MeV/u}) = 0.429\ \text{MeV}.$$

The second reaction releases

$$\Delta E_2 = [m_p + m_D - M(^3{}_2He)]c^2 = (1.00428 - 2.01355 - 3.01493)c^2$$

$$= (.0059\ u)(931.49\ MeV/u) = 5.50\ MeV.$$

T: The final step releases

$$\Delta E_3 = [2M(^3{}_2He) - m_\alpha - 2m_p]c^2 = (6.02986 - 4.00151 - 2.01456)c^2$$

$$= (0.1379\ u)(931.49\ MeV/u) = 12.85\ MeV.$$

S: So $\Delta E_{tot} = \Delta E_1 + \Delta E_2 + \Delta E_3 = 18.78$ MeV for the whole cycle. How do we get the mass of hydrogen involved in the sun from this?

T: Oops! Hold it a second! You have to make *two* $^3{}_2He$ nuclei for the third step to happen, which means that steps 1 and 2 have to happen *twice*. So the total energy release to make *one* $^4{}_2He$ is

$$\Delta E_{tot} = 2(\Delta E_1 + \Delta E_2) + \Delta E_3 = 24.7\ MeV.$$

Notice that this is the same as the *net* reaction $4(^1{}_1p) \rightarrow \alpha$ + energy. For part (b.), the atomic mass of hydrogen is 1, so 1 gram of protons is about 6×10^{23} of them. You know 4 of them can produce 24.7 MeV.

S: Wow! One gram would produce

$$(6 \times 10^{23}/4)(24.7\ MeV) = 3.7 \times 10^{24}\ MeV.$$

T: And the power output of the sun is 4×10^{26} joules/sec, or

$$4 \times 10^{26}\ J/s \times (1\ MeV/1.6 \times 10^{-13}\ J) = 2.5 \times 10^{39}\ MeV/sec\ !!$$

S: So it would take

$$(2.5 \times 10^{39}\ MeV/s)/(3.7 \times 10^{24}\ Mev/gm) = 6.8 \times 10^{14}\ gm/sec.$$

T: Since 1000 kg weighs 2200 pounds (on earth), or 1 metric ton, this is *680 million metric tons of hydrogen per second*!

Nuclear Fission--In the case of some nuclei, bombardment by neutrons causes the original nucleus to *split* into two *major fragments*, releasing energy in a process called *nuclear fission*. Examples of nuclei for which the fission reaction occurs with significant probability are $^{233}{}_{92}U$, $^{235}{}_{92}U$, and $^{239}{}_{94}Pu$. Such nuclei are termed *fissile*.

Breeding Reactions--Of the three fissile nuclei above, only $^{235}{}_{92}U$ occurs in nature, making up 0.7% of natural uranium. The other two have to be manufactured from other naturally occurring isotopes by nuclear processes called *breeding reactions*. These reactions are:

$$^1{}_0n + {}^{238}{}_{92}U \rightarrow {}^{239}{}_{92}U \rightarrow {}^{239}{}_{93}Np + {}^0{}_{-1}e;$$

$$^{239}{}_{93}Np \rightarrow {}^{239}{}_{94}Pu + {}^0{}_{-1}e$$

and

$$^1{}_0n + {}^{232}{}_{90}Th \rightarrow {}^{233}{}_{90}Th \rightarrow {}^{233}{}_{91}Pa + {}^0{}_{-1}e;$$

$$^{233}_{91}Pa \rightarrow {}^{233}_{92}U + {}^{0}_{-1}e$$

Important Observations

1. Other heavy nuclei can occasionally fission when struck by a neutron, but the probability of it happening is much lower than for a fissile nucleus.
2. When fission occurs, 2 to 3 neutrons are released and are thus available to produce subsequent fissions. This multiplication allows the rate of fissioning in a sample to increase exponentially, resulting in a *chain reaction* and potentially explosive behavior. The size and shape of the sample, along with the presence of neutron absorbing material, determine how fast the reactions can proceed.
3. The fission fragments are not of equal mass number, nor are they always the same nuclei. They are statistically distributed around a 40%-60% split of the original mass.
4. The fragments share the same neutron/proton ratio that the fissile nucleus originally had, which makes them far too neutron rich to be stable in the middle of the periodic table of elements. Thus they are *unstable* against β^- decay, which essentially exchanges a neutron for a proton. Most of the radioactivity associated with nuclear fallout and power plants comes from these highly radioactive fission fragments.

Example 29-7. Suppose a particular fission event is as follows:

$$^{1}_{0}n + {}^{235}_{92}U \rightarrow {}^{133}_{53}I + {}_{39}Y + 3({}^{1}_{0}n).$$

(a.) Find the atomic mass nuber of the yttrium isotope.

(b.) The fission fragments are highly unstable, undergoing successive β^- decays until they reach a stable configuration. Using a table of stable isotopes, find the final form of these fragments.

S: Part (b.) just requires conserving nucleon number. We have $A_{tot} = 236$. Thus $A_Y = 236 - 133 - 3 = 100$. We thus have $^{100}_{39}Y$.

T: Good. A table of isotopes shows the *stable* forms of iodine and yttrium to be $^{127}_{53}I$ and $^{89}_{39}Y$. What happens to A and Z as a result of β^- decay?

S: Z increases by 1, and A stays the same.

T: That's right. So what happens to $^{133}_{53}I$?

S: When Z changes, the *element* changes: $^{133}_{53}I \rightarrow {}^{133}_{54}Xe + {}^{0}_{-1}e$. I don't see this isotope listed as stable in the table.

T: It must decay further. See what another β^- decay would do.

S: $^{133}_{54}Xe \rightarrow {}^{133}_{55}Cs + {}^{0}_{-1}e$. This *is* stable.

T: As a matter of fact, it is the *only* stable nucleus with A = 133, and forms 100% of naturally occurring Cs, as can be seen from the table.

S: From the yttrium, we get ${}^{100}_{39}Y \rightarrow {}^{100}_{40}Zr + {}^{0}_{-1}e$, unstable, then ${}^{100}_{40}Zr \rightarrow {}^{100}_{41}Nb + {}^{0}_{-1}e$, still unstable, and finally ${}^{100}_{41}Nb \rightarrow {}^{100}_{42}Mo + {}^{0}_{-1}e$.

T: This is *one* of the stable isotopes of ${}_{42}Mo$, occurring with about 10% abundance in nature.

Example 29-8. Knowing that ${}^{235}_{92}U$ has an atomic mass of 235.044 and those of ${}^{100}_{42}Mo$ and ${}^{133}_{55}Cs$ are 99.0935 and 132.906 respectively, calculate the energy released in this fission and compare with Table 29.3 (text).

S: The mass before fission was $m_n + M({}^{235}_{92}U) = 1.0087 + 235.044 = 236.053$ u. Afterwards, it is $3m_n + M({}^{133}_{55}Cs) + M({}^{100}_{42}Mo)$.

T: *Plus* the mass of 4 electrons which left the picture in the β^- decay. So the total final mass is

$$3(1.0087) + 132.906 + 99.9035 + 4(.00055) = 235.838 \text{ u.}$$

S: Then $\Delta m = .215$ u, which gives us

$$\Delta E = (.215 \text{ u})(931.49 \text{ MeV/u}) = 200 \text{ MeV.}$$

T: This is close to the average fission energy of 215 MeV quoted in your text.

QUESTIONS ON FUNDAMENTALS

1. What is the size of the nucleus of ^{238}U ?

(a.) 286 fm (b.) 7.43 fm (c.) 95.2 fm
(d.) .194 fm (e.) 6.59 fm

2. A deuterium nucleus consists of a proton and neutron bound together, having a mass of 2.014102 u. Using the mass values of the free proton and neutron as given in Example 29-2, calculate the energy with which the proton and neutron are bound together.

(a.) 1.84 MeV (b.) 937.8 MeV (c.) 14.8 MeV
(d.) 1.71 MeV (e.) 1.84 keV

3. The half-life of ^{47}Ca is 4.54 days. How much of an original sample of ^{47}Ca would be left after a time of 30 days?

(a.) 1% (b.) 15% (c.) 0.13% (d.) 85% (e.) 2.5%

4. Approximately what fraction of the original sample of ^{47}Ca decays in the first hour?

(a.) .19 (b.) .205 (c.) .00512 (d.) .0064
(e.) .27

5. $^{47}_{20}Ca$ decays by beta emission. What isotope is formed by this decay?

(a.) $^{43}_{18}Ar$ (b.) $^{47}_{19}K$ (c.) $^{47}_{21}Sc$ (d.) $^{46}_{21}Sc$

(e.) $^{46}_{19}K$

GENERAL PROBLEMS

Problem 29-1. Which requires more energy: the separation of a proton or a neutron from ^4_2He?

Q1: What nuclei would result from these separations?
Q2: What are the atomic masses of these nuclei?
Q3: What is the proton separation energy?
Q4: What is the neutron separation energy?

Problem 29-2. $^{226}_{88}Ra$ decays into $^{226}_{86}Rn$ with a half-life of 1590 years. Rn in turn has a half-life of 3.82 days as it decays into $^{218}_{84}Po$. Calculate the mass of Rn that is in *equilibrium* with 1 gm of Ra.

Q1: What precisely does equilibrium mean here?
Q2: What is the rate of formation of Rn?
Q3: Doesn't this change as the sample ages?
Q4: What is the rate of decay of Rn?
Q5: What is the mathematical statement of equilibrium?
Q6: What is this in terms of masses?

Problem 29-3. You have a sample of radioactive sodium, $^{24}_{11}Na$ that has an activity of .01 mCi. 24 hours later, you observe the activity has decreased to .00325 mCi. What is the half-life $^{24}_{11}Na$, and how much was in the original sample?

Q1: What determines this decrease in activity?
Q2: What does the *ratio* of activities tell us?
Q3: How do we get A from this?
Q4: How is the half-life related to this?
Q5: How is the original activity related to the original sample amount?

Problem 29-4. In the previous problem, the $^{24}_{11}Na$ β^--decays into $^{24}_{12}Mg$. The masses are: $M(^{24}_{11}Na) = 23.990962$ u, $M(^{24}_{12}Mg) = 23.985042$ u, and $m_e = .00549$ u.

(a.) What is the *power* release of the sample with the original activity?
(b.) How much energy has been given off during the entire 24 hours?

Q1: How much energy does *each* decay release?
Q2: How many events per second occur in the original sample?
Q3: How many decays have occurred in the 24 hours?
Q4: What is N(24 hrs)?

ANSWERS TO QUESTIONS

FUNDAMENTALS:

1. (b.) 2. (d.) 3. (a.) 4. (d.)
5. (c.)

GENERAL PROBLEMS:

Problem 29-1.

A1: ${}^{4}{}_{2}He \rightarrow {}^{1}{}_{1}p + {}^{3}{}_{1}H$, and ${}^{4}{}_{2}He \rightarrow {}^{1}{}_{0}n + {}^{3}{}_{2}He$.

A2: ${}^{3}{}_{1}H$: 3.016050 u, ${}^{3}{}_{2}He$: 3.016029 u, ${}^{4}{}_{2}He$: 4.002604 u.

A3: $\Delta E_p = [m_p + M({}^{3}{}_{1}H) - M({}^{4}{}_{2}He)]c^2$.

A4: $\Delta E_n = [m_n + M({}^{3}{}_{2}He) - M({}^{4}{}_{2}He)]c^2$.

Problem 29-2.

A1: The amount of Rn for which the rate of decay is equal to its rate of formation.

A2: The rate of decay, or activity, of its parent, Ra. This is $\lambda_{Ra}N_{Ra}$.

A3: Yes, but *very slowly* compared to the activity of Rn. This slowly changing equilibrium is called *secular equilibrium*.

A4: $\lambda_{Rn}N_{Rn}$.

A5: $\lambda_{Rn}N_{Rn} = \lambda_{Ra}N_{Ra}$, or $N_{Rn} = (\lambda_{Ra}/\lambda_{Rn})N_{Ra}$.

A6: Since number is equal to mass divided by atomic mass, this would become

$$M_{Rn} = (M_{Rn}/M_{Ra})(\lambda_{Ra}/\lambda_{Rn})M_{Ra}$$

Problem 29-3.

A1: Activity is $\lambda N(t)$, and $N(t) = N_0 e^{-\lambda t}$.

A2: The ratio is

$$\lambda N(t = 24 \text{ hrs})/\lambda N_0 = e^{-\lambda(24 \text{ hrs})} = .00325/.01 = .325$$

A3: Take the natural log of both sides: $-\lambda(24 \text{ hrs}) = -1.124$.

A4: $T_{\frac{1}{2}} = .693/\lambda$.

A5: Original activity is λN_0. So find N from $\lambda N_0 = .01$ mCi, where 1 Ci = 3.7×10^{10} Bq.

Problem 29-4.

A1: $\Delta E = [M({}^{24}{}_{11}Na) - M({}^{24}{}_{12}Mg) - m_e]c^2 = 5.003$ MeV.

A2: .01 mCi = 3.7×10^5 Bq.

A3: $N_0 - N(24 \text{ hrs})$. You have N from Problem 29-3.

A4: $N(24 \text{ hrs}) = N_0 e^{-\lambda(24 \text{ hrs})} = .325\, N_0$.

CHAPTER 30. SOLID STATE PHYSICS

BRIEF SUMMARY AND RESUME

Major types of solid *crystalline geometries* are discussed and calculations are made of typical inter-atomic separations. *Thermal* conductivity is observed to be proportional to *electrical* conductivity, as expressed in the *Wiedeman-Franz Law*. The *Dulong-Petit* and *Debye* theories of molar specified heats are discussed. The *energy band theory* of solids is introduced, and differences between conductors, semi-conductors, and insulators are investigated. Both *intrinsic* and *extrinsic* semi-conductors are discussed in more detail, and a number of solid state devices based on their properties are noted. *Superconductivity* and the *Josephson effect* are mentioned.

Material from other chapters relevant to Chapter 30 includes:

1. Calculations involving densities, molecular mass number, and Avogadro's number.
2. Kinetic Theory of Gases.
3. Thermal and electrical conductivity.
4. Molar specific heat.
5. Ground state atomic configuration.
6. The Hall effect.

NEW DATA AND PHYSICAL CONSTANTS

1. Josephson Effect Constant. The constant of proportionality between voltage applied to a Josephson junction and the frequency of a.c. current across the junction:

$$2e/h = 4.8361 \times 10^{14} \text{ Hz/volt}$$

NEW CONCEPTS AND PRINCIPLES

Thermal and Electrical Conductivities of Solids

Understanding this section should enable you to do the following:

1. *Calculate the relationship between the electrical and thermal conductivities of metallic solids.*

The Wiedeman-Franz Law--For metals, there is theoretically a direct proportionality between their thermal and electrical conductivities at any temperature, given by

$$K = (\pi^2 k^2 T)\sigma/3e^2$$

where K = the thermal conductivity and σ = the electrical conductivity, which is the reciprocal of resistivity.

Important Observations

1. Notice the correctness of the units: $k2T/e^2$ has units of $(J^2\text{-}K)/(K^2\text{-}C^2\text{-}\Omega\text{-m})$ which can be untangled to yield W/K-m, which are the SI units of thermal conductivity.

2. The proportionality between K and σ is not a constant, but is a linearly increasing function of temperature.

<u>Example 30-1</u>. Predict the thermal conductivity at T = 20°C for tungsten, given that at 0°C, its electrical resistivity is $\rho = 4.9 \times 10^{-8}\ \Omega\text{-m}$.

<u>S</u>: To get electrical conductivity σ, I have to take the reciprocal of ρ, right?

$$\sigma = 1/\rho = 1/(4.9 \times 10^{-8}\ \Omega\text{-m}) = 2.04 \times 10^{7}/\Omega\text{-m}$$

<u>T</u>: We can calculate the constant $\pi^2k^2/3e^2$ once and for all:

$$\pi^2k^2/3e^2 = \pi^2(1.38 \times 10^{-23}\ \text{J/K})^2/3(1.6 \times 10^{-19}\ \text{C})^2$$

$$= 2.45 \times 10^{-8}\ \text{J}^2/\text{K}^2\text{C}^2.$$

<u>S</u>: What temperature do we use, 20°C?

<u>T</u>: T must always be in kelvins in the Wiedeman-Franz law. Since the value of σ is for 0°C, the correct temperature to use is T = 273 K. This will give you the thermal conductivity K at 0°C.

<u>S</u>: All right. Then

$$K = (2.45 \times 10^{-8}\ \text{J}^2/\text{K}^2\text{-C}^2)(2.04 \times 10^{7}/\Omega\text{-m})$$

$$= 136\ \text{J}^2/\text{K}^2\text{-C}^2\text{-}\Omega\text{-m}.$$

These units are really mixed up!

<u>T</u>: Notice that Ω = volts/amp = volt-sec/coul, and that coul-volts = joules. Then after cancelling, you come out with W/K-m.

<u>S</u>: In Table 30.2 (text), K is <u>169</u> W/m-K for tungsten.

<u>T</u>: Yes. First, the W-F law is only approximate, and secondly, we calculated K for T = 0°C, not 20°C.

<u>S</u>: So we should have K(20°C) = (136 W/m-K)(293 K/273 K) = 146 W/m-K?

<u>T</u>: To be really consistent, you should also put in the value of conductivity σ at 20°C:

$$\rho_{20} = \rho_0(1 + \alpha\ \ T) = (4.9 \times 10^{-8}\ \Omega\text{-m})(1 + 4.8 \times 10^{-3}/\text{C}^\circ \times 20\ \text{C}^\circ)$$

$$= 5.4 \times 10\text{-}8\ \Omega\text{-m}.$$

So

$$\sigma_{20} = 1.86 \times 10^{7}/\Omega\text{-m},$$

giving

$$K(20^\circ\text{C}) = K(0^\circ\text{C})(1.86 \times 10^{7})/(2.04 \times 10^{7}) \times (293)/(273) = 131\ \text{W/m-K}$$

The Band Theory of Solids and Conductivity of Materials

Understanding this section should enable you to do the following:

1. *Identify the difference between conductors, semi-conductors, and insulators on the basis of the band theory of solids.*
2. *Define conduction by "holes" and state how the Hall effect can determine the sign of the charges actually producing the current.*
3. *Identify the difference between intrinsic, n-type extrinsic, and p-type extrinsic semi-conductors.*
4. *Calculate the electrical conductivity by knowing the density of free charges and their drift velocity.*

Band Theory of Electron Energy Levels--As atoms become crowded, to the extent that their outer electrons influence each other, the sharply defined electron energy levels of the single atoms coalesce into *bands* of closely spaced energy levels. When a very large number of atoms have formed into a crystalline solid, these levels are so closely spaced that they represent a *continuous* range of allowed energies within each band. The bands are separated by gaps of energy "forbidden" to the electrons. This situation is shown in Figure 30.6 (text). The Pauli Exclusion Principle will determine the extent to which any of the allowed energy bands are filled in the ground state of the crystal. When an electrical field is applied to the crystal, the electrons must be able to gain in energy in order to respond to the field as a current. If no empty levels are available within a band, no current can flow, since the field ordinarily is not strong enough to give the electrons sufficient energy to jump the gap to a higher band where unoccupied states might exist.

Valence Band--The energy band occupied by the *valence* electrons of the crystal. All lower energy bands are occupied completely by the inner electrons.

Conduction Band--The energy band just higher in energy than the valence band. It is unoccupied by electrons when the material is in its ground state.

Insulator--A material whose electrons completely fill its valence band, and whose conduction band is separated from the valence band by a forbidden energy gap of several eV. Electrons in such a material have no way of increasing their energy in response to an applied electric field, since there are no vacant energy states *slightly* higher in energy. Thus their electrical conductivity is extremely low. See Figure 30.7 (text).

Conductor--A material whose valence band is not completely filled with electrons, or whose valence band *overlaps* its conduction band in energy. In either case, vacant energy states *are* available, so that electrons can gain small increments of energy, creating a current in response to an applied electric field. See Figure 30.8 (text).

Semi-conductor--A material whose valence band is filled, but whose conduction band is separated from it by a *narrow* ($\approx$ 1 eV or less) energy gap. With a relatively small amount of excitation, either thermally at an elevated temperature, or by photon absorption, electrons can be raised across this gap into the empty conduction band. In this state such materials have conductivities intermediate between conductors and insulators. See Figure 30.9 (text).

Holes--When an electron is promoted across an energy gap into the conduction band in a semi-conductor, it leaves a *vacant* state behind in the valence band. This "hole" allows an electron adjacent in energy to gain an increment of energy and fill it, which in turn vacates its former state, etc. Thus each hole allows electrons in the valence band to advance a little in energy in response to an applied field. The net effect is that the hole acts like a *positive* charge moving oppositely to the electron flow, representing a *positive* current. See Figure 30.10 (text).

Important Observations

1. Materials which, in the *pure* state, have a narrow energy gap between valence and conduction bands are called *intrinsic* semi-conductors.
2. Intrinsic semi-conductors obviously have a large *negative* temperature coefficient of resistivity.
3. In intrinsic semi-conductors, the number densities of electrons in the conduction band and holes in the valence band are *equal*, since each promoted electron produces one hole. Thus both contribute *equally* to the current in these materials.

Extrinsic Semi-conductors--Introducing even a small amount of certain impurity atoms into an intrinsic semi-conductor (a process referred as doping) greatly increases its conductivity by establishing impurity electron energy levels *very* close ($\approx$ .01 eV) to the valence or conduction bands. There are two types of extrinsic semi-conductors.

n-type Semi-conductors--The impurity atoms *donate* additional electrons occupying energy states (*donor levels*) just *below* the *conduction* band, from which they can easily be raised into that band by thermal or optical energy *without* creating holes in the valence band. These semi-conductors have a predominance of *negative* charges carrying the current.

P-type Semi-conductors--The impurity atoms produce *vacant* energy states (*acceptor levels*) just above the *valence* band. These states can *accept* electrons excited out of the valence band by small amounts of thermal or photon energy, creating *holes* in the valence band. Since no *conduction* electrons are produced, these semi-conductors have a predominance of *positive* charges carrying the current.

Important Observation

1. The *Hall effect* (see Chapter 19) enables us to determine the sign of the predominant current carrier and their *drift* velocity, $v_d = V_H/Bw$. Ideally pure *intrinsic* semi-conductors thus exhibit *no* Hall voltage, while n-type, along with conductors, exhibit a Hall voltage showing the negative sign of their free charges. p-type semi-conductors show a Hall effect which verifies the theory of positive holes being responsible for their currents.

Example 30-2. Suppose 500 grams of silicon has 1 μg of added boron. Assuming each boron atom provides one charge carrier for the semi-conductor, answer the following:

(a.) What density of free charges are there in the crystal?

(b.) In a Hall measurement as pictured in Figure 19.13 (text), which edge of the crystal, top or bottom, would be at higher potential?

(c.) What Hall voltage would occur in a 0.5 T magnetic field if the crystal was 10 cm long, .1 cm thick, and 1 cm wide, and was carrying 10^{-4} amps?

S: For part (a.), we first have to find out how many atoms of Si and B there are. What are their atomic mass numbers?

T: For Si, M = 28.1, and for B, M = 10.8, to three figures.

S: Then we can say that 500 grams of Si contains

$$(500 \text{ gm})(6.02 \times 10^{23} \text{ atoms/mole})/(28.1 \text{ gm/mole}) = 1.07 \times 10^{25} \text{ atoms},$$

and 1 g of boron contains

$$(10^{-6} \text{ gm})(6.02 \times 10^{23} \text{ atoms/mole})/(10.8 \text{ gm/mole}) = 5.57 \times 10^{16} \text{ atoms}.$$

T: So the fraction of boron to silicon atoms is:

$$N_B/N_{Si} = (5.57 \times 10^{16})/(1.07 \times 10^{25}) = 5.21 \times 10^{-9}.$$

S: Now we need to convert these to number per *volume*. How do we find the volume of this sample?

T: With as small a fraction of B as this, we can assume the sample will have essentially the density of pure Si, namely 2.33 gm/cm^3.

S: O.K., and the atomic mass will still be 28.1, so we have for the number density:

$$(2.33 \text{ gm/cm}^3)(6.02 \times 10^{23} \text{ atoms/mole})/(28.1 \text{ gm/mole})$$

$$= 4.99 \times 10^{22} \text{ atoms/cm}^3 = 4.99 \times 10^{28} \text{ atoms/m}^3.$$

T: From above, about 5 billionths of these will produce free charges. More precisely, the number density of free charges will be

$$n = (5.21 \times 10^{-9})(4.99 \times 10^{28} \text{ atoms/m}^3) = 2.60 \times 10^{20}/\text{m}^3.$$

S: How do we tell what sign these charges will have?

T: As discussed in the text, the boron creates *acceptor* levels just above the valence band. These are *not conduction electrons*, but when filled, *holes* are left in the valence band as *positive* charge carriers.

S: In Figure 19.13 (text), these would be moving to the right and be deflected *upward* by the magnetic field. This would create a positive, *higher potential upper* edge, with $V_H = v_d B w$.

T: That's right -- good observation! . We obviously need to know v_d.

S: In Chapter 19, we used $j = nev_d$, where j is the current density I/A.

T: You're certainly sharp today! Applying that to the prsent situation gives us $n = I/v_d eA$.

S: So

$$v_d = I/neA = (10^{-4} \text{ amps})/(2.60 \times 10^{20}/m^3)(1.6 \times 10^{-19} \text{ C})(10^{-5} \text{ m}^2)$$

$$= .24 \text{ m/s} = 24 \text{ cm/sec}.$$

T: Then finally

$$V_H = (.24 \text{ m/s})(.5 \text{ T})(10^{-2} \text{ m}) = 3.45 \text{ mV}.$$

Superconductivity

Understanding this section should enable you to do the following:

1. *Define superconductivity and the approximate temperature requirement for it to occur in typical metals and alloys.*
2. *Define the Josephson effect and calculate the relation between voltage and a.c. current across a Josephson junction.*

Critical Temperature--In many metals and alloys the electrical resistivity becomes *zero* when they are cooled below a *critical* temperature, which is different for each substance. Until recently, the critical temperatures were all below 23 K. Now we know that superconductivity occurs well above 100 K for a number of ceramic substances. A partial list of critical temperatures is given in Table 30.4 (text).

The Nature of Superconductivity--Resistance in a conductor is the result of individual electrons making uncoordinated collisions with lattice atoms. At temperature below critical, the electrons become bound in a *coordinated* motion throughout the crystal which the collisions with lattice atoms are unable to break, since the thermal energy is so low. This is a remarkable example of a *macroscopic* quantum state, in which there is a definite minimum energy and hence minimum lattice temperature needed to break the coordination of the electrons. Unless this temperature is exceeded, electrons cannot lose energy by individual collisions, and the resistivity is *zero*.

The Josephson Effect--One of the more dramatic and useful phenomena associated with superconductivity is the *Josephson effect*, in which high frequency a.c. currents can be produced by applying a small voltage across a thin (≈ 2 nm) insulating oxide layer between two superconductors. The a.c. frequency is independent of the materials being used and proportional to the d.c. voltage applied. Mathematically,

$$\nu = (2V)(e/h),$$

where V = the applied voltage.

Important Observations

1. The resistivity below the critical temperature is not just very small, it is *exactly* zero.
2. Achievement of superconducting temperatures depends on the availability of a liquid coolant in which to immerse the metal or alloy. So far that has meant liquid He, since no other material stays liquid to such low temperatures as most critical temperatures. The recent discovery of materials with higher critical temperatures allows the more common liquid hydrogen or even nitrogen to be used as a coolant.

Example 30-3. If 10 mV was applied across a Josephson junction, what frequency would the a.c. current produced have?

S: The expression gives

$$\nu = (2V)(e/h) = 2(10^{-2}\ \text{v})(1.6 \times 10^{-19}\ \text{C})/(6.62 \times 10^{-34}\ \text{J-s})$$

$$= 4.83 \times 10^{12}\ \text{Hz}.$$

T: The *period* of this variation is $T = 1/\nu = .207 \times 10^{-12}$ sec. These extremely high current frequencies, difficult to achieve previously, but rather routinely attainable by this method, can serve as high speed switches, with switching times of less than *one-trillionth* of a second!

ANSWERS TO GENERAL PROBLEMS

Chapter 1.

1-5: (a.) 45 mi/hr ; 33.3 mi/hr, N.E. (b.) 54 mi/hr; 0

1-6: $v = 71.3$ km/hr, $\theta = 82.4^\circ$,
$\langle a \rangle = 9.9\ m/s^2$, 52.6° S of W.

Chapter 2.

2-1: 7 m

2-3: $t_1 = 7.93$ s; $t_2 = 5.95$ s

Chapter 3.

3-1: They pass at s = 5 m in both cases.
(a.) t = 1.75 s (b.) t = 1.43 s

3-2: $a_2 = (m_2 g)/(m_2 + 4m_1)$; $a_2 = (5/13)g$

3-3: $g_{mars} = (.363)g_{earth}$

3-4: (a.) F = 9N (b.) $F_N = 6N$ (c.) F = 20.8 N ,
$F_N = 13.8$ N

3-5: (a.) $a_{max} = 4.9\ m/s^2$; $F_{max} = 16.3$ N

(b.) $a_{max} = 2.45\ m/s^2$; $F_{max} = 7.35$ N

Chapter 4.

4-1: $\theta = 64.3^\circ$; y = 1093 ft

4-2: v = 17.8 m/s; F_N(driver) = 1097 N = 246 pounds

4-3: t = 44.3 s; x = 3.01×10^4 m; $v_y = 72.4$ m/s and
$v_x = 680$ m/s at impact, so v = 684 m/s.
Impact angle: $\theta = 6.1^\circ$ rel. to horizontal

4-4: $v_{max}^2 = gR(\sin\theta + \mu_s\cos\theta)/(\cos\theta - \mu_s\sin\theta)$

4-5: (a.) $\omega = .3132$ rad/s (b.) g/2 (c.) 31.3 m/s

Chapter 5.

5-1: 37°

5-2: F = 112.3 N; $\theta = 14.9^\circ$ above horizontal

5-3: $l_1 = 5.5$ ft

5-4: $\langle X \rangle = .464$ m, $\langle Y \rangle = .464$ m

Chapter 6.

6-1: Zero acceleration at $x = mg/k$

6-2: $P = 9 \times 10^5$ watts

6-3: (a.) $(F_{net})_P = 3000$ N; $(F_{net})_N = 2(193 + 6s)$ N

(b.) $\Delta KE = 6 \times 10^5$ J (c.) $v_{max} = 37.3$ m/s

6-4: (a.) $W_{fric} = -3.78 \times 10^5$ J

(b.) $W_{fric} = -4.9 \times 10^5$ J

6-5: (a.) IMA = 503; AMA = 100 (b.) Q = 18,120 J

Chapter 7.

7-1: e = .376 ; $\langle F \rangle = 344$ N

7-2: t = 2.65 s

7-3: KE_2/KE = .0392, .331, .331, and .0392. The ratio is a maximum (=1) for k = 1 and decreases symmetrically for larger and smaller values.

Chapter 8.

8-2: $\omega_2 = 2.71$ rad/s

8-3: $(\Delta m/\Delta t) = .204$ kg/s

8-4: (a.) $\omega = 7.79$ rad/s (b.) t = 1.62 s

8-5: $\omega = 6.88$ rad/s

Chapter 9.

9-1: (a.) $F_{net} = 138$ N to the right (b.) L = 64.6 cm from the left

9-2: .094 moles is vented

9-3: $m/V = 64.2$ kg/m^3 ; $v_{rms} = 652$ m/s

9-4: $P_{gauge} = 2625$ torr; $M(O_2) = 46.4$ gm;

$M(N_2) = 51.2$ gm

Chapter 10.

10-1: r = 9.88 cm

10-2: specific grav. = $(w - w_{fl})/(w - w_w)$

10-3: h' = 744 mm Hg

10-4: $t_{max} = L/46.8$

10-5: $r_2 = 3.3$ cm

Chapter 11.

11-1: $T = 5.09 \times 10^3$ s = 84.8 minutes

11-2: (b.) $T = 1.64$ s, $L_{eq} = .668$ m

(c.) $T = 3.48$ s, $L_{eq} = 3.01$ m

11-3: $f = 4.21$ Hz

Chapter 12.

12-1: (a.) $f = 275$ Hz, (b.) maximum

12-2: $L = 1.84$ m

12-3: $f = 118.8$ Hz

Chapter 13.

13-1: 5.5 kcal

13-2: $\Delta V_{turp} = 13.9$ cm^3; $\Delta V_{net\ cont} = .43$ cm^3

13-3: $\Delta V_{tot} = -1.84$ cm^3. 34% due to pressure, 66% due to temperature

13-4: $T_f = 32.4$ °C

13-5: (a.) 770 °C (b.) $F = 3.34 \times 10^5$ N

Chapter 14.

14-1: $T_{rad} = 256$ K

14-2: $\Delta T = 0.375$ C°

14-3: $(Q/At)_{rad} = 39.7$ W/m^2; $(Q/At)_{cond.} = 30.1$ W/m^2

Chapter 15.

15-2: $W_{AB} = 0$; $Q_{AB} = \Delta U_{AB} = +7445$ J;

$Q_{BC} = 0$; $W_{BC} = -\Delta U_{BC} = +3623$ J;

$W_{CA} = -2561$ J: $Q_{CA} = -6383$ J; $\Delta U_{CA} = -3822$ J

15-3: $(e_1)_{actual} = .143$; $(e_1)_{actual}/(e_1)_{ideal} = .286$

15-4: $(e_1)_{actual} = .134$; less efficient.

15-5: $P = 7.2$ watts.

Chapter 16.

16-1: $E_x = +18 \times 10^5$ N/C ; $E_y = +3.14 \times 10^3$ N/C

16-2: (a.) $E = k_e q/r^2$ for $r<R_i$ (b.) $E = 0$ for $R_i<r<R_o$

(c.) $E = k_e q/r^2$ for $r>R_o$

Chapter 17.

17-1: $C = R/k_e$
17-2: (a.) $\Delta V = 900$ volts $= V_1 - V_2$ (b.) $Q_1' = -1.33$ nC

$Q_2' = -.67$ nC (c.) $V_1' = V_2' = -120$ volts

17-3: (a.) $C = .53$ nF (b.) $Q = 12.7$ nC

(c.) $Q' = 12.7$ nC, $C' = .088$ nF, $V' = 144$ v.

Chapter 18.

18-1: $R_{tot} = 9.6\ \Omega$
18-2: $R_{tot} = 1.07\ \Omega$

18-4: (a.) $C = 2 \times 10^{-8}$ F (b.) $C'_{tot} = 1.43 \times 10^{-8}$ F;
$V' = 100$ v.; $Q' = 1.43\ \mu C$

Chapter 19.

19-1: $V = 4.9 \times 10^4$ m/s

19-2: $I = 1.40 \times 10^3$ A

19-3: $B = 3.14 \times 10^{-4}$ T

19-4: $m = 3.2 \times 10^{-26}$ kg ≈ 20 u. The ions are ${}^{20}{}_{10}Ne$.

Chapter 20.

20-1: (a.) $Q(50\text{ ms}) = .415\ Q_{max}$

(b.) $Q_{max} = (4 \times 10^{-6}\text{ F}) \times \mathcal{E}$.
20-2: N = 291 turns
20-3: $I_T = 122$ A ; $v_T = 61.2$ m/s

20-4: $M = 4.93 \times 10^{-5}$ H (b.) $(\Delta I/\Delta t)_{min} = 2.03 \times 10^{-2}$ A/s

Chapter 21.

21-1: (a.) P = 50 watts (b.) P = 46.7 watts
21-2: (a.) I = 2.01 A (b.) $C_{res} = 10.1\ \mu F$
21-3: $C = 2.54\ \mu F$; $R = 52.9\ \Omega$
21-4: $C = 14.3\ \mu F$

Chapter 22.

22-1: $I(t = .2\text{ ms}) = 1.78$ mA ; $B = 3.95 \times 10^{-9}$ T

22-2: $E_{max} = 63.9$ mV ; $B_{max} = 2.13 \times 10^{-10}$ T

Chapter 23.

23-1: 48.8° on either side of vertical, total angle = 97.6°
23-3: R = 24.8 cm
23-4: f = +187.5 cm

Chapter 24.

24-1: (a.) θ_1 = 59.3° (b.) θ_3 = 48.8°
24-2: Angular dispersion = 10.2°
24-3: ΔT = 11.1 C°
24-4: (a.) t = .019 mm

Chapter 25.

25-1: f = 215 cm
25-2: f = 2.67 cm
25-3: M = 1.56
25-4: (a.) M_{tot} = 738 (b.) M_{tot} = 790

Chapter 26.

26-1: v = 2.25×10^8 m/s = .75c
26-2: v' = .68c
26-3: $\Delta t = 5.33 \times 10^{-6}$ s ; $\Delta t' = 1.07 \times 10^{-10}$ s

26-4: T = 2.42×10^8 K

Chapter 28.

28-1: n_i = 6, n_f = 2 ; E_6 = -.378 eV
28-2: N_{tot} = 32
28-3: n = 4.86×10^{47}. The precise integer is impossible to determine from the limited significant figures of the data.

$\Delta r = 2.70 \times 10^{-41}$ meters. This is enormously smaller than a nucleus, so the "allowed" orbits are not measurably discrete, but take on essentially a continuum of values.

Chapter 29.

29-1: $E_n > E_p$

29-2: $M_{Rn} = 6.5 \times 10^{-6}$ gm

29-3: $N_o = 2.84 \times 10^{10}$ atoms

29-4: (a.) P = 2.96×10^{-7} watts

(b.) $E_{tot} = 9.59 \times 10^{10}$ MeV = 1.53×10^{-2} J